Horiba International Conference - The 19th International Conference on the Application of High Magnetic Fields in Semiconductor Physics and Nanotechnology 2010

Journal of Physics: Conference Series Volume 334

Fukuoka, Japan
1-6 August 2010

ISBN: 978-1-62276-681-9
ISSN: 1742-6588

Printed from e-media with permission by:

Curran Associates, Inc.
57 Morehouse Lane
Red Hook, NY 12571

Some format issues inherent in the e-media version may also appear in this print version.

Printed by Curran Associates, Inc. (2013)

For permission requests, please contact the Institute of Physics
at the address below.

Institute of Physics
Dirac House, Temple Back
Bristol BS1 6BE UK

Phone: 44 1 17 929 7481
Fax: 44 1 17 920 0979

techtracking@iop.org

Additional copies of this publication are available from:

Curran Associates, Inc.
57 Morehouse Lane
Red Hook, NY 12571 USA
Phone: 845-758-0400
Fax: 845-758-2634
Email: curran@proceedings.com
Web: www.proceedings.com

Papers

19th International Conference on the Application of High Magnetic Fields in Semiconductor Physics and Nanotechnology (HMF-19)

Koji Muraki and Shojiro Takeyama

PREFACE

This volume contains invited and contributed papers from the 19th International Conference on the Application of High Magnetic Fields in Semiconductor Physics and Nanotechnology (HMF-19) held in Fukuoka, Japan, from 1–6 August 2010. This conference was mainly sponsored by the Tokyo University-'Horiba International fund', which was donated by Dr Masao Horiba, the founder of Horiba Ltd. The scientific program of HMF-19 consisted of 37 invited talks, 24 contributed talks, and 83 posters, which is available from the conference homepage http://www.hmf19.iis.u-tokyo.ac.jp/index.html. Each manuscript submitted for publication in this volume has been independently reviewed. The Editor is very grateful to all the reviewers for their quick responses and helpful reports and to all the authors for their submissions and patience for the delay in the editorial process. Finally, the Editor would like to express his sincere gratitude to all the individuals involved in the conference organization and all the attendees, who made this conference so successful.

Koji Muraki

Committees

Chair

Conference chair	S Takeyama	(ISSP-UT)
Conference secretary	T Machida	(IIS-UT)
Program chair	K Muraki	(NTT)
Local organizing chair	K Oto	(Chiba Univ.)

Advisory Committee

International		Domestic	
L Brey	(ES)	T Ando	(TIT)
Z H Chen	(CN)	Y Hirayama	(Tohoku Univ.)
S Das Sarma	(US)	G Kido	(NIMS)
L Eaves	(GB)	N Miura	(JP)
J P Eisenstein	(US)	J Nitta	(Tohoku Univ.)
K Ensslin	(CH)	T Takamasu	(NIMS)

J Furdyna	(US)
G M Gusev	(BR)
I Kukushkin	(RU)
Z D Kvon	(RU)
G Landwehr	(DE)
J C Maan	(NL)
A H MacDonald	(US)
N F Oliveira Jr	(BR)
A Pinczuk	(US)
J C Portal	(FR)
A Sachrajda	(CA)
M K Sanyal	(IN)
R Stepniewski	(PL)

Program Committee

Chair: K Muraki (NTT)

International		**Domestic**	
G Bauer	(AU)	H Ajiki	(Osaka Univ.)
G Boebinger	(US)	H Aoki	(Hongo, UT)
S Ivanov	(RU)	K Nomura	(RIKEN)
K von Klitzing	(DE)	T Okamoto	(Hongo, UT)
R Nicholas	(GB)	T Osada	(ISSP-UT)
M Potemski	(FR)		
N Studart	(BR)		
U Zeitler	(NL)		

Local Organizing Committee

Chair: K Oto	(Chiba Univ.)
Y H Matsuda	(ISSP-UT)
H Yokoi	(Kumamoto Univ.)
M Itoh	(IIS-UT)
M Noda	(ISSP-UT)
H Sawabe	(ISSP-UT)

Sponsors

Horiba International Conference (Dr Masao Horiba's

Donation)

The University of Tokyo

Fukuoka City

The Institute for Solid State Physics, The University of Tokyo

The Global Center of Excellence for Physical Sciences Frontier, The University of Tokyo

Spin-orbit coupling and the Landau level spectrum of ABA-stacked trilayer graphene

Edward McCann[1] **and Mikito Koshino**[2]

[1]Department of Physics, Lancaster University, Lancaster, LA1 4YB, UK
[2]Department of Physics, Tokyo Institute of Technology, 2-12-1 Ookayama, Meguro-ku, Tokyo 152-8551, Japan

E-mail: ed.mccann@lancaster.ac.uk

Abstract. We review the tight-binding model of ABA-stacked trilayer graphene. The presence of mirror reflection symmetry of the crystal structure leads to a decomposition of the band structure into separate monolayerlike and bilayerlike parts. Owing to a lack of spatial inversion symmetry, next-nearest layer coupling induces gaps in the monolayerlike and the bilayerlike parts of the spectrum, and also shifts the monolayerlike and bilayerlike bands with respect to each other. We calculate the Landau level spectrum of ABA-stacked trilayer graphene taking into account terms within the single-particle picture that break level degeneracy including next-nearest layer coupling and spin-orbit coupling.

1. Introduction

Soon after the isolation of individual graphene flakes [1], observation of the integer quantum Hall effect in monolayers [2, 3] and bilayers [4] led to an explosion of interest in the nature of chiral charge carriers in them. The electronic spectrum of monolayer graphene [5] consists of two Dirac cones with linear dispersion supporting chiral quasiparticles with Berry's phase π. The Dirac cones are located at two inequivalent corners of the Brillouin zone, denoted K_+ and K_- points, and also referred to as valleys [6]. In bilayer graphene [4, 7, 8], the band structure is also gapless at the K points with a parabolic dispersion supporting chiral quasiparticles corresponding to Berry's phase 2π.

In graphene monolayers and bilayers, both space and time inversion operations connect electronic states at K_+ to those at K_- [9, 10, 11]. Thus, the energy spectrum is degenerate between the two valleys, even when time reversal symmetry is broken by the presence of a magnetic field [12], as long as spatial inversion symmetry is not broken by an external field produced, for example, by an external gate or doping [7, 8, 13, 14]. This holds for general ABA-stacked (Bernal) multilayer graphene composed of N layers as long as N is even (bilayer graphene corresponds to $N = 2$). The situation is different in ABA-stacked multilayer graphene with N odd, beginning with trilayer graphene $N = 3$, because their crystal structures do not obey spatial inversion symmetry [15, 11, 12, 16].

Instead of spatial inversion symmetry, odd-N multilayers possess mirror reflection symmetry of their crystal structure in the plane of their central layer. It can not guarantee valley degeneracy when time reversal symmetry is broken by a magnetic field, because it does not connect states at the two valleys, but it does allow the electronic Hamiltonian to be decomposed into separate

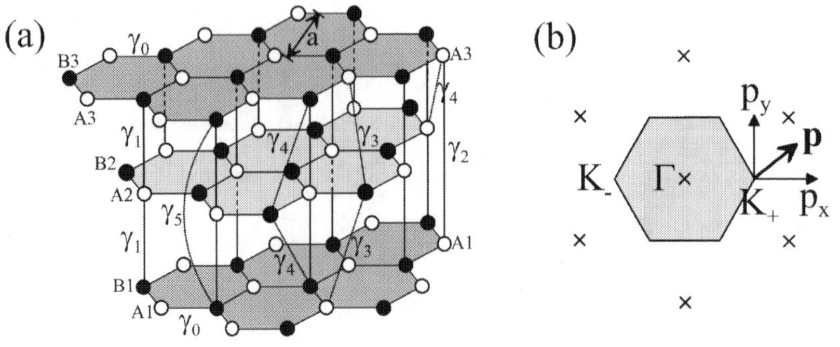

Figure 1. (a) Schematic of the crystal structure of ABA-stacked trilayer graphene with six atomic sites in the unit cell, A (white circles) and B (black circles) on each layer, showing the Slonczewski-Weiss-McClure parameterization [25] of couplings γ_0-γ_5. (b) Brillouin zone of trilayer graphene (shaded hexagon) showing momentum $\mathbf{p} = (p_x, p_y)$ measured with respect to the corner of the Brillouin zone K_+. Crosses indicate reciprocal lattice points.

monolayer- and bilayerlike parts [17, 18, 19, 20, 21, 22, 23, 24]. Each part approximately satisfies an effective inversion symmetry corresponding to that of the real lattice of monolayer or bilayer graphene, respectively, and this ensures approximate valley degeneracy of the electronic spectrum of odd-N multilayers even for finite magnetic field [12]. Mirror symmetry is broken by interlayer asymmetry induced by an external electric field, which breaks valley degeneracy of the Landau level spectrum [12]. In this paper, we focus instead on intrinsic trilayer graphene ($N = 3$), when symmetry is not broken by an external electric field, and describe terms in the single-particle picture that break the effective inversion symmetry and the approximate valley degeneracy of the Landau level spectrum. They include next-nearest layer coupling and spin-orbit coupling. We begin with a review of the tight-binding model of ABA-stacked trilayer graphene in Section 2, followed by a discussion of the Landau level spectrum in the presence of next-nearest layer coupling in Section 3 and in the presence of spin-orbit coupling in Section 4.

2. Tight-binding model of ABA-stacked trilayer graphene

We consider ABA-stacked trilayer graphene to consist of three layers of carbon atoms, each arranged with a honeycomb crystal structure (as in monolayer graphene), as shown in Fig. 1. The unit cell consists of six atomic sites: pairs $\{A1, B1\}$, $\{A2, B2\}$, and $\{A3, B3\}$ in the bottom, center, and top layers, respectively. The relative arrangement of the layers is Bernal (A-B) stacking [25], in that half of the atomic sites have a counterpart directly above or below them in the adjacent layer: $B1$ is directly below $A2$, and $A2$ is directly below $B3$.

Our tight-binding model [7, 17, 24] takes into account a single p_z orbital per atomic site. There are six orbitals per unit cell and six bands in the electronic band structure. We adapt the Slonczewski-Weiss-McClure parameterization [25, 26] of couplings between atomic sites including γ_0 which describes nearest-neighbor (Ai-Bi for $i = \{1, 2, 3\}$) coupling within each layer, γ_1 which describes strong nearest-layer coupling between sites ($B1$-$A2$ and $A2$-$B3$) that lie directly above or below each other, and γ_3 (γ_4) which describes weaker nearest-layer coupling between sites $A1$-$B2$ and $B2$-$A3$ ($A1$-$A2$, $B1$-$B2$, $A2$-$A3$, and $B2$-$B3$). If the model would include only these couplings, it would satisfy an effective inversion symmetry (as explained in detail later) guaranteeing the presence of a degeneracy point at each of two inequivalent corners, $K_\pm$, of the hexagonal Brillouin zone [6]. However, the effective inversion symmetry and the degeneracy tends to be broken by next-nearest-layer coupling γ_2 (between $A1$ and $A3$) and γ_5 (between $B1$

Journal of Physics: Conference Series **334** (2011) 012001 doi:10.1088/1742-6596/334/1/012001

and $B3$), as well as parameter δ describing a difference of on-site energy between sites $A1$, $B2$, $A3$, and $B1$, $A2$, $B3$. Parameter Δ that is often used in models of bulk graphite [25] is related to δ by $\Delta = \delta + \gamma_2 - \gamma_5$.

In a basis with components ψ_{A1}, ψ_{B1}, ψ_{A2} , ψ_{B2}, ψ_{A3}, ψ_{B3}, the ABA-stacked trilayer Hamiltonian [17, 24, 27, 28] is

$$\widetilde{H} = \begin{pmatrix} U_1 & v\pi^\dagger & -v_4\pi^\dagger & v_3\pi & \gamma_2/2 & 0 \\ v\pi & U_1+\delta & \gamma_1 & -v_4\pi^\dagger & 0 & \gamma_5/2 \\ -v_4\pi & \gamma_1 & U_2+\delta & v\pi^\dagger & -v_4\pi & \gamma_1 \\ v_3\pi^\dagger & -v_4\pi & v\pi & U_2 & v_3\pi^\dagger & -v_4\pi \\ \gamma_2/2 & 0 & -v_4\pi^\dagger & v_3\pi & U_3 & v\pi^\dagger \\ 0 & \gamma_5/2 & \gamma_1 & -v_4\pi^\dagger & v\pi & U_3+\delta \end{pmatrix}, \quad (1)$$

where $\pi = \xi p_x + ip_y$ and $\pi^\dagger = \xi p_x - ip_y$ are operators related to the in-plane momentum $\mathbf{p} = (p_x, p_y)$ measured with respect to the K point [6], and $\xi = \pm 1$ is the valley index. Effective velocities are defined as $v = (\sqrt{3}/2)a\gamma_0/\hbar$, $v_3 = (\sqrt{3}/2)a\gamma_3/\hbar$, and $v_4 = (\sqrt{3}/2)a\gamma_4/\hbar$, and $U_i = -eV_i$ describe different on-site energies in each layer $i = \{1, 2, 3\}$ corresponding to different layer potentials V_i created by external fields.

Mirror reflection symmetry of the lattice in the plane of its central layer suggests that a basis involving linear combinations of orbitals from the top and bottom layers would be convenient [20, 21, 22, 24]. There are a pair that are odd with respect to mirror reflection symmetry,

$$\phi_{1,m} = (\psi_{A1} - \psi_{A3})/\sqrt{2}, \quad (2)$$
$$\phi_{2,m} = (\psi_{B1} - \psi_{B3})/\sqrt{2}, \quad (3)$$

and four that are even with respect to mirror reflection symmetry,

$$\phi_{1,b} = (\psi_{A1} + \psi_{A3})/\sqrt{2}, \quad (4)$$
$$\phi_{2,b} = \psi_{B2}, \quad (5)$$
$$\phi_{3,b} = \psi_{A2}, \quad (6)$$
$$\phi_{4,b} = (\psi_{B1} + \psi_{B3})/\sqrt{2}. \quad (7)$$

Then, in a basis with components $\phi_{1,m}$, $\phi_{2,m}$, $\phi_{1,b}$, $\phi_{2,b}$, $\phi_{3,b}$, $\phi_{4,b}$, the ABA-stacked trilayer Hamiltonian [24] is

$$H = \begin{pmatrix} H_m & D \\ D^\dagger & H_b \end{pmatrix}, \quad (8)$$
$$H_m = \begin{pmatrix} \Delta_2 - \gamma_2/2 & v\pi^\dagger \\ v\pi & \Delta_2 - \gamma_5/2 + \delta \end{pmatrix},$$
$$H_b = \begin{pmatrix} \Delta_2 + \gamma_2/2 & \sqrt{2}v_3\pi & -\sqrt{2}v_4\pi^\dagger & v\pi^\dagger \\ \sqrt{2}v_3\pi^\dagger & -2\Delta_2 & v\pi & -\sqrt{2}v_4\pi \\ -\sqrt{2}v_4\pi & v\pi^\dagger & -2\Delta_2+\delta & \sqrt{2}\gamma_1 \\ v\pi & -\sqrt{2}v_4\pi^\dagger & \sqrt{2}\gamma_1 & \Delta_2 + \gamma_5/2 + \delta \end{pmatrix},$$
$$D = \begin{pmatrix} \Delta_1 & 0 & 0 & 0 \\ 0 & 0 & 0 & \Delta_1 \end{pmatrix}.$$

In this Hamiltonian we use two parameters to take into account differences in the on-site energies U_1, U_2, U_3 of the three layers [24]:

$$\Delta_1 = (U_1 - U_3)/2, \quad (9)$$
$$\Delta_2 = (U_1 - 2U_2 + U_3)/6, \quad (10)$$

and we set the average on-site energy $[U_1+U_2+U_3]/3$ to zero. The first parameter, Δ_1, describes the average energy difference between the outer layers [29, 17, 30, 24, 12], while the second, Δ_2, describes the difference between the energy of the central layer as compared to the average energy of the outer layers. Parameter Δ_1 (Δ_2) is odd (even) with respect to mirror reflection.

The Hamiltonian Eq. (8) has a two by two block H_m written in a basis $\phi_{1,m}$, $\phi_{2,m}$ of orbitals that are odd with respect to mirror reflection. It is similar to the Dirac-like Hamiltonian of monolayer graphene, and it contributes two monolayerlike bands near zero energy. There is a four by four block H_b with components $\phi_{1,b}$, $\phi_{2,b}$, $\phi_{3,b}$, $\phi_{4,b}$ that are even with respect to mirror reflection. It is similar to the Hamiltonian of bilayer graphene [7] except that terms proportional to γ_1, γ_3, and γ_4 describing interlayer coupling appear with a factor $\sqrt{2}$. Block H_b gives four bilayerlike bands: two bands near zero energy and two bands split away from zero by energy $\epsilon \approx \pm\sqrt{2}\gamma_1$ at the K point. Parameters γ_3 and γ_4 have a similar effect as in bilayer graphene [7], with γ_3 giving trigonal warping of the bands and γ_4 contributing to electron-hole asymmetry.

The only parameter in the Hamiltonian Eq. (8) that is odd with respect to mirror reflection symmetry is the asymmetry Δ_1 between the outer layers [29, 17, 30, 24, 12]. Therefore, it is the only parameter that appears in the off-diagonal blocks D. It connects the odd and even parts of the Hamiltonian, H_m and H_b, and hybridizes the monolayerlike and bilayerlike bands. The electronic spectrum for finite Δ_1 has been discussed previously for zero [29, 17, 30, 24, 27, 28] and finite magnetic field [12], and we refer the reader there for further details.

For $\Delta_1 = 0$, the odd and even parts of the Hamiltonian Eq. (8) are separate giving superimposed monolayerlike and bilayerlike bands. Furthermore, for $\gamma_2 = \gamma_5 = \delta = \Delta_2 = 0$, the monolayerlike part and the bilayerlike part assume an effective inversion symmetry corresponding to that of the real lattice of monolayer or bilayer graphene, respectively. This symmetry, combined with time-reversal symmetry, guarantees degeneracy of the spectrum at the K point. Finite γ_2, γ_5, δ, and Δ_2 break the effective inversion symmetry, and their effect is to induce gaps in the monolayerlike and the bilayerlike parts of the spectrum, but also to shift the monolayerlike and bilayerlike bands with respect to each other. At the K point ($p = 0$), and for $\Delta_1 = 0$, the band energies are $\epsilon = \Delta_2 - \gamma_2/2$ and $\epsilon = \Delta_2 - \gamma_5/2 + \delta$ for the monolayer part, and $\epsilon = \Delta_2 + \gamma_2/2$ and $\epsilon = -2\Delta_2$ for the bilayer part (as well as the two split bands at $|\epsilon| \approx \sqrt{2}|\gamma_1|$).

3. Landau level spectrum in the presence of next-nearest layer coupling

In order to focus on the low-energy electronic spectrum, we perform a Schrieffer-Wolff transformation [31, 7] to write an effective four-component Hamiltonian describing the four bands that approach zero energy. This eliminates components $\phi_{3,b}$, $\phi_{4,b}$ corresponding to atomic sites that are strongly coupled by γ_1 and produce the split bands at $\epsilon \approx \pm\sqrt{2}\gamma_1$ in the bilayerlike spectrum. Then, in basis with components $\phi_{1,m}$, $\phi_{2,m}$, $\phi_{1,b}$, $\phi_{2,b}$, the approximate four-component ABA-stacked trilayer Hamiltonian [12, 32] is

$$h = \begin{pmatrix} h_m & d \\ d^\dagger & h_b \end{pmatrix}, \tag{11}$$

$$h_m = \begin{pmatrix} a_m + b_m & v\pi^\dagger \\ v\pi & a_m - b_m \end{pmatrix}, \tag{12}$$

$$h_b = \begin{pmatrix} a_b + b_b & 0 \\ 0 & a_b - b_b \end{pmatrix} + \frac{v^2}{2\gamma_1^2}\begin{pmatrix} (r_b + s_b)\,\pi^\dagger\pi & 0 \\ 0 & (r_b - s_b)\,\pi\pi^\dagger \end{pmatrix} \tag{13}$$

$$-\frac{(v^2 + 2v_4^2)}{\sqrt{2}\gamma_1}\begin{pmatrix} 0 & \left(\pi^\dagger\right)^2 \\ \pi^2 & 0 \end{pmatrix} + \sqrt{2}v_3\begin{pmatrix} 0 & \pi \\ \pi^\dagger & 0 \end{pmatrix}, \tag{14}$$

$$d = \Delta_1\begin{pmatrix} 1 & 0 \\ v_4\pi/\gamma_1 & -v\pi^\dagger/\sqrt{2}\gamma_1 \end{pmatrix}, \tag{15}$$

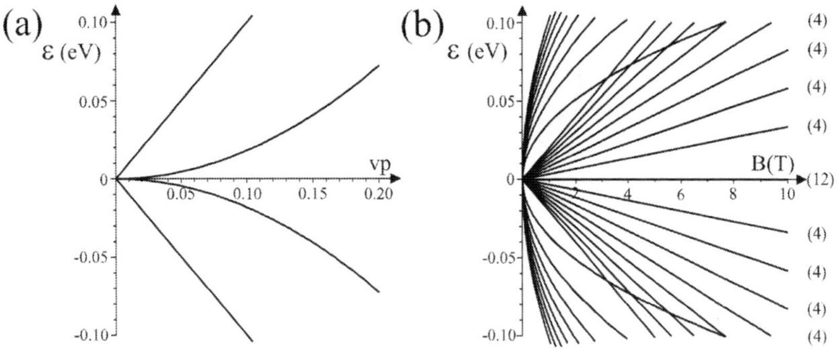

Figure 2. The low-energy spectrum of ABA-stacked trilayer graphene for $\gamma_2 = \gamma_5 = 0$ showing (a) energy bands in the vicinity of the K_+ point at zero magnetic field and (b) Landau level spectrum, taking into account both valleys. Numbers in parenthesis indicate level degeneracy, including spin and valley degeneracy. In both plots, other parameter values are $v = 1.0 \times 10^6 \mathrm{ms}^{-1}$, $\gamma_1 = 0.39\mathrm{eV}$, $\gamma_3 = \gamma_4 = \Delta_1 = \Delta_2 = \delta = 0$. Plots were made using approximations Eqs. (22), (23) and Eqs. (24-28) including, for clarity, only levels up to $n = 8$.

where

$$a_m = \Delta_2 - (\gamma_2 + \gamma_5)/4 + \delta/2\,, \quad (16)$$
$$b_m = (\gamma_5 - \gamma_2)/4 - \delta/2\,, \quad (17)$$
$$a_b = -\Delta_2/2 + \gamma_2/4\,, \quad (18)$$
$$b_b = 3\Delta_2/2 + \gamma_2/4\,, \quad (19)$$
$$r_b = (\gamma_5 - \gamma_2)/4 + \delta + 4\gamma_1 v_4/v\,, \quad (20)$$
$$s_b = -3\Delta_2 - (\gamma_2 + \gamma_5)/4\,. \quad (21)$$

We assume $\{|\varepsilon|, vp, |\Delta_1|, |a_m|, |b_m|, |a_b|, |b_b|\} \ll |\gamma_1|$ and generally keep terms in the four-component Hamiltonian h up to order $1/\gamma_1$ [26]. Terms in the two-component bilayerlike part h_b are analogous to those in the two-component Hamiltonian of bilayer graphene [7, 33]. The first term in Eq. (13) indicates the role of next-layer coupling in shifting the bilayer bands with respect to the monolayer bands (a_b term) and in opening a gap in the bilayerlike spectrum (b_b term). Although it is of order $1/\gamma_1^2$, the second term in Eq. (13) is kept because it describes electron-hole asymmetry (r_b term) and weak degeneracy breaking of the bilayerlike Landau levels (s_b term). The quadratic term in Eq. (14) dominates over a broad range of energy $|\epsilon| \ll |\gamma_1|$ and the linear term Eq. (14) describes trigonal warping of the bands. For simplicity, we neglect γ_3 and γ_4 in the following.

As mentioned above, the spectrum for finite Δ_1 has been discussed previously for zero [29, 17, 30, 24, 27, 28] and finite magnetic field [12]. Here, we focus on $\Delta_1 = 0$. The odd and even (with respect to mirror reflection) parts of the Hamiltonian produce separate monolayerlike ϵ_m and bilayerlike ϵ_b bands. At zero magnetic field, valley degeneracy is preserved [32],

$$\epsilon_m = a_m \pm \sqrt{b_m^2 + v^2p^2}\,, \quad (22)$$
$$\epsilon_b \approx a_b \pm \sqrt{b_b^2 + v^4p^4/(2\gamma_1^2)}\,, \quad (23)$$

where we neglected small contributions from the terms containing r_b and s_b. The parameters $a_{m/b}$ produce shifts of the monolayerlike and bilayerlike spectra with respect to each other,

Journal of Physics: Conference Series **334** (2011) 012001 doi:10.1088/1742-6596/334/1/012001

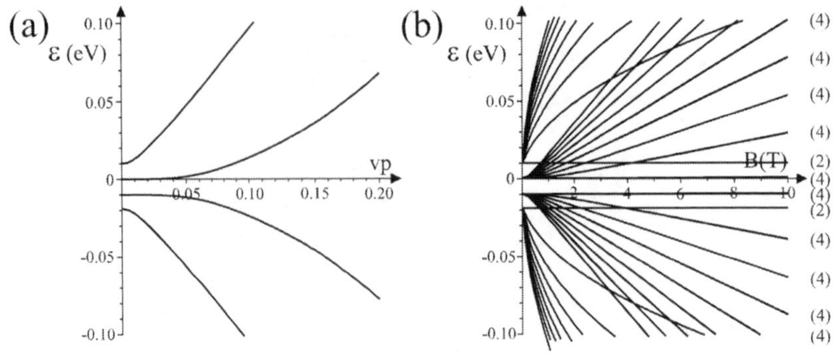

Figure 3. The low-energy spectrum of ABA-stacked trilayer graphene for $\gamma_2 = -0.020\text{eV}$, $\gamma_5 = +0.038\text{eV}$ showing (a) energy bands in the vicinity of the K_+ point at zero magnetic field and (b) Landau level spectrum, taking into account both valleys. Numbers in parenthesis indicate level degeneracy, including spin and valley degeneracy. In both plots, other parameter values are $v = 1.0 \times 10^6\text{ms}^{-1}$, $\gamma_1 = 0.39\text{eV}$, $\gamma_3 = \gamma_4 = \Delta_1 = \Delta_2 = \delta = 0$. Plots were made using approximations Eqs. (22), (23) and Eqs. (24-28) including, for clarity, only levels up to $n = 8$.

whereas parameters $b_{m/b}$ cause gaps to occur within the monolayerlike and bilayerlike spectra. This can be seen by plotting the band structure for different next-layer coupling parameters γ_2 and γ_5. These parameters not been measured in trilayer graphene, so we use different numerical values to illustrate their effect. The zero-magnetic-field low-energy band structure for $\gamma_2 = \gamma_5 = 0$ is plotted in Fig. 2(a), for $\gamma_2 = -0.020\text{eV}$, $\gamma_5 = +0.038\text{eV}$ in Fig. 3(a), and for $\gamma_2 = -0.020\text{eV}$, $\gamma_5 = -0.038\text{eV}$ in Fig. 4(a). Thus, different γ_2, γ_5 values can produce energy gaps or an overlap between bands. Note that using parameters from bulk graphite [15, 18, 24] tends to produce an overlap as in Fig. 4(a).

In the presence of a finite perpendicular magnetic field of magnitude B, the Landau level spectrum has levels $n \geq 1$ given by

$$\epsilon_m^{(n)} = a_m \pm \sqrt{b_m^2 + n\Gamma^2}, \tag{24}$$

$$\epsilon_b^{(n)} \approx a_b + \xi\frac{s_b\Gamma^2}{4\gamma_1^2} \pm \sqrt{b_b^2 + \frac{n(n+1)\Gamma^4}{2\gamma_1^2}}, \tag{25}$$

that are spin degenerate. The monolayerlike levels $\epsilon_m^{(n)}$ are valley degenerate, but parameter s_b introduces weak splitting of the bilayerlike levels between valleys K_+ ($\xi = +1$) and K_- ($\xi = -1$). Here, $\Gamma = \sqrt{2\hbar v^2 eB} = \sqrt{2}\hbar v/\lambda_B$ is an energy scale related to the inverse of the magnetic length, $\lambda_B = \sqrt{\hbar/(eB)}$, and we consider $\{|\epsilon|, \sqrt{n}\Gamma, , |a_{m/b}|, |b_{m/b}|, |r_b|, |s_b|\} \ll |\gamma_1|$. There are three "zero" energy levels per valley and spin component, hence twelve in total [17]. In the presence of next-layer coupling, we find their level splitting to be given by

$$\epsilon_m^{(0)} = a_m + \xi b_m, \tag{26}$$

$$\epsilon_b^{(0)} = a_b + \xi b_b + (r_b + \xi s_b)\Gamma^2/(2\gamma_1^2), \tag{27}$$

$$\epsilon_b^{(-1)} = a_b + \xi b_b. \tag{28}$$

Valley degeneracy is broken more strongly for the "zero" energy levels, Eqs. (26-28), by parameters $b_{m/b}$, than it is for the higher levels. As in bilayer graphene [7], there is also a

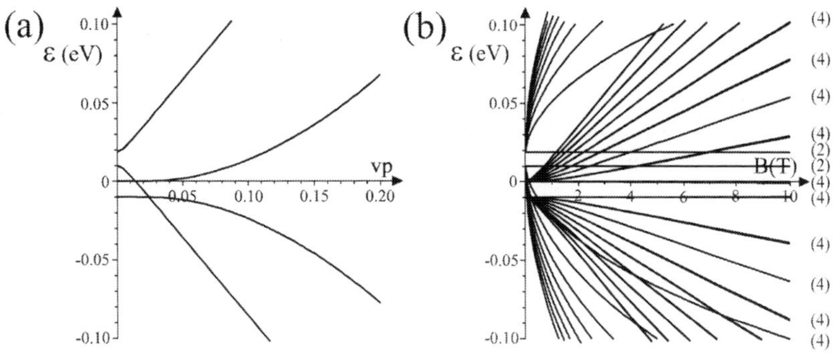

Figure 4. The low-energy spectrum of ABA-stacked trilayer graphene for $\gamma_2 = -0.020$eV, $\gamma_5 = -0.038$eV showing (a) energy bands in the vicinity of the K_+ point at zero magnetic field and (b) Landau level spectrum, taking into account both valleys. Numbers in parenthesis indicate level degeneracy, including spin and valley degeneracy. In both plots, other parameter values are $v = 1.0 \times 10^6$ms^{-1}, $\gamma_1 = 0.39$eV, $\gamma_3 = \gamma_4 = \Delta_1 = \Delta_2 = \delta = 0$. Plots were made using approximations Eqs. (22), (23) and Eqs. (24-28) including, for clarity, only levels up to $n = 8$.

weak splitting of the bilayerlike levels $\epsilon_b^{(0)}$ and $\epsilon_b^{(-1)}$. The spectrum Eqs. (24-28) is plotted in Fig. 2(b) for $\gamma_2 = \gamma_5 = 0$ (and $a_m = b_m = a_b = b_b = 0$) where numbers in parenthesis indicate level degeneracy. Fig. 3(b) shows the spectrum for $\gamma_2 = -0.020$eV, $\gamma_5 = +0.038$eV (the zero-field band structure has a gap, Fig. 3(a)) and fig. 4(b) shows the spectrum for $\gamma_2 = -0.020$eV, $\gamma_5 = -0.038$eV (the zero-field band structure is gapless, Fig. 4(a)). In Figs. 3 and 4, valley degeneracy splitting of the "zero" energy levels can be seen for finite γ_2 and γ_5, but level splitting due to s_b and r_b is not resolved. The energies of the "zero" energy levels correspond to the band energies at the K point ($p = 0$) for zero field. The monolayerlike levels are only two-fold (spin) degenerate, reading $\epsilon_m^{(0)} = a_m + b_m = \Delta_2 - \gamma_2/2$ at valley K_+, and $\epsilon_m^{(0)} = a_m - b_m = \Delta_2 - \gamma_5/2 + \delta$ at K_-. The "zero" energy bilayerlike levels are almost four-fold degenerate, $\epsilon_b^{(0)} \approx \epsilon_b^{(-1)} = a_b + b_b = \Delta_2 + \gamma_2/2$ at K_+, and $\epsilon_b^{(0)} \approx \epsilon_b^{(-1)} = a_b - b_b = -2\Delta_2$ at K_-.

4. Landau level spectrum in the presence of spin-orbit coupling

In this section, we consider the influence of intrinsic spin-orbit terms that exist in samples in the absence of external electric field, *e.g.* we neglect the influence of a transverse electric field that breaks reflection symmetry and produces an additional, Rashba-like term [34, 35, 36, 37, 38]. In monolayer graphene, the only spin-orbit term that exists at the center of each valley in the absence of external fields was introduced by Kane and Mele [34],

$$\mathcal{H}_{KM} = \alpha \Pi_z \sigma_z S_z \,, \tag{29}$$

where α is a parameter and Pauli matrices Π_z, σ_z, S_z act in K_+/K_- valley, A/B lattice, and $\uparrow/\downarrow$ spin space, respectively. This terms couples the out-of-plane component of electronic spin with the out-of-plane component of pseudospin. It doesn't break spin and valley degeneracy, but it breaks the A/B sublattice symmetry, opening a band gap and realizing a new, topological state of matter, a quantum spin Hall insulator [34].

In trilayer graphene, in addition to the Kane-Mele term Eq. (29), there is a second type of intrinsic spin-orbit coupling present at low-energy at the center of the valley [32],

$$\mathcal{H}_{ABA} = \beta \Pi_z S_z \,. \tag{30}$$

It is allowed because of the lack of spatial inversion symmetry of the lattice. It couples the out-of-plane component of electronic spin with the out-of-plane component of the valley 'spin' degree of freedom. This "valley-Zeeman" effect produces Zeeman-like spin splitting of the energy bands at each valley, of magnitude proportional to the parameter β, with an opposite sign of the effective Zeeman field in the two valleys.

In the presence of these spin-orbit terms, the approximate four-component ABA-stacked trilayer Hamiltonian Eq. (11) is modified [32]. Per spin and valley, the monolayerlike block, in basis $\phi_{1,m}$, $\phi_{2,m}$, becomes

$$h_m = \begin{pmatrix} a_m + b_m + s\xi\alpha_m + s\xi\beta_m & v\pi^\dagger \\ v\pi & a_m - b_m - s\xi\alpha_m + s\xi\beta_m \end{pmatrix}, \tag{31}$$

while the bilayerlike block of Eq. (11), in basis $\phi_{1,b}$, $\phi_{2,b}$, is

$$h_b = \begin{pmatrix} a_b + b_b + s\xi\alpha_b + s\xi\beta_b & -v^2\left(\pi^\dagger\right)^2/(\sqrt{2}\gamma_1) \\ -v^2\pi^2/(\sqrt{2}\gamma_1) & a_b - b_b - s\xi\alpha_b + s\xi\beta_b \end{pmatrix}, \tag{32}$$

where $s = \pm 1$ labels spin components, $\xi = \pm 1$ valleys. Parameters $\alpha_{m/b}$ indicate the strength of the Kane-Mele term Eq. (29) in the monolayerlike and bilayerlike parts of the Hamiltonian, respectively, and $\beta_{m/b}$ indicate the strength of the "valley-Zeeman" term Eq. (30). For simplicity, we neglect parameters r_b and s_b in Eq. (32). At zero magnetic field [32],

$$\epsilon_m = a_m + s\xi\beta_m \pm \sqrt{(b_m + s\xi\alpha_m)^2 + v^2p^2}\,, \tag{33}$$

$$\epsilon_b = a_b + s\xi\beta_b \pm \sqrt{(b_b + s\xi\alpha_b)^2 + v^4p^4/(2\gamma_1^2)}\,, \tag{34}$$

where $\{\epsilon_{m/b}, vp, |a_{m/b}|, |b_{m/b}|, |\alpha_{m/b}|, |\beta_{m/b}|\} \ll \gamma_1$. The Kane-Mele spin-orbit term, responsible for factors $\alpha_{m/b}$, has been discussed in detail elsewhere [34, 39, 40, 36, 37, 41, 42, 43, 44, 45, 46, 47, 48]. Here, we note that, in trilayers, it can conspire with the $b_{m/p}$ parameters to break spin degeneracy even at zero magnetic field.

In the presence of a finite perpendicular magnetic field of magnitude B, the Landau levels $n \geq 1$ are given by

$$\epsilon_m^{(n)} = a_m + s\xi\beta_m \pm \sqrt{(b_m + s\xi\alpha_m)^2 + n\Gamma^2}\,, \tag{35}$$

$$\epsilon_b^{(n)} = a_b + s\xi\beta_b \pm \sqrt{(b_b + s\xi\alpha_b)^2 + \frac{n\,(n+1)\,\Gamma^4}{2\gamma_1^2}}\,, \tag{36}$$

where $\{\epsilon_{m/b}, \sqrt{n}\Gamma, |a_{m/b}|, |b_{m/b}|, |\alpha_{m/b}|, |\beta_{m/b}|\} \ll \gamma_1$. There are three "zero" energy levels per valley and spin component (hence twelve in total) given by

$$\epsilon_m^{(0)} = a_m + \xi b_m + s\alpha_m + s\xi\beta_m\,, \tag{37}$$

$$\epsilon_b^{(-1)} = \epsilon_b^{(0)} = a_b + \xi b_b + s\alpha_b + s\xi\beta_b\,. \tag{38}$$

Parameters $\beta_{m/b}$ arise from the "valley-Zeeman" spin-orbit term unique to odd-N multilayers and they describe a Zeeman-like spin splitting of the energy bands at each valley, with an opposite sign of the effective Zeeman field in the two valleys. A crude estimate, using the tight-binding model, found $|\alpha_{m/b}| = |\beta_{m/b}| \sim 0.1$meV [32]. Using $\alpha_m = \alpha_b = \beta_m = -\beta_b \equiv \alpha$ (with a large value $\alpha = 2$meV chosen for illustrative purposes), the electronic spectrum in the presence of spin-orbit in ABA-stacked trilayer graphene is plotted in Fig. 5, neglecting the effect of next-layer coupling that was discussed in Section 3.

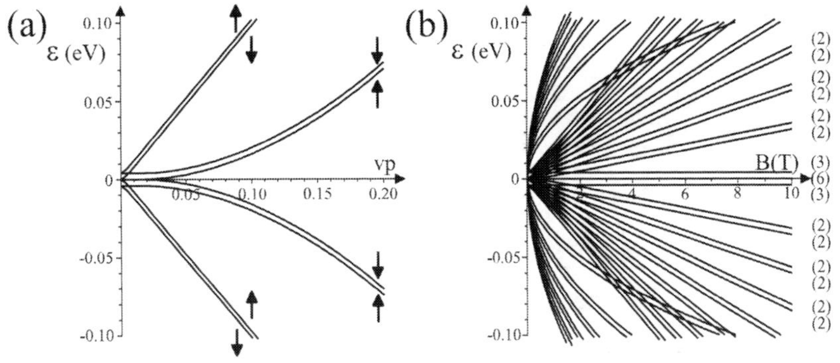

Figure 5. The low-energy spectrum in the presence of spin-orbit coupling showing (a) energy bands in the vicinity of the K_+ point at zero magnetic field and (b) Landau level spectrum, taking into account both valleys. Numbers in parenthesis indicate level degeneracy, including spin and valley degeneracy. In both plots, $a_m = a_b = b_m = b_b = 0$, $\alpha_m = \alpha_b = \beta_m = -\beta_b \equiv \alpha$ with the value $\alpha = 2\text{meV}$ chosen for illustrative purposes. Other parameter values are $v = 1.0 \times 10^6 \text{ms}^{-1}$, $\gamma_1 = 0.39\text{eV}$, $\gamma_3 = \gamma_4 = \Delta_1 = \Delta_2 = \delta = 0$. Plots were made using the analytic approximations Eqs. (33), (34) and Eqs. (35-38) including, for clarity, only levels up to $n = 8$.

5. Conclusions

In this paper, we calculated the Landau level spectrum of ABA-stacked trilayer graphene, showing how degeneracy of the levels is affected by next-nearest layer coupling and spin-orbit coupling. We considered effects within the single-particle picture, neglecting spontaneous symmetry breaking driven by electron-electron interactions that is likely to occur in quantum Hall states of trilayer graphene in a similar fashion as monolayer [10, 49, 50, 51, 52, 53, 54] and bilayer [55, 56] graphene.

Acknowledgments

This project has been funded by JST-EPSRC Japan-UK Cooperative Programme Grant EP/H025804/1.

References

[1] Novoselov K S, Geim A K, Morozov S V, Jiang D, Zhang Y, Dubonos S V, Grigorieva I V and 2004 *Science* **306** 666
[2] Novoselov K S, Geim A K, Morozov S V, Jiang D, Katsnelson M I, Grigorieva I V, Dubonos S V and Firsov A A 2005 *Nature* **438** 197
[3] Zhang Y B, Tan Y W, Stormer H L and Kim P 2005 *Nature* **438** 201
[4] Novoselov K S, McCann E, Morozov S V, Fal'ko I V, Katsnelson M I, Zeitler U, Jiang D, Schedin F and Geim A K 2006 *Nature Phys.* **2** 177
[5] Castro Neto A H, Guinea F, Peres N M R, Novoselov K S, and Geim A K 2009 *Rev. Mod. Phys.* **81** 109
[6] Corners of the hexagonal Brillouin zone are located at wave vector $\mathbf{K}_\xi = \xi(\frac{4}{3}\pi a^{-1}, 0)$, where $\xi = \pm 1$ and a is the lattice constant.
[7] McCann E and Fal'ko V I 2006 *Phys. Rev. Lett.* **96** 086805
[8] Ohta T, Bostwick A, Seyller T, Horn K and Rotenberg E 2006 *Science* **313** 951
[9] Gusynin V P, Sharapov S G and Carbotte J 2007 *Int. J. Mod. Phys. B* **21** 4611
[10] Zhang Y, Jiang Z, Small J P, Purewal M S, Tan Y-W, Fazlollahi M, Chudow J D, Jaszczak J A, Stormer H L and Kim P 2006 *Phys. Rev. Lett.* **96** 136806
[11] Manes J L, Guinea F and Vozmediano M A H 2007 *Phys. Rev. B* **75** 155424
[12] Koshino M and McCann E 2010 *Phys. Rev. B* **81** 115315

[13] McCann E 2006 *Phys. Rev. B* **74** 161403(R)
[14] Castro E V, Novoselov K S, Morozov S V, Peres N M R, Lopes dos Santos J M B, Nilsson J, Guinea F, Geim A K and Castro Neto A H 2007 *Phys. Rev. Lett.* **99** 216802
[15] Latil S and Henrard L 2006 *Phys. Rev. Lett.* **97** 036803
[16] We consider ABA-stacked trilayer graphene in this paper. Note that ABC-stacked trilayer graphene has inversion symmetry, guaranteeing valley degeneracy of its Landau levels in the single-particle picture.
[17] Guinea F, Castro Neto A H and Peres N M R 2006 *Phys. Rev. B* **73** 245426
[18] Partoens B and Peeters F M 2006 *Phys. Rev. B* **74** 075404
[19] Partoens B and Peeters F M 2007 *Phys. Rev. B* **75** 193402
[20] Koshino M and Ando T 2007 *Phys. Rev. B* **76** 085425
[21] Koshino M and Ando T 2008 *Phys. Rev. B* **77** 115313
[22] Koshino M and Ando T 2009 *Solid State Commun.* **149** 1123
[23] Min H and MacDonald A H 2008 *Phys. Rev. B* **77** 155416
[24] Koshino M and McCann E 2009 *Phys. Rev. B* **79** 125443
[25] Dresselhaus M S and Dresselhaus G 2002 *Adv. Phys.* **51** 1
[26] The tight-binding parameters are defined as $\gamma_0 = -\langle\Phi_{Ai}|H|\Phi_{Bi}\rangle = -\langle\Phi_{Bi}|H|\Phi_{Ai}\rangle$ where Φ_{Ai} is an atomic orbital located on site Ai, $i = \{1,2,3\}$, $\gamma_1 = \langle\Phi_{A2}|H|\Phi_{B1}\rangle = \langle\Phi_{A2}|H|\Phi_{B3}\rangle$, $\gamma_3 = -\langle\Phi_{A1}|H|\Phi_{B2}\rangle = -\langle\Phi_{A3}|H|\Phi_{B2}\rangle$, and $\gamma_4 = \langle\Phi_{A1}|H|\Phi_{A2}\rangle = \langle\Phi_{B1}|H|\Phi_{B2}\rangle$. Next layer couplings contain an additional factor of two $\gamma_2 = 2\langle\Phi_{A1}|H|\Phi_{A3}\rangle$, $\gamma_5 = 2\langle\Phi_{B1}|H|\Phi_{B3}\rangle$, to make their definitions consistent with those in bulk graphite [25] where each layer has two next-nearest layers. In Ref. [32], γ_2 and γ_5 did not contain this factor of two.
[27] Avetisyan A A, Partoens B and Peeters F M 2009 *Phys. Rev. B* **79** 035421
[28] Avetisyan A A, Partoens B and Peeters F M 2009 *Phys. Rev. B* **80**, 195401
[29] Lu C L, Chang C P, Huang Y C, Chen R B and Lin M L 2006 *Phys. Rev. B* **73** 144427
[30] Aoki M and Amawashi H 2007 *Solid State Commun.* **142** 123
[31] Schrieffer J R and Wolff P A 1966 *Phys. Rev.* **149** 491
[32] McCann E and Koshino M 2010 *Phys. Rev. B* **81** 241409(R)
[33] Mucha-Kruczyński M, McCann E and Falko V I 2010 *Semicond. Sci. Technol.* **25** 033001
[34] Kane C L and Mele E J 2005 *Phys. Rev. Lett.* **95** 226801
[35] Bychkov Y A and Rashba E I 1984 *J. Phys. C* **17** 6039
[36] Huertas-Hernando D, Guinea F and Brataas A 2006 *Phys. Rev. B* **74** 155426
[37] Min H, Hill J E, Sinitsyn N A, Sahu B R, Kleinman L and MacDonald A H 2006 *Phys. Rev. B* **74** 165310
[38] Rashba E I 2009 *Phys. Rev. B* **79** 161409
[39] Sinitsyn N A, Hill J E, Min H, Sinova J and MacDonald A H 2006 *Phys. Rev. Lett.* **97** 106804
[40] Wang X F and Chakraborty T 2007 *Phys. Rev. B* **75** 033408
[41] Yao Y, Ye F, Qi X-L, Zhang S-C and Fang Z 2007 *Phys. Rev. B* **75** 041401(R)
[42] Boettger J C and Trickey S B 2007 *Phys. Rev. B* **75** 121402(R)
[43] Gmitra M, Konschuh S, Ertler C, Ambrosch-Draxl C and Fabian J 2009 *Phys. Rev. B* **80** 235431
[44] van Gelderen R and Morais Smith C 2010 *Phys. Rev. B* **81** 125435
[45] Guinea F 2010 *Preprint* arXiv:1003.1618
[46] Liu H-W, Xie X C and Q-f Sun 2010 *Preprint* arXiv:1004.0881
[47] Cortijo A, Grushin A G and Vozmediano M A H 2010 *Preprint* arXiv:1007.3704
[48] Prada E, San-Jose P, Brey L and Fertig H A 2010 *Preprint* arXiv:1007.4910
[49] Jiang Z, Zhang Y, Stormer H L and Kim P 2007 *Phys. Rev. Lett.* **99** 106802
[50] Giesbers A J M, Ponomarenko L A, Novoselov K S, Geim A K, Katsnelson M I, Maan J C and Zeitler U 2009 *Phys. Rev. B* **80** 201430(R)
[51] Jiang Z, Zhang Y, Tan Y-W, Stormer H L and Kim P 2007 *Solid State Commun.* **143** 14
[52] Yang K 2007 *Solid State Commun.* **143** 27
[53] Checkelsky J G, Li L and Ong N P 2008 *Phys. Rev. Lett.* **100** 206801
[54] Checkelsky J G, Li L and Ong N P 2009 *Phys. Rev. B* **79** 115434
[55] Feldman B E, Martin J and Yacoby A 2009 *Nature Phys.* **5** 889
[56] Zhao Y, Cadden-Zimansky P, Jiang Z and Kim P 2010 *Phys. Rev. Lett.* **104** 066801

Journal of Physics: Conference Series **334** (2011) 012002 doi:10.1088/1742-6596/334/1/012002

Spontaneous Quantum Hall States and Novel Luttinger Liquids in Chiral Graphene

Fan Zhang, Jeil Jung, Allan H. MacDonald
Department of Physics, University of Texas at Austin, Austin TX 78712, USA
E-mail: zhangfan@physics.utexas.edu

Abstract. Chirally stacked neutral N-layer graphene systems with $N \geq 2$ exhibit spontaneous inversion symmetry breaking and interaction induced charge gaps due to interlayer exchange interactions. A variety of distinct broken symmetry states are distinguished by their charge, spin, and valley Hall conductivities, by their orbital magnetizations, and by their edge state properties. In the spinless case, quantum anomalous and valley Hall states are favored by a weak magnetic field and by an electric field between the graphene layers, respectively. At interfaces between different phases one dimensional gapless modes exhibit novel Luttinger liquid behaviors.

1. Introduction

Broken symmetry and topological order[1, 2] in reduced dimensions have been important themes of condensed matter physics for decades. Recent success in isolating monolayer and few-layer graphene sheets[3, 4] from bulk graphite, combined with progress in epitaxial growth methods, has provided us a new family of two-dimensional electron systems that can host broken symmetries and topological orders. In chirally stacked graphene systems[5, 6, 7], a new type of broken symmetry states[8, 9] with momentum space vortices and real-space inversion symmetry breaking, first proposed theoretically within a Hartree Fork mean-field theory[10], appears likely on the basis of a perturbative renormalization group analysis[8], and recent experiments.[11, 12]. These states have recently attracted a great deal of attention [8, 9, 10, 11, 12, 13, 15, 14, 16, 17]. These spontaneous quantum Hall states[9] are characterized by spin and valley dependent spontaneous layer polarization, driven by interlayer electron-electron interactions. A family of closely related broken symmetry states exhibit different orbital magnetizations and topological edge states[9], and are distinguished by their quantized charge, spin, and valley Hall conductances[9].

Quasiparticles in graphene are described by the massless Dirac fermion model[3, 4] and the dispersion of the band remains linear down to the charge neutrality point. When these honeycomb layers are stacked, electronic properties are strongly modified in a way that is controlled by the specific stacking arrangement[5, 6, 7]. Among all stacking possibilities, only the ABC arrangement[6, 7], also referred to chiral stacking[9] arrangements, extend the following features that make Bernal bilayer[5] ($N = 2$) electronic structure particularly interesting to larger N-layer films: (i) two low-energy sublattice sites are localized in the outermost layers, and any inversion asymmetry, i.e. an electric field perpendicular to the film, is able to induce a energy gap;[5, 6, 7] (ii) hopping between low-energy sites via high energy sites is an N-step

Published under licence by IOP Publishing Ltd

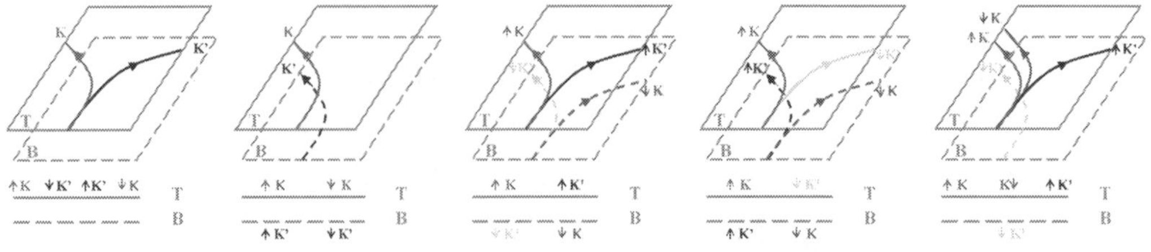

(a) QVH (b) QAH (c) LAF (d) QSH (2D TI) (e) "ALL"

Figure 1. For cases (a-e) the lower panel describes the sense of layer polarization for each spin-valley combinations while the upper panel schematically indicates the corresponding Hall conductivity contributions[9]. (a) a valley Hall insulator with a net layer polarization and a mass $m\sigma_z$; (b) an anomalous Hall insulator with a valley-dependent mass $m\tau_z\sigma_z$; (c) a layer-antiferromagnetic insulator with a spin-dependent mass $m\, s_z\sigma_z$; (d) a quantum spin Hall (or 2D Topological) insulator with a valley and spin dependent mass term $m\tau_z s_z\sigma_z$; (e) an exotic Hall state with a valley and spin dependent mass term $m(\frac{1+\tau_z}{2}+\frac{1-\tau_z}{2}s_z)\sigma_z$.

process which leads to $p^{\rm N}$ dispersion at Dirac points, sublattice pseudospin chirality N and Berry phase $N\pi$;[6, 7] (iii) the density-of-state $D(E)\sim E^{(2-{\rm N})/{\rm N}}$ diverges[7] as E approaches zero for $N>2$ whereas it remains finite for $N=2$ and vanishes for $N=1$. Because of the flat bands and the large pseudospin chirality, many-body effects become dominant[7] in the low-energy physics of a charge neutral chiral multilayer systems.

When these band states are described in a pseudospin language[10], the broken symmetry state is characterized by a momentum-space vortex with vorticity N and a vortex-core which is polarized in the top-or-bottom layers. For a chiral (Bernal) bilayer, for example, interactions lead to a broken symmetry ground state with a spontaneous gap in which charge is transferred between top and bottom layers. As discussed in Ref.[8, left=,right=], the chiral bilayer can be viewed as a two-dimensional analog of the Luttinger liquid; however, the mean-field CDW-like instability is enhanced by quantum fluctuations[8] in bilayers, whereas it is canceled by the weak attractive BCS instability in Luttinger liquids.

In the rest of the current paper, we will first classify the broken symmetry states, which can be characterized by the signs of the valley and spin dependent mass terms[9] which appear in their mean-field Hamiltonians and by their quantized charge, valley and spin Hall conductances[9]. In section III we will explain why quantum anomalous[18, 19, 9, 16, 11, 12] and valley Hall[20, 21, 9] states are favored by a weak magnetic field and an electric field across the graphene layers, respectively. When the spin-degree of freedom is accounted for a topological insulator phase[22, 23, 9, 24] can also appear. In section IV we discuss the Luttinger liquid systems which appear at boundaries between different phases. We conclude with a brief summary.

2. Classification of Broken Symmetry States

We discuss the electronic properties of N-layer ABC-stacked systems in terms of ordered state mean-field Hamiltonians of the form[9],

$$\mathcal{H}_{\rm N}=\frac{(v_0 p)^{\rm N}}{(-\gamma_1)^{\rm N-1}}[\cos(N\phi_{\boldsymbol p})\sigma_{\rm x}+\sin(N\phi_{\boldsymbol p})\sigma_{\rm y}]+m\sigma_{\rm z}\,. \qquad (1)$$

We have used the notation $\cos\phi_{\boldsymbol p}=\tau_z p_{\rm x}/p$ and $\sin\phi_{\boldsymbol p}=p_{\rm y}/p$ where $\tau_z=\pm1$ labels valleys K and K', the two inequivalent Brillouin zone corners. The Pauli matrices $\vec\sigma$ act on a *which-layer* pseudospin degree-of-freedom. In Eq. (1) the first term is the low-energy $\vec{k}\cdot\vec{p}$ band Hamiltonian for a single valley[5, 7]. Weak remote hopping processes have been dropped with the view that they do not play an essential role in the broken symmetry states. The second term is an interlayer interaction induced gap term which defines the direction of layer polarization in the momentum space vortex core[10, 8]. For each spin and valley, symmetry is broken by choosing a sign for m[8, 10]. We have dropped the momentum dependence of m because, as we see below, it does not play an essential role. In the mean-field Hamiltonians $2|m|$ is the size of the spontaneous gap, v_0 is the Fermi velocity in monolayer graphene, and $\gamma_1 \sim 0.4$ eV is the inter-layer hopping energy. The p^{N} dispersion is a consequence of the N-step process in which electrons hop between low-energy sites in top and bottom layers via high-energy states.

When spin and valley degrees-of-freedom are taken into account, the system has sixteen distinct broken symmetry states in which the sign of m is chosen separately for $(K\uparrow)$, $(K\downarrow)$, $(K'\uparrow)$ and $(K'\downarrow)$ flavors[9]. We take the view that any of these states could be stable, depending on details[17] that are beyond current knowledge and might be tunable experimentally. The sixteen states can be classified according to their total layer-polarization which is proportional to the sum over spin-valley of the sign of m. Six of the sixteen states have no net charge transfer between top and bottom layers and are likely to be lowest in energy in the absence of an external electric field. These six states can be separated into three doublets which differ only by layer inversion in every spin-valley. Thus three essentially distinct states compete for the broken symmetry ground state[9]: the anomalous Hall state in which the sign of m is valley-dependent but not spin-dependent ($m\sigma_z \to m\tau_z\sigma_z$), the layer-antiferromagnetic state in which m is only spin-dependent ($m\sigma_z \to ms_z\sigma_z$) and the topological insulator (TI) state in which m is both spin and valley dependent ($m\sigma_z \to m\tau_z s_z\sigma_z$). Other interesting states[9] with net layer charge transfer between top and bottom layers can be favored by a potential difference between the layers: for example the valley Hall state in which m is neither valley nor spin dependent and the "All" state in which m depends on both valley and spin ($m\sigma_z \to m(\frac{1+\tau_z}{2} + \frac{1-\tau_z}{2}s_z)\sigma_z$).[9] These states are distinguished[9] by their spin and valley dependent Hall conductivities and orbital magnetizations indicated schematically in Fig. 1 and summarized in Table 1. The broken symmetry states are distinguished by the signs of the Berry curvatures contributions[9] from near the K and K' valleys of $\uparrow$ and $\downarrow$ spin bands.

The Berry curvatures[25, 9] are non-zero only when inversion symmetry is spontaneously broken and act like an effective magnetic field in momentum space. In a gapped monolayer graphene, this effective magnetic field around one valley can be understood as a field generated by a magnetic monopole located at the valley center. The Berry curvature has opposite sign[9] near the two inequivalent Brillouin zone corners for a quantum valley Hall state as we see

Table 1. Summary[9] of spin-valley layer polarizations (T or B) and corresponding charge, spin, and valley Hall conductivities (e^2/h units) and insulator types for the three distinct states (b-d) with no overall layer polarization, for a state in which every spin-valley is polarized toward the top layer (a), and for a state with partial layer polarization (e).

Fig.	$K\uparrow$	$K\downarrow$	$K'\uparrow$	$K'\downarrow$	$\sigma^{(SH)}$	$\sigma^{(VH)}$	$\sigma^{(CH)}$	$\sigma^{(SVH)}$	Insulator
1(b)	T	T	B	B	0	0	$2N$	0	QAH
1(c)	T	B	T	B	0	0	0	$2N$	LAF
1(d)	T	B	B	T	$2N$	0	0	0	QSH
1(a)	T	T	T	T	0	$2N$	0	0	QVH
1(e)	T	T	T	B	N	N	N	N	All

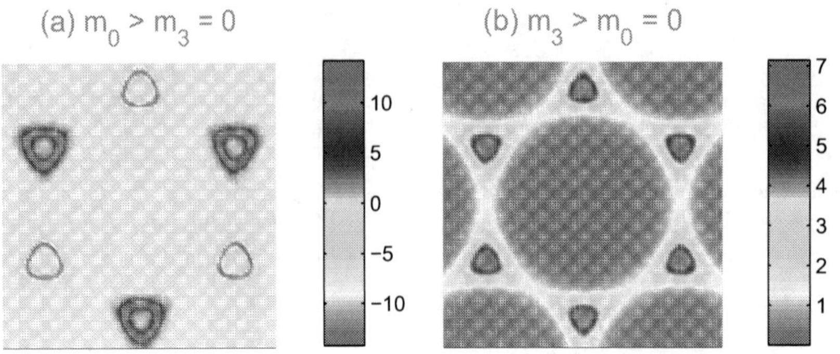

Figure 2. (Color online) Berry curvature (an effective magnetic field in momentum space) as a function of momentum-space position in the bilayer graphene Brillouin zone[9]. (a) For a quantum valley Hall state; (b) for a quantum anomalous Hall state. Large energy gaps are chosen deliberately in order to visualize the shape of the peaks and the trigonal warping effect.

in Fig. 2(a) but has the same sign[9] at both corners for a quantum anomalous Hall state as illustrated in Fig. 2(b). The Berry curvature not only has sharp peaks near all the Brillouin zone corners but at these large mass values exhibits large trigonal warping effect as well, reflecting the C_3 symmetry of the sublattice. In the presence of an in-plane electric field, an electron acquires an anomalous transverse velocity proportional to the Berry curvature, giving rise to an intrinsic Hall conductivity[25, 19]. Provided that the Fermi level lies in the mass gap, we find that the intrinsic Hall conductivity contribution[9] from a given valley and spin is

$$\sigma_{\mathrm{H}}(\tau_{\mathrm{z}}, s_{\mathrm{z}}) = \tau_{\mathrm{z}}\mathrm{sgn}(\mathrm{m})\frac{\mathrm{Ne}^2}{2\mathrm{h}}, \tag{2}$$

each spin and valley contributes $Ne^2/2h$ to the total Hall conductivity, with the sign given by $\tau_{\mathrm{z}}\mathrm{sgn}(\mathrm{m})$.

3. Competing Orders

In the spinless case, there are only two broken symmetry states, namely, the anomalous Hall state and the valley Hall state. In the case of a valley Hall state, valley K and K' have the same mass $m_0\sigma_{\mathrm{z}}$, and each valley is layer polarized in the same sense[9]. Mass terms with the valley Hall form can be generated simply by a potential difference between the layers, so the valley Hall state is an easily generated non-interacting electron state. The total Hall conductivity of this state is zero, i.e. $\nu = 0$ for bilayer graphene, with the K and K' valleys making Hall conductivity and magnetization contributions of opposite sign, preserving time reversal symmetry. The valley Hall states can be favored experimentally by applying a perpendicular electric field[9, 21], as shown in Fig. 3(a).

In the anomalous Hall state, on the other hand, one valley has a positive mass $m_3\sigma_{\mathrm{z}}$ while the other has a negative mass $-m_3\sigma_{\mathrm{z}}$. Correspondingly the two valleys spontaneously choose different layer polariztions[9]. Therefore the Hall conductivity and orbital magnetization contributions have the same sign for each valley[9]. This state breaks time reversal symmetry. The total Hall conductivity has the quantized value $\pm 2Ne^2/h$, i.e. $\nu = \pm 4$ in the case of bilayer graphene. In addition to its anomalous Hall effect, this state has a substantial orbital magnetization. The anomalous Hall state is probably most simply identified experimentally[11, 12] by observing a $\nu = 2N$ QHE which persists to zero magnetic field[9, 11, 12, 16], as depicted in Fig. 3(b).

The spectra of translationally invariant two-dimensional electron systems are quantized into Landau levels in the presence of a finite perpendicular magnetic field. In a gapless chirally stacked graphene N-layer, the zero energy Landau level (Fig. 3(c)) is $4N$ fold degenerate. In the case of a valley Hall insulator, the states around K and K' valleys have very large orbital magnetic moments with opposite sign, which couple to the out-of-plane magnetic field and gives rise to the relative energy shift between valley K and K'. Consequently, the two-fold valley degeneracy is lifted and there is an asymmetry between those Landau levels around two Dirac points, as described in Fig. 3(d). The energy spectrum of a quantum valley Hall insulator is adiabatically connected to that of the $\nu = 0$ quantum Hall state. In contrast, as shown in Fig. 3(e)(f), the valley degeneracy is unbroken in the presence of magnetic field, because the nature of the orbital magnetic moments near the two Dirac points are the same. The energy gap in a quantum anomalous Hall state is adiabatically connected to that of the $\nu = \pm 4$ quantum Hall state where the sign is a choice of whether the magnetic field is in $\hat{z}$ or $-\hat{z}$ direction relative

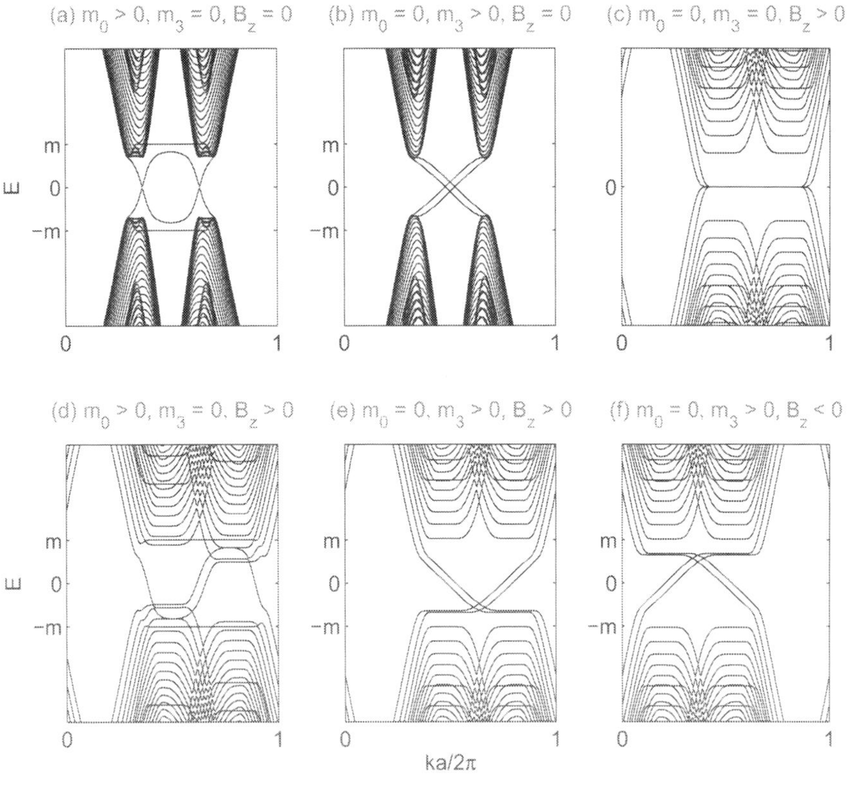

Figure 3. (Color online) (a) Ribbon quantum valley Hall states and (b) ribbon quantum anomalous Hall states in the absence of magnetic fields; (c) Ribbon quantum Hall states for a gapless bilayer graphene; (d) $\nu = 0$ quantum Hall states for a gapped bilayer graphene in QVH phase; (e) and (f) $\nu = \pm 4$ quantum Hall states for a gapped bilayer graphene in QAH phase. To visualize the edge states, we use a zigzag ribbon, the intralayer and interlayer nearest neighbor hoppings are chosen as $\gamma_0 = 1$ and $\gamma_1 = 0.3$, respectively, and nonzero m is fixed as 0.15.

to m_3.

When spin is taken into account, the quantized Landau levels of the three additional phases[9] are obtained by each spin choosing to be a quantum valley Hall state or a quantum anomalous Hall state. For the case of a quantum spin Hall state, one spin flavor is the $\nu = 2$ QAH state and the other flavor is the $\nu = -2$ QAH state. For the layer-antiferromagnetic state, each spin flavor is a $\nu = 0$ QVH state but with the opposite layer polarization to the other flavor. In contrast, one spin flavor is a QVH state while the other is a QAH state in the "All" state.

By continuously lowering the magnetic field to zero, an anomalous Hall state is adiabatically identified[11, 12]. When a perpendicular electric field is switched on, the quantum valley Hall phase starts to compete with the quantum anomalous Hall phase. The sizes of the gap at the two valleys are $|m_0 + m_3|$ and $|m_0 - m_3|$, respectively. The state is still within the anomalous Hall phase but with two unequal gaps at K and K' for $|m_0| < |m_3|$ as seen in Fig. 4(a). Beyond the critical point where $|m_0| = |m_3|$ as depicted in Fig. 4(c), the state jumps to the valley Hall phase instead. In the quantum phase transition region as described by Fig. 4(b), the energy gap is enhanced at one valley while it closes at the other valley, where quantum anomalous and valley Hall effects coexist.

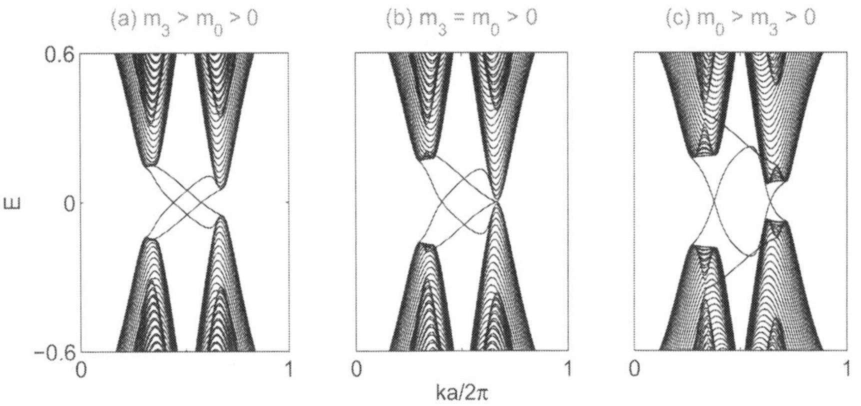

Figure 4. (Color online) The competition between the quantum anomalous Hall state (intra-valley edge state) and the quantum valley Hall (inter-valley edge state) state in broken symmetry bilayer graphene. (a) a quantum anomalous Hall state with unequal gaps at K and K'; (b) a critical point with an enhanced gap at K and a closed gap at K', where quantum anomalous and valley Hall effects coexist; (c) a quantum valley Hall state with a reopened gap at K'. To visualize the edge states, we use a zigzag ribbon with the same γ_0 and γ_1 values as in Fig. 3.

In the quantum spin Hall state[9], the helical edge modes are likely to localize in a N-even system, due to the possible backscattering process in which N left movers and N right movers scatter into each other allowed by time reversal symmetry. Therefore, the QSH phase is topologically protected in chirally stacked odd number of graphene layers. Besides the effective spin-orbit coupling induced by interlayer interaction, Rashba interaction $\lambda_{\rm R}(\tau_{\rm z}\sigma_{\rm x}s_{\rm y} - \sigma_{\rm y}s_{\rm x})$ is possibly induced by inversion symmetry breaking, and a small staggered potential $m_{\rm S}\sigma_{\rm z}$ is also present due to coupling to substrates. These interactions lift the degenerate zero energy to $m_{\rm SH} \pm \lambda_{\rm S}$, $-m_{\rm SH} \pm \sqrt{4\lambda_{\rm R}^2 + \lambda_{\rm S}^2}$. The topological-nontrivial phase persists as long as the effective spin-orbit gap is not closed, or in other words,

$$\left(\frac{\lambda_{\rm R}}{m_{\rm SH}}\right)^2 + \left|\frac{\lambda_{\rm S}}{m_{\rm SH}}\right| \leq 1\,, \qquad (3)$$

which is independent of layer number N and determines the phase diagram. Trigonal warping and other remote hopping terms are time reversal invariant, not disfavoring the TI phase.

4. Edge States at Domain Walls

At zero temperature, in a clean chirally stacked few-layer graphene system, there are 16 possible broken symmetry states and they are classified as 5 distinct phases, as discussed in the section II. In the presence of disorder or thermal fluctuations, different phase are likely to appear locally in different parts of the system. There are 240 possible domain walls and they can be classified into 16 distinct types. In the spinless case, as only valley Hall phase and anomalous Hall phase are allowed, there are 2 types of intra-phase and 1 type of inter-phase domain walls. At each type of domain wall, we argue that a unique Luttinger liquid emerges; the spinless bilayer examples of which are illustrated in Fig. 5.

In a quantum valley Hall state, the Chern numbers[2, 9] of the two valleys are $\pm N/2$, respectively. For the case of a quantum anomalous Hall state, the Chern number[2, 9] is $N/2$ at both valleys with a uniform sign. These features are equivalently shown in Fig. 2. At the domain wall separating two quantum valley Hall regions with the opposite layer polarization, the Chern numbers change by $\pm N$ for a single valley and hence N parallel zero modes appear at each valley as depicted in Fig. 5(a). These one dimensional zero modes form N copies of full Luttinger liquids and the valley-pseudospin becomes exactly the left-or-right chirality. This QVH-QVH domain walls and the generated Luttinger liquids are likely to be formed in the electron-hole puddles, and can be easily realized and tuned by an external electric field[26].

At a domain wall separating two quantum anomalous Hall regions with opposite total Hall conductance, the change of Chern number is N (neglecting spin) for both valleys. Therefore, we expect that N parallel zero modes appear at each valley as seen in Fig. 5(b). Each valley has a copy of a purely chiral "spin"-$\frac{N-1}{2}$ like Luttinger liquid. At the domain wall between a quantum valley Hall and a quantum anomalous Hall regions, the chern number is changed by N for one valley while it is preserved for the other. Thus the edge states at the interface are purely chiral at one valley while they completely disappear at the other. We expect the Luttinger liquid behaviors of the one dimensional zero modes at the spinfull domain walls are much more exotic.

5. Discussion

The trigonal warping effect is sometimes ignored in theoretical efforts to identify the broken symmetry physics based on mean field theory or renormalization group analysis. This is reasonably justified in the bilayer case for the following two reasons. (i) The trigonal warping effect dominates only below 1 meV[5], compared to relevant band broadening not much smaller than 1 meV. Trigonal warping effect are therefore likely to be smeared and become inessential, due to residue disorder and fluctuations at higher energies. (ii) The infrared cutoff of the RG flows can be reasonably set at where the quadratic band dispersion overwhelms the trigonal warping effect. For chirally stacked graphene with the layer number $N > 2$, the trigonal warping energy scale is increased by almost a factor of 10[7] while the even flatter gapless bands are much more unstable to interlayer interactions. Interaction effects are expected to dominate at low carrier densities[7] and to drive the spontaneous inversion symmetry breaking, but samples that are clean enough to reveal its interaction physics have not yet been studied. We comment that the broken symmetry is likely to occur only in low disorder and high quality samples, since large disorder can destroy the perfect nesting conditions and reduce the parameter space of the RG flows.

Recent experiments[11, 12] in bilayers appear to provide definitive proof that, at $\nu = \pm 4$, the ground state at very weak external magnetic fields is the quantized anomalous Hall state. (At $\nu = 0$, LAF state (or called SDW state) and QSH state are competing for the ground state at zero magnetic field.) Even though a QAH state does not have a finite spin-polarization, the orbital

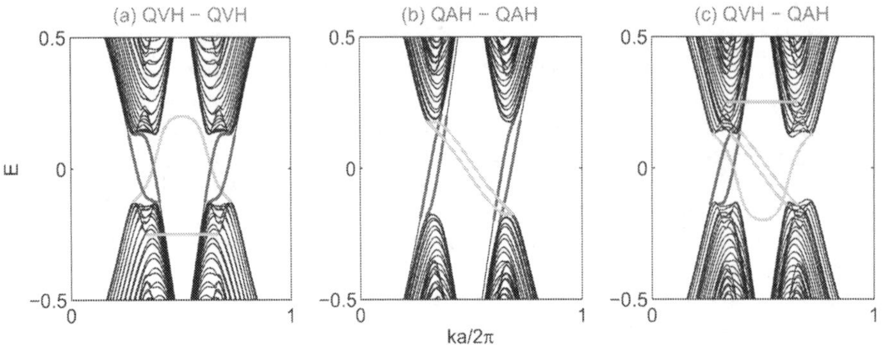

Figure 5. (Color online) Three distict Luttinger liquids at domain walls in spinless bilayer graphene with broken symmetry. The red lines denotes the zero modes localized at domain walls between (a) two quantum valley Hall regions with opposite layer polarization; (b) two quantum anomalous Hall regions with opposite total Hall conductance; (c) a quantum valley Hall region and a quantum anomalous Hall region. The green lines represent the edge states on the outermost zigzag boundaries and note that they are doubly degenerate in (a) and (b) for the two zigzag boundaries. To visualize the edge states, we use a zigzag ribbon with the same γ_0 and γ_1 values as in Fig. 3 and $|m| = 0.25$.

magnetic moment close to each Dirac point has a symmetric sharp peak[9] at which individual states carry moments twenty times larger than μ_B, and a total orbital magnetization per area goes as $\sim (N\lambda m_e/2\pi\hbar^2)\ln(\gamma_1/|\lambda|)\mu_B$, that is $\sim 0.002\mu_B$ per carbon atom for $|m| = 10$ meV[9]. Thus the energy of the quantized anomalous Hall state will be lowered by a perpendicular external magnetic field. Lattice mean field theory shows the size of the gap can be as large as 30 meV[17] but the actual ground state is dependent on subtle correlation and microscopic physics issues. We estimate that a magnetic field of the order of 0.004 T is sufficient to favor the QAH state over the QVH state[17]. Increasing the magnetic field further results in quantum Hall ferromagnetism[27, 28, 29]. The fully layer polarized QVH state will be favored by an external electric field which produces a potential difference between the layers. The turning point of the band gap is approximately 7 mV/nm for a perpendicular electric filed[17].

The quantum spin Hall effect we discuss in this paper is in several respects different from that discussed in the well known papers[22, 23] which foreshadowed the identification of topological insulators. (i) The broken symmetry occurs only for $N \geq 2$, rather than in the single-layer systems[22, 23]. (ii) The quantum spin Hall effect is driven by broken symmetries produced by weak repulsive interaction instability, rather than by spin-orbit interactions[22, 23]. the interaction induced effective spin-orbit coupling $\lambda\tau_z s_z \sigma_z$ can be 10^4 times larger[9, 17] than the intrinsic one[30].

We close this discussion by pointing out that the edge states of a broken symmetry state have physical significance[9]. (i) The edge states for QAH phase are inter-valley one dimensional gapless modes while the zero modes are intra-valley like for QVH phase[9]. (ii) The states with anomalous Hall effects have N topologically protected robust chiral edge states associated with the QHE[9]. (iii) For the valley Hall effects, in general we expect $[N/2]$ chiral edge state branches at each valley in an N-layer stack[9]; the full e^2/h unit of Hall conductance requires the two valleys to act in concert; the additional half quantum Hall effect from each valley in the N-odd layers is insufficient to produce a new chiral edge state branch. This is a novel manifestation of the chiral anomaly in condensed matter systems. Of course valley Hall edge states are topologically protected only when the edge-direction projections of K and K' valleys are not coincident and inter-valley scattering due to disorder is absent.

Acknowledgement
This work has been supported by Welch Foundation under grant TBF1473, NRI-SWAN, DOE grant Division of Materials Sciences and Engineering DE-FG03-02ER45958, and ARO grant W911NF-09-1-0527. We acknowledge helpful discussions with C. Lau, B. Halperin, A. Yacoby, K. Novoselov, W. Bao, J. Velasco and C. Schönenberger. When this manuscript is about to publish, some experiments in graphene bilayers[31, 32] and trilayers[33] appear online, and a gapped $\nu = 0$ state has been observed in bilayer graphene at zero magnetic field[31].

References

[1] K. Von Klitzing, G. Dorda, and M. Pepper, Phys. Rev. Lett. **45**, 494 (1980).
[2] D. J. Thouless, M. Kohmoto, M. P. Nightingale, and M. Dennijs, Phys. Rev. Lett. **49**, 405 (1982).
[3] A. K. Geim and A. H. MacDonald, Phys. Today **60**(8), 35(2007).
[4] A. H. Castro Neto *et al.*, Rev. Mod. Phys. **81**, 109 (2009).
[5] E. McCann and V. I. Fal'ko, Phys. Rev. Lett. **96**, 086805 (2006).
[6] Hongki Min and A. H. MacDonald, Phys. Rev. B **77**, 155416 (2008).
[7] F. Zhang, B. Sahu, H. Min, A. H. MacDonald, Phys. Rev. B **82**, 035409 (2010).
[8] F. Zhang, H. Min, M. Polini and A. H. MacDonald, Phys. Rev. B **81**, 041402(R) (2010).
[9] F. Zhang, J. Jung, G. A. Fiete, Q. Niu and A. H. MacDonald, Phys. Rev. Lett. **106**, 156801 (2011).
[10] H. Min, G. Borghi, M. Polini and A. H. MacDonald, Phys. Rev. B **77**, 041407(R) (2008).
[11] J. Martin *et al.*, Phys. Rev. Lett. **105**, 256806 (2010).
[12] R. T. Weitz *et al.*, Science **330**, 812 (2010).
[13] R. Nandkishore and L. Levitov, Phys. Rev. Lett. **104**, 156803 (2010).
[14] O. Vafek and K. Yang, Phys. Rev. B **81**, 041401(R) (2010).
[15] F. Guinea, Physics **3**, 1 (2010).
[16] R. Nandkishore and L. Levitov, Phys. Rev. B **82**, 115124 (2010).
[17] J. Jung, F. Zhang, A. H. MacDonald, Phys. Rev. B **83**, 115408 (2011).
[18] F. D. M. Haldane, Phys. Rev. Lett. **61**, 2015 (1988).
[19] N. Nagaosa, J. Sinova, S. Onoda, A. H. MacDonald and N. P. Ong, Rev. Mod. Phys. **82**, 1539 (2010).
[20] D. Xiao, W. Yao and Q. Niu, Phys. Rev. Lett. **99**, 236809 (2007).
[21] J. Li, I. Martin, M. Buttiker, A. F. Morpurgo, Nature Phys. **7**, 38 (2011).
[22] C. L. Kane and E. J. Mele, Phys. Rev. Lett. **95**, 226801 (2005).
[23] S. Raghu, X. Qi, C. Honerkamp, and S. Zhang, Phys. Rev. Lett. **100**, 156401 (2008).
[24] E. Prada, P. San-Jose and L. Brey, arXiv:1007.4910 (2010).
[25] D. Xiao, M. Chang and Q. Niu, Rev. Mod. Phys. **82**, 1959 (2010).
[26] I. Martin, Y. M. Blanter, and A. F. Morpurgo, Phys. Rev. Lett. **100**, 036804 (2008).
[27] Y. Barlas, R. Cote, K. Nomura and A. H. MacDonald, Phys. Rev. Lett. **101**, 097601(2008).
[28] B. Feldman, J. Martin and A. Yacoby, Nature Phys. **5**, 889 (2009).
[29] Y. Zhao, P. Cadden-Zimansky, Z. Jiang and P. Kim, Phys. Rev. Lett. **104**, 066801 (2010).
[30] H. Min *et al.*, Phys. Rev. B **74**, 165310 (2006).
[31] J. Velasco Jr. *et al.*, to appear soon (2011).
[32] F. Freitag *et al.*, arXiv:1104.3816 (2011).
[33] W. Bao *et al.*, arXiv:1103.6088 (2011).

Journal of Physics: Conference Series **334** (2011) 012003 doi:10.1088/1742-6596/334/1/012003

DMRG study of fractional quantum Hall effect and valley skyrmions in graphene

Naokazu Shibata

Department of Physics, Tohoku University, Aoba, Aoba-ku, Sendai 980-8578, Japan

E-mail: shibata@cmpt.phys.tohoku.ac.jp

Abstract. The ground state and low-energy excitations of graphene and its bilayer are investigated by the density matrix renormalization group (DMRG) method. We analyze the effect of Coulomb interaction between the electrons including valley degrees of freedoms. The obtained results show finite charge excitation gap at various fractional fillings $\nu_n = 1/3, 2/5, 2/3$ in the $n = 0$ and 1 Landau levels of single-layer graphene (SLG) and $n = 2$ Landau level of bilayer graphene (BLG). The lowest charge excitations at $\nu = 1/3$, and 1 in SLG are valley skyrmions.

1. Introduction

Graphene is an atomic scale two-dimensional electron system with low-energy electrons in two valleys, K and K' in the Brillouin zone. The K and K' are inequivalent Brillouin zone corners defined by triangular sublattices of hexagonal structure of carbon atoms. In contrast to conventional two-dimensional systems, the single-electron wavefunction has two sublattice components reflecting two triangular sublattices of graphene, which are connected by nearest-neighbor hopping element t. Such two components wavefunction and the valley degrees of freedom are unique properties of graphene [1, 2, 3, 4]. In the present paper, we investigate how these characteristics affect fractional quantum Hall states, which is expected to be realized in high magnetic field [5, 6].

In magnetic field, the presence of two valleys is observed in half-integer Hall conductivity[7, 8]

$$\sigma_{xy} = (n + \frac{1}{2})\frac{4e^2}{h}, \tag{1}$$

at $\nu = \pm 2, \pm 6, \pm 10, \cdots$ with factor 4 for single-layer graphene (SLG), which reflects 4-fold Landau level (LL) degeneracy, resulting from spin and valley symmetry in graphene.

The two components single-particle Hamiltonian for the K-valley of SLG[4] is written as

$$\mathcal{H}_{\mathrm{K}}^{(\mathrm{SLG})} = \frac{\sqrt{2}\hbar v_F}{\ell_B}\begin{pmatrix} 0 & a \\ a^\dagger & 0 \end{pmatrix}, \tag{2}$$

where $a \equiv (\pi_x - i\pi_y)\ell_B/\sqrt{2}\hbar$ is the magnetic ladder operator and $\pi = \mathbf{p} - e\mathbf{A}$ is the kinetic momentum. v_F is the Fermi velocity of SLG, $\ell_B = \sqrt{\hbar/eB}$ is the magnetic length, and 2×2 matrices represent sublattice degrees of freedom in graphene. $\mathcal{H}_{K'}^{(\mathrm{SLG})}$ is the transpose of $\mathcal{H}_{K}^{(\mathrm{SLG})}$. The eigenenergy is given by

$$\epsilon_n = \pm\hbar v_F \sqrt{2n}/\ell_B. \tag{3}$$

and the eigenvectors are $|0\rangle_K^{(\mathrm{SLG})} = (0, |0\rangle)^t$ for $n = 0$ and $|n, \pm\rangle_K^{(\mathrm{SLG})} = (|n-1\rangle, \pm|n\rangle)^t/\sqrt{2}$ for $n \geq 1$. Here, $|n\rangle$ is the eigenvector of the number operator $N \equiv a^\dagger a$ with an eigenvalue n.

The bilayer graphene (BLG) has an ordinary parabolic spectrum in the vicinity of the neutrality point. In a magnetic field, the effective Hamiltonian for BLG is written in the form [9, 10]

$$\mathcal{H}_{\mathrm{K}}^{(\mathrm{BLG})} = \hbar\omega_c \begin{pmatrix} 0 & a^2 \\ (a^\dagger)^2 & 0 \end{pmatrix}, \tag{4}$$

and its eigenenergy is

$$\epsilon_n = \pm\hbar\omega_c\sqrt{n(n-1)}. \tag{5}$$

Here, $\omega_c = eB/m$ with m being the effective mass of BLG. The eigenvectors in eq. (4) are $|0,0\rangle_K^{(\mathrm{BLG})} = (0, |0\rangle)^t$, $|0,1\rangle_K^{(\mathrm{BLG})} = (0, |1\rangle)^t$ for the zero energy level and $|n, \pm\rangle_K^{(\mathrm{BLG})} = (|n-2\rangle, \pm|n\rangle)^t/\sqrt{2}$ for $n \geq 2$.

The relevant energy scales in graphene in a magnetic field are as follows: (i) LL separation around the neutrality point, $\sqrt{2}\hbar v_F/\ell_B \simeq 400\sqrt{B[T]}[K]$ for SLG, while $\sqrt{2}\hbar\omega_c \simeq 30 \times (B[\mathrm{T}])[\mathrm{K}]$ for BLG, (ii) Zeeman coupling, $\Delta_z \equiv g\mu_B|\mathbf{B}| \simeq 1.5 \times (B[T])[K]$, and (iii) the Coulomb energy $e^2/\epsilon\ell_B \simeq 100\sqrt{B[T]}[K]$.

We focus mainly on many-body states in nonzero LLs of SLG and BLG. We introduce the effective filling factors of the topmost partially filled nth LL,

$$\nu_n^{(\mathrm{SLG})} = \nu - 4(n - 1/2) \tag{6}$$

for SLG and

$$\nu_n^{(\mathrm{BLG})} = \nu - 4(n-1) \tag{7}$$

for BLG. Here, $\nu_n \leq 4$ and $n \geq 2$ for BLG.

2. Model and Method

Since the LL separation is larger than typical Coulomb energy $e^2/\epsilon\ell_B$ in high magnetic fields, we first project Coulomb interaction Hamiltonian onto a certain LL. The projected Hamiltonian onto the nth LL is written as[11]

$$H = \frac{1}{L^2}\sum_{i<j}\sum_{\mathbf{q}} V(q)e^{-q^2/2}[F_n(q)]^2 e^{i\mathbf{q}\cdot(\mathbf{R}_i - \mathbf{R}_j)}, \tag{8}$$

where $\mathbf{R}_i$ is the guiding center coordinate of the ith particle. The *relativistic* form factor in the nth LL of SLG is written as [12]

$$F_0^{(\mathrm{SLG})}(q) = L_0\left(q^2/2\right) \tag{9}$$

and

$$F_{n\geq 1}^{(\mathrm{SLG})}(q) = (1/2)\left[L_{|n|}\left(q^2/2\right) + L_{|n|-1}\left(q^2/2\right)\right]. \tag{10}$$

For the $n(\geq 2)$th LL in BLG,

$$F_{n\geq 2}^{(\mathrm{BLG})}(q) = (1/2)\left[L_{|n|}\left(q^2/2\right) + L_{|n|-2}\left(q^2/2\right)\right]. \tag{11}$$

Here, $L_n(x)$ are the Laguerre polynomials. In the spherical geometry, it is convenient to write the Hamiltonian as

$$H^{(n)} = \sum_{i<j}\sum_{m} V_m^{(n)} P_{ij}[m], \tag{12}$$

where $P_{i,j}[m]$ projects onto states in which particles i and j have the relative angular momentum $\hbar m$, and $V_m^{(n)}$ is their interaction energy in the nth LL.[11] Using the above form factors, the pseudopotentials[13, 11] are given by

$$V_m^{(n)} = \int_0^\infty \frac{dq}{2\pi} qV(q)e^{-q^2}[F_n(q)]^2 L_m(q^2). \tag{13}$$

The corresponding integrals for electrons on the surface of a sphere that are used in the present work are described in refs. [11, 14, 15]. Note that our Hamiltonian eqs. (8) and (12) have SU(2) symmetry in the valley degrees of freedom.

In the following, we treat the valley degrees of freedom K and K' in the language of the pseudospin, while real spin degrees of freedom are supposed to be frozen by Zeeman splitting. LL mixing is also neglected.

We calculate the exact wave functions of the ground and low-energy excited states in SLG and BLG, using the density matrix renormalization group (DMRG) method[16, 17, 18], which is applied to torus[17, 18] and spherical geometries[11, 13, 14, 15, 19, 21, 22]. The accuracy of the results is systematically controlled by the truncation error, which is smaller than 10^{-4} in the present calculation. We investigate systems of various sizes with up to 40 electrons in the unit cell keeping 1400 basis states in each block[17, 18].

3. Results

3.1. $\nu_n = 1$ *and* $1/3$ *states*

At $\nu_n = 1$ integer fillings, the pseudospins of valley degrees of freedom are completely polarized through the exchange Coulomb interaction. Since fully pseudospin-polarized quasiparticle excitations require energy to the next higher LL, we calculate the unpolarized and partially polarized excitations within the same LL by the DMRG method.

The pseudospin polarization and the energy of the lowest charge excitation are shown in Fig. 1 for systems in spherical geometry. This figure shows that the pseudospin-unpolarized state is the lowest excited state when the system size is sufficiently large in SLG ($n = 0$, 1, and 2) and BLG ($n = 2$). The extrapolated values of the pseudospin-unpolarized excitation gaps Δ_s in units of $e^2/(\epsilon\ell_B)$ are 0.63, 0.28, and 0.42 in the $n = 0, 1,$ and 2 LLs of SLG, and 0.48 in the $n = 2$ LL of BLG[22].

In the $n = 2$ LL, the lowest quasiparticle excitation in small systems ($N_e \lesssim 20$) has a large pseudospin polarization $P = 1 - 2/N_e$ both for SLG and BLG, where $P \equiv (N_K - N_{K'})/(N_K + N_{K'})$ with N_K ($N_{K'}$) being the number of electrons in the K (K') valley. These results show the instability of the unpolarized $P = 0$ excitations in small systems. This is consistent with the fact that the unpolarized excitations are unstable in higher LLs because of the long-range nature of the effective exchange interaction, which increases the energy of the skyrmion-like pseudospin-unpolarized state.

The elementally charge excitation gaps at the fractional fillings $\nu_n = 1/3$ are shown in Fig. 2. The pseudospin (valley)-polarized $P = 1$ excitation gap Δ_c for $n = 0$ SLG is 0.101 in units of

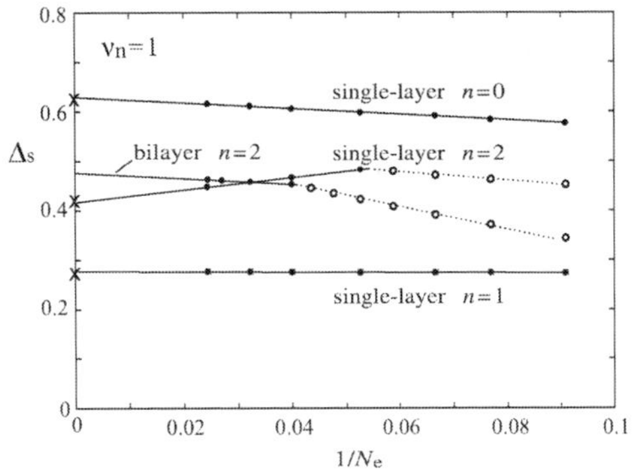

Figure 1. Lowest charge excitation gaps in SLG and BLG at $\nu_n = 1$ in the $n = 0$, 1 and 2 LLs in the spherical geometry. N_e is the number of electrons in the unit cell. Closed circles represent the pseudospin-unpolarized $P = 0$ (skyrmion) excitation gaps and open circles represent nearly pseudospin-polarized $P = 1 - 2/N_e$ excitation gaps. The crosses on the vertical axis represent results obtained by Hartree-Fock calculations[28].

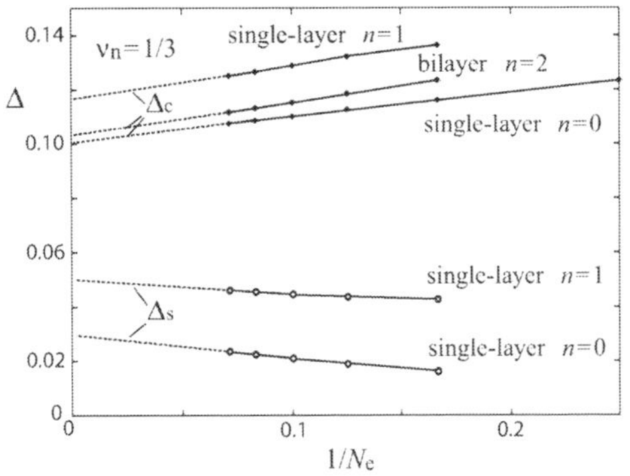

Figure 2. Pseudospin (valley)-polarized excitation gap Δ_c and the pseudospin (valley)-unpolarized (skyrmion) excitation gap Δ_s at $\nu_n = 1/3$ in SLG ($n = 0$ and 1 LLs) and in BLG ($n = 2$ LL).

$e^2/(\epsilon\ell_B)$ in the thermodynamic limit, which is in good agreement with the previous work[15]. In the $n = 1$ LL of SLG, Δ_c is 0.115, which is larger than 0.101 in the $n = 0$ LL of SLG. Such enhancement of the excitation energy in $n = 1$ LL is obtained in the energy spectra of finite system[5].

In BLG, $\Delta_c = 0.103$ in the $n = 2$ LL, which is also slightly larger than that in the $n = 0$ LL of SLG. This increase in Δ_c in higher LLs is consistent with the increase in the difference in Haldane's pseudopotential between $m = 1$ and 3; $V_1^n - V_3^n$ are 0.168 in units of $e^2/(\epsilon\ell_B)$ for BLG in the $n = 2$ LL, 0.198 for SLG in the $n = 1$ LL, and 0.166 for SLG in the $n = 0$ LL.

To study the pseudospin structure in the unpolarized excited states, we compare our numerical results with the Hartree-Fock (HF) trial states of skyrmions [23, 24, 25, 26, 27, 28].

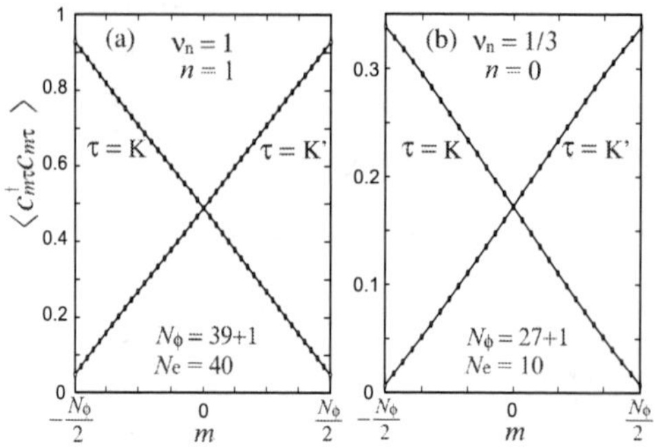

Figure 3. Expectation values, $\langle c^\dagger_{m\tau} c_{m\tau} \rangle$, in pseudospin (valley) unpolarized excited states with one extra flux at ν_n =1 and 1/3. N_ϕ is the total flux.

The HF trial state of quasihole skyrmions at $\nu_n = 1$ is written in the form[10]

$$|\Psi_{sk}\rangle = \prod_{m=-N_\phi/2}^{N_\phi/2} [\alpha_m c^\dagger_{mK} + \beta_m c^\dagger_{m+1K'}]|0\rangle, \tag{14}$$

where $\langle c^\dagger_{mK} c_{mK} \rangle = |\alpha_m|^2$ and $\langle c^\dagger_{mK'} c_{mK'} \rangle = |\beta_{m-1}|^2$. We have calculated the expectation values $\langle c^\dagger_{m\tau} c_{m\tau} \rangle$ from the wave function obtained in the present DMRG study. The results for $\nu_1 = 1$ in SLG indicate that they are approximately given by $\langle c^\dagger_{m\tau} c_{m\tau} \rangle = 1/2 \mp m/N_\phi$ for $\tau = K$ and $\tau = K'$ as shown in Fig. 3(a). Similar results are also obtained for $\nu_n = 1/3$ in the $n = 0$ and 1 LLs of SLG as shown in Fig. 3(b), which shows that the pseudospin-unpolarized elementally charge excitations at these fillings are pseudospin (valley) skyrmions.

As seen in conventional 2DEGs, $\nu = 1/3$ (valley or spin)-unpolarized excited states can be lower in energy than the polarized excited states. Indeed, as shown in Fig. 2, the unpolarized excitation gaps Δ_s in SLG are 0.03 in the $n = 0$ LL and 0.05 in the $n = 1$ LL, which are much smaller than the polarized excitation gaps Δ_c in the $n = 0$ and 1 LLs, respectively. In the case of BLG, however, we find that valley unpolarized excited states have higher energies than polarized excited states. Consequently, the activation energy at $\nu_2 = 1/3$ in BLG is 2-3 times larger than that in SLG.

3.2. $\nu_n = 2/5$ *and 2/3 states*

According to the composite fermion theory[29], $\nu = m/(1+2m)$ fractional quantum Hall states are mapped onto the $\nu_{\text{eff}} = m$ integer quantum Hall states. The $\nu = 2/5$ fractional quantum Hall state is then mapped onto the $n = 2$ integer quantum Hall state, where the lowest LL is doubly occupied by pseudospin-up and pseudospin-down electrons. We therefore expect a pseudospin-unpolarized ground state at $\nu_n = 2/5$. In the $n = 0$ LL of SLG, this is confirmed by the DMRG calculation as shown in Table 1. On the other hand, in the $n = 1$ LL of SLG, the ground state is pseudospin-polarized state.

The pseudospin (valley)-unpolarized excitation gap Δ_s at $\nu_n = 2/5$ is also shown in Table 1. The Δ_s from the unpolarized ground state in the $n = 0$ LL of SLG is 0.04 $e^2/(\epsilon\ell_B)$, and that from the polarized ground state in the $n = 1$ LL of SLG is 0.035 $e^2/(\epsilon\ell_B)$. In BLG, the polarized excitation gap is smaller than the unpolarized excitation gap, and $\Delta_c = 0.052$.

We finally study the valley polarization in the ground state and the elementally charged excitations at $\nu_n = 2/3$. the ground state at $\nu_n = 2/3$ in the $n = 0$ LL of SLG is pseudospin

Table 1. Pseudospin (valley) polarization in the ground state and the lowest charge excited state, and the unpolarized excitation gap Δ_s and the polarized excitation gap Δ_c extrapolated to the thermodynamic limit at the effective filling ν_n in the nth LL of single-layer graphene (SLG) and bilayer graphene (BLG).

ν_n		LL	ground state	excited state	Δ_s	Δ_c
1	SLG	0	polarized	skyrmion	0.63	
1	SLG	1	polarized	skyrmion	0.28	
1	SLG	2	polarized	skyrmion	0.42	
1	BLG	2	polarized	skyrmion	0.48	
1/3	SLG	0	polarized	skyrmion	0.03	0.101
1/3	SLG	1	polarized	skyrmion	0.05	0.115
1/3	BLG	2	polarized	polarized		0.103
2/5	SLG	0	unpolarized	unpolarized	0.04	(0.050)[30]
2/5	SLG	1	polarized	skyrmion	0.035	0.058
2/5	BLG	2	polarized	polarized		0.052
2/3	SLG	0	unpolarized	unpolarized	0.08	(0.101)[30]
2/3	SLG	1	polarized	unpolarized	0.10	0.115
2/3	BLG	2	polarized	polarized		0.103

unpolarized. However, in the $n = 1$ LL of SLG and the $n = 2$ LL of BLG, the size dependence of the polarization energy shows that the pseudospins are fully polarized in the ground state.

The valley unpolarized excitation gap Δ_s from the unpolarized ground state in the $n = 0$ LL of SLG is 0.08 $e^2/(\epsilon\ell_B)$. In the $n = 1$ LL of SLG, we find that Δ_s is 0.10. The valley unpolarized elementally charge excitations at $\nu_n = 2/3$ in the $n = 1$ LL of SLG are not skyrmion-like pseudospin-textured states.

4. Summary

Our DMRG calculation confirms various types of quantum Hall states in graphene at $\nu_n = 1, 1/3, 2/5$, and 2/3 in the $n = 0$ and 1 LLs of SLG and in the $n = 2$ LL of BLG. These results are summarized in Table 1, where the pseudospin polarizations for the ground state and the lowest charge excited state are listed with the elementally charge excitation energies Δ_c and Δ_s. The elementally charge excitations are obtained by increasing or decreasing the flux quantum number N_ϕ by 1 for $\nu = 1$, 1/3, and 2/5, and by increasing or decreasing the flux quantum number by adding or removing a single electron in the system for $\nu = 2/3$. We have studied both (a) pseudospin-polarized excitations (Laughlin's quasiholes and quasiparticles) and (b) pseudospin-unpolarized excitations (quasihole and quasiparticle skyrmions at $\nu_n = 1$ and 1/3 in the $n = 0$ and 1 LLs of SLG, and at $\nu_n = 1$ in the $n = 2$ LL of BLG). The activation energies obtained in finite systems are extrapolated to the thermodynamic limit, which give theoretical predictions for future experimental studies of the fractional quantum Hall states in graphene.

References

[1] Geim A K and Novoselov K S 2007 *Nat. Mater.* **6** 183
[2] Castro Neto A H, Guinea F, Peres N M R, Novoselov K S and Geim A K 2009 *Rev. Mod. Phys.* **81** 109
[3] Novoselov K S, Geim A K, Morozov S V, Jiang D, Katsnelson M I, Grigorieva I V, Dubonos S V and Firsov A A 2005 *Nature* **438** 197

[4] Zhang Y, Tan Y -W, Stormer H L and Kim P 2005 *Nature* **438** 201
[5] Apalkov V M and Chakraborty T 2006 *Phys. Rev. Lett.* **97** 126801
[6] Abergel D S L, Apalkov V, Berashevich J, Ziegler K and Chakraborty T 2010 *Adv. Phys.* **59** 261
[7] Zheng Y and Ando T 2002 *Phys. Rev. B* **65** 245420
[8] Zhang Y, Jiang Z, Small J P, Purewal M S, Tan Y -W, Fazlollahi M, Chudow J D, Jaszczak J A, Stormer H L, and Kim P 2006 *Phys. Rev. Lett.* **96** 136806; Jiang Z, Zhang Y, Stormer H L and Kim P 2007 *Phys. Rev. Lett.* **99** 106802
[9] McCann E and Fal'ko V I 2006 *Phys. Rev. Lett.* **96** 086805
[10] Girvin S M and MacDonald A H 1997 *Perspectives in Quantum Hall Effects*, ed S Das Sarma and A Pinczuk (New York: John Wiley and Sons)
[11] Haldane F D M 1983 *Phys. Rev. Lett.* **51** 605; Haldane F D M and Rezayi E H 1985 *Phys. Rev. Lett.* **54** 237
[12] Nomura K and MacDonald A H 2006 *Phys. Rev. Lett.* **96** 256602
[13] Haldane F D M 1987 *The Quantum Hall Effect*, ed R E Prange and S M Girvin (New York: Springer Verlag)
[14] Fano G, Ortolani F and Colombo E 1986 *Phys. Rev. B* **34** 2670
[15] Morf R H, d'Ambrumenil N and S. Das Sarma S 2002 *Phys. Rev. B* **66** 075408
[16] White S R 1992 *Phys. Rev. Lett.* **69** 2863; White S R 1993 *Phys. Rev. B* **48** 10345
[17] Shibata N 2003 *J. Phys. A* **36** R381
[18] Shibata N and Yoshioka D 2001 *Phys. Rev. Lett.* **86** 5755; Shibata N and Yoshioka D 2006 *J. Phys. Soc. Jpn.* **75** 043712; Shibata N and Nomura K 2007 *J. Phys. Soc. Jpn.* **76** 103711
[19] Toke C, Lammert P E, Crespi V H and Jain J K 2006 *Phys. Rev. B* **74** 235417
[20] The extrapolated gaps for non-zero Landau levels in ref. [19] are contradict with our results and the Hartree-Fock calculation in ref. [28].
[21] Feiguin A E, Rezayi E, Nayak C and Das Sarma S 2008 *Phys. Rev. Lett.* **100** 166803
[22] Shibata N and Nomura K 2008 *Phys. Rev. B* **77** 235426; Shibata N and Nomura K 2009 *J. Phys. Soc. Jpn.* **78** 104708
[23] Sondhi S L, Karlhede A, Kivelson S and Rezayi E H 1993 *Phys. Rev. B* **47** 16419
[24] Fertig H A, Brey L , Cote R and MacDonald A H 1994 *Phys. Rev. B* **50** 11018
[25] Moon K, Mori H, Yang K, Girvin S M, MacDonald A H, Zheng L, Yoshioka D and Zhang S C 1995 *Phys. Rev. B* **51** 5138
[26] Rezayi E H 1991 *Phys. Rev. B* **43** 5944
[27] Yoshioka D 1998 *J. Phys. Soc. Jpn.* **67** 3356
[28] Yang K, Das Sarma S and MacDonald A H 2006 *Phys. Rev. B* **74** 075423
[29] Jain J K 1989 *Phys. Rev. Lett.* **63** 199
[30] Δ_c for $\nu_n = 2/5$ and $2/3$ in the $n = 0$ LL of SLG is the gap from the lowest polarized state.

Journal of Physics: Conference Series **334** (2011) 012004 doi:10.1088/1742-6596/334/1/012004

Topological aspect of graphene physics

Y. Hatsugai

Institute of Physics, University of Tsukuba, Tsukuba, 305-8571 Japan

E-mail: y.hatsugai@gmail.com
hatsugai@sakura.cc.tsukuba.ac.jp

Abstract. Topological aspects of graphene are reviewed focusing on the massless Dirac fermions with/without magnetic field. Doubled Dirac cones of graphene are topologically protected by the chiral symmetry. The quantum Hall effect of the graphene is described by the Berry connection of a manybody state by the filled Landau levels which naturally possesses non-Abelian gauge structures. A generic principle of the topologically non trivial states as the bulk-edge correspondence is applied for graphene with/without magnetic field and explain some of the characteristic boundary phenomena of graphene.

1. Introduction

Graphene as a playground of massless Dirac fermions has a long history of theoretical study[1, 2, 3]. However its experimental realization is a real surprise since the massless Dirac fermion itself is topologically non trivial especially under a magnetic field[4, 5]. It implies that we have a chance to observe exotic topological phenomena in a real world, which have been confirmed experimentally for the past few years.

The quantized Hall conductance is one of the most well-known topological objects as the Chern numbers of the Berry connection which describes the bulk (without boundaries). It is also described by the edge states, supplemented by the Laughlin argument, which gives a different topological number. These two topological quantities are closely related as the bulk-edge correspondence[6]. This bulk-edge relation has been realized as a universal feature of topological states in several different quantum systems such as superconductors, quantum (spin) Hall effects[7, 8, 9], gapped quantum magnets and photonic crystals[10]. The edge states in a magnetic field are chiral fermions in a sense, they have specific direction, and can not be destroyed as far as the bulk is gapped. It is the topological stability of the edge states. Graphene under a magnetic field is one of the typical topological states where the bulk-edge correspondence of the Dirac fermions plays a fundamental role. Localized zero modes near the zigzag boundaries (Fujita states)[11], which have been measured by the STM experiments[12], and their generalization under a magnetic field, are also governed by this topological principle[6]. Further appearance of two Dirac fermions with opposite chirality at the K and K' points are not accidental but is topologically protected by the chiral symmetry which is a two dimensional analogue of the Nielsen-Ninomiya theorem in four dimensions[13].

Also the Landau level of graphene is quite special reflecting its singular dispersion as of the Dirac fermions. Then the quantum Hall effects of it is anomalous[14, 15, 16]. Further Landau degeneracy of the characteristic zero energy Landau level are topologically protected against for some class of randomness by the index theorem. It is fundamental for free-standing single layer graphene where its intrinsic disorder is a gauge field fluctuation[17, 18].

Based on our works, these topological phenomena in graphene physics will be explained intuitively.

2. Topological stability of the Dirac cones and the $n = 0$ Landau level

2.1. Chiral symmetry and doubled Dirac cones

Let us consider a single orbital tight-binding hamiltonian of graphene with magnetic field[16] ($\boldsymbol{j} = (j_1, j_2)$ is a two dimensional coordinate with the unit translations $\boldsymbol{e}_{1,2}$ and ϕ is a total flux per hexagon in flux quantum $\Phi_0 = h/e$, See Fig.1)

$$H = t\sum_{\boldsymbol{j}}\left[c_\bullet^\dagger(\boldsymbol{j})c_\circ(\boldsymbol{j}) + e^{2\pi\phi j_1}c_\bullet^\dagger(\boldsymbol{j})c_\circ(\boldsymbol{j}-\boldsymbol{e}_2) + c_\bullet^\dagger(\boldsymbol{j}+\boldsymbol{e}_1)c_\circ(\boldsymbol{j}) + h.c.\right] \quad \text{(zigzag)}$$

$$= t\sum_{\boldsymbol{j}}\left[c_\bullet^\dagger(\boldsymbol{j})c_\circ(\boldsymbol{j}) + e^{2\pi\phi(j_1+1)}c_\bullet^\dagger(\boldsymbol{j})c_\circ(\boldsymbol{j}+\boldsymbol{e}_1-\boldsymbol{e}_2) + c_\bullet^\dagger(\boldsymbol{j}-\boldsymbol{e}_1)c_\circ(\boldsymbol{j}) + h.c.\right] \quad \text{(bearded)}$$

where we give two expressions using two different unit cells that are compatible for zigzag and bearded edges. When $\phi = 0$, it is written in the momentum representation as $H = \int \frac{d^2k}{(2\pi)^2}\boldsymbol{c}^\dagger(\boldsymbol{k})\boldsymbol{h}(\boldsymbol{k})\boldsymbol{c}(\boldsymbol{k})$,

$$\boldsymbol{h}_{Z,B}(\boldsymbol{k}) = \begin{pmatrix} 0 & \Delta_{Z,B}(\boldsymbol{k}) \\ \Delta^*_{Z,B}(\boldsymbol{k}) & 0 \end{pmatrix}, \quad \begin{cases} \Delta_Z(\boldsymbol{k}) = t(1+e^{-ik_1}+e^{-ik_2}) & : \text{(zigzag)} \\ \Delta_B(\boldsymbol{k}) = t(1+e^{i(k_1-k_2)}+e^{ik_1}) & : \text{(bearded)} \end{cases}$$

where $\boldsymbol{c}^\dagger(\boldsymbol{k}) = (c_\circ^\dagger(\boldsymbol{k}), c_\bullet^\dagger(\boldsymbol{k}))$, $c_\alpha(\boldsymbol{j}) = \int \frac{d^2k}{(2\pi)^2}e^{i\boldsymbol{k}\cdot\boldsymbol{j}}c_\alpha(\boldsymbol{k})$. The energy dispersion is given as $\epsilon(\boldsymbol{k}) = \pm|\Delta(\boldsymbol{k})|$ and vanishing momenta $\boldsymbol{k}_D$, $\Delta(\boldsymbol{k}_D) = 0$ give massless Dirac cones if exist.

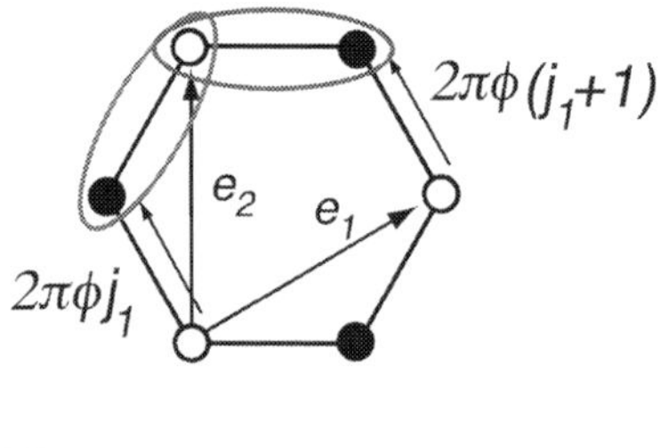

Figure 1. The unit cell of the graphene. The red ellipse is consistent with the zigzag edges and the blue one is for the bearded one. Two primitive translation vectors are also shown.

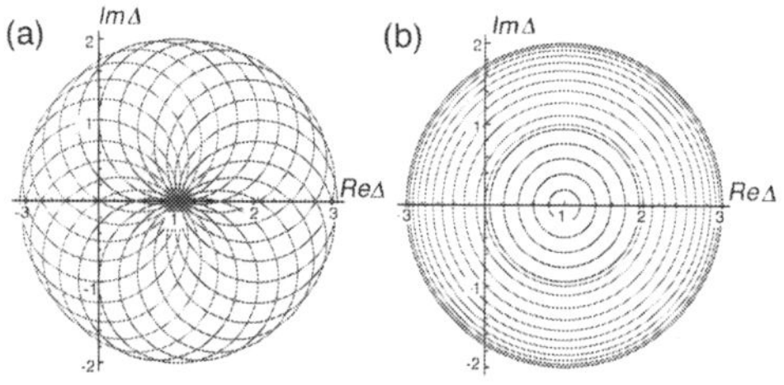

Figure 2. Trajectories of $\Delta(\boldsymbol{k})$ parametrized by k_1 for various values of k_2. (a) for $\Delta = \Delta_Z$ and (b) $\Delta = \Delta_B$ ($t = -1$). Curves that encloses the origin is drawn in red.

Since the Brillouin zone, $T^2 = \{(k_1, k_2)|k_1, k_2 \in [0, 2\pi]\}$, is a two-torus by identifying $k_1 = 0 \leftrightarrows k_1 = 2\pi$ and $k_2 = 0 \leftrightarrows k_2 = 2\pi$, it does not have boundaries. Then considering k_1 as a parameter, $\Delta(\boldsymbol{k})$ makes a closed loop in the complex plane for each fixed k_2. This loop is modified by changing k_2 from 0 to 2π and comes back to the original one (Fig.2). When the loop cuts the origin, it gives the Dirac cone. Then once the loop cuts the origin, the situation is stable against for small but finite perturbation. This is the topological stability of the Dirac cones. Here we need the 2×2 hamiltonian is characterized by single complex parameter Δ (diagonal parts are zero). This is the *chiral symmetry*, that is, there exists $\{\boldsymbol{h}, {}^\exists\gamma\} = 0$, $\gamma^2 = \sigma_0$: 2×2

unit matrix. (In this case, $\gamma = \sigma_3$.) Also the doubling of the Dirac cones is again clear since the origin cutting the loop from the outside to the inside inevitably followed by cutting from inside to the out (See Fig.2 and Fig.4). This simple observation guarantees that the number of Dirac cones are always even. Note here that the chiral symmetry is related to the bipartite structure of the honeycomb lattice, that is, the hoppings are only between the two sub-lattices $\circ$ and $\bullet$

On the other hand, as for generic semiconductors including graphene, let us assume the effective hamiltonian of the valence and the conduction bands, $\boldsymbol{h}$, is chiral symmetric, that is, it does anti-commutes with some chiral matrix γ, $\{\boldsymbol{h}, \gamma\} = 0$. It implies $\boldsymbol{h}$ is traceless and thus one can expand it by the Pauli matrices as $\boldsymbol{h}(\boldsymbol{k}) = \boldsymbol{R}(\boldsymbol{k}) \cdot \boldsymbol{\sigma}$ with three real parameters $R_1(\boldsymbol{k})$, $R_2(\boldsymbol{k})$ and $R_3(\boldsymbol{k})$ that makes a vector $\boldsymbol{R}(\boldsymbol{k}) = \begin{pmatrix} R_1(\boldsymbol{k}) \\ R_2(\boldsymbol{k}) \\ R_3(\boldsymbol{k}) \end{pmatrix}$. The momentum dependent energy gap is $E_g(\boldsymbol{k}) = 2|\boldsymbol{R}(\boldsymbol{k})|$. Then the zero gap condition is $\boldsymbol{R} = \boldsymbol{0}$. Writing the chiral matrix as, $\gamma = \boldsymbol{n}_\gamma \cdot \boldsymbol{\sigma}$ ($\boldsymbol{n}_\gamma^2 = 1$), the chiral symmetric condition $\{\boldsymbol{h}, \gamma\} = (\boldsymbol{R} \cdot \boldsymbol{n}_\gamma)\sigma_0 + i(\boldsymbol{R} \times \boldsymbol{n}_\gamma) \cdot \boldsymbol{\sigma} + (\boldsymbol{R} \rightleftarrows \boldsymbol{n}_\gamma) = 2(\boldsymbol{R} \cdot \boldsymbol{n}_\gamma)\sigma_0 = 0$, implies $\boldsymbol{R} \cdot \boldsymbol{n}_\gamma = 0$, that is, $\boldsymbol{R}$ is always in the plane $\mathbb{R}^2(\boldsymbol{n}_\gamma)$ whose normal vector is $\boldsymbol{n}_\gamma$. Since the Brilluine zone T^2 is a 2-dimensional closed surface, the image $\boldsymbol{R}(T^2)$ is also a closed surface (like a balloon) in 3-dimensions. Then the condition $\boldsymbol{R}(T^2) \subset \mathbb{R}^2(\boldsymbol{n}_\gamma)$ means the closed surface is collapsed on the plane (a rubber balloon without air on the desk). When the collapsed image includes the origin, it gives the Dirac cones. Then the topological stability of the Dirac cones and the doubling of them are clear[16, 19, 20]. (See Figs3.) It is the 2-dimensional analogue of the Nielsen-Ninomiya theorem in the 4-dimensional lattice gauge theory [13]. A relation between the 4D graphene and chiral fermions is recently discussed as well[21].

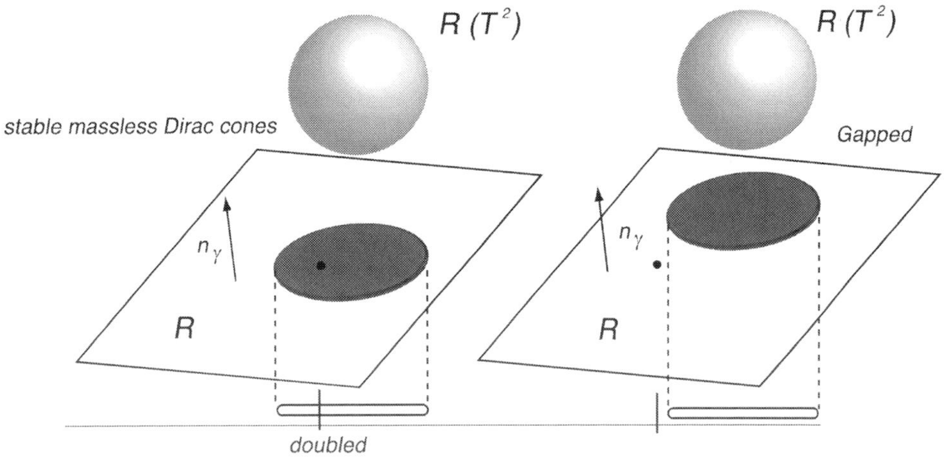

Figure 3. Topological stabilities of the Dirac cones. The image of the Brilloine zone $\boldsymbol{R}(T^2)$ is generically a surface in three dimensional space $\boldsymbol{R}$, which is collapsed into the plane which is normal to the $\boldsymbol{n}_\gamma$, when the hamiltonian is chiral symmetric as $\{H, \gamma\} = 0$, $\gamma = \boldsymbol{\sigma} \cdot \boldsymbol{n}_\gamma$.

2.2. Chiral symmetry and the $n = 0$ Landau level

Now let us expand the hamiltonian $\boldsymbol{h}(\boldsymbol{k})$ with chiral symmetry ($\{\boldsymbol{h}, \gamma\}$) near one of the doubled zero gap momentum, say, $\boldsymbol{k}_D$. Then the effective hamiltonian in the lowest order in $\delta\boldsymbol{k} = \boldsymbol{k} - \boldsymbol{k}_D$ is written as $\boldsymbol{h} \approx \boldsymbol{h}_C = \delta k_1 \boldsymbol{\sigma}\cdot\boldsymbol{X} + \delta k_2 \boldsymbol{\sigma}\cdot\boldsymbol{Y}$ where $\boldsymbol{X} = \partial_{k_1}\boldsymbol{R}$ and $\boldsymbol{Y} = \partial_{k_2}\boldsymbol{R}$ which are perpendicular to the $\boldsymbol{n}_\gamma$. It implies the chiral symmetry of the effective hamiltonian, $\{\boldsymbol{h}_C, \gamma\} = 0$. One may define the chirality of the effective hamiltonian $\chi_D = \pm 1$ as $\boldsymbol{n}_\gamma = \chi_D \boldsymbol{X}\times\boldsymbol{Y}/(c\hbar)^2$ where the effective *light velocity* $(c > 0)$ is defined as $c^2 \equiv |\boldsymbol{X}\times\boldsymbol{Y}|/\hbar^2$. When the vector, $(\boldsymbol{X}, \boldsymbol{Y}, \boldsymbol{n}_\gamma)$, forms a right handed triple, the chirality, χ_D is $+1$ and, if left-handed, $\chi_D = -1$. Then the chiral matrix is written as $\boldsymbol{\gamma} = \chi_D \boldsymbol{\sigma}\cdot(\boldsymbol{X}\times\boldsymbol{Y})/(c\hbar)^2$.

By the inverse procedure of the quantization, $\hbar\delta\boldsymbol{k} \to \boldsymbol{p} - e\boldsymbol{A} \equiv \boldsymbol{\pi}$, $p_\alpha = -i\hbar\partial_\alpha$, $(\alpha = x, y)$, we have a real space form of the effective hamiltonian with a magnetic field, $B = \partial_x A_y - \partial_y A_x > 0$, $[\pi_x, \pi_y] = i\hbar eB$ as

$$\boldsymbol{h}_C = \hbar^{-1}\big[(\boldsymbol{\sigma}\cdot\boldsymbol{X})\pi_x + (\boldsymbol{\sigma}\cdot\boldsymbol{Y})\pi_y\big]$$

The spectrum and the wave functions are determined by considering its square as

$$\begin{aligned}\boldsymbol{h}_C^2 &= \hbar^{-2}\big[\boldsymbol{X}^2\pi_x^2 + \boldsymbol{Y}^2\pi_y^2 + (\boldsymbol{X}\cdot\boldsymbol{Y})(\pi_x\pi_y + \pi_y\pi_x) + i(\boldsymbol{X}\times\boldsymbol{Y})\cdot\boldsymbol{\sigma}[\pi_x, \pi_y]\big] \\ &= c^2\,(\boldsymbol{\pi}^\dagger\Xi\,\boldsymbol{\pi})\boldsymbol{\sigma}_0 - \chi_D(eB\hbar c^2)\boldsymbol{\gamma}\end{aligned}$$

where

$$\begin{aligned}\Xi &= \frac{1}{(\hbar c)^2}(\boldsymbol{X}, \boldsymbol{Y})^\dagger(\boldsymbol{X}, \boldsymbol{Y}) = \frac{1}{(\hbar c)^2}\begin{pmatrix}\boldsymbol{X}\cdot\boldsymbol{X} & \boldsymbol{X}\cdot\boldsymbol{Y} \\ \boldsymbol{X}\cdot\boldsymbol{Y} & \boldsymbol{Y}\cdot\boldsymbol{Y}\end{pmatrix} \\ \det\Xi &= \frac{1}{(\hbar c)^4}\big(|\boldsymbol{X}|^2|\boldsymbol{Y}|^2 - (\boldsymbol{X}\cdot\boldsymbol{Y})^2\big) = \frac{|\boldsymbol{X}\times\boldsymbol{Y}|^2}{(\hbar c)^4} = 1\end{aligned}$$

By identifying as $\frac{1}{2m} \equiv c^2$, the first term of $\boldsymbol{h}_C$ is considered as a hamiltonian of the standard parabolic electrons with anisotropic masses (See Appendix A). Here the cyclotron frequency ω_C is identified as $\omega_C = eB/m = 2eBc^2$. Also note that the chiral matrix, $\boldsymbol{\gamma}$, commutes with the squared hamiltonian $\boldsymbol{h}_C^2$.

Then writing a normalized standard Landau state with energy, $\hbar\omega_C(n+1/2) = 2\hbar eBc^2(n+1/2)$, $n = 0, 1, 2, \cdots$, as $\psi_n(x, y)$, the eigen state of $\boldsymbol{h}_C^2$ is written as $\boldsymbol{\Psi}_n^\pm = \boldsymbol{\varphi}_\pm\psi_n(x, y)$ where $\boldsymbol{\varphi}_\pm$ is an eigen state of $\boldsymbol{\gamma}$ with eigen values ± 1 respectively. It satisfies $\boldsymbol{h}_C^2\boldsymbol{\Psi}_n^\pm = (eB\hbar c^2)(2n + 1 \mp \chi_D)\boldsymbol{\Psi}_n^\pm$. It implies that the effective hamiltonian $\boldsymbol{h}_C$ always has a zero energy Landau level as

$$\boldsymbol{\Psi}_{\chi_D}^0(x, y) \equiv \boldsymbol{\varphi}_{\chi_D}\psi_0(x, y) \quad : \quad \boldsymbol{h}_C\boldsymbol{\Psi}_{\chi_D}^0(x, y) = 0$$

which has the chirality χ_D.

As for the non zero energy Landau levels, $\boldsymbol{h}_C^2$ is doubly degenerated at ϵ_n^2, $\epsilon_n = c\sqrt{2neB\hbar}$, spanned by $\boldsymbol{\Psi}_n^{+\chi_D} = \boldsymbol{\varphi}_{+\chi_D}\psi_n(x, y)$ and $\boldsymbol{\Psi}_{n-1}^{-\chi_D} = \boldsymbol{\varphi}_{-\chi_D}\psi_{n-1}(x, y)$, which diagonalize $\boldsymbol{h}_C$ without degeneracy with the energies $\pm\epsilon_n$.

2.3. Chiral symmetry and Fermion doubling

Up to this point, we have discussed the effective theory of the chiral symmetric zero gap semiconductor near one of the gapless momentum.

As discussed the chiral symmetry of the global hamiltonian $\boldsymbol{h}$, $\{\boldsymbol{h}(\boldsymbol{k}), \boldsymbol{\gamma}\} = 0$, $(\boldsymbol{h} = \boldsymbol{R}\cdot\boldsymbol{\sigma})$ requires that the three dimensional vector $\boldsymbol{R}({}^\forall\boldsymbol{k})$ is always perpendicular to the $\boldsymbol{n}_\gamma$ and on the plane $\mathbb{R}^2(\boldsymbol{n}_\gamma)$ cutting the origin $\boldsymbol{R} = 0$ ($\boldsymbol{R}(\boldsymbol{k}) \in \mathbb{R}^2(\boldsymbol{n}_\gamma)$). Then one can repeat the discussion for $\Delta(\boldsymbol{k})$ in Fig.2 for the contour $\boldsymbol{R}()$ on the plane $\mathbb{R}^2(\boldsymbol{n}_\gamma)$, that is, considering a closed curve characterized by k_2, $C(k_2) = \{\boldsymbol{k}|k_1 \in [0, 2\pi]\}$. This curve is on the plane $\mathbb{R}^2(\boldsymbol{n}_\gamma)$ and comes back to the original one by changing k_1 form 0 to 2π. When it cuts the origin, it gives the Dirac

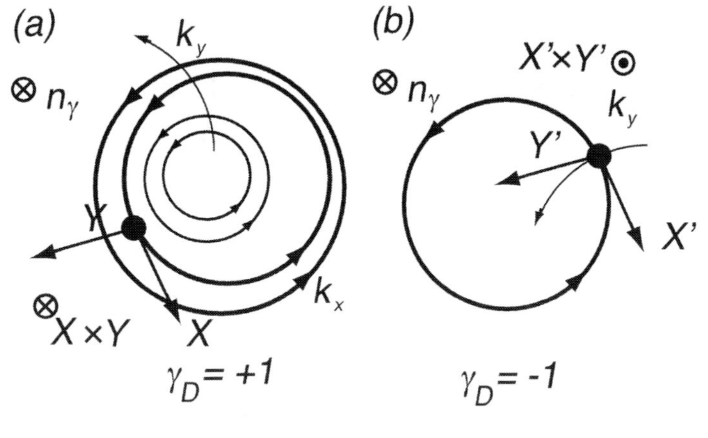

Figure 4. $\boldsymbol{R}(\boldsymbol{k})$ on the plane $\mathbb{R}^2(\boldsymbol{n}_\gamma)$. (a) When the curve cuts the origin which defines the Dirac cone. (b) Another Dirac cone which is paired with the one in (a). The chirality is reversed.

cones. The doubling of the Dirac cones is also clear as before (Figs.2 and 4). Also as for the chirality of the Dirac cones, χ_D, the paired Dirac cones, D and D', have a reversed chirality as

$$\chi_D + \chi_{D'} = 0$$

which is also clear from the Fig.4. Then the chirality of the $n = 0$ Landau level is also reversed. As for the graphene, the Dirac fermions at K and K' do have the reversed chirality.

2.4. *Aharonov-Casher argument*

As discussed the $n = 0$ Landau level is an eigen state of the chiral operator for each Dirac fermion which is a characteristic feature of the chiral symmetric massless Dirac fermions on lattice. It is realized in graphene. This complete degeneracy of the Landau level is not only for the pure system and it does persist for some class on the randomness[18]. There is well-known discussion by Aharonov-Casher[22] which is considered as a direct demonstration of the index theorem where the degeneracy is only determined by the total flux passing through the system as far as the effective description by the Dirac fermions is allowed. Here the magnetic field B may not be uniform. We compactly describe it extending the discussion to our anisotropic case. Since the anisotropic mass matrix Ξ discussed before is real symmetric and its determinant unity, it is diagonalized by the orthogonal matrix $\boldsymbol{V}$ as (See also appendix), $\boldsymbol{\Xi} = \boldsymbol{V}^\dagger \begin{pmatrix} \xi_X & \\ & \xi_Y \end{pmatrix} \boldsymbol{V}$, where $\xi_X, \xi_Y > 0$, $\xi_X \xi_Y = 1$. Then defining canonical momenta $\boldsymbol{\Pi} = \boldsymbol{V}\boldsymbol{\pi}$, $[\Pi_X, \Pi_Y] = [\pi_x, \pi_y] = i\hbar e B]$, $\boldsymbol{h}_C^2$ is factorized as

$$\begin{aligned} \boldsymbol{h}_C^2 &= c^2(\xi_X \Pi_X^2 + \xi_Y \Pi_Y^2) + i\chi_D \gamma [\Pi_X, \Pi_Y]) = c^2 \mathcal{D}^\dagger \mathcal{D} \\ \mathcal{D} &= \sqrt{\xi}_X \Pi_X + i\chi_D \gamma \sqrt{\xi}_Y \Pi_Y \end{aligned}$$

Then writing the vector potential as $A_X = -\xi_Y \partial_Y \phi$, $A_Y = \xi_X \partial_X \phi$ and defining $\tilde{X} = X/\sqrt{\xi}_X$ and $\tilde{Y} = Y/\sqrt{\xi}_Y$ ($\sqrt{\xi}_X \partial_X = \partial_{\tilde{X}}$ and $\sqrt{\xi}_Y \partial_Y = \partial_{\tilde{Y}}$), the zero mode $\Psi^0_{\chi_D} = \varphi_{\chi_D} \psi$ satisfies, $0 = \mathcal{D}\psi^0_{\chi_D} = -i\hbar \left[\partial_{\tilde{X}} + i\frac{2\pi}{\phi_0} \partial_{\tilde{Y}} \phi + i\partial_{\tilde{Y}} + \frac{2\pi}{\phi_0} \partial_{\tilde{X}} \phi \right] \psi^0_{\chi_D}$. Then writing as $\psi_0 = e^{-2\pi \frac{\phi}{\phi_0}} f$, the function

f satisfies $(\partial_{\tilde{X}} + i\partial_{\tilde{Y}})f = 0$. This is written by defining $z = \tilde{X} + i\tilde{Y}$ as $\partial_{\bar{z}} f = 0$. Thus the function f is an entire function of z in the whole complex plane $z \in \mathbb{C}$, that is, polynomials. Also note that the function ϕ needs to satisfy $B = \partial_X A_Y - \partial_Y A_X = (\xi_X \partial_{\tilde{X}}^2 + \xi_Y \partial_{\tilde{Y}}^2)\phi = (\partial_{\tilde{X}}^2 + \partial_{\tilde{Y}}^2)\phi$. It implies $\phi(\tilde{X},\tilde{Y}) = \int d\tilde{X}d\tilde{Y} G(\tilde{X}-\tilde{X}', \tilde{Y}-\tilde{Y}')B(\tilde{X}',\tilde{Y}')$ where $G(\tilde{X},\tilde{Y}) = \frac{1}{2\pi}\log\frac{r}{r_0}$, $r^2 = \tilde{X}^2+\tilde{Y}^2$ and r_0 is a constant. When assuming the magnetic field is only nonzero in the finite region, we have for the asymptotic behavior in the limit $r \to \infty$ as $\phi \overset{r\to\infty}{\longrightarrow} \frac{\Phi}{2\pi}\log\left(\frac{r}{r_0}\right)$ where $\Phi = \int d\tilde{X}d\tilde{Y}\, B = \int dXdY\, B$ is a total flux. Then we have for the asymptotic behavior of the zero mode as $\psi \overset{r\to\infty}{\longrightarrow} f(z)\left(\frac{r}{r_0}\right)^{-\frac{\Phi}{\phi_0}}$. It implies that the number of the degeneracy is $\frac{\Phi}{\phi_0}$.

3. Hall conductance and Berry connection of Dirac sea

Geometrical phase of the quantum mechanical states are of fundamental interest for physical society today. The quantum Hall effect and its time reversal invariant analogue as the quantum spin Hall effect are the typical examples.

Now let us start from the Niu-Thouless-Wu formula for the Hall conductance[23]. It reads that the Hall conductance of the manybody state is given by the average over the twisted boundary condition as

$$\sigma_{xy} = \frac{e^2}{h}C, \quad C = \frac{1}{2\pi i}\int_0^{2\pi} d\phi^1 \int_0^{2\pi} d\phi^2 \left[\langle \partial_1 G|\partial_2 G\rangle - \langle \partial_2 G|\partial_1 G\rangle\right]$$

where $\partial_\mu = \frac{\partial}{\partial\phi^\mu}$, $(\mu = x, y)$ and where $|G\rangle = |G(\phi^1,\phi^2)\rangle$ is a unique gapped ground state of the hamiltonian, $H(\phi^1,\phi^2)$, with a twisted boundary condition specified by $e^{i\phi^1}$ and $e^{i\phi^2}$ for each of the two translational directions as $H|G\rangle = E|G\rangle$.

This C is the topological number (Chern number) and is intrinsically integer as explained below. This topological property is clearly demonstrated by defining the Berry connection (a formal form of the vector potential) as

$$\mathcal{A} = \langle G|dG\rangle = \langle G|\partial_\mu G\rangle d\phi^\mu = d\boldsymbol{\phi}\cdot\boldsymbol{A}, \qquad \boldsymbol{A} = \langle G|\boldsymbol{\nabla} G\rangle$$

where $\boldsymbol{\phi} = (\phi^1,\phi^2,0)$ and $\boldsymbol{\nabla} = (\partial_1,\partial_2,\partial_3)$. Note that, since the phase of the state $|G\rangle$ is completely arbitrary, one may take a different choice as $|G'\rangle = |G\rangle\omega$, $\omega = e^{i\theta}$, $\theta \in \mathbb{R}$, which gives a different connection $\mathcal{A}' = \langle G'|dG'\rangle = d\boldsymbol{\phi}\cdot\boldsymbol{A}'$. They are related with each other as [24]

$$\mathcal{A} = \mathcal{A}' + \omega^{-1}d\omega = \mathcal{A}' + id\theta, \qquad \boldsymbol{A} = \boldsymbol{A}' + i\boldsymbol{\nabla}\theta$$

This is the gauge transformation similar to the case of the electromagnetic field governed by the Maxwell equations. (As for the differential form, see [25].) This is called Abelian since ω is just a complex number $|\omega| = 1$.

Since the twisted boundary condition of $e^{i2\pi}$ is equivalent to the periodic boundary condition without any twist, the integral region $T^2 = \{(\phi^1,\phi^2)|\phi^\mu \in [0,2\pi]\}$ is identified as the two-torus. Then the Chern number of the manybody state is written as

$$C = \frac{1}{2\pi i}\int_{T^2} d\boldsymbol{S}\cdot\mathrm{rot}\,\boldsymbol{A} = \frac{1}{2\pi i}\int_{T^2} d\mathcal{A} \equiv \frac{1}{2\pi i}\int_{T^2}\mathcal{F} = \frac{1}{2\pi i}\int_{T^2}\langle dG|dG\rangle$$

where $d\boldsymbol{S} = d\phi^1 d\phi^2(0,0,1)$ and the field strength is defined as $\mathcal{F} = d\mathcal{A} = \langle dG|dG\rangle$. The connection $\mathcal{A}$ is gauge dependent but the Abelian field strength $\mathcal{F}$ is gauge invariant as $\mathcal{F}' = d\mathcal{A}' = d\mathcal{A}$ since $d(\omega^{-1}d\omega) = -\omega^{-1}d\omega\omega^{-1}\omega = -(\omega^{-1}d\omega)^2 = 0$ (See Appendix B).

To understand the quantization of the Hall conductance, we need to specify the gauge[26]. Here we discuss this gauge fixing generically[27]. Since the two-torus T^2 is boundaryless,

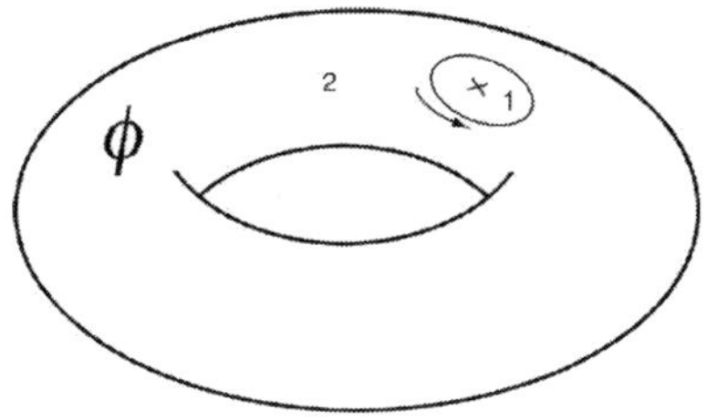

Figure 5. Singularity of the gauge and the two patches to avoid it.

the Chern number is always vanishing if one is allowed to use the Stokes theorem globally $C = \int_{T^2} d\mathcal{A} = \int_{\partial T^2=\emptyset} \mathcal{A} = 0$. To have a nonzero Chern number, the single global gauge is not allowed and one needs to use two different gauges, at least, say, $\mathcal{A}_1$ and $\mathcal{A}_2$. Let us first mention that the state $|G\rangle$ is gauge dependent but the projection into the ground state space $P = |G\rangle\langle G| = |G'\rangle\langle G'|$ is gauge invariant. Then taking arbitrary (fixed) state $|T_1\rangle$, a normalized and gauge fixed state $|G_1\rangle$ is given as $|G_1\rangle = P|T_1\rangle/\sqrt{N_1}$ where $N_1 = \langle T_1|P|T_1\rangle = |\langle G|T_1\rangle|^2 \geq 0$ is a gauge invariant normalization constant. It is clear that this gauge becomes singular when the normalization N_1 is vanishing at some point in T^2 ($N_1 = 0$). Near this singular point ϕ_1, one needs to use the other gauge, say $|G_2\rangle$ using another arbitrary state $|T_2\rangle$ that is generically regular in the region R_1 that includes ϕ_1, $N_2(\phi) \neq 0, \phi \in R_1$.

The gauge transformation between the two, $\mathcal{A}_1 = \mathcal{A}_2 + i\theta_{21}$ is given as $|G_1\rangle = |G_2\rangle\omega_{21}$, $\omega_{21} = \langle T_2|P|T_1\rangle/\sqrt{N_1N_2} = e^{i\theta_{21}}$. When these two gauges are enough to span the whole ϕ space, the Chern number is written as

$$C = \frac{1}{2\pi i}\left(\int_{R_1} d\mathcal{A}_2 + \int_{T^2\setminus R_1} d\mathcal{A}_1\right) = \frac{1}{2\pi i}\int_{\partial R_1}(\mathcal{A}_1 - \mathcal{A}_2) = \frac{1}{2\pi}\oint_{\partial R_1} d\theta_{21}$$

Here the reason the Chern number is integer is clear since the gauge transformation ω_{21} is single valued on the boundary ∂R_1.

When the system is non-interacting, $V = 0$, one may go further. In this case, the ground state of the many body system, (M particle state), $|G\rangle$ is constructed from one particle states. It is explicitly expressed by writing the hamiltonian as $H = \boldsymbol{c}^\dagger(\boldsymbol{h})\boldsymbol{c}$ where $\boldsymbol{c}^\dagger = (c_1, \cdots, c_N)$ and N is a total number of system sites. The ground state is given by the filling the one particle states below the fermi energy ϵ_F as ($\epsilon_M \leq \epsilon_F$)

$$|G\rangle = |\boldsymbol{\Psi}\rangle = \prod_{\ell=1}^{M} \boldsymbol{c}^\dagger\boldsymbol{\psi}_\ell|0\rangle, \quad \boldsymbol{h}\boldsymbol{\psi}_\ell = \epsilon_\ell\boldsymbol{\psi}_\ell, \quad \epsilon_\ell \leq \epsilon_{\ell'}, (\ell < \ell'), \quad \boldsymbol{\psi}_\ell^\dagger\boldsymbol{\psi}_{\ell'} = \delta_{\ell\ell'}$$

where the many body state is labeled by the filled one particle state, that form a multiplet, $\boldsymbol{\Psi} = (\boldsymbol{\psi}_1, \cdots, \boldsymbol{\psi}_M)$[28, 29, 30]. The Berry connection of this non-interacting ground state has an interesting form by the non-Abelian Berry connection $\boldsymbol{A}_M$ which is defined as

$$\mathcal{A} = \langle g|\sum_\ell (\boldsymbol{c}^\dagger\boldsymbol{\psi}_1)\cdots(\boldsymbol{c}^\dagger d\boldsymbol{\psi}_\ell)\cdots\cdot(\boldsymbol{c}^\dagger\boldsymbol{\psi}_M)|0\rangle = \sum_\ell \det_M \boldsymbol{\Psi}^\dagger(\boldsymbol{\psi}_1, \cdots, d\boldsymbol{\psi}_\ell, \cdots, \boldsymbol{\psi}_M) = \mathrm{Tr}\,\boldsymbol{A}_M$$

$$\boldsymbol{A}_M = \boldsymbol{\Psi}^\dagger d\boldsymbol{\Psi} = \begin{bmatrix} \boldsymbol{\psi}_1^\dagger d\boldsymbol{\psi}_1 & \boldsymbol{\psi}_1^\dagger d\boldsymbol{\psi}_2 & \cdots \\ \boldsymbol{\psi}_2^\dagger d\boldsymbol{\psi}_1 & \boldsymbol{\psi}_2^\dagger d\boldsymbol{\psi}_2 & \cdots \\ \vdots & \vdots & \ddots \end{bmatrix}.$$

Journal of Physics: Conference Series **334** (2011) 012004 doi:10.1088/1742-6596/334/1/012004

Here the choice of the one particle states below the fermi energy is arbitrary and one may further allow to mix up among them as

$$\boldsymbol{\Psi} = \boldsymbol{\Psi}'\boldsymbol{\omega}, \quad \boldsymbol{\omega}^\dagger\boldsymbol{\omega} = E_M$$

which results in the same many body state $|\boldsymbol{\Psi}\rangle = |\boldsymbol{\Psi}\rangle' \det\boldsymbol{\omega}$ since $\boldsymbol{\omega} \in U(M)$ and $\det\boldsymbol{\omega} \in U(1)$. This $\boldsymbol{A}_M$ defines a non-Abelian Berry connection. Its property under the gauge transformation is given as[31, 27]

$$\boldsymbol{A}_M = \boldsymbol{\omega}^{-1}\boldsymbol{A}'_M\boldsymbol{\omega} + \boldsymbol{\omega}^{-1}d\boldsymbol{\omega}$$

This is the gauge transformation of the non Abelian matrix valued connection spanned by the filled one particle states. Correspondingly the field strength (matrix) is defined as

$$\boldsymbol{F}_M \equiv d\boldsymbol{A}_M + \boldsymbol{A}_M^2 = \sum_{\mu<\nu} \boldsymbol{F}_{\mu\nu} d\phi^\mu d\phi^\nu, \quad \boldsymbol{F}_{\mu\nu} = \partial_\mu \boldsymbol{A}_\nu - \partial_\nu \boldsymbol{A}_\mu + [\boldsymbol{A}_\mu, \boldsymbol{A}_\nu]$$

It transforms as $\boldsymbol{F}_M = \boldsymbol{\omega}^{-1}\boldsymbol{F}'_M\boldsymbol{\omega}$. This non Abelian field strength is directly related to the field strength of the manybody state

$$\mathcal{F} = \mathrm{Tr}\, \boldsymbol{F}_M$$

which gives the TKNN formula when each Landau levels are well separated[32]. It transforms as $\mathcal{F} = \mathrm{Tr}\, \boldsymbol{F}_M = \mathcal{F}' = \mathrm{Tr}\, \boldsymbol{F}'_M$, which is consistent with the physical observable σ_{xy} is independent of the choice of the one particle state.

The advantage of this non-Abelian formulation is that we do not need to sum up all Chern numbers of the filled one particle states. We just need to evaluated the matrix values field strength. This is especially useful for the graphene in a weak magnetic field. Since we are mainly interested in the Hall conductance when the fermi energy is near zero. Then one need to evaluate all Chern numbers of Landau levels with negative one particle states energies (Landau levels of Dirac sea). The final result is expected to be of the order of unity. It implies most of the Chern numbers of the filled Dirac sea are canceled. That is, making one error in evaluating each Chern numbers causes a serious problem on the final results.

Also we mentioned here a numerical technique developed in the lattice gauge theory to evaluate lattice topological invariants are quite useful for the evaluation of the $U(1)$ part of the non Abelian Berry connection, which is a two-dimensional analogue of the King Smith-Vanderbilt formula for the polarization, this is a lattice version of the one-dimensional line integral[33]. This non Abelian method is applied for calculation of the Hall conductance of graphene with realistic multi using non orthogonal bases[34].

4. Bulk edge correspondence

Based on the decades of studies for topological states in condensed matter physics, it has been widely understood that although the bulk of topologically non trivial states is featureless and does not show any fundamental symmetry breaking, near their boundaries and defects as local perturbation, there are characteristic local physics governed by the edge states. This is the *bulk-edge correspondence* that can be useful to investigate topological phases both theoretically and experimentally[6]. As for the graphene, there have been two types of interesting boundary phenomena. One is the boundary states at the zigzag edge with/without magnetic field and the other is quantum Hall edge states of the Dirac fermions. Both of them are fundamentally controlled by the bulk which we demonstrate here.

To investigate, let us put the graphene on the cylinder (Fig.6) with zigzag-bearded edges. Let us then take a momentum representation only in tangential $\boldsymbol{e}_2$ direction as $c_\alpha(\boldsymbol{j}) =$

Journal of Physics: Conference Series **334** (2011) 012004 doi:10.1088/1742-6596/334/1/012004

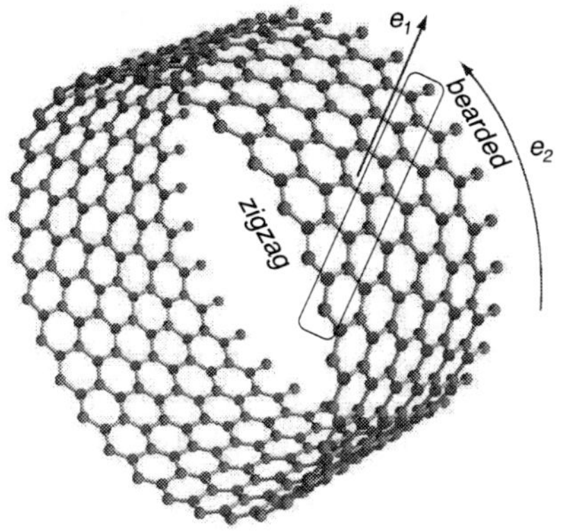

Figure 6. Graphene on a cylinder with zigzag and bearded edges. Unit cell of the dimensional system for the momentum representation by k_2 is shown.

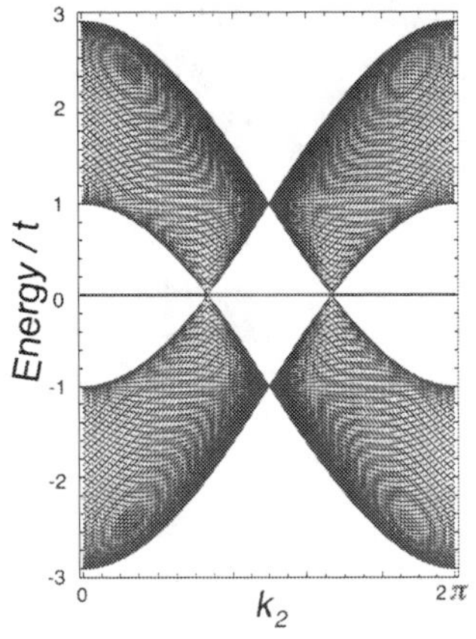

Figure 7. One particle energy spectrum of the graphene without magnetic field labelled by k_2. There are zero mode boundary states. The red ones are localized at the zigzag boundary and the blue ones are localized at the bearded one[11, 35].

$\int \frac{dk_2}{d2\pi} e^{ik_2 j_2} c_\alpha(j_1, k_2)$. Now the total hamiltonian is decomposed as $H = \int \frac{dk_2}{d2\pi} H_{cyl}(k_2)$ by a one dimensional hamiltonian $H_{cyl}(k_2)$ with k_2 as a parameter. $H_{cyl} = \left[t_{\bullet\circ}(j_1) c_\bullet^\dagger(j_1+1) c_\circ(j_1) + t_{\circ\bullet}(j_1) c_\circ^\dagger(j_1) c_\bullet(j_1) + h.c.\right]$ where $c_\alpha(j_1) = c_\alpha(j_1, k_2)$, $t_{\bullet\circ}(j_1) = t$ and $t_{\circ\bullet}(j_1) = t(1 + e^{i(k_2 - 2\pi\phi j_1)})$. The spectrum of H without magnetic field is shown in Fig.7. There is characteristic zero energy states which are localized near zigzag and bearded boundaries[11].

One particle Schrodinger equation for each k_2, $H_{cyl}(k_2)|E(k_2)\rangle = E(k_2)|E(k_2)\rangle$ by $|E\rangle = \sum_{j_1}(\psi_\bullet(j_1) c_\bullet^\dagger(j_1, k_2) + \psi_\circ(j_1) c_\circ^\dagger(j_1, k_2))|0\rangle$ is reformulated using a transfer matrix as ($\phi = p/q$, p and q are mutually prime.)

$$\begin{pmatrix} \psi_\bullet(q\ell+1) \\ \psi_\circ(q\ell) \end{pmatrix} = M^\ell \begin{pmatrix} \psi_\bullet(1) \\ \psi_\circ(0) \end{pmatrix}$$

where M is a 2×2 matrix (see Appendix C for the precise definition). It implies that the boundary condition for zigzag-bearded boundaries is then $\psi_\circ(q\ell) = 0$ (bearded) and $\psi_\circ(0) = 0$ (zigzag). Then the spectrum $\epsilon_\ell(k_2)$ of the edge states is determined by $M_{21}(\epsilon_\ell) = 0$ and their position is specified as

$$\begin{array}{rlll} \text{for zigzag-bearded: } |M_{11}(\epsilon_\ell)| & > & 1: & \text{right } j_1 \approx \infty : \text{(bearded edges)} \\ |M_{11}(\epsilon_\ell)| & < & 1: & \text{left } \ j_1 \approx 1 \ : \text{(zigzag edges)} \end{array}$$

When one consider the bearded-zigzag boundaries, the condition is $\psi_\bullet(q\ell+1) = 0$ (zigzag) and $\psi_\bullet(1) = 0$ (bearded). Then the spectrum is determined by $M_{12}(\epsilon_\ell) = 0$ and

$$\begin{array}{rlll} \text{for bearded-zigzag: } |M_{22}(\epsilon_\ell)| & > & 1: & \text{right } j_1 \approx \infty : \text{(zigzag edges)} \\ |M_{22}(\epsilon_\ell)| & < & 1: & \text{left } \ j_1 \approx 1 \ : \text{(bearded edges)} \end{array}$$

4.1. Zero mode boundary states

Graphene with zigzag edges without magnetic field has a characteristic boundary states (Fujita states) which has been also confirmed by a realistic first principle calculation and also experimentally[11, 36, 12]. By the general discussion above, we have ($\phi = 0$) the zero modes localized at the zigzag edge when $|M_{11}| = |1 + e^{ik_2}| < 1$ $(2\pi/3 < k_2 < 4\pi/3)$ and at the bearded edge when $|M_{11}| = |1 + e^{ik_2}| > 1$ $(0 < k_2 < 2\pi/3,\ 4\pi/3 < k_2 < 2\pi)$[11, 35] as shown in Fig7. Actually the existence of the zero localize states is guaranteed by the bulk[35]. Let us first define the Berry phase (Zak phase) $\gamma(k_2)$ for each k_2 using a bulk hamiltonian $\boldsymbol{h}(\boldsymbol{k})$ as

$$\gamma_{Z,B}(k_2) \ = \ -i \int_0^{2\pi} dk_1\, \boldsymbol{\psi}^\dagger_{Z,B} \partial_{k_1} \boldsymbol{\psi}_{Z,B} = 0 \text{ or } \pi \pmod{2\pi}$$

where $\boldsymbol{\psi}_{Z,B}$ is an eigen state of the hamiltonian of the periodic systems $\boldsymbol{h}_{Z,B}$ (compatible with the zigzag or bearded edges. This quantization as Z_2 Berry phase is due to the chiral symmetry[35, 29]. The zero modes which is compatible for the unit cell of the periodic system only exist when $\gamma(k_2) = \pi$.

$$\gamma_{Z,B}(k_2) = \begin{cases} \pi \pmod{2\pi} & : \text{zero modes exist at the zigzag (bearded) edges for } k_2 \\ 0 \pmod{2\pi} & : \text{no reason to have the zero mode for } k_2 \end{cases}$$

The condition for the zigzag boundary states, $|M_{11}| = |1 + e^{-ik_2}| < 1$, is the condition for the $\Delta_Z(\boldsymbol{k}) = 1 + e^{-ik_2} + e^{-ik_1}$ encloses the origin by changing $k_1 :\rightarrow 2\pi$. Since the curve is a unit circle centered at $1 + e^{-ik_2}$, the distance between the origin and center $|1 + e^{-ik_2}|$ needs to be less than unity for the origin to be enclosed. *It establishes a bulk-edge correspondence of the zero mode boundary states for the zigzag edges.*

With a magnetic field, there still exist zero mode boundary states though the $n = 0$ Landau level also co-exists. Even though the bulk and the edge are both at the same energy, the local density of state at $E = 0$ shows characteristic behavior which is consistent with the localized boundary modes[37].

4.2. Quantum Hall edge states of Dirac fermions

Anomalous quantum Hall effect of graphene as of the Dirac fermions is also discussed with the boundaries on the cylindrical geometry. Then counting the light and left edges modes between the energy gap where the fermi energy lies, one can assign the Hall conductance using the Laughlin argument. Using the transfer matrix formulation given before, we can assign a winding number of the edge state on the complex energy surface, which is, in our case, a Riemann surface of genus g where g is a number of energy gaps (Landau gaps). Using the setup, one may prove a direct relation between the two topological numbers, the one is the Chern number and the other is a winding number of the edge state [38, 6, 16]. It reads physically

$$\sigma_{xy}^{\text{bulk}} \ = \ \sigma_{xy}^{\text{edge}}.$$

That is, the Hall conductance described by the bulk and the edges are the same. It should be like this. This is the bulk-edge correspondence of the Dirac fermions in graphene. In Fig.8, energy spectrum with the zigzag-bearded boundaries are shown as a function of k_2 and the corresponding Hall conductance. It confitms the principle.

Acknowledgement

We thank discussion with H. Aoki, T. Kawarabayashi, T. Fukui, M. Arai, T. Morimoto and H. Watanabe. The work is supported in part by Grants-in-Aid for Scientific Research, No.203400984 from JSPS and No.22014002 (Novel States of Matter Induced by Frustration) on Priority Areas from MEXT (JAPAN).

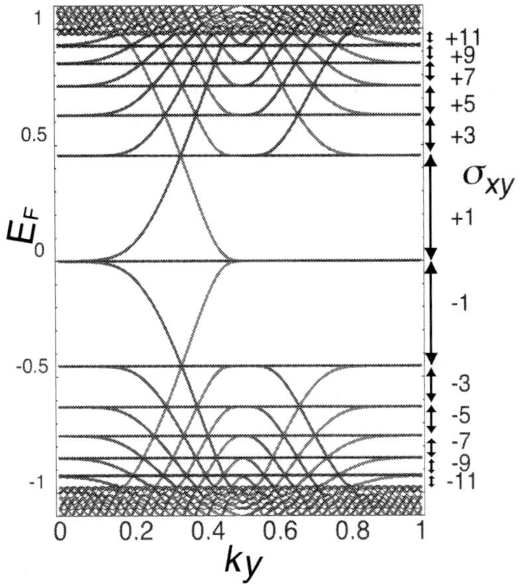

Figure 8. One particle energy spectrum of the graphene on a cylinder. The blue (red) lines are edge states localized near the left (right) edges. k_y is a momentum along the edges and the σ_{xy}'s when the fermi energy sit there are shown in unit $-e^2/h$. The numbers are the Chern numbers of the bulk and, at the same time, numbers of the blue (red) lines in the gap ($\phi = 1/51$). (See also [2])

Appendix A. Landau Level with anisotropic mass

Let us summarize the standard Landau quantization of electrons with parabolic dispersion with (effective) anisotropic mass described by the following hamiltonian $\boldsymbol{H} = \boldsymbol{\pi}^\dagger \frac{1}{2m} \Xi_L \boldsymbol{\pi}$, ($\mathrm{rot}\, \boldsymbol{A} = B\hat{z}$) where $\boldsymbol{\pi} = \begin{pmatrix} \pi_x \\ \pi_y \end{pmatrix}$, $\pi_i = p_i - eA_i = \pi_i^\dagger$ and $\Xi_L = \begin{pmatrix} \xi_x & \xi_{xy} \\ \xi_{xy} & \xi_y \end{pmatrix}$ is a real symmetric anisotropy matrix. It satisfies $[\pi_x(\ell_B/\hbar), \pi_y(\ell_B/\hbar)] = i$ ($\ell_B = \sqrt{\frac{\hbar}{eB}}$). Since the matrix Ξ_L is real symmetric, it is diagonalized by the orthogonal matrix as $\Xi_L = \boldsymbol{V}^\dagger \Xi_D \boldsymbol{V} = \boldsymbol{V}^\dagger \Xi_D \boldsymbol{V}$ $\Xi_D = \mathrm{diag}(\xi_X, \xi_Y)$, $\xi_X \xi_Y = \det \Xi_L$, $\xi_X + \xi_Y = \mathrm{Tr}\, \Xi_L$. $\boldsymbol{V} = \begin{pmatrix} \cos\theta & -\sin\theta \\ \sin\theta & \cos\theta \end{pmatrix}$, ${}^\exists\theta \in \mathbb{R}$. Then we have $H = \boldsymbol{\Pi}^\dagger \Xi_D \boldsymbol{\Pi}$, $\boldsymbol{\Pi} \equiv \boldsymbol{V}\boldsymbol{\pi}$, $[\Pi_X(\ell_B/\hbar), \Pi_Y(\ell_B/\hbar)] = i$.

Now defining a bosonic operator, ($[a, a^\dagger] = 1$) $a = (\ell_B/\hbar)(\Pi_X + i\Pi_Y)/\sqrt{2}$, the hamiltonian is written as $H = \frac{\hbar\omega}{4}\Big(\xi_X(a + a^\dagger)^2 - \xi_Y(a - a^\dagger)^2\Big)$ where $\omega = \dfrac{eB}{m}$ is a cyclotron frequency.

Now we define a new bosonic operator ($[b, b^\dagger] = 1$) as $a = ub + v^* b^\dagger$ requiring $[a, a^\dagger] = [ub + v^* b^\dagger, u^* b^\dagger + vb] = |u|^2 - |v|^2 = 1$. Here noting that $\xi_X(a + a^\dagger)^2 - \xi_Y(a - a^\dagger)^2 = b^2\{\xi_X(u + v)^2 - \xi_Y(u - v)^2\} + h.c. + (bb^\dagger + b^\dagger b)\{\xi_X|u + v|^2 + \xi_Y|u - v|^2\}$, we choose as $\xi_X(u + v)^2 = \xi_Y(u - v)^2$, $u + v = C\sqrt{\xi_Y}$, $u - v = \pm C\sqrt{\xi_X}$. Assuming $\xi_X, \xi_Y > 0$ and imposing $|u|^2 - |v|^2 = 1$, we have $|C|^2 = 1/\sqrt{\xi_X \xi_Y} = 1/(\det \Xi_L)^{1/2}$, $u = \frac{\sqrt{\xi_X} + \sqrt{\xi_Y}}{2(\det \Xi_L)^{1/4}}$, $v = \frac{-\sqrt{\xi_X} + \sqrt{\xi_Y}}{2(\det \Xi_L)^{1/4}}$.

Finally the hamiltonian is written as

$$H = \frac{1}{2}\hbar\omega(bb^\dagger + b^\dagger b)|C|^2(\xi_X \xi_Y) = \hbar\omega_\Xi(b^\dagger b + \frac{1}{2})$$

where $\omega_\Xi = \omega\sqrt{\det \Xi_L} = \frac{eB}{m}\sqrt{\xi_X \xi_Y}$. As for the Landau degeneracy, we stressed that it is

Journal of Physics: Conference Series **334** (2011) 012004 doi:10.1088/1742-6596/334/1/012004

independent of the anisotropy since the magnetic length is independent of the mass.

Appendix B. Some notations of differential form

In this appendix, we supplement some details for the differential forms which may not be so popular as the standard vector analysis.

- We have omitted the wedge product ($\wedge$) which is anti-commuting and useful for the integral over the oriented surface. For example, $d\phi^1 \wedge d\phi^2 = -d\phi^2 \wedge d\phi^1$ and $d\phi^1 \wedge d\phi^1 = -d\phi^1 \wedge d\phi^1 = 0$. It implies

$$\langle dG|dG\rangle \equiv \langle dG| \wedge |dG\rangle = d\phi^\mu \langle \partial_\mu G| \wedge d\phi^\nu |\partial_\nu G\rangle = d\phi^1 \wedge d\phi^2 \big[\langle \partial_1 G|\partial_2 G\rangle - \langle \partial_2 G|\partial_1 G\rangle\big]$$

- For any (matrix valued) one form $\boldsymbol{X} = \boldsymbol{X}_\mu d\phi^\mu$, $\mathrm{Tr}\,\boldsymbol{X}^2 = 0$

$$\mathrm{Tr}\,\boldsymbol{X}^2 = \mathrm{Tr}\,\boldsymbol{X}_\mu \boldsymbol{X}_\nu d\phi^\mu d\phi^\nu = \mathrm{Tr}\,\boldsymbol{X}_\nu \boldsymbol{X}_\mu d\phi^\mu d\phi^\nu = \mathrm{Tr}\,\boldsymbol{X}_\mu \boldsymbol{X}_\nu (-d\phi^\nu d\phi^\mu) = -\mathrm{Tr}\,\boldsymbol{X}^2$$

- As for a matrix valued function $\boldsymbol{\omega}$, $d\boldsymbol{\omega}^{-1} = -\boldsymbol{\omega}^{-1} d\boldsymbol{\omega}\boldsymbol{\omega}^{-1}$. It obeys from $0 = d\boldsymbol{E} = d(\boldsymbol{\omega}\boldsymbol{\omega}^{-1}) = (d\boldsymbol{\omega})\boldsymbol{\omega}^{-1} + \boldsymbol{\omega} d\boldsymbol{\omega}^{-1}$.
- Gauge transformation of the non-Abelian gauge strength $\boldsymbol{F}_M = d\boldsymbol{A}_M + \boldsymbol{A}_M^2$ where $\boldsymbol{A}_M = \boldsymbol{\omega}^{-1}\boldsymbol{A}'_M\boldsymbol{\omega} + \boldsymbol{\omega}^{-1}d\boldsymbol{\omega}$.

$$\begin{aligned}\boldsymbol{F}_M &= d(\boldsymbol{\omega}^{-1}\boldsymbol{A}'_M\boldsymbol{\omega}) + d(\boldsymbol{\omega}^{-1}d\boldsymbol{\omega}) + \boldsymbol{\omega}^{-1}{\boldsymbol{A}'_M}^2\boldsymbol{\omega} + \boldsymbol{\omega}^{-1}\boldsymbol{A}'_M d\boldsymbol{\omega} + \boldsymbol{\omega}^{-1}d\boldsymbol{\omega}\boldsymbol{\omega}^{-1}\boldsymbol{A}'_M\boldsymbol{\omega} + (\boldsymbol{\omega}^{-1}d\boldsymbol{\omega})^2 \\ &= -\boldsymbol{\omega}^{-1}d\boldsymbol{\omega}\boldsymbol{\omega}^{-1}\boldsymbol{A}'_M\boldsymbol{\omega} + \boldsymbol{\omega}^{-1}d\boldsymbol{A}'_M\boldsymbol{\omega} - \boldsymbol{\omega}^{-1}\boldsymbol{A}'_M d\boldsymbol{\omega} - \boldsymbol{\omega}^{-1}d\boldsymbol{\omega}\boldsymbol{\omega}^{-1}d\boldsymbol{\omega} \\ &\quad + \boldsymbol{\omega}^{-1}{\boldsymbol{A}'_M}^2\boldsymbol{\omega} + \boldsymbol{\omega}^{-1}\boldsymbol{A}'_M d\boldsymbol{\omega} + \boldsymbol{\omega}^{-1}d\boldsymbol{\omega}\boldsymbol{\omega}^{-1}\boldsymbol{A}'_M\boldsymbol{\omega} + (\boldsymbol{\omega}^{-1}d\boldsymbol{\omega})^2 = \boldsymbol{\omega}^{-1}\boldsymbol{F}'_M\boldsymbol{\omega}\end{aligned}$$

Appendix C. Transfer matrix formulation

[38, 16] The Schrodinger equation of one particle state with fixed k_2 is written as

$$\begin{aligned} t_{\bullet\circ}(j_1-1)\psi_\circ(j_1-1) + t^*_{\circ\bullet}(j_1)\psi_\circ(j_1) &= E\psi_\bullet(j_1) \\ t^*_{\bullet\circ}(j_1)\psi_\bullet(j_1+1) + t_{\circ\bullet}(j_1)\psi_\bullet(j_1) &= E\psi_\circ(j_1) \end{aligned}$$

which are expressed in the transfer matrix form as

$$\begin{aligned}\begin{pmatrix}\psi_\circ(j_1)\\ \psi_\bullet(j_1)\end{pmatrix} &= M_{\circ\bullet}(j_1)\begin{pmatrix}\psi_\bullet(j_1)\\ \psi_\circ(j_1-1)\end{pmatrix}, \quad M_{\circ\bullet}(j_1) = \begin{pmatrix}\frac{E}{t^*_{\circ\bullet}(j_1)} & -\frac{t_{\bullet\circ}(j_1-1)}{t^*_{\circ\bullet}(j_1)}\\ 1 & 0\end{pmatrix} \\ \begin{pmatrix}\psi_\bullet(j_1+1)\\ \psi_\circ(j_1)\end{pmatrix} &= M_{\bullet\circ}(j_1)\begin{pmatrix}\psi_\circ(j_1)\\ \psi_\bullet(j_1)\end{pmatrix}, \quad M_{\bullet\circ}(j_1) = \begin{pmatrix}\frac{E}{t^*_{\bullet\circ}(j_1)} & -\frac{t_{\circ\bullet}(j_1)}{t^*_{\bullet\circ}(j_1)}\\ 1 & 0\end{pmatrix}\end{aligned}$$

where $t_{\bullet\circ}(j_1) = t$ and $t_{\circ\bullet}(j_1) = t(1 + e^{i(k_2 - 2\pi\phi j_1)})$. Then the total transfer matrix is given as

$$\begin{pmatrix}\psi_\bullet(j_1+1)\\ \psi_\circ(j_1)\end{pmatrix} = M_t(j_1)\begin{pmatrix}\psi_\bullet(j_1)\\ \psi_\circ(j_1-1)\end{pmatrix}, \quad M_t(j_1) = M_{\bullet\circ}(j_1)M_{\circ\bullet}(j_1)$$

As for $\phi = p/q$, we finally have

$$\begin{pmatrix}\psi_\bullet(q\ell+1)\\ \psi_\circ(q\ell)\end{pmatrix} = M^\ell\begin{pmatrix}\psi_\bullet(1)\\ \psi_\circ(0)\end{pmatrix}, \quad M = M_t(q)\cdots M_t(1),$$

where $\det M = (-)^q \prod_{j_1=1}^q \frac{t_{\circ\bullet}}{t^*_{\circ\bullet}}\frac{t_{\bullet\circ}}{t^*_{\bullet\circ}}$, $|\det M| = 1$.

References

[1] Wallace P R 1947 *Phys. Rev.* **71** 622
[2] Lomer W H 1955 *Proc. Roy. Soc. (London)* **A227** 330
[3] McClure J W 1956 *Phys. Rev.* **104** 666
[4] Novoselov K S, Geim A K, Morozov S V, Jiang D, Katsnelson M I, Grigorieva I V, Dubonos S V and Firsov A A 2005 *Nature* **438** 197
[5] Zhang Y, Tan Y, Stormer H and Kim P 2005 *Nature* **438** 201
[6] Hatsugai Y 1993 *Phys. Rev. Lett.* **71** 3697
[7] Qi X L and Zhang S C 2010 *Physcs today* **63** 33
[8] Moore J E 2010 *Nature* **464** 194
[9] Hasan M Z and Kane C L *arXiv:1002.3895*
[10] Wang Z, Chong Y, Joannopoulos J D and Soljacic M 2009 *Nature* **461** 772
[11] Fujita M, Wakabayashi K, Nakada K and Kusakabe K 1996 *J. Phys. Soc. Jpn.* **65** 1920
[12] Kobayashi Y, Fukui K, Enoki T, Kusakabe K and Kaburagi Y 2005 *Phys. Rev. B* **71** 193406
[13] Nielsen H B and Ninomiya M 1981 *Nucl. Phys. B* **185** 20
[14] Zheng Y and Ando T 2005 *Phys. Rev. B* **65** 245420
[15] Gusynin V P and Sharapov S 2005 *Phys. Rev. Lett.* **95** 146801
[16] Hatsugai Y, Fukui T and Aoki H 2006 *Phys. Rev. B* **74** 205414
[17] Guinea F, Horovitz B, and Doussal P L 2008 *Phys. Rev. B* **77** 205421
[18] Kawarabayashi T, Hatsugai Y and Aoki H 2009 *Phys. Rev. Lett.* **103** 156804
[19] Hatsugai Y, Fukui T and Aoki H 2007 *Eur. Phys. J. Special topics* **148** 133
[20] Hatsugai Y 2010 *New J. Phys.* **12** 065004
[21] Creutz M 2008 *JHEP* **04** 017
[22] Aharonov Y and Casher A 1979 *Phys. Rev. B* **19** 2461
[23] Niu, Thouless D J, and Wu Y S 1985 *Phys. Rev. B* **31** 3372
[24] Berry M V 1984 *Proc. R. Soc.* **A392** 45
[25] Flanders H 1989 *Differential Forms with Applications to the Physical Sciences* (Dover Publ.)
[26] Kohmoto M 1985 *Ann. Phys. (N. Y.)* **160** 355
[27] Hatsugai Y 2004 *J. Phys. Soc. Jpn.* **73** 2604
[28] Hatsugai Y, Ryu S and Kohmoto M 2004 *Phys. Rev. B* **70** 054502
[29] Hatsugai Y 2006 *J. Phys. Soc. Jpn.* **75** 123601
[30] Hatsugai Y 2005 *J. Phys. Soc. Jpn.* **74** 1374
[31] Wilczek F and Zee A 1984 *Phys. Rev. Lett.* **52** 2111
[32] Thouless D J, Kohmoto M, Nightingale P and den Nijs M 1982 *Phys. Rev. Lett.* **49** 405
[33] Fukui T, Hatsugai Y and Suzuki H 2005 *J. Phys. Soc. Jpn.* **74** 1674
[34] Arai M and Hatsugai Y 2009 *Phys. Rev. B* **79** 075429
[35] Ryu S and Hatsugai Y 2002 *Phys. Rev. Lett.* **89** 077002
[36] Okada S and Oshiyama A 2001 *Phys. Rev. Lett.* **87** 146803
[37] M Arikawa Y H and Aoki H 2008 *Phys. Rev. B* **78** 205401
[38] Hatsugai Y 1993 *Phys. Rev. B* **48** 11851

HMF-19 IOP Publishing
Journal of Physics: Conference Series **334** (2011) 012005 doi:10.1088/1742-6596/334/1/012005

Orbital magnetism of Dirac systems

Mikito Koshino and Tsuneya Ando

Department of Physics, Tokyo Institute of Technology, 2-12-1 Ookayama, Meguro-ku, Tokyo 152-8551, Japan

E-mail: koshino@stat.phys.titech.ac.jp

Abstract. Orbital magnetic susceptibility is calculated for massive Dirac systems, such as graphene with a band gap, and bismuth. Gap opening allows us to relate the singular orbital magnetism of massless Dirac system to the usual magnetism of conventional metal. In gapped graphene, we show that the valley degree of freedom produces Zeeman splitting as a real spin does, and causes Pauli paramagnetism which dominates over Landau diamagnetism. Singular susceptibility of intrinsic graphene can be understood as divergence of those contributions in zero-mass limit. A similar analysis is applied to three-dimensional Dirac systems to explain the strong diamagnetism of bismuth.

1. Introduction

In condensed matter systems, the orbital magnetism sensitively depends on the detail of the electronic band structure, and sometimes largely deviates from the conventional Landau diamagnetism. Particularly, narrow gap materials such as graphite [1, 2, 3] or bismuth [4, 5, 6] exhibit a singular behavior in the orbital susceptibility near the energy gap. Graphene monolayer [7, 8, 9] is a zero-gap system in which the conduction and valance bands stick touch at K and K' points in the Brillouin zone, called valleys, and the low-energy physics is characterized by chiral quasiparticles analogous to massless Dirac fermions. [1, 10, 11, 12, 13, 14, 15, 16, 17] At the Dirac point where two bands cross each other, the magnetic susceptibility has a singularity expressed as a delta function in Fermi energy ε_F, i.e., the susceptibility is negative infinite at $\varepsilon_F = 0$, while vanishes when ε_F enters the conduction or valence band. [1, 18, 19, 20, 21, 22, 23, 24, 25] The orbital magnetism was also studied for related materials, such as graphite intercalation compounds,[26, 27, 28, 29] carbon nanotube,[30, 31, 32, 33] few-layer graphenes,[34, 35, 36] and organic compounds having Dirac-like spectrum.[37]

In this paper, we calculate the orbital magnetism of *massive* Dirac systems having an energy gap, to understand a singular diamagnetism of massless Dirac system as a zero-mass (i.e., zero-gap) limit of a conventional system. In Sec. 2, we calculate the susceptibility of a gapped monolayer graphene. We find that the pseudo-spin degree of freedom associated with valleys produces Zeeman splitting as a real spin does, and gives rise to Pauli paramagnetism in analogous way to the usual system. In zero-gap graphene, the singular susceptibility change in varying ε_F i.e., disappearance of the diamagnetism off the Dirac point, is understood as a contribution of valley Pauli paramagnetism give by band electrons. In Sec. 3 we extend the analysis to a three-dimensional Dirac system which models bismuth.

2. Monolayer graphene

Electronic states of graphene in the vicinity of K and K' points in the Brillouin zone are well described by the effective mass approximation.[1, 10] Let $|A\rangle$ and $|B\rangle$ be the Bloch functions at the K point, corresponding to A and B sublattices, respectively. In a basis $(|A\rangle, |B\rangle)$, the Hamiltonian for the monolayer graphene around the K point becomes

$$\mathcal{H}^K = \begin{pmatrix} \Delta & v\pi_- \\ v\pi_+ & -\Delta \end{pmatrix}, \tag{1}$$

where $v \approx 1 \times 10^6$ m/s is the band velocity [8, 9], $\pi_\pm = \pi_x \pm i\pi_y$, and $\boldsymbol{\pi} = -i\hbar\boldsymbol{\nabla} + (e/c)\mathbf{A}$ with vector potential $\mathbf{A}$ giving external magnetic field $\mathbf{B} = \boldsymbol{\nabla} \times \mathbf{A}$. In the following, we neglect the spin Zeeman energy because the spin splitting is much smaller than Landau-level separations. The Hamiltonian at the K' point is obtained by exchanging $\pi_\pm$ in Eq. (1).

The diagonal terms $\pm\Delta$ represent the potential asymmetry between A and B sites, which opens an energy gap at the Dirac point. This can arise in graphene placed on a certain substrate material, where the interaction between the graphene and the substrate lattice produces different potentials between A and B. Indeed, a band gap of the order of 0.1eV has been observed in graphene epitaxially grown on a SiC substrate. [38, 39] From a theoretical point of view, the singular behavior in ideal graphene with vanishing gap is better understood by taking the limit $\Delta \to 0$, as will be shown below. We can safely assume $\Delta \geq 0$ without loss of generality. In a magnetic field, the eigenenergy becomes

$$\begin{aligned} \varepsilon_n^K &= \mathrm{sgn}_-(n)\sqrt{(\hbar\omega_B)^2|n| + \Delta^2}, \\ \varepsilon_n^{K'} &= \mathrm{sgn}_+(n)\sqrt{(\hbar\omega_B)^2|n| + \Delta^2} \end{aligned} \tag{2}$$

with $n = 0, \pm1, \pm2, \cdots$, $\hbar\omega_B = \sqrt{2}\hbar v/l_B$, $l_B = \sqrt{\hbar/eB}$, and

$$\mathrm{sgn}_\pm(n) = \begin{cases} +1 & (n > 0); \\ \pm1 & (n = 0); \\ -1 & (n < 0). \end{cases} \tag{3}$$

The Landau levels of $n \neq 0$ are doubly degenerate between the K and K' valleys, while those of $n = 0$ are not. Figure 2(a) shows an example of energy levels at $\hbar\omega_B = 2\Delta$.

The thermodynamical potential at temperature T then becomes

$$\Omega = -\frac{1}{\beta}\frac{g_v g_s}{2\pi l_B^2}\sum_{s=\pm}\sum_{n=0}^{\infty}\varphi\left[\varepsilon_s((\hbar\omega_B)^2 n)\right]\left(1 - \frac{\delta_{n0}}{2}\right), \tag{4}$$

where $\beta = 1/k_BT$, $\varepsilon_s(x) = s\sqrt{x + \Delta^2}$, $\varphi(\varepsilon) = \log\left[1 + e^{-\beta(\varepsilon-\zeta)}\right]$ with ζ being the chemical potential, and $g_s = 2$ and $g_v = 2$ represent the degrees of freedom associated with spin and valley, respectively. The magnetic susceptibility is defined by

$$\chi = -\left(\frac{\partial^2\Omega}{\partial B^2}\right)_\zeta\Big|_{B=0}. \tag{5}$$

In weak magnetic field, using the Euler-Maclaurin formula, the summation in n in Eq. (4) can be written as an integral in continuous variable x and a residual term proportional to B^2. At zero temperature, we have [25]

$$\chi(\varepsilon_F) = -g_v g_s \frac{e^2 v^2}{6\pi c^2}\frac{1}{2|\Delta|}\theta(|\Delta| - |\varepsilon_F|). \tag{6}$$

In the limit of $\Delta \to 0$, the susceptibility approaches a delta function

$$\chi(\varepsilon_F) = -g_v g_s \frac{e^2 v^2}{6\pi c^2} \delta(\varepsilon_F), \tag{7}$$

in agreement with the previous result.[1, 26, 21] The susceptibility Eq. (6) and the density of states are shown in Fig. 2 (b). The susceptibility is not zero in the gap, because the completely filled valence band gives a constant diamagnetic susceptibility. When the Fermi energy enters the conduction band, the susceptibility jumps downs to zero, resulting in zero total magnetism.

Because the Hamiltonian is equivalent to that of a Dirac electron with a nonzero mass, the magnetic susceptibility around the band edge should correspond to that of a conventional electron. This is clearly illustrated by the effective Hamiltonian expanded in the vicinity of $\mathbf{k} = 0$. For the conduction band, the effective Hamiltonian for the A site near the band bottom ($\varepsilon = \Delta$) is written apart from the constant energy as [25]

$$\mathcal{H}^K \approx \frac{v^2}{2\Delta} \pi_- \pi_+ = \frac{\boldsymbol{\pi}^2}{2m^*} + \frac{1}{2} g^* \mu_B B, \tag{8}$$

$$\mathcal{H}^{K'} \approx \frac{v^2}{2\Delta} \pi_+ \pi_- = \frac{\boldsymbol{\pi}^2}{2m^*} - \frac{1}{2} g^* \mu_B B, \tag{9}$$

where $\mu_B = e\hbar/(2mc)$ is the Bohr magneton with m being the free electron mass, and we defined $m^* = \Delta/v^2$, $g^* = 2m/m^*$. For instance, the g factor is estimated at $g^* \sim 60$ at $\Delta = 0.1$ eV, and diverges as $\propto \Delta^{-1}$ as the gap decreases. The last term in each Hamiltonian can be regarded as the pseudo-spin Zeeman term, where the different valleys K and K' serve as pseudo-spin up ($\xi = +1$) and down ($\xi = -1$), respectively. This agrees with the Zeeman energy expected for an intrinsic magnetic moment, that originates from the self-rotation of the wave packet in Bloch electron. [40, 41]

Obviously, the pseudo-spin Zeeman term gives the Pauli paramagnetism and the first term containing $\boldsymbol{\pi}^2$ gives the Landau diamagnetism in the usual form as

$$\chi_P(\varepsilon) = \left(\frac{g^*}{2}\right)^2 \mu_B^2 D(\varepsilon), \tag{10}$$

$$\chi_L(\varepsilon) = -\frac{1}{3}\left(\frac{m}{m^*}\right)^2 \mu_B^2 D(\varepsilon), \tag{11}$$

with density of states $D(\varepsilon) = g_v g_s m^*/(2\pi\hbar^2)\,\theta(\varepsilon)$. The total susceptibility $\chi_P + \chi_L$ actually agrees with the amount of the jump at the conduction band bottom in χ of Eq. (6). Because $g = 2m/m^*$ in the present case, we have $\chi_L = -\chi_P/3 \propto 1/m^*$ as in the free electron, giving the paramagnetic susceptibility in total. Therefore the susceptibility exhibits a discrete jump toward the paramagnetic direction when the Fermi energy moves off the Dirac point.

In the original Hamiltonian, the Landau-level energies in Eq. (2) can be rewritten as

$$\varepsilon_{\xi,s,n'} = s\sqrt{(\hbar\omega_B)^2\left(n' + \frac{1}{2} + \frac{\xi s}{2}\right) + \Delta^2} \quad (n' = 0, 1, 2, \cdots). \tag{12}$$

Figure 2(a) shows energy levels for $\hbar\omega_B = 2\Delta$ and the relationship between the different labeling schemes of Eqs. (2) and (12). For the conduction band, the levels of the same n' with opposite pseudo-spins $\xi = \pm 1$ share the same Landau level function labeled by n' on the A site, on which the states near the conduction-band bottom ($\varepsilon = \Delta$) have most of the amplitude. For the valence band, similarly, n' describes the index of the Landau-level function at the B site.

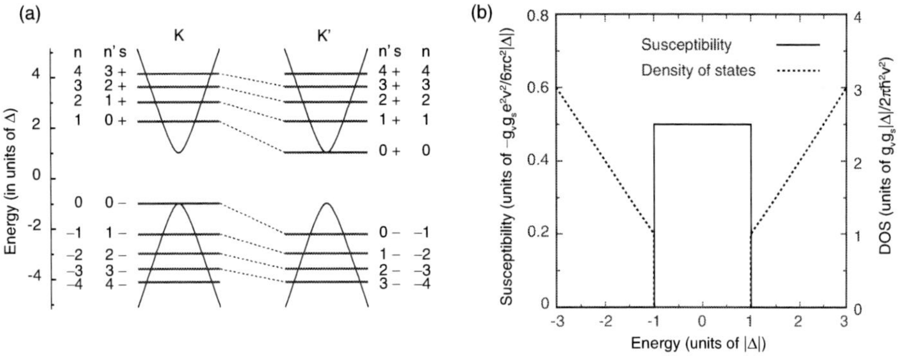

Figure 1. (a) Landau-level energies of gapped monolayer graphene for $\hbar\omega_B = 2\Delta$. (b) Orbital susceptibility (solid) and density of states (dashed) of monolayer graphene with asymmetric potential Δ.

3. Three dimensional Dirac system

The results of graphene can be directly extended to three-dimensional Dirac Hamiltonian, which approximately describes the electronic structure of bismuth. [4, 5, 6, 42] The Hamiltonian is given by

$$\mathcal{H} = \begin{pmatrix} \Delta & 0 & v\pi_z & v\pi_- \\ 0 & \Delta & v\pi_+ & -v\pi_z \\ v\pi_z & v\pi_- & -\Delta & 0 \\ v\pi_+ & -v\pi_z & 0 & -\Delta \end{pmatrix}, \tag{13}$$

where four bases correspond to two orbital and two spin degrees of freedom. The anisotropy of the velocity present in bismuth is ignored for simplicity, as we focus on the parallel argument to its 2D counterpart. The orbital susceptibility was previously calculated for realistic Hamiltonian retaining the anisotropy and other factors. [6]

The density of states at zero magnetic field is

$$D(\varepsilon) = \frac{g_v g_s}{\pi^2\hbar^3 v^3}|\varepsilon|\sqrt{\varepsilon^2-\Delta^2}\,\theta(\varepsilon^2-\Delta^2), \tag{14}$$

where g_v is the valley degeneracy allowing the presence of different k points described by the above Hamiltonian in the first Brillouin zone. The Landau levels in a uniform magnetic field in z direction are given by [25]

$$\varepsilon_{s,n,\sigma} = s\sqrt{(\hbar\omega_B)^2\Big(n+\frac{1}{2}+\frac{\sigma}{2}\Big)+v^2p_z^2+\Delta^2} \quad (n = 0,1,2,\cdots), \tag{15}$$

with $\hbar\omega_B = \sqrt{2}\hbar v/l_B$, $s = \pm 1$, and $\sigma = \pm 1$. This is equivalent to the two-dimensional Dirac system, Eq. (12), when the term Δ^2 is replaced with $\Delta^2 + v^2p_z^2$. The susceptibility $\chi(\varepsilon)$ is

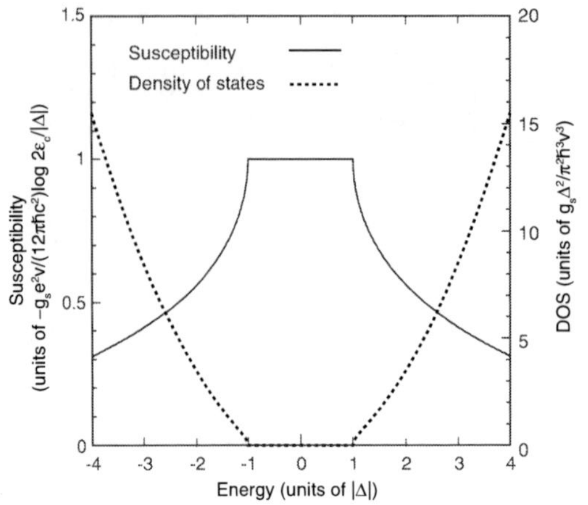

Figure 2. Orbital susceptibility and density of states of three-dimensional Dirac electron.

calculated by integrating Eq. (6) in p_z as

$$\begin{aligned}\chi(\varepsilon) &= -\frac{g_v g_s e^2 v^2}{6\pi c^2}\int \frac{dp_z}{2\pi\hbar}\frac{\theta(\Delta^2+v^2p_z^2-\varepsilon^2)}{2\sqrt{\Delta^2+v^2p_z^2}} \\ &= -\frac{g_v g_s e^2 v}{12\pi^2\hbar c^2}\begin{cases}\log\dfrac{2\varepsilon_c}{|\Delta|} & (|\varepsilon|<|\Delta|); \\ \log\dfrac{2\varepsilon_c}{|\varepsilon|+\sqrt{\varepsilon^2-\Delta^2}} & (|\varepsilon|>|\Delta|),\end{cases}\end{aligned} \tag{16}$$

where ε_c is a cut-off energy. In the limit of $\Delta \to 0$, the susceptibility at zero energy logarithmically diverges.

At an energy ε just above the band bottom $|\Delta|$, we obtain the paramagnetic contribution

$$\chi(\varepsilon) - \chi(0) \approx \frac{2}{3}\left(\frac{m}{m^*}\right)^2 D(\varepsilon)\mu_B^2, \tag{17}$$

where $D(\varepsilon) = (g_s g_v/4\pi^2)(2m^*/\hbar^2)^{3/2}\sqrt{\varepsilon}$ with $m^* = \Delta/v^2$. This is nothing but the magnetic susceptibility, dominated by the Pauli paramagnetism, of a three-dimensional metal with mass m^* and g factor $g^* = 2m/m^*$. Figure 2 shows the susceptibility and the density of states in the present system. The singular decrease of the susceptibility at the band edges is understood in terms of the appearance of the dominant spin paramagnetism inside the band.

We note that in bismuth the index σ in Eq. (15) represents real spin, while it was valley pseudo-spin in Eq. (12) for graphene. The Pauli component included in Eq. (17) thus describes the real spin paramagnetism enhanced by the strong spin-orbit coupling, apart from the bare electron paramagnetism.

4. Conclusion

We have calculated the orbital magnetism of massive Dirac systems. In an analogous way to conventional metal, the susceptibility near the band edge can be expressed by the Landau diamagnetism and the Pauli paramagnetism associated with pseudo-spin, such as valleys in graphene. The diverging magnetism of zero-gap graphene can be intuitively interpreted as a result of vanishing the effective mass. The similar analysis is also applied to three dimensional Dirac systems which models bismuth. In any cases, the susceptibility near the energy gap is typically of the order of $(m/m^*)^2\mu_B^2 D(\varepsilon)$, and overwhelms the real-spin Pauli paramagnetism of bare electrons, owing to small effective mass m^*. We expect that the singular susceptibility in graphene is observed by employing the experimental techniques used for two-dimensional electron systems on semiconductor. [43, 44]

This work was supported in part by Grant-in-Aid for Scientific Research on Priority Area "Carbon Nanotube Nanoelectronics," by Grant-in-Aid for Scientific Research, and by Global Center of Excellence Program at Tokyo Tech "Nanoscience and Quantum Physics" from Ministry of Education, Culture, Sports, Science and Technology Japan.

[1] J. W. McClure, Phys. Rev. **104**, 666 (1956).
[2] J. W. McClure, Phys. Rev. **119**, 606 (1960).
[3] M. P. Sharma, L. G. Johnson, and J. W. McClure, Phys. Rev. B **9**, 2467 (1974).
[4] P. A. Wolff, J. Phys. Chem. Solids **25**, 1057 (1964).
[5] H. Fukuyama and R. Kubo, J. Phys. Soc. Jpn. **27**, 604 (1969).
[6] H. Fukuyama and R. Kubo, J. Phys. Soc. Jpn. **28**, 570 (1970).
[7] K. S. Novoselov, A. K. Geim, S. V. Morozov, D. Jiang, Y. Zhang, S. V. Dubonos, I. V. Grigorieva, and A. A. Firsov, Science **306**, 666 (2004).
[8] K. S. Novoselov, A. K. Geim, S. V. Morozov, D. Jiang, M. I. Katsnelson, I. V. Grigorieva, S. V. Dubonos, and A. A. Firsov, Nature **438**, 197 (2005).
[9] Y. Zhang, Y.-W. Tan, H. L. Stormer, and P. Kim, Nature **438**, 201 (2005).
[10] J. C. Slonczewski and P. R. Weiss, Phys. Rev. **109**, 272 (1958).
[11] D. P. DiVincenzo and E. J. Mele, Phys. Rev. B **29**, 1685 (1984).
[12] G. W. Semenoff, Phys. Rev. Lett. **53**, 2449 (1984).
[13] N. H. Shon and T. Ando, J. Phys. Soc. Jpn. **67**, 2421 (1998).
[14] Y. Zheng and T. Ando, Phys. Rev. B **65**, 245420 (2002).
[15] T. Ando, J. Phys. Soc. Jpn. **74**, 777 (2005).
[16] V. P. Gusynin and S. G. Sharapov, Phys. Rev. Lett. **95**, 146801 (2005).
[17] N. M. R. Peres, F. Guinea, and A. H. Castro Neto, Phys. Rev. B **73**, 125411 (2006).
[18] S. G. Sharapov, V. P. Gusynin, and H. Beck, Phys. Rev. B **69**, 075104 (2004).
[19] H. Fukuyama, J. Phys. Soc. Jpn. **76**, 043711 (2007).
[20] M. Nakamura, Phys. Rev. B **76**, 113301 (2007).
[21] M. Koshino and T. Ando, Phys. Rev. B **75**, 235333 (2007).
[22] A. Ghosal, P. Goswami, and S. Chakravarty, Phys. Rev. B **75**, 115123 (2007).
[23] T. Ando, Physica E **40**, 213 (2007).
[24] M. Koshino, Y. Arimura, and T. Ando, Phys. Rev. Lett. **102**, 177203 (2009).
[25] M. Koshino and T. Ando, Phys. Rev. B **81**, 195431 (2010).
[26] S. A. Safran and F. J. DiSalvo, Phys. Rev. B **20**, 4889 (1979).
[27] S. A. Safran, Phys. Rev. B **30**, 421 (1984).
[28] J. Blinowski and C. Rigaux, J. Phys. (Paris) **45**, 545 (1984).
[29] R. Saito and H. Kamimura, Phys. Rev. B **33**, 7218 (1986).
[30] H. Ajiki and T. Ando, J. Phys. Soc. Jpn. **62**, 1255 (1993).
[31] H. Ajiki and T. Ando, J. Phys. Soc. Jpn. **62**, 2470 (1993); J. Phys. Soc. Jpn. **63**, 4267 (1994) (Erratum).
[32] H. Ajiki and T. Ando, J. Phys. Soc. Jpn. **64**, 4382 (1995).
[33] M. Yamamoto, M. Koshino, and T. Ando, J. Phys. Soc. Jpn. **77**, 084705 (2008).
[34] M. Koshino and T. Ando, Phys. Rev. B **76**, 085425 (2007).
[35] M. Nakamura and L. Hirasawa, Phys. Rev. B **77**, 045429 (2008).
[36] A. H. Castro Neto, F. Guinea, N. M. Peres, K. S. Novoselov, and A. K. Geim, Rev. Mod. Phys. **81**, 109 (2009).
[37] A. Kobayashi, Y. Suzumura, and H. Fukuyama, J. Phys. Soc. Jpn. **77**, 064718 (2008).
[38] S. Y. Zhou, G.-H. Gweon, A. V. Fedorov, P. N. First, W. A. de Heer, D.-H. Lee, F. Guinea, A. H. Castro Neto, and A. Lanzara, Nature Mater. **6**, 770 (2007).
[39] S. Y. Zhou, D. A. Siegel, A. V. Fedorov, F. El Gabaly, A. K. Schmid, A. H. Castro Neto, D.-H. Lee, and A. Lanzara, Nature Mater. **7**, 259 (2008).
[40] M.-C. Chang and Q. Niu, Phys. Rev. B **53**, 7010 (1996).
[41] D. Xiao, W. Yao, and Q. Niu, Phys. Rev. Lett. **99**, 236809 (2007).
[42] Y. Fuseya, M. Ogata, and H. Fukuyama, Phys. Rev. Lett. **102**, 066601 (2009).

[43] H. L. Stormer, T. Haavasoja, V. Narayanamurti, A. C. Gossard, and W. Wiegmann, J. Vac. Sci. Technol. B **1**, 423 (1983).
[44] J. P. Eisenstein, H. L. Stormer, V. Narayanamurti, A. Y. Cho, A. C. Gossard, and C. W. Tu, Phys. Rev. Lett. **55**, 875 (1985).

Tilt-Induced Anisotropic Phases in Wide GaAs Quantum Wells

D R Luhman[1,2] , W Pan[3], T M Lu[2], D C Tsui[2], L N Pfeiffer[2], K W Baldwin[2], and K W West[2]

[1] Department of Physics and Astronomy, Carleton College, Northfield, MN 55057, USA
[2] Department of Electrical Engineering, Princeton University, Princeton, NJ 08544, USA
[3] Sandia National Laboratories, Albuquerque, NM 87185, USA

dluhman@carleton.edu, wpan@sandia.gov

Abstract. Tilt induced anisotropic transport is observed in a series of three wide GaAs quantum wells. In the absence of tilt the transport is isotropic and well developed quantum Hall states are observed. For large values of tilt (θ>60°) the quantum Hall states at ν=5 and 7 are replaced by strongly anisotropic states in the two widest wells, occurring at slightly lower values of θ for the widest well. A similar transition is also seen at ν=9 in the widest quantum well. This transition is not observed in the narrowest quantum well up to the largest tilt angle measured.

1. Introduction

Two-dimensional electron systems (2DESs) in high mobility GaAs quantum wells produce a variety of interesting quantum phases with increased magnetic field. Among them are the anisotropic states in the N≥2 Landau levels which occur at half values of the filling factor, such as ν=9/2 [1,2]. These observed anisotropic states are believed to be due to the formation of a unidirectional charge-density wave state, also known as a stripe phase [3,4]. The orientation of these anisotropic phases has been observed to be sensitive to the application of an in-plane magnetic field [5,6].

Much of the experimental work involving stripe phases has been carried out in single layer systems. However in bilayer systems the presence of interlayer coupling may allow for the formation of more exotic phases. Indeed, the possibility of coherent stripe phases has been studied theoretically [7-13]. In particular, a tilt-induced transition from a quantum Hall state to various isospin stripe phases at odd values of filling factor in the N≥1 Landau levels has been predicted to occur for certain system parameters [9-11]. While experiments [14] with low mobility bilayer samples have shown some preliminary indications for anisotropy at odd ν, unequivocal evidence is still lacking.

2. Summary of Results

In recent experiments we have examined anisotropic transport in three $Al_{0.24}Ga_{0.76}As/GaAs/Al_{0.24}Ga_{0.76}$ wide single quantum wells (WSQWs), with well widths L=58, 64, and 70 nm respectively. For sufficiently large values of L, the Coulomb interaction causes the electrons to collect near the walls of the quantum well forming an effective bilayer system. In the absence of an in-plane magnetic field, isotropic quantum Hall states are observed. When the two widest quantum wells are adequately tilted, the quantum Hall states at ν=5 and ν=7 disappear and are replaced by anisotropic phases. These data clearly demonstrate a tilt-induced transition from a quantum Hall state to a compressible anisotropic state at ν=5 and 7. For the widest quantum well, at the largest tilt angle studied, the quantum Hall states at ν=9 and 11 are also replaced by compressible states with apparent anisotropy at ν=9. Furthermore, in all samples investigated, anisotropic transport is also seen to emerge at half filling of the total filling factor and is particularly prominent at ν=9/2 and ν=11/2.

3. Samples

The WSQWs are modulation doped at a distance of 80 nm from both sides of the quantum well. The samples were cleaved into 5 mm ×5 mm squares and eight equally spaced ohmic contacts were positioned around the perimeter of each specimen to make contact to the conducting channel. The electron densities of each sample varied slightly but were all near $n_e \sim 2.7\times10^{11}$ cm^{-2} and each had a mobility approaching $\mu \sim 10^7$ cm^2/Vs. Each sample was mounted in the mixing chamber of a dilution refrigerator with a base temperature of $T \approx 20$ mK. Transport coefficients were measured as a function of the applied magnetic field, *B*, using a lock-in amplifier technique with an excitation frequency of ~ 7 Hz and current of 10 nA. The samples were tilted with respect to the direction of the applied magnetic field using an *in situ* rotator. The tilt angle, θ, is defined as the angle between the normal vector perpendicular to the 2DEG and *B* and is experimentally determined by aligning the quantum Hall effect minima at integer filling factors. The perpendicular (parallel) component of magnetic field is denoted by $B_\perp$ ($B_{//}$).

In the data presented below we focus on results from the sample with a quantum well width of *L*=70 nm. The results of similar experiments conducted with *L*=64 and 58 nm samples will be discussed following the presentation of the *L*=70 nm data.

4. Experimental Results and Discussions

Figure 1 shows the diagonal resistance, R_{xx}, and the Hall resistance, R_{xy}, for the *L*=70 nm QW with θ=0°. Clear minima are seen at integer values of filling factor, both even and odd, for ν≥4 and ν=2. At ν=3 the quantum Hall state is very weak and at ν=1 the quantum Hall state is missing entirely. In addition, the fractional quantum Hall states surrounding ν=1 all have even numerators. This behavior is very similar to previous observations in double quantum wells [15] and wide single quantum wells [16] and provides strong evidence that the *L*=70 nm QW is effectively a bilayer system.

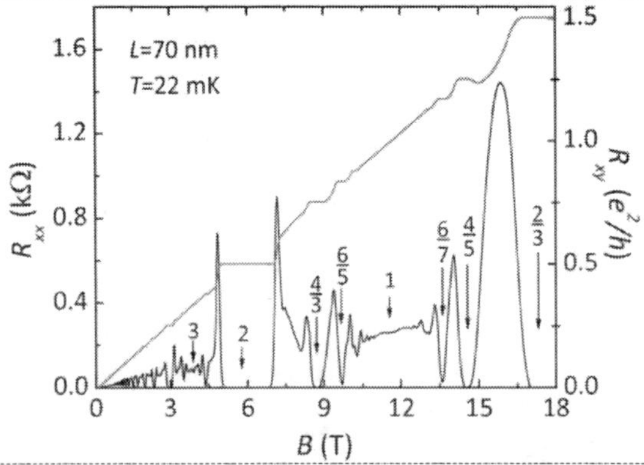

Figure 1. R_{xx} and R_{xy} as function of *B* for the *L*=70 nm quantum well at *T*=22 mK. Several integer and fractional filling factors are labeled.

To detect the anisotropy, the diagonal resistance was measured with the current oriented in two perpendicular directions for each sample and each value of θ. The resistance R_{xx} is defined with the excitation current parallel to $B_{//}$ and R_{yy} is defined with the current perpendicular to the direction of $B_{//}$. In the L=70 nm case only, R_{xx} was multiplied by a scaling factor of 0.4 to make the amplitude at low field coincide with R_{yy} at θ =0°. This discrepancy was most likely due to the specific contacts used, and does not effect the conclusions drawn from the data, though, it actually reduces the observed anisotropy.

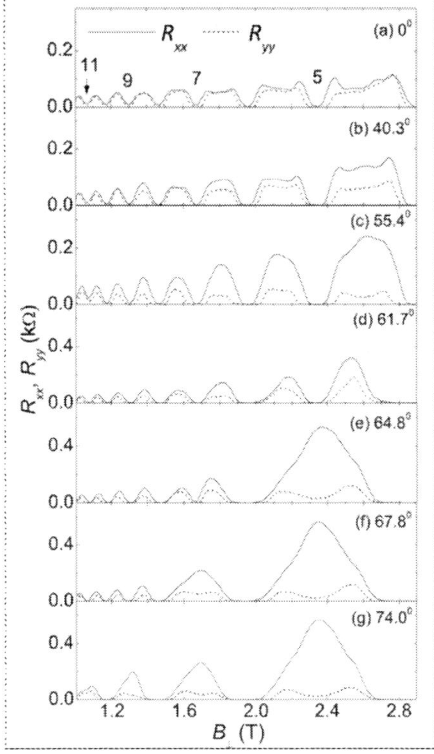

Figure 2. R_{xx} (solid line) and R_{yy} (dashed line) as function of $B_{\perp}$ at various values of θ for the L=70 nm sample at T=20 mK. The value of θ is listed for each trace, and several filling factors are labeled. The vertical axes in panels (d) through (g) are twice that of the other panels.

Figure 2 displays typical data showing both R_{xx} and R_{yy} as a function of $B_{\perp}$ over a large range of θ for the L=70 nm sample. Initially when θ=0°, very little anisotropy is evident and well defined integer quantum Hall minima are present in the displayed range of $B_{\perp}$. As tilt increases to θ=40.3°, a slight disparity between R_{xx} and R_{yy} is discernable at half total fillings factors ν=9/2 and ν=11/2 and weakens with increasing ν. This anisotropy grows with θ and is strongest at θ=55.4°. With further increase in tilt to θ=61.7° the anisotropy weakens and the respective peaks and dips in R_{xx} and R_{yy} move slightly toward adjacent odd filling factors. With another increase in the θ, by a mere *3.1* degrees, to θ=64.8° a sudden transition occurs at ν=5 where the minimum corresponding to the quantum Hall state is replaced by a strongly anisotropic state. Another small increase in tilt to θ=67.8°causes a similar transition at ν=7. When the tilt angle is increased to θ=74.0° the quantum Hall states at ν=9 and ν=11 also disappear with ν=9 exhibiting some anisotropy. The anisotropic states, especially at ν=5 and 7, are characterized by a peak in R_{xx} and a minimum in R_{yy} or, in other words, the easy transport direction is perpendicular to $B_{//}$ and the hard transport direction is parallel to $B_{//}$. The disappearance of the quantum Hall states and onset of transport anisotropy also corresponds to the disappearance of the quantized plateaus in the Hall resistance, R_{xy}. This is shown in Figure 3 where R_{xy} is plotted as a function of $B_\perp$ near the transition to anisotropic states for the R_{yy} current configuration. As the data show, the plateau at ν=5 is absent for θ=64.8° and both the ν=5 and 7 plateaus are absent when θ=67.8°. Well defined plateaus at ν=9 and 11 are not present when θ=74.0°, however, a visible minimum remains in the derivative of R_{xy} at these two filling factors (not shown) in contrast to the situation at ν=5 and 7. No anisotropy is observed in the Hall resistance.

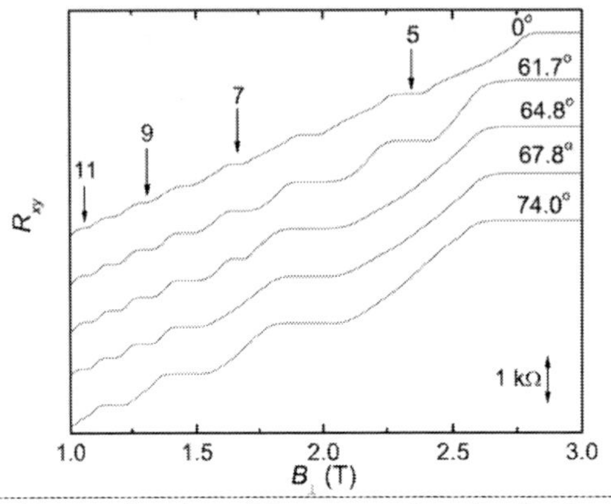

Figure 3. R_{xy} as a function of $B_\perp$ at several angles for the L=70 nm sample. The traces are offset for clarity and the odd filling factors are labeled to highlight the disappearance of the Hall plateaus with increasing angle.

The strong anisotropy at ν=5 and 7 is presented in Figure 4 (a-d) at several different temperatures with θ=71.2°. The anisotropy is strongest at the lowest obtainable temperature of T=22 mK and rapidly weakens with increasing T. The easy direction (R_{yy}) increases while the hard direction (R_{xx}) decreases with increasing T, similar to the anisotropy observed at ν=9/2 in single layer systems [1,2]. In Fig.4 (e) the ratio R_{xx}/R_{yy} is plotted versus T for both filling factors. Interestingly, the dramatic weakening of the anisotropy occurs for T<300 mK which may indicate the importance of correlations.

Before discussing the origin of the anisotropic transport, we argue that it is bilayer in nature. As discussed above, the transport in the absence of tilt suggests that the L=70 nm QW is effectively a bilayer system (see Fig.1). In addition to this sample, data were also acquired from L=64 nm and L=58 nm WSQWs. Magneto-transport data in the absence of tilt for the L=64 nm QW show that the bilayer characteristics appear to be weakened compared to the L=70 nm data [17]. Anisotropic transport at ν=5 and 7 was observed in the L=64 nm sample but at a slightly larger value of tilt angle (θ ≈ 70°) than for L=70 nm. In contrast, for the L=58 nm sample the transport [17] in the absence of tilt was very similar to what is expected for a single layer and no anisotropy was observed at odd values of ν up to the largest investigated angle of θ=80.4°. The presence of anisotropy at odd ν in the widest two samples and the absence of anisotropy in the narrowest sample suggest that the anisotropy is a bilayer phenomenon. We also note that a similar type of transition was observed at the even filling factors of ν=4, 6 and 8 at larger tilt angles (θ>80°) in a narrower QW with L=35 nm by Pan *et al* [18]. This is in contrast to our results which show a tilt-induced transition to anisotropic states at odd values of ν. The major distinction between our samples and that of Pan *et al* [18] is the relatively narrow QW width in the case of the latter whose L=35 nm sample is basically single layer in nature, albeit with two occupied subbands.

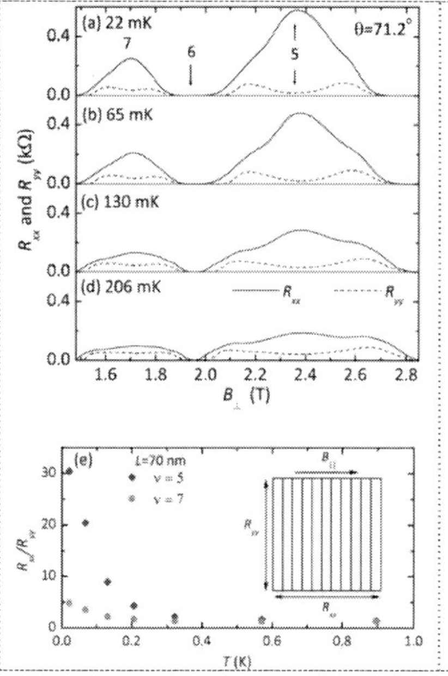

Figure 4. R_{xx} and R_{yy} at four different temperatures for θ=71.2°. The anisotropy at ν=5 and 7 decreases dramatically with increased T. In (e) the ratio R_{xx}/R_{yy} is shown versus T. The inset illustrates the definition of R_{xx} and R_{yy} relative to $B_{//}$. The vertical lines indicate the possible presence of a stripe phase with R_{xx} the hard direction and R_{yy} the easy direction. All data are from the L=70 nm sample.

Odd integer quantum Hall states in bilayer systems depend on coherent interlayer interaction and can be described by the so-called {111} state [19]. For sufficiently large values of $B_{//}$ in WSQWs, increases in $B_{//}$ result in reduced interlayer interaction and tunneling [20] as well as a modest increase in the interlayer separation [21]. Consequently, the {111} is suppressed and eventually destroyed with increased $B_{//}$, and the system behaves as two independent single layer systems each with filling factor ν/2. In single layer systems, in the presence of a large parallel magnetic field, the transport at ν=5/2 and 7/2 is anisotropic [5,6]. Therefore it is possible that the onset of anisotropy in our samples at total filling factors ν=5 and 7 may be the result of a transition to two independent layers each with filling factor 5/2 and 7/2 for ν=5 and 7 respectively. In this regard, the easy axis of our observed anisotropic state is perpendicular to the applied field (see the inset in Fig.4 (e), similar to that observed at ν=5/2 and 7/2 in single layer systems [5,6]. In our samples, the transition to anisotropic transport is observed at $B_{//}\approx$4.5 T for ν=5, 7 and 9 in the 70 nm QW and $B_{//}\approx$ 6.0 T for ν=5 and 7 in the 64 nm QW. If we consider the ratio of interlayer spacing [22], d, to the magnetic length due to the parallel field, $l_{//}=\hbar/(eB_{//})^{1/2}$, the transitions occur at a similar value of $d/l_{//}\approx$ 4.2 in each sample.

If the above scenario, i.e., a transition from a {111} quantum Hall state to two independent layers, is correct, a similar transition would be expected for larger odd filling factors, as well as for the L=58 nm sample, provided $B_{//}$ is large enough. For the L=70 nm QW, possible evidence for this transition was observed at ν=11 (see Fig.2). On the other hand, no anisotropy was observed in the L=58 nm quantum well up to θ=80.4°, corresponding to a value of $d/l_{//}$ beyond that of $d/l_{//}\approx$ 4.2 observed in the other samples.

Another possible explanation for the observed anisotropy at odd ν is the tilt-induced onset of a stripe phase. Theoretical work [9-11] has considered the possibility of different stripe phases at odd integer quantum Hall states in WSQWs and bilayer systems. In particular Wang *et al.* [11] have shown that in a wide quantum well an isospin Skyrmion stripe phase can occur at odd ν when $B_{//}$ is large. A quantitative comparison between our experimental results and the theoretical calculations is not available at the present time, but the calculations at ν=1 indicate that the transition from an integer quantum Hall state to a stripe phase occurs at large values of $B_{//}$ as well width is decreased, consistent with our experimental results. Further detailed numerical calculations taking into account our specific sample parameters are needed to shed more light on the observed anisotropy.

5. Acknowledgments

We would like to thank G. Jones, E. Palm and T. Murphy for experimental assistance. A portion of this work was performed at the National High Magnetic Field Laboratory, which is supported by NSF Cooperative Agreement No. DMR-0084173, by the State of Florida, and by the DOE. The work at Sandia was supported by the Division of Materials Sciences and Engineering, Office of Basic Energy Sciences, U.S. Department of Energy. Sandia National Laboratories is a multi-program laboratory operated by Sandia Corporation, a wholly owned subsidiary of Lockheed Martin company, for the U.S. Department of Energy's National Nuclear Security Administration under contract DE-AC04-94AL85000. The work at Princeton University was funded by DOE grant No. DE-FG-02-98ER45683.

References

[1] Lilly M P, *et al.* 1999, Phys. Rev. Lett. **82** 394.
[2] Du R R, *et al.* 1999, Solid State Commun. **109**, 389.
[3] Fogler M M, Koulakov A A, and Shklovskii B I 1996, Phys. Rev. B **54**, 1853.
[4] Moessner R and Chalker J T 1996, Phys. Rev. B **54**, 5006.
[5] Pan W, *et al.* 1999 Phys. Rev. Lett. **83** 820.
[6] Lilly M P, *et al.* 1999, Phys. Rev. Lett. **83** 824.
[7] Brey L and H.A. Fertig H A 2000, Phys. Rev. B **62**, 10268.
[8] Côté R and H.A. Fertig H A 2002, Phys. Rev. B **65**, 85321.
[9] Demler E, *et al.* 2002, Solid State Commun. **123**, 243.
[10] Côté R, *et al.* 2002, Phys. Rev. B **66**, 205315.
[11] Wang D W, Demler E, and S. Das Sarma S 2003, Phys. Rev. B **68**, 165303
[12] Papa E, *et al.* 2003, Phys. Rev. B **67**, 115330.
[13] Doiron C B, Côté R, and H.A. Fertig H A 2005, Phys. Rev. B **72**, 115336.
[14] Gusev G M, *et al.* 2007, Phys. Rev. Lett. **99**, 126804.
[15] Boebinger G S, *et al.* 1990, Phys. Rev. Lett. **64**, 1793.
[16] Suen Y W, *et al.* 1991, Phys. Rev. B **44**, 5947.
[17] Luhman D R, *et al.* 2008, Physica E **40**, 1059.
[18] Pan W, *et al.* Phys. Rev. B **64**, 121305.
[19] Halperin B I, 1983, Helv. Phys. Acta **56**, 75.
[20] Hu J and MacDonald A H, 1992, Phys. Rev. B **46**, 12554.
[21] T.S. Lay T S, *et al.* 1997, Phys. Rev. B **56**, R7092.
[22] The value of *d* was determined from the peak separation in the calculated electron density profile across the quantum well. Finite values of magnetic field were not taken into account, see Ref. [17].

Nonequilibrium transport in very high Landau levels

M. A. Zudov[1], A. T. Hatke[1], H.-S. Chiang[1], L. N. Pfeiffer[2], K. W. West[2], and J. L. Reno[3]

[1] School of Physics and Astronomy, University of Minnesota, Minneapolis, MN 55455, USA
[2] Department of Electrical Engineering, Princeton University, Princeton, NJ 08544, USA
[3] Sandia National Laboratories, Albuquerque, NM 87185, USA

E-mail: zudov@physics.umn.edu

Abstract. Low temperature transport properties of high mobility two-dimensional electron systems placed in a weak perpendicular magnetic field can be modified dramatically by microwave or dc electric fields. This paper surveys recent experimental developments which include zero-differential resistance states, Hall field-induced resistance oscillations in tilted magnetic fields, nonlinear response of the Shubnikov-de Haas Oscillations, and a novel microwave photoconductivity peak near the second harmonic of the cyclotron resonance.

1. Introduction

Over the past decade it was realized that high mobility two-dimensional electron systems (2DESs) exhibit an array of fascinating transport phenomena occurring in very high Landau levels where the Shubnikov-de Haas oscillations (SdHOs) are not yet resolved. Among these are several classes of magnetooscillations, namely, microwave-(MIROs) [1; 2], phonon- [3–11], Hall field-(HIROs) [12–15] induced resistance oscillations, and their combinations [16–25]. In a clean 2DES the minima of MIROs can evolve into zero-resistance states (ZRSs) [26–39], which are currently explained in terms of the absolute negative resistance and its instability with respect to formation of current domains [40–43]. Transport in irradiated 2DESs was extensively studied both experimentally [44–67] and theoretically [68–105] but open issues remain. Among these are experimental confirmation of domains, apparent insensitivity of MIROs to the sense of circular polarization of microwaves [37], and suppression of MIROs [33] by in-plane magnetic fields.

More recent experiments on non-irradiated 2DESs have revealed further intriguing phenomena which remain poorly understood. Among these are zero-differential resistance states (ZdRSs), which emerge from the SdHO maxima in quantizing magnetic fields ($B \sim 10$ kG) [106–108] and from the HIRO minimum [109] in much lower magnetic fields ($B \sim 1$ kG). It was suggested [106] that ZdRSs are analogous to the radiation-induced ZRSs in a sense that they can also be explained by the domain model. As such, ZdRSs offer an alternative playground for exploring instabilities which proved to be difficult to access in irradiated 2DESs.

This paper describes some of the recent and ongoing experimental developments in the field of nonlinear transport in very high Landau levels and is organized as follows. After a brief review of MIROs and HIROs (Sec. 2), we discuss the temperature evolution of the HIRO-originated ZdRS (Sec. 3), HIROs in tilted magnetic fields (Sec. 4), nonlinear transport in the SdHO regime (Sec. 5), and the anomalously large microwave photoconductivity effect near the second harmonic of the cyclotron resonance which appears in addition to MIROs (Sec. 6).

2. Microwave- and Hall field-induced resistance oscillations

MIROs appear in magnetoresistivity under microwave irradiation and are controlled by a dimensionless parameter $\epsilon_{\rm ac} = \omega/\omega_{\rm c}$, where $\omega = 2\pi f$ is the microwave frequency and $\omega_{\rm c} = eB/m^*$ is the cyclotron frequency of an electron with an effective mass m^*. The resistivity of an irradiated 2DES can be expressed as $\rho_\omega = \rho + \delta\rho_\omega$, where ρ is the resistivity of the non-irradiated 2DES and $\delta\rho_\omega(\epsilon_{\rm ac})$ is the photoresistivity which is a sign-alternating function of $\epsilon_{\rm ac}$. According to the displacement model [68; 69; 71; 79; 101; 104], $\delta\rho_\omega(\epsilon_{\rm ac})$ originates from the microwave-induced impurity-assisted transitions between the disorder-broadened Landau levels. In another mechanism, known as inelastic [75; 91; 101; 104], microwaves create a nonequilibrium distribution of electron states which, in turn, translates to the oscillating $\delta\rho_\omega(\epsilon_{\rm ac})$. Which mechanism is dominant depends on temperature and type of disorder in the 2DES [61; 104] but in the regime of overlapping Landau levels both models predict

$$\frac{\delta\rho_\omega}{\rho} \simeq -\eta\mathcal{P}_\omega\epsilon_{\rm ac}\lambda^2 \sin 2\pi\epsilon_{\rm ac}. \tag{1}$$

Here, $\lambda = \exp(-\pi/\omega_{\rm c}\tau_{\rm q})$ is the Dingle factor, $\tau_{\rm q}$ is the quantum lifetime, η is the model-dependent scattering parameter, and $\mathcal{P}_\omega$ is the dimensionless parameter proportional to the microwave power. Away from the cyclotron resonance $\mathcal{P}_\omega$ decays rapidly with increasing frequency, roughly as ω^{-4}. According to Eq. (1), maxima$^+$ and minima$^-$ of MIROs should occur at $\epsilon_{\rm ac}^{n\pm} = n \mp 1/4$ $(n = 1, 2, 3, \dots)$. However, in many experiments using high mobility 2DES the condition of overlapping Landau levels is violated at sufficiently high B and the maxima and minima are pushed towards cyclotron resonance harmonics and occur at $\epsilon_{\rm ac}^{n\pm} \simeq n \mp \delta$, where δ can be considerably smaller than 1/4 [31; 79; 91].

HIROs appear in differential magnetoresistivity $r \equiv dV/dI$ when a direct current I is passed through a Hall bar-shaped 2DES. HIROs are explained by the displacement model [22; 24; 110; 111] owing to the impurity-mediated transitions between Landau levels tilted by the Hall electric field, $\mathcal{E}_{\rm dc} = \rho_H j$, where ρ_H is the classical Hall resistivity and $j = I/w$ (w is the sample width). In this scenario, the dominant scattering process involves an electron which is backscattered off an impurity. The guiding center of such an electron is displaced by a distance equal to the cyclotron diameter, $2R_{\rm c}$, in a single scattering event. When $2R_{\rm c}$ matches an integral multiple of the real-space Landau level separations the probability of such events is enhanced. This enhancement leads to a maximum in the differential resistivity occurring whenever the dimensionless parameter $\epsilon_{\rm dc} \equiv e\mathcal{E}_{\rm dc}(2R_{\rm c})/\hbar\omega_{\rm c}$ is equal to an integer [14; 110]. At $2\pi\epsilon_{\rm dc} \gtrsim 1$, HIROs are described by

$$\frac{r}{\rho} \simeq \frac{16\lambda^2}{\pi}\frac{\tau_{\rm tr}}{\tau_\pi}\cos 2\pi\epsilon_{\rm dc}, \tag{2}$$

where $\tau_{\rm tr}$ is the transport lifetime and τ_π is the time describing electron backscattering [22; 110]. At weak dc fields, $2\pi\epsilon_{\rm dc} \ll 1$, the theory[110] predicts another source of nonlinearities originating from the inelastic [91] mechanism. In this approach a dc field creates a nonequilibrium distribution of electron states which leads to a resistance drop. Since both models were developed in the limit of strongly overlapping Landau levels, a condition which is not always satisfied in experiments [14; 106; 107; 112], the relative importance of these mechanisms at $2\pi\epsilon_{\rm dc} \ll 1$ remains poorly understood.

Both MIROs and HIROs are best observed at $T \lesssim 1$ K and quickly decay at higher T. In high mobility 2DESs, the decay originates primarily from the electron-electron interaction effects which modify the single particle lifetime $\tau_{\rm q}$ entering the Dingle factor λ [15; 61; 104; 113]. However, in 2DESs with larger impurity contribution to $1/\tau_{\rm q}$, with different type of disorder, and/or at sufficiently low temperatures, the decay of MIROs can also originate from $\eta \propto \tau_{\rm in} \simeq E_F/T^2$ [66; 91; 104] within a framework of the inelastic mechanism. Here, $\tau_{\rm in}$ is the inelastic relaxation time and E_F is the Fermi energy.

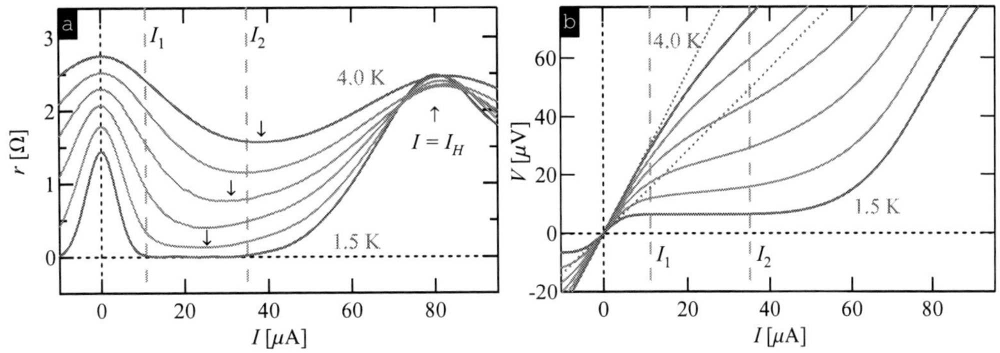

Figure 1. (a) r and (b) V vs I measured at $B = 1.5$ kG and at different T from 1.5 K to 4.0 K, in step of 0.5 K. ZdRS is observed at $I_1 \lesssim I \lesssim I_2$ where I_1 and I_2 can be associated with currents inside the domains [114]. These data were obtained on a Hall bar sample (width $w = 100$ μm) with density $n_e \simeq 3.8 \times 10^{11}$ cm^{-2} and mobility $\mu \simeq 1.0 \times 10^7$ cm^2/Vs.

3. Zero-differential resistance state in high Landau levels

Recent experiments [109] have shown that when a high mobility 2DES is subject to weak electric and magnetic fields, its current-voltage characteristic exhibits a plateau signaling a formation of a state with zero differential resistance. In contrast to the ZdRS originating from the SdHO maxima (and thus appearing only at the discrete values of B corresponding to odd filling factors) [106], the ZdRS discussed here originates from a HIRO minimum and is observed over a continuous range of magnetic fields extending well below the onset of the SdHOs.

In Fig. 1(a) we present differential resistivity $r(I)$ measured at T from 1.5 K to 4.0 K, in step of 0.5 K. At $T = 1.5$ K, $r(I)$ decays all the way to zero with increasing I, showing a well developed ZdRS, which is formed between $I_1 \simeq 11$ μA and $I_2 \simeq 35$ μA (cf. dashed lines). As discussed in Ref. [109], I_1 is roughly B-independent, whereas I_2 increases linearly with B tracing the fundamental HIRO peak (cf., ↑) at $I \simeq I_H \equiv en_e w\omega_c/2k_F \propto B$, where $k_F = \sqrt{2\pi n_e}$. With increasing T, the ZdRS disappears evolving into a well defined HIRO minimum. The differential resistivity at this minimum increases monotonically with T, while its position shifts to higher I (cf., ↓). The linear response resistivity $\rho \equiv r(0)$ also monotonically increases with T. This increase is likely due to increased scattering on acoustic phonons which reduces the mobility of clean 2DESs. Integrating the data in Fig. 1(a) yields the $V(I)$ curves which are presented in Fig. 1(b). At $T = 1.5$ K, concurrent with the drop of differential resistivity [cf. Fig. 1(a)] the longitudinal voltage V departs from Ohm's law (cf., lower dotted line) and saturates to a plateau which extends over a finite current range, $I_1 \lesssim I \lesssim I_2$. Inside this range, V is independent of I, so that $dV/dI = 0$. With increasing T, the plateau disappears and the nonlinearity becomes progressively weaker. The increase in $\rho = r(0)$ with T is illustrated by dotted lines which represent the linear response regime (Ohm's law) at 1.5 K and 4.0 K.

The exact mechanism responsible for the observed T-dependence remains a subject of future work. On one hand, recent studies [7] have shown that electron-electron interactions modify the quantum lifetime entering the Dingle factor and thus play an important role in the T-dependence of HIROs (from which ZdRS originates). On the other hand, the observed growth of the linear-response resistivity due to electron-phonon scattering may also contribute to the ZdRS disappearance and the subsequent increase of the differential resistivity. Finally, identifying the dominant mechanism responsible for the nonlinearity leading to the ZdRS is difficult because the effect appears in the separated Landau levels, a regime which remains poorly understood.

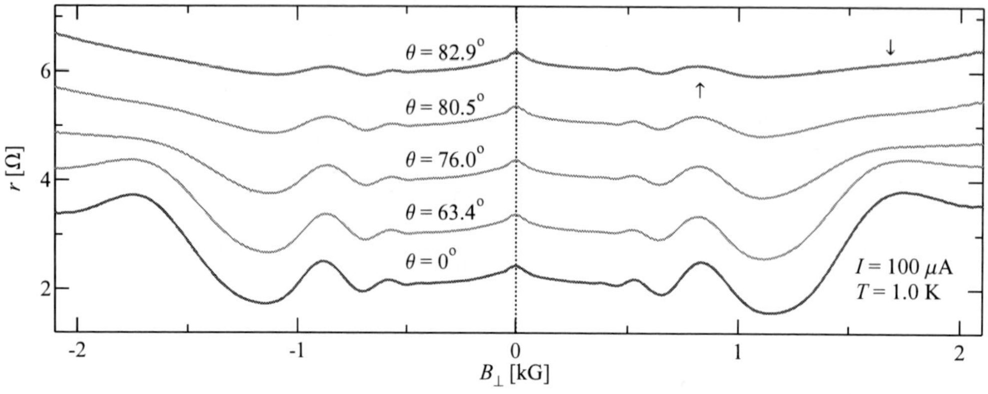

Figure 2. Differential magnetoresistivity $r(B)$ at $I = 100\ \mu$A and $T = 1.0$ K measured at different tilt angles θ (as marked). HIRO peaks at $\epsilon_{\rm dc} = 1$ and $\epsilon_{\rm dc} = 2$ are marked by ↓ and ↑, respectively. These data were obtained on a Hall bar sample ($w = 100\ \mu$m) with $n_e \simeq 3.6 \times 10^{11}$ cm^{-2} and $\mu \simeq 1.1 \times 10^7$ cm^2/Vs.

4. Hall field-induced resistance oscillations in tilted magnetic fields

The effect of an in-plane magnetic field, $B_{\parallel}$, on MIROs has been studied in two experiments [33; 34]. In one experiment [34] the 2DES was tilted with respect to the total field B by an angle θ as B was varied. In this setup the ratio of $B_{\parallel} = B\sin\theta$ to $B_{\perp} = B\cos\theta$ remains fixed, $B_{\parallel}/B_{\perp} = \tan\theta$, and, as a result, lower order (higher $B_{\perp}$) MIROs experience larger $B_{\parallel}$ than the higher order (lower $B_{\perp}$) MIROs. Another experiment [33] employed a two-axis magnet so that MIROs of all orders, occurring at different $B_{\perp}$, were subject to the same $B_{\parallel}$. While both experiments agreed that the oscillation period is governed by $B_{\perp}$, Ref. [33] found that MIROs disappear under $B_{\parallel} \simeq 5$ kG while in Ref. [34] MIROs were essentially unchanged up to $\theta \simeq 80^{\circ}$ (or $B_{\parallel} \lesssim 12$ kG). Obviously, the role of $B_{\parallel}$ remains poorly understood and it is interesting and timely to examine how other classes of magneto-oscillations respond to $B_{\parallel}$. In what follows, we present the results of our experiment studying the effect of $B_{\parallel}$ on HIROs.

Figure 2 shows differential resistivity r vs $B_{\perp}$ measured at $I = 100\ \mu$A and $T = 1.0$ K for various tilt angles up to $\theta = 82.9^{\circ}$ (as marked). Similar to MIROs [33; 34], Fig. 2 clearly demonstrates that the HIRO period is controlled by $B_{\perp}$, as the corresponding peak positions coincide with each other for all tilt angles. At $\theta = 0$ (bottom curve) the data reveal at least three well developed oscillations which gradually decay away with increasing θ. Careful examination of the data reveals that the the decay is nonuniform; the strongest, fundamental HIRO peak (cf., ↓) is much more sensitive to the tilt than the higher order peaks; it disappears completely at $\theta \simeq 82.9^{\circ}$ ($B_{\parallel} \simeq 1.3$ T), while the second (cf., ↑) and the third HIRO peaks are still clearly observed. The second HIRO peak eventually also disappears, at $\theta \simeq 85.9^{\circ}$, while the third HIRO peak still remains visible at this θ.

Our analysis [115] of the HIRO decay with increasing tilt angle shows that it can be understood in terms of a $B_{\parallel}$-induced correction to the single particle scattering rate which scales roughly as $B_{\parallel}^2$. While the exact mechanism of such an increase remains a subject of future theoretical and experimental studies, this scenario is qualitatively consistent with the observations of Ref. [33] where $B_{\parallel}$ was held constant and MIROs of *higher* orders decayed faster than the lower orders. Therefore, it appears plausible that suppressions of MIROs [33] and HIROs originate from the same physical mechanism.

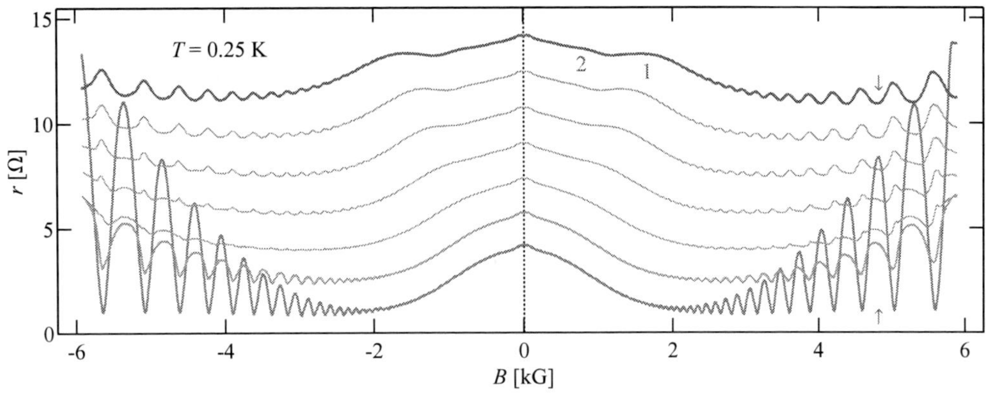

Figure 3. Differential resistance r vs. magnetic field B for different dc currents I, from $0\,\mu$A (bottom) to 6 μA (top), in step of 1 μA. Traces are vertically offset for clarity. Integers at the top trace mark HIRO maxima, $\epsilon_{dc} = 1$ and $\epsilon_{dc} = 2$. These data were obtained on a Hall bar sample ($w = 10$ μm) with density $n_e \simeq 2.4 \times 10^{11}$ cm^{-2} and mobility $\mu \simeq 6 \times 10^6$ cm^2/Vs.

5. Effect of dc electric field on Shubnikov-de Haas oscillations

When a direct current is passed through a 2DES, one can naively expect that SdHOs will smoothly decay with increasing current due to increased Joule heating. Recent experiments [116], however, revealed far more intriguing evolution of the SdHOs with current which could not be explained by conventional heating effects. In particular, it was found that the SdHO frequency doubles at certain currents suggesting significant redistribution between the first and the second Fourier components of the SdHOs. This behavior was analyzed and interpreted within a framework of the inelastic mechanism [91; 110] which leads to a filling factor-dependent spectral diffusion of electrons.

In Fig. 3 we illustrate another experimental aspect of the nonlinear transport in the SdHO regime. Here, we plot the differential magnetoresistivity $r(B)$ measured at different currents I, from $0\,\mu$A (bottom) to 6 μA (top), in step of 1 μA. As one would expect, SdHOs initially decay with increasing I. However, at still higher I SdHOs reappear but with the opposite phase. Indeed, comparison of the bottom ($I = 0$) trace to the top ($I = 6$ μA) trace shows that the SdHO maxima have evolved into the minima and vice versa. At such low currents, HIROs remain confined to lower magnetic fields (cf. peaks marked by 1 and 2 at the top trace) so that the SdHOs appear in the regime of $\epsilon_{dc} < 1$. Therefore one cannot explain the observed phase reversal of the SdHOs by the inter-Landau level scattering processes which give rise to HIROs.

As discussed in Sec. 2, both the inelastic and the displacement contributions may play an important role in the regime of $2\pi\epsilon_{dc} \lesssim 1$. However, in both cases calculations predict that the characteristic current at which SdHOs would change a phase should increase roughly linearly with the magnetic field. As evidenced by Fig. 3 this is clearly not the case as *all* of the SdHOs evolve together; the sudden phase change occurs simultaneously at roughly the same I regardless of the magnetic field (which varies by a factor of three in this measurement). One should note, however, that existing calculations assumed overlapping Landau levels and therefore might not be directly applicable to experiments. Unfortunately, a comprehensive theory of nonlinear transport in the separated Landau levels is not currently available. To explain the observed phase change of the SdHOs, the theory might also have to examine if and how the applied electric field modifies relative contributions of short- and long-range disorder potentials.

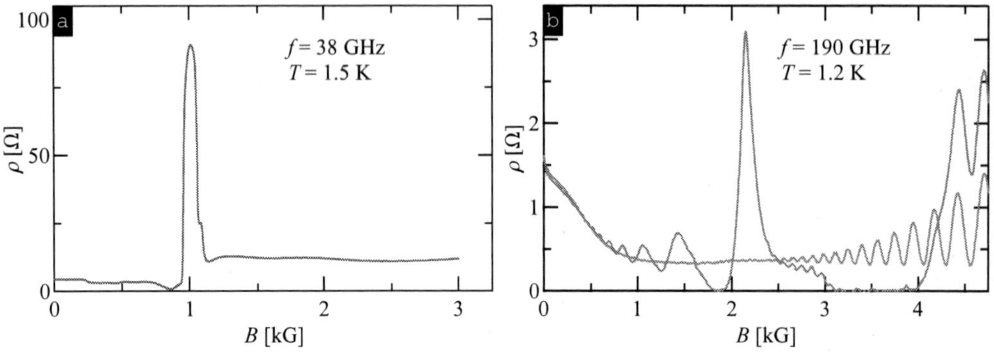

Figure 4. (a)[(b)] $\rho(B)$ measured with microwave irradiation of $f = 38\,[190]$ GHz at $T \simeq 1.5\,[1.2]$ K in a 2DES with $n_e \simeq 2.4\,[3.8] \times 10^{11}$ cm^{-2} and $\mu \simeq 0.6\,[1.0] \times 10^7$ cm^2/Vs. Panel (b) also shows $\rho(B)$ without microwave irradiation.

6. Photoresistance near the second harmonic of the cyclotron resonance

As discussed in the introduction, MIRO minima evolve into the radiation-induced ZRSs in sufficiently clean 2DESs and therefore the negative photoresistivity never exceeds the dark resistivity by absolute value. In contrast, positive photoresistivity responsible for the MIRO maxima has no associated instabilities and was routinely found to exceed the dark resistivity (but usually no more than 2-3 times). In addition to MIROs, Smet *et al.* [37] reported a remarkably strong and narrow photoresistivity peak in close proximity to the cyclotron resonance, which showed threshold-like dependence on the radiation intensity. In Fig. 4(a) we present the magnetoresistivity $\rho(B)$ measured in a moderate mobility 2DES under microwave irradiation of frequency $f = 38$ GHz at $T \simeq 1.5$ K which shows a giant photoresistivity peak. Similar to the peak reported in Ref. [37], the peak shown in Fig. 4(a) occurs near the cyclotron resonance and is only observed at microwave intensities above some critical value. While this peak can be phenomenologically linked to the bolometric effect [37], due to the resonant heating at the cyclotron resonance, its exact nature was not theoretically considered.

In Fig. 4(b) we present another example of an unusually large microwave photoresistance which is observed in addition to MIROs but occurs only near the *second harmonic* of the cyclotron resonance [117–119]. Here we plot magnetoresistivity $\rho(B)$ measured under microwave irradiation of frequency $f = 190$ GHz at $T \simeq 1.2$ K. The data show a very sharp peak at $B \simeq 2$ kG which appears superimposed on the second MIRO maximum. More detailed analysis of the oscillation waveform [31] reveals that the peak appears *between* the second harmonic of the cyclotron resonance and the second MIRO maximum. Frequency dependence studies shows that in contrast to MIROs, which are known to decay with microwave frequency, this peak is observed only above certain microwave frequency (about 100 GHz in our 2DES). We also find that the peak exhibits roughly linear dependence on microwave intensity, disappears with increasing temperature in a way similar to MIROs [118], and can be more than an order of magnitude stronger than MIROs. Due to an unusually strong negative magnetoresistance effect in our 2DES, which has yet to be understood, the peak can also be more than two orders of magnitude larger than the dark resistivity. Based on the above, we conclude that such a dramatic effect cannot be explained by existing theories of microwave photoconductivity and understanding its nature remains a subject of future theoretical and experimental work. In particular, it would be interesting to examine possible role of quasiclassical memory effects [84; 102] and to investigate the effects of dc electric field and in-plane magnetic field.

7. Summary

Owing to both experimental and theoretical advances over the past few years our understanding of nonequilibrium transport in high Landau levels has improved dramatically but questions remain. Among the long standing issues are apparent immunity of MIROs to the sense of circular polarization [37], suppression of MIROs/HIROs by modest in-plane magnetic fields [33; 115], the nonlinear resistivity in the SdHO regime [108; 116] and in the regime preceeding zero-differential resistance states [109], and the novel photoresistivity peak recently discovered near the second harmonic of the cyclotron resonance [117; 118].

Acknowledgments

We thank A. Bykov, I. Dmitriev, R. Du, M. Dyakonov, M. Khodas, K. von Klitzing, I. Kukushkin, A. Mirlin, D. Polyakov, J. Smet, B. Shklovskii, M. Vavilov, and S. Vitkalov for discussions and D. Smirnov for technical assistance. Part of this work was performed at the NHMFL, supported by NSF Cooperative Agreement No. DMR-0654118, by the State of Florida, and by the DOE. The work in Secs. 3, 4 (5, 6) was supported by the NSF Grant No. DMR-0548014 (DOE Grant No. DE-SC00025670). The work at Princeton was partially funded by the Gordon and Betty Moore Foundation as well as the NSF MRSEC Program through the Princeton Center for Complex Materials (DMR-0819860).

References

[1] Zudov M A, Du R R, Simmons J A and Reno J L 2001 *Phys. Rev. B* **64** 201311(R)
[2] Ye P D, Engel L W, Tsui D C, Simmons J A, Wendt J R, Vawter G A and Reno J L 2001 *Appl. Phys. Lett.* **79** 2193
[3] Zudov M A, Ponomarev I V, Efros A L, Du R R, Simmons J A and Reno J L 2001 *Phys. Rev. Lett.* **86** 3614
[4] Ponomarev I V and Efros A L 2001 *Phys. Rev. B* **63** 165305
[5] Yang C L, Zudov M A, Zhang J, Du R R, Simmons J A and Reno J L 2002 *Physica E (Amsterdam)* **12** 443
[6] Bykov A A, Kalagin A K and Bakarov A K 2005 *JETP Lett.* **81** 523
[7] Hatke A T, Zudov M A, Pfeiffer L N and West K W 2009 *Phys. Rev. Lett.* **102** 086808
[8] Lei X L 2008 *Phys. Rev. B* **77** 205309
[9] Raichev O E 2008 *Phys. Rev. B* **78** 125304
[10] Bykov A A and Goran A V 2009 *JETP Lett.* **90** 578
[11] Zudov M A, Chiang H S, Hatke A T, Zhang W, Pfeiffer L N and West K W 2009 *Int. J. of Mod. Phys. B* **23** 2684
[12] Yang C L, Zhang J, Du R R, Simmons J A and Reno J L 2002 *Phys. Rev. Lett.* **89** 076801
[13] Bykov A A, Zhang J, Vitkalov S, Kalagin A K and Bakarov A K 2005 *Phys. Rev. B* **72** 245307
[14] Zhang W, Chiang H S, Zudov M A, Pfeiffer L N and West K W 2007 *Phys. Rev. B* **75** 041304(R)
[15] Hatke A T, Zudov M A, Pfeiffer L N and West K W 2009 *Phys. Rev. B* **79** 161308(R)
[16] Zhang W, Zudov M A, Pfeiffer L N and West K W 2008 *Phys. Rev. Lett.* **100** 036805
[17] Zhang W, Zudov M A, Pfeiffer L N and West K W 2007 *Phys. Rev. Lett.* **98** 106804
[18] Zhang W, Zudov M, Pfeiffer L and West K 2008 *Physica E (Amsterdam)* **40** 982
[19] Lei X L 2007 *Appl. Phys. Lett.* **91** 112104

[20] Hatke A T, Chiang H S, Zudov M A, Pfeiffer L N and West K W 2008 *Phys. Rev. B* **77** 201304(R)

[21] Hatke A T, Chiang H S, Zudov M A, Pfeiffer L N and West K W 2008 *Phys. Rev. Lett.* **101** 246811

[22] Khodas M and Vavilov M G 2008 *Phys. Rev. B* **78** 245319

[23] Lei X L 2009 *Phys. Rev. B* **79** 115308

[24] Khodas M, Chiang H S, Hatke A T, Zudov M A, Vavilov M G, Pfeiffer L N and West K W 2010 *Phys. Rev. Lett.* **104** 206801

[25] Dmitriev I, Gellmann R and Vavilov M G 2010 *Phys. Rev. B* **82** 201311(R)

[26] Mani R G, Smet J H, von Klitzing K, Narayanamurti V, Johnson W B and Umansky V 2002 *Nature (London)* **420** 646

[27] Zudov M A, Du R R, Pfeiffer L N and West K W 2003 *Phys. Rev. Lett.* **90** 046807

[28] Yang C L, Zudov M A, Knuuttila T A, Du R R, Pfeiffer L N and West K W 2003 *Phys. Rev. Lett.* **91** 096803

[29] Du R R, Zudov M A, Yang C L, Yuan Z Q, Pfeiffer L N and West K W 2004 *Int. J. of Mod. Phys. B* **18** 3465

[30] Du R R, Zudov M A, Yang C L, Pfeiffer L N and West K W 2004 *Physica E (Amsterdam)* **22** 7

[31] Zudov M A 2004 *Phys. Rev. B* **69** 041304(R)

[32] Mani R G 2004 *Appl. Phys. Lett.* **85** 4962

[33] Yang C L, Du R R, Pfeiffer L N and West K W 2006 *Phys. Rev. B* **74** 045315

[34] Mani R G 2005 *Phys. Rev. B* **72** 075327

[35] Zudov M A, Du R R, Pfeiffer L N and West K W 2006 *Phys. Rev. B* **73** 041303(R)

[36] Zudov M A, Du R R, Pfeiffer L N and West K W 2006 *Phys. Rev. Lett.* **96** 236804

[37] Smet J H, Gorshunov B, Jiang C, Pfeiffer L, West K, Umansky V, Dressel M, Meisels R, Kuchar F and von Klitzing K 2005 *Phys. Rev. Lett.* **95** 116804

[38] Bykov A A, Bakarov A K, Islamov D R and Toropov A I 2006 *JETP Lett.* **84** 391

[39] Bykov A A 2010 *JETP Lett.* **91** 361

[40] Andreev A V, Aleiner I L and Millis A J 2003 *Phys. Rev. Lett.* **91** 056803

[41] Anderson P W and Brinkman W F unpublished *arXiv:cond-mat/0302129*

[42] Auerbach A, Finkler I, Halperin B I and Yacoby A 2005 *Phys. Rev. Lett.* **94** 196801

[43] Finkler I G and Halperin B I 2009 *Phys. Rev. B* **79** 085315

[44] Dorozhkin S I 2003 *JETP Lett.* **77** 577

[45] Willett R L, Pfeiffer L N and West K W 2004 *Phys. Rev. Lett.* **93** 026804

[46] Mani R G, Smet J H, von Klitzing K, Narayanamurti V, Johnson W B and Umansky V 2004 *Phys. Rev. Lett.* **92** 146801

[47] Studenikin S A, Potemski M, Coleridge P T, Sachrajda A S and Wasilewski Z R 2004 *S. S. Commun.* **129** 341

[48] Kovalev A E, Zvyagin S A, Bowers C R, Reno J L and Simmons J A 2004 *S. S. Commun.* **130** 379

[49] Studenikin S A, Potemski M, Sachrajda A, Hilke M, Pfeiffer L N and West K W 2005 *Phys. Rev. B* **71** 245313

[50] Dorozhkin S I, Smet J H, Umansky V and von Klitzing K 2005 *Phys. Rev. B* **71** 201306(R)

[51] Bykov A A, Bakarov A K, Kalagin A K and Toropov A I 2005 *JETP Lett.* **81** 284

[52] Yuan Z Q, Yang C L, Du R R, Pfeiffer L N and West K W 2006 *Phys. Rev. B* **74** 075313

[53] Wirthmann A, McCombe B D, Heitmann D, Holland S, Friedland K J and Hu C M 2007 *Phys. Rev. B* **76** 195315

[54] Dorozhkin S I, Smet J H, von Klitzing K, Pfeiffer L N and West K W 2007 *JETP Lett.* **86** 543

[55] Dorozhkin S I, Bykov A A and and AK Bakarov I P 2007 *JETP Lett.* **85** 543

[56] Bykov A A, Islamov D R, Goran A V and Bakarov A K 2007 *JETP Lett.* **86** 779

[57] Wiedmann S, Gusev G M, Raichev O E, Lamas T E, Bakarov A K and Portal J C 2008 *Phys. Rev. B* **78** 121301

[58] Bykov A A 2008 *JETP Lett.* **87** 551

[59] Bykov A A 2008 *JETP Lett.* **87** 233

[60] Wiedmann S, Gusev G M, Raichev O E, Bakarov A K and Portal J C 2009 *Phys. Rev. B* **80** 035317

[61] Hatke A T, Zudov M A, Pfeiffer L N and West K W 2009 *Phys. Rev. Lett.* **102** 066804

[62] Pechenezhskii I V, Dorozhkin S I and Dmitriev I A 2007 *JETP Lett.* **85** 86

[63] Tung L C, Yang C, Smirnov D, Pfeiffer L, West K, Du R and Wang Y J 2009 *Solid State Commun.* **149** 1531

[64] Fedorych O M, Potemski M, Studenikin S A, Gupta J A, Wasilewski Z R and Dmitriev I A 2010 *Phys. Rev. B* **81** 239903

[65] Bykov A A 2009 *JETP Lett.* **89** 575

[66] Studenikin S A, Sachrajda A S, Gupta J A, Wasilewski Z R, Fedorych O M, Byszewski M, Maude D K, Potemski M, Hilke M, West K W and Pfeiffer L N 2007 *Phys. Rev. B* **76** 165321

[67] Andreev I V, Muravev V M, Kukushkin I V, Schmult S and Dietsche W 2011 *Phys. Rev. B* **83** 121308(R)

[68] Ryzhii V I 1970 *Sov. Phys. Solid State* **11** 2078

[69] Ryzhii V I, Suris R A and Shchamkhalova B S 1986 *Sov. Phys. Semicond.* **20** 1299

[70] Phillips J C 2003 *S. S. Commun.* **127** 233

[71] Durst A C, Sachdev S, Read N and Girvin S M 2003 *Phys. Rev. Lett.* **91** 086803

[72] Shi J and Xie X C 2003 *Phys. Rev. Lett.* **91** 086801

[73] Lei X L and Liu S Y 2003 *Phys. Rev. Lett.* **91** 226805

[74] Koulakov A A and Raikh M E 2003 *Phys. Rev. B* **68** 115324

[75] Dmitriev I A, Mirlin A D and Polyakov D G 2003 *Phys. Rev. Lett.* **91** 226802

[76] Ryzhii V and Vyurkov V 2003 *Phys. Rev. B* **68** 165406

[77] Ryzhii V 2003 *Phys. Rev. B* **68** 193402

[78] Ryzhii V, Chaplik A and Suris R 2004 *JETP Lett.* **80** 363

[79] Vavilov M G and Aleiner I L 2004 *Phys. Rev. B* **69** 035303

[80] Joas C, Raikh M E and von Oppen F 2004 *Phys. Rev. B* **70** 235302

[81] Park K 2004 *Phys. Rev. B* **69** 201301

[82] Lee D H and Leinaas J M 2004 *Phys. Rev. B* **69** 115336

[83] Mikhailov S A 2004 *Phys. Rev. B* **70** 165311

[84] Dmitriev I A, Mirlin A D and Polyakov D G 2004 *Phys. Rev. B* **70** 165305

[85] Dietel J, Glazman L I, Hekking F W J and von Oppen F 2005 *Phys. Rev. B* **71** 045329
[86] Joas C, Dietel J and von Oppen F 2005 *Phys. Rev. B* **72** 165323
[87] Torres M and Kunold A 2005 *Phys. Rev. B* **71** 115313
[88] Alicea J, Balents L, Fisher M P A, Paramekanti A and Radzihovsky L 2005 *Phys. Rev. B* **71** 235322
[89] Lei X L and Liu S Y 2005 *Phys. Rev. B* **72** 075345
[90] Lei X L and Liu S Y 2005 *Appl. Phys. Lett.* **86** 262101
[91] Dmitriev I A, Vavilov M G, Aleiner I L, Mirlin A D and Polyakov D G 2005 *Phys. Rev. B* **71** 115316
[92] Toyoda T, Fujita M, Koizumi H and Zhang C 2005 *Phys. Rev. B* **71** 033313
[93] Ng T K and Dai L 2005 *Phys. Rev. B* **72** 235333
[94] Kashuba A 2006 *Phys. Rev. B* **73** 125340
[95] Mikhailov S A and Savostianova N A 2006 *Phys. Rev. B* **74** 045325
[96] Dietel J 2006 *Phys. Rev. B* **73** 125350
[97] Lei X L and Liu S Y 2006 *Appl. Phys. Lett.* **89** 182117
[98] Lei X L and Liu S Y 2006 *Appl. Phys. Lett.* **88** 212109
[99] Lei X L 2006 *Phys. Rev. B* **73** 235322
[100] Volkov V A and Takhtamirov E E 2007 *JETP* **104** 602
[101] Dmitriev I A, Mirlin A D and Polyakov D G 2007 *Phys. Rev. B* **75** 245320
[102] Dmitriev I A, Evers F, Gornyi I V, Mirlin A D, Polyakov D G and Wolfle P 2008 *Phys. St. Sol. B* **245** 239
[103] Dmitriev I A, Dorozhkin S I and Mirlin A D 2009 *Phys. Rev. B* **80** 125418
[104] Dmitriev I A, Khodas M, Mirlin A D, Polyakov D G and Vavilov M G 2009 *Phys. Rev. B* **80** 165327
[105] Raichev O E 2010 *Phys. Rev. B* **81** 165319
[106] Bykov A A, Zhang J Q, Vitkalov S, Kalagin A K and Bakarov A K 2007 *Phys. Rev. Lett.* **99** 116801
[107] Romero N, McHugh S, Sarachik M P, Vitkalov S A and Bykov A A 2008 *Phys. Rev. B* **78** 153311
[108] Vitkalov S 2009 *Int. J. of Mod. Phys. B* **23** 4727
[109] Hatke A T, Chiang H S, Zudov M A, Pfeiffer L N and West K W 2010 *Phys. Rev. B* **82** 041304(R)
[110] Vavilov M G, Aleiner I L and Glazman L I 2007 *Phys. Rev. B* **76** 115331
[111] Lei X L 2007 *Appl. Phys. Lett.* **90** 132119
[112] Zhang J Q, Vitkalov S, Bykov A A, Kalagin A K and Bakarov A K 2007 *Phys. Rev. B* **75** 081305(R)
[113] Hatke A, Zudov M, Pfeiffer L and West K 2010 *Physica E (Amsterdam)* **42** 1081
[114] A. A. Bykov *et al.*, Phys. Rev. Lett. **99**, 116801 (2007); N. Romero *et al.*, Phys. Rev. B **78**, 153311 (2008).
[115] Hatke A T, Zudov M A, Pfeiffer L N and West K W 2011 *Phys. Rev. B* **83** 081301(R)
[116] Kalmanovitz N R, Bykov A A, Vitkalov S A and Toropov A I 2008 *Phys. Rev. B* **78** 085306
[117] Dai Y, Du R R, Pfeiffer L N and West K W 2010 *Phys. Rev. Lett.* **105** 246802
[118] Hatke A T, Zudov M A, Pfeiffer L N and West K W 2011 *Phys. Rev. B* **83** 121301(R)
[119] Hatke A T, Zudov M A, Pfeiffer L N and West K W 2011 *Phys. Rev. B* **83** 201301(R)

Journal of Physics: Conference Series **334** (2011) 012008 doi:10.1088/1742-6596/334/1/012008

Real-space mapping of a disordered two-dimensional electron system in the quantum Hall regime

K. Hashimoto[1,2], J. Wiebe[3], T. Inaoka[4], Y. Hirayama[1,2], R. Wiesendanger[3], and M. Morgenstern[5,6]

[1] Department of Physics, Tohoku University, Sendai 980-8578, Japan
[2] ERATO Nuclear Spin Electronics Project, Sendai 980-8578, Japan
[3] Institute of Applied Physics, Hamburg University, Jungiusstraße 11, D-20355 Hamburg, Germany
[4] Department of Physics and Earth Sciences, University of the Riyuukyu, 1 Senbaru, Nishihara,Okinawa, 903-0213, Japan
[5] II. Institute of Physics B, RWTH Aachen University, Aachen, D-52056, Germany
[6] JARA-Fundamentals of Future Information Technology

hashi@m.tains.tohoku.ac.jp

Abstract. By using scanning tunnelling spectroscopy, we study the influence of potential disorder on an adsorbate-induced two-dimensional electron system in the integer quantum Hall regime. The real-space imaged local density of states exhibits transition from localized drift states encircling the potential minima to another type of localized drift states encircling the potential maxima. While the former states show regular round shapes, the latter have irregular-shaped patterns. This difference is induced by different sources for the potential minima and maxima, i.e., substrate donors and an inhomogeneous distribution of the adsorbates, respectively.

1. Introduction

The interplay between a high magnetic field (B) and electrostatic disorder is a fundamental physical aspect of the integer quantum Hall effect [1]. At a high magnetic field, an electron of a two dimensional electron system (2DES) performs cyclotron motion leading to discrete Landau levels (LLs). An additional electric field caused by potential disorder leads to drift of the guiding center of the cyclotron motion. The resulting quantum mechanical states, so-called drift states, meander along equipotential lines with a width of about the cyclotron radius $r_c = \sqrt{(2n+1)\cdot\hbar/(|e|B)}$ (n, LL index; $\hbar$, Dirac's constant; e, electron charge) [2]. When the drift states are trapped around a potential minimum, such localized states split off from the degenerate LL in energy, composing the low-energy tail of the LL. In contrast, the drift states localized around a potential maximum form the high-energy tail. The transition between these two types of localized states through the critical state, i.e. the extended state, is the so-called quantum Hall transition, which we had previously probed [3] and which was recently observed also within a graphene 2DES [4] and a quasi-2DES on the graphite surface [5].

This paper presents a scanning tunnel spectroscopy (STS) study of a disordered 2DES, focusing on the influences of the disorder potential. The 2DES is induced on the *n*-InSb(110) surface. At $B = 0$ T, the local density of states (LDOS) shows wave patterns scattered by the weak potential disorder with a wave length according to a non-parabolic dispersion relation which is in agreement with a self-consistent calculation. At $B = 12$ T, we observe the evolution from localized drift states encircling the potential minima to localized states encircling the potential maxima by mapping the energy dependent LDOS of the lowest Landau level (LL0). While the former localized states show round-shaped LDOS patterns, the latter localized states exhibit irregular-shaped patterns. To clarify the cause of the difference, we map the potential landscape by measuring the spatial peak-energy fluctuations of LL0. Comparison between the potential landscape and the topographic scanning tunneling microscopy (STM) image of the same area reveals that the difference is induced by different sources of the potential minima and maxima, i.e., substrate donors and an inhomogeneous distribution of Cs atoms, respectively.

2. Methods

2-1. Experiments

The 2DES was prepared by depositing 0.01 - 0.02 monolayer Cs on an atomically smooth, clean *n*-InSb(110) surface prepared in ultrahigh vacuum ($\sim 1\times10^{-10}$ mbar) by cleavage. The donor (acceptor) density, n_d (n_a), of the *n*-InSb is determined to be 9×10^{21} m^{-3} (5×10^{21} m^{-3}) from compensation electron density ($n_d - n_a$) and compensation ratio (n_a/n_d) measured by transport and STM images [6]. Cs was deposited from a well out-gassed Cs dispenser [7] at a pressure of about 10^{-10} mbar. The Cs induces downward band bending at the InSb surface and, thereby, creates the 2DES [8]. To minimize an inhomogeneous distribution of Cs atoms due to the formation of chain structures [9], the deposition was carried out onto the cold surface at $T \sim 30$ K and, subsequently, the Cs atoms were stabilized by annealing the sample at a temperature of $T = 100$ - 300 K for about 5 - 30 min. We obtained a homogeneous distribution of Cs with less than 4 % in chain structures.

We performed STS at $T = 0.3$ K by using an ultra-high vacuum scanning tunnelling microscope [10] homebuilt at Hamburg University. The distance between tip and sample was stabilized at a current I_{stab} and a voltage V_{stab} prior to opening the feedback. Then, the $dI/dV(V_s)$ curves were recorded at a sample voltage V_s using lock-in technique with an ac modulation of amplitude V_{mod} (modulation frequency ~ 1.5 kHz). We used an ex-situ etched W-tip which is subsequently prepared in-situ by flash annealing and voltage pulses on W(110) [10].

2-2 Calculations

The self-consistent calculation used to determine the adsorbate induced band bending is described elsewhere in detail [11]. As input parameters, the experimentally determined compensation electron density $n_c = n_d - n_a = 4\times10^{21}$ m^{-3}, the standard values for InSb of static dielectric constant $\varepsilon = 17.9$, band gap $E_{Gap} = 235$ meV, and spin-orbit splitting $E_{SO} = 850$ meV as well as a surface band-shift of $\Delta E = 278$ meV as estimated at band-bending saturation by photoemission spectroscopy [8] are used. The calculation does not use further fitting parameters and considers the non-parabolicity of the conduction band.

3. Results and discussion

First we sketch the influence of a potential disorder on the Cs-induced 2DES at $B = 0$ T. The solid curve in Fig. 1(a) is the spatially averaged dI/dV curve which represents the energy dependence of the macroscopically averaged LDOS, i.e. the DOS [12]. The three steplike features marked by arrows represent the first (E_1), second (E_2), and third (E_3) subband edges [13]. They are in good agreement with the subband energies $E_1 = -116$ meV, $E_2 = -51$ meV, and $E_3 = -22$ meV (measured from the Fermi energy E_F) resulting from the self-consistent calculation [Fig. 1(b)]. The band edge, however, locally fluctuates in energy owing to the potential disorder [12]. Indeed, local dI/dV curves A and B obtained by locally averaging dI/dV curves (diameter of about 20 nm) show lower and higher band edge energies, respectively, than the one in the DOS, indicating that they are influenced by a lower and higher electrostatic disorder potential. The measured areas A and B are marked in the STM image [inset of Fig. 1(a)]. In the area A, the Cs atoms are homogeneously distributed as in the surrounding area. However, in the area B, the Cs atoms form chain structures, suggesting that the inhomogeneous distribution of Cs due to the chain formation induces a higher potential. Notice that the band edge of the d*I*/d*V* curve A shows an additional peak structure which can be understood as a shallow bound state due to an attractive potential [14, 15].

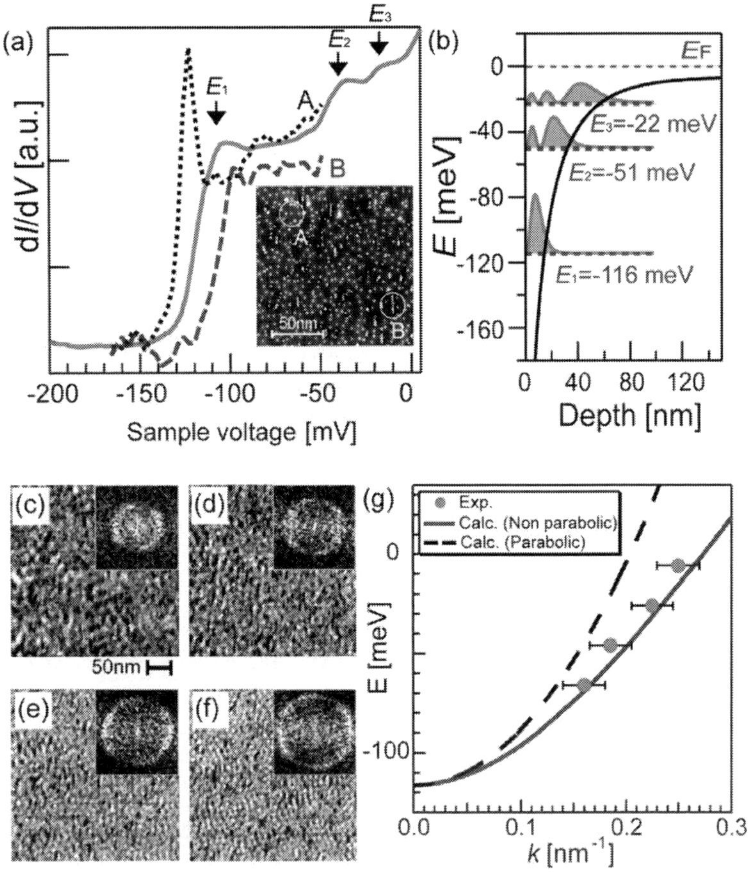

Fig. 1 (colour on line) Zero-field property of the Cs-induced disordered 2DES. (a) Dotted (A) and dashed (B) d*I*/d*V* curves are locally averaged in area A and B (diameter of about 20 nm) marked in the STM image (inset). Solid d*I*/d*V* curve is averaged in a large area (48 x 48 points, 350 x 350 nm^2). Arrows mark subband energies E_i (i =1, 2, 3). (b) Self-consistently calculated band bending at the InSb surface (solid curve) and electron density distribution of *i*th subband indicated by E_i. (c)-(f) d*I*/d*V*(*x*,*y*) images recorded in the same surface area at V_s = -66 mV (c),-46 mV (d), -26 mV (e), and -6 mV (f); V_{stab} = 100 mV, I_{stab} = 0.2 nA, V_{mod} = 1.5 mV, insets show Fourier transforms of the real-space images. (g) Dispersion of electron energy E plotted as a function of wave number k. The dots represent the dominant k values extracted from (c)-(f). The solid curve represents a dispersion curve of the lowest subband (E_1 = -116 meV) self-consistently calculated taking full account of the non-parabolic conduction-band dispersion of InSb, in comparison with the parabolic dispersion using the band-edge effective mass (dashed curve).

Journal of Physics: Conference Series **334** (2011) 012008 doi:10.1088/1742-6596/334/1/012008

Figures 1(c)-(f) show dI/dV images of the 2DES at different V_s values. The wave length of the visible standing wave patterns apparently decreases with increasing energy. The corresponding Fourier transforms of the real-space images are shown in the insets. In each Fourier transform, the dominating feature is a ring which is increasing in diameter for increasing energy. This feature reflecting the 2DES dispersion cannot be resolved within a strongly disordered system [15]. The dots in Fig. 1(g) show the extracted wave numbers k of the rings as a function of the energy. We perform a self-consistent calculation which takes full account of the non-parabolic conduction-band dispersion of InSb [11]. The resulting dispersion curve of the lowest subband (E_1 = -116 meV) is plotted as a solid line. For the sake of comparison, the dispersion curve assuming a parabolic conduction band with the band-edge effective mass is also plotted as a dashed line. The experimentally obtained results are in agreement with the calculated non-parabolic dispersion, evidencing that the underlying potential disorder acts as a weak disorder which does not disrupt general properties of the 2DES on InSb. Similar results are also obtained in a 2DES induced by Cs on p-InSb(110) [16] and by Fe on n-InAs(110) [12].

Next, we examine the disordered 2DES in the quantum Hall regime. Figures 2 (a)-(f) present the real-space dI/dV images recorded at different V_s values in the lowest spin-down LL (LL0↓) at B = 12 T [3]. The spatially averaged dI/dV curve in the corresponding voltage region is shown in the right panel of Fig. 2(g). In the low-energy tail of the LL0↓ [Fig. 2(a)], we observe round disk/loop patterns as marked by arrows. With increasing V_s [Fig. 2(b), (c)], they increase in size, indicating that the localized drift states probes a longer equipotential line at higher energy within the potential valley. In contrast, loop patterns marked by arrowheads in the high-energy tail [Fig. 2(e), (f)] shrink in size with increasing energy. They are, therefore, attributed to localized drift states around the potential maxima. The shapes of the localized states around the potential maxima tend to be irregular in contrast to the round-shaped localized drift states around the potential minima. At the LL center [Fig. 2(d)], the drift states delocalize in areas between the potential valleys and hills, corresponding to the extended state. The observed energy dependencies can be confirmed by the energy-position plot shown in the left panel of Fig. 2(g), which is determined by dI/dV(V_s) curves recorded along the dashed line marked in Fig. 2 (a) or (f). The LL0↓ level fluctuates in energy with a dip (peak) structure near x = 130 nm (95 nm) deviating from the LL center (-101 mV) by about 19 mV (16 mV). Such LL fluctuations cause broadened LLs in the measured DOS as shown in the right panel. The accompanying low-energy LL0↑ energetically overlaps with the high-energy LL0↓ around x = 130 nm, causing overlapping localized LL0↑ states around the potential minima to appear at -92.4 mV as marked by arrows in Fig. 2 (f).

The LL fluctuation additionally provides information of the local 2DES potential analogous to the tip-induced quantum dot state [12, 17]. Since the drift states track the potential fluctuation down to a length scale of r_c, the LL0↓ fluctuation at B = 12 T monitors the potential landscape with a resolution of r_c = 7.4 nm. The corresponding fluctuation amplitude with respect to the energy of the LL center is plotted in Fig. 2(h) as a function of position within the same area shown in Figs. 2(a)-(f). One can see five potential valleys with a depth of 18±3 meV marked by arrows, and three potential hills with a height of 15±2 meV marked by arrowheads. An STM image of the same area is given in Fig. 2(i). It shows an almost homogenous distribution of the Cs atoms with only 4 % chain structures. The position of the chain structures is marked by crosses on the potential map [Fig. 2(h)]. The three large potential hills as well as a number of small hills appear at the positions of the chain structures in agreement with the result at B = 0 T [Fig. 1(a)]. This indicates that the Cs chains dominantly induce the potential maxima in this area, although the corresponding surface-2DES region with a depth of $d \approx$ 15 nm is expected to contain a finite number of substrate acceptors (~2 in a volume of 170×170×15 nm^3) which may also contribute to the formation of the potential maxima. In Fig. 2(i), we superimpose particular equipotential line extracted from the potential hills onto the STM image. This reveals that the irregular-shaped equipotential line encircles an area where the density of the single Cs atoms is apparently lower than in the surrounding area. If we assume that the single Cs chain, analogous to a Co-island on InAs(110) [18], acts as a single donor possessing the same charge as a single Cs atom, these areas have negative electrostatic potential due to the lower density of positively charged donors than areas without chains. We therefore conclude that the irregular-shaped localized states are induced by the inhomogeneous distribution of the Cs atoms due to the Cs chain formation. On the other hand, there is no correspondence between the position of the Cs chains and the potential valley as visible in Fig. 2(h). Additionally, there is no other specific inhomogeneity of the Cs distribution in the valley area. The most possible source for the potential minima are, thus, point charges such as single substrate donors [12, 17], which may induce the round-shaped electrostatic potential and thus round-shaped localized drift states. Indeed, the potential depth of 18±3 meV is comparable to the attractive potential of a single donor averaged over the extension of the 2DES [12]. Moreover, the number of the valleys is also nearly equal to the number of substrate donors expected in the surface-2DES region (~ 4).

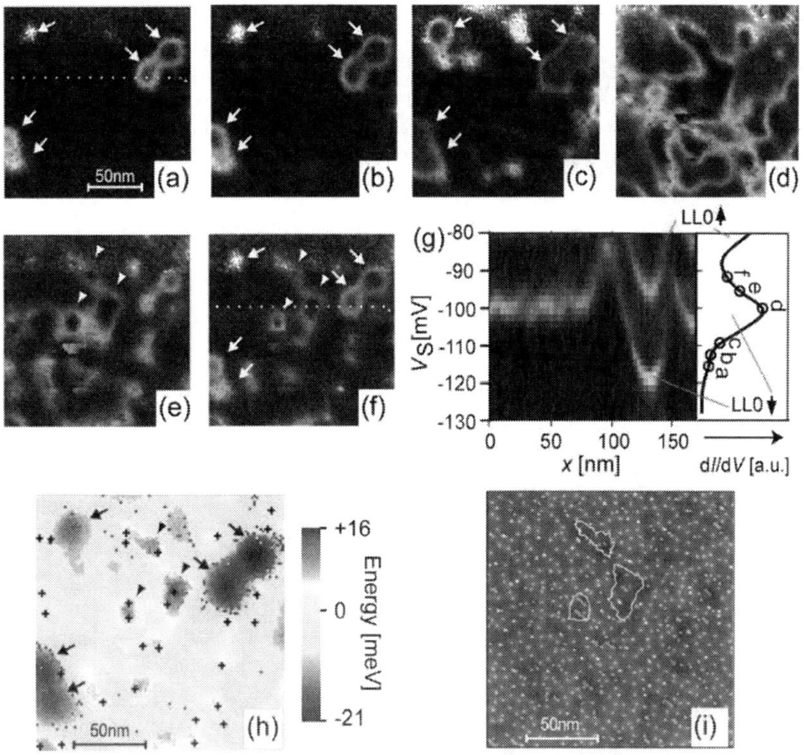

Fig. 2 (colour on line) (a) – (f) dI/d$V(x, y)$ images; B = 12 T, V_s = -116.3 mV (a), -114.6 mV (b), -109.5 mV (c), -100.9 mV (d), -95.8 mV (e), and -92.4 mV (f); V_{stab} = 150 mV, I_{stab} = 0.10 nA, V_{mod} = 1.0 mV; each arrow (arrowhead) in (a) - (f) marks drift states encircling a potential minimum (maximum). (g) left: dI/d$V(V_s, x)$ measured along the dashed line marked in (a) or (f); right: spatially averaged dI/dV curve with circles at the V_s values used in (a) – (f). (h) Potential landscape of the same surface area as (a) – (f) determined from the energetic position of the LL0↓ peak. Crosses mark positions of Cs chains as determined in the corresponding STM image (i); arrows (arrowheads) mark the potential minima (maxima). Equipotential lines at an energy extracted from the three potential hills in (h) are marked by solid lines in (i).

4. Conclusions

We have shown the evolution of drift states between two different types of localized states in both the low- and high-energy tails of a LL0. While the former states have round shapes, the latter states form irregular-shaped patterns. The difference is induced by different sources of the potential disorder, i.e., substrate donors and an inhomogeneous distribution of adsorbates, which induce rather round potential valleys and irregular-shaped potential hills, respectively.

Acknowledgements

We thank F. Meier, D. Haude, F. Marczinowski, S. v. Oehsen, and G. Meier for helpful discussions and acknowledge financial support via the priority program ''Quantum-Hall systems'' as well as SFB 508-B4 of the German science foundation. Part of the numerical calculations was performed at Cyberscience Center, Tohoku University and Supercomputer Center, the Institute for Solid State Physics, University of Tokyo.

References

[1] B. Krame, T. Ohtsuki, and S. Kettemann, Phys. Rep. **417**, 211 (2005).
[2] T. Ando, J. Phys. Soc. Jpn. **53**, 3101 (1984).
[3] K. Hashimoto *et al.*, Phys. Rev. Lett. **101**, 256802 (2008).
[4] D. L. Miller *et al.*, Nature Phys. **6**, 811 (2010).
[5] Y. Niimi, H. Kambara, and H. Fukuyama, Phys. Rev. Lett. **102**, 026803 (2009).
[6] The electron density and compensation ratio are measured to be $n_d - n_a = 4\times10^{21}$ m^{-3} and n_a/n_d = 0.56.
[7] The Cs dispencer was purchased from SAES Getters
[8] M. G. Betti *et al.*, Phys. Rev. B **63**, 155315 (2001).
[9] L. J. Whitman *et al.*, Phys. Rev. Lett. **66**, 1338 (1991).
[10] J. Wiebe *et al.*, Rev. Sci. Instrum. **75**, 4871 (2004).
[11] S. Abe, T. Inaoka, and M. Hasegawa, Phys. Rev. B **66**, 205309 (2002).
[12] M. Morgenstern *et al.*, Phys. Rev. Lett. **89**, 136806 (2002).
[13] The larger step increase of the dI/dV curve at the lowest subband is caused by the higher probability density of the lowest subband nearer to the surface due to its stronger confinement induced by the triangular potential [Fig. 1(b)]. This is different from results on a surface quantum well [14].
[14] S. Perraud *et al.*, Phys. Rev. Lett. **100**, 056806 (2008).
[15] J. Wiebe *et al.*, Phys. Rev. B **68**, 041402 (2003).
[16] S. Becker *et al.*, Phys. Rev. B **81**, 155308 (2010).
[17] R. Dombrowski *et al.*, Phys. Rev. B **59**, 8043 (1999).
[18] M. Morgenstern *et al.*, Phys. Rev. B **65**, 155325 (2002).

Coherent control of two individual electron spins and influence of hyperfine coupling in a double quantum dot

S Tarucha[1-3], T Obata[1,2], M Pioro-Ladriere[1,4], R Brunner[1,5], Y -S Shin[1], T Kubo[1,6], and Y Tokura[1,6]

[1] Quantum Spin Information Project, ICORP, Japan Science and Technology Agency, Atsugi-shi, Kanagawa, Japan
[2] QPEC & Department of Applied Physics, University of Tokyo, Hongo, Bunkyo-ku, Tokyo, Japan
[3] Institute for Nano Quantum Information Electronics, the University of Tokyo, Komaba, Meguro-ku, Tokyo, Japan
[4] Department de Physique, Universit'e de Sherbrooke, Sherbrooke, Quebec, J1K-2R1, Canada
[5] Institute for Physics, University of Leoben, 8700, Austria
[6] NTT Basic Research Laboratories, NTT Corporation, Atsugi-shi, Kanagawa, Japan

E-mail: tarucha@ap.t.u-tokyo.ac.jp

Abstract. Electric dipole spin resonance of two individual electrons and the influence of hyperfine coupling on the spin resonance are studied for a double quantum dot equipped with a micro-magnet. The spin resonance occurs by oscillating the electron in each dot at microwave (MW) frequencies in the presence of a micro-magnet induced stray field. The observed continuous wave (CW) and time-resolved spin resonances are consistent with calculations in which the MW induced AC electric field and micro-magnet induced stray field are taken into account. The influence of hyperfine coupling causes an increase and broadening of the respective CW spin resonance peaks through dynamical nuclear polarization when sweeping up the magnetic field. This behaviour appears stronger for the larger of the two spin resonance peaks and in general becomes more pronounced as the MW power increases, both reflecting that the electron-nuclei interaction is more efficient for the stronger spin resonance. In addition the hyperfine coupling effect only becomes pronounced when the MW induced AC magnetic field exceeds the fluctuating nuclear field.

1. Introduction

Following the Loss and DiVincenzo's proposal on spin-based quantum computing with quantum dots (QDs) [1], and motivated by the realization of single and double QDs holding just one or two electrons [2-4] and the observation of a robust spin degree of freedom [5-7], considerable effort has been devoted to make electron spin qubits. Consequently spin qubits utilising the spin up and down state in single QDs have been demonstrated using electron spin resonance (ESR) [8-11] and qubits utilising the spin singlet and triplet state have been demonstrated by manipulating the inter-dot exchange

Published under licence by IOP Publishing Ltd

coupling in a double QD [12,13]. ESR is a well-known technique but its application to spin qubits is still a challenge, because a sufficiently strong AC magnetic field has to be operated on single electrons in single QDs. Such an AC field has been prepared using a micro-coil placed on top of a QD with an AC current flowing through it [8] and an electric dipole spin resonance (EDSR) technique. In the latter a local AC magnetic field for a QD is generated by application of a MW electric field to the QD through a spin-orbit interaction [9,14,15] and an inhomogeneous magnetic field [16]. We have recently used a micro-magnet to generate the inhomogeneous magnetic field [10,17] and demonstrated two qubits for the first time [11], and in addition proposed a scheme for implementing a multiple qubit system. The micro-magnet when externally laterally magnetized, produces an out-of-plane stray field across a QD with a gradient of ~ T/μm. Spatial oscillation of an electron inside the dot driven by the MW field can generate an AC magnetic field local to the dot (see Fig. 1a). In addition, the micro-magnet induced stray field has an in-plane component along the magnetization direction, which is added to the external magnetic field to provide a total Zeeman field to the dot. Spin resonance localized to the dot occurs when the MW frequency is consistent with the total Zeeman energy.

Here we present CW and time-resolved EDSR experiments for two individual electrons in a double QD using the micro-magnet technique and then we discuss the influence of the hyperfine interaction observed in the CW EDSR. This influence occurs through dynamical nuclear polarization (DNP) [18,19] and statistical fluctuation of the Overhauser field [20], and has been argued as an emerging problem in spin-based quantum computing. We find that the DNP appears stronger for the stronger of the two spin resonances.

2. Electron spin resonance

Figure 1b shows a scanning electron micrograph of the GaAs double QD used in the experiment. Two dots, Dot 1 and 2, are made in a two-dimensional electron gas by applying relevant voltages to the gate electrodes (Fig. 1c). A current I_{dot} flowing through the double dot is initially measured as a function of two side gate voltages to define the charge state with the electron number N_1 and N_2, for Dot 1 and 2, respectively of $(N_1,N_2)=(1,1)$ and (2,0) under an external in-plane magnetic field B_0. When the Co micro-magnet on top of the gated structure is externally magnetized in-plane (along z), a stray magnetic field is generated at the double dot with a slanted out-of-plane field $B_\perp(z)$ ($dB_\perp/dz$ ~ 0.8 T/μm) and an inhomogeneous in-plane field $B_\cdot(x)$ providing a Zeeman field offset, $\Delta B_0 \approx B_{0L}-B_{0R}$ (= 20 to 30 mT) between the two dots.

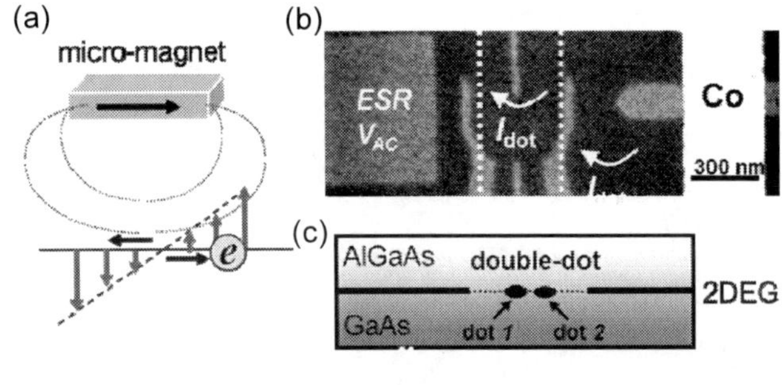

Figure 1. (a) Oscillating electron in a micro-magnet induced stray field. (b) Scanning electron micrograph of the device. The Co micro-magnet is intentionally displaced but usually placed between the dotted lines. (c) Schematic device cross-section, indicating formation of a double dot.

EDSR occurs for each QD by spatially displacing the electron in the presence of B. (z) by applying high frequency AC voltage to the MW gate electrode to satisfy the condition $hf_{a.c.}=g\mu_B B_{0v=1,2}$, where $f_{a.c.}$ denotes the MW frequency, μ_B is the Bohr's magneton, $B_{0v=1,2}$ corresponds to the total in-plane magnetic fields at the left and right dot, respectively (see Fig. 1b). To observe the EDSR for each QD we operate a double dot in the Pauli spin blockade (PSB) [21] established by the formation of a spin triplet state ($T_+=|\uparrow\rangle|\uparrow\rangle$ or $T_-=|\downarrow\rangle|\downarrow\rangle$) for (N_1,N_2)=(1,1) (Fig. 2a). Figure 2b shows the current I_{dot} through the double dot vs. the external magnetic field B_0 measured for $f_{a.c.}$ = 25.6 GHz. A sufficiently large source-drain bias of 1.0 to 1.5 mV is applied to avoid the influence of photon assisted tunneling (PAT) through the outer barriers. In this figure, we see two peaks separated by 30 mT, reflecting the EDSR for the individual electrons in the two dots. The current through the double dot is initially blocked by Pauli exclusion once a spin triplet state has been formed in the double dot. Spin-flip induced by EDSR in either dot lifts the blockade, giving rise to a finite leakage current. The EDSR peak height, I_{EDSR}, is proportional to the square of the MW induced magnetic field, B_{MW}, and therefore $I_{EDSR} \propto E_{MW}^2\,(dB_\perp/dz)^2$, where E_{MW} is the amplitude of the MW electric field at each dot [11]. From calculation of E_{MW} and $dB_\perp/dz$ we assign the larger peak (smaller peak) at the lower (higher) $f_{a.c.}$ to EDSR for Dot 1 (2).

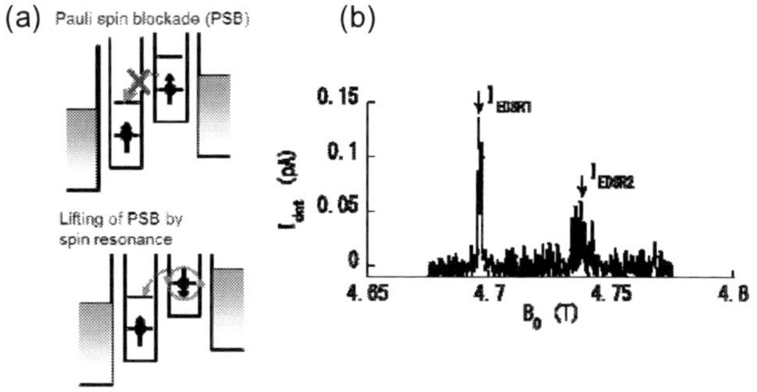

Figure 2. (a) Measurement scheme of EDSR current. (b) CW EDSR current measured for two individual electrons in the two dots (Dot 1 and 2).

After the CW EDSR experiment, we readjusted the device conditions and performed a pump and probe experiment to observe time-resolved EDSR (Fig. 3). We varied the detuning ε with gate voltage pulses to switch the operation stage between A (initialization to either $T_-(1,1)$ or $T_+(1,1)$ in the Pauli spin blockade regime with $\varepsilon = 0$ and $J \sim t$, inter-dot tunnel coupling) and B (spin rotation in either dot with $\varepsilon >> 0$, $J \sim 0$). The EDSR is only pumped in stage B for various MW burst time τ_b and then probed in stage A. The result of the pump and probe experiment performed for the right dot is shown in Fig. 4 as a function of MW burst time τ_b. The oscillatory current with τ_b indicates Rabi oscillations with a frequency ν_{Rabi} of 8.9 MHz in the sinusoidal curve fit. The Rabi frequency ν_{Rabi} is proportional to B_{MW}, i.e. $\nu_{Rabi} = g\mu_B B_{MW}/2h = g\mu_B/2h \times (dB_\perp/dz)\, eE_{MW} l_{orb}^2/\Delta$ [17]. Here, the orbital spread, l_{orb}, and the QD confinement energy, Δ, are 48 nm, and 0.5 meV, respectively obtained from experiments performed separately. Using a calculated value of E_{MW} we derive ν_{Rabi} of 11 MHz, fairly consistent with the experiment.

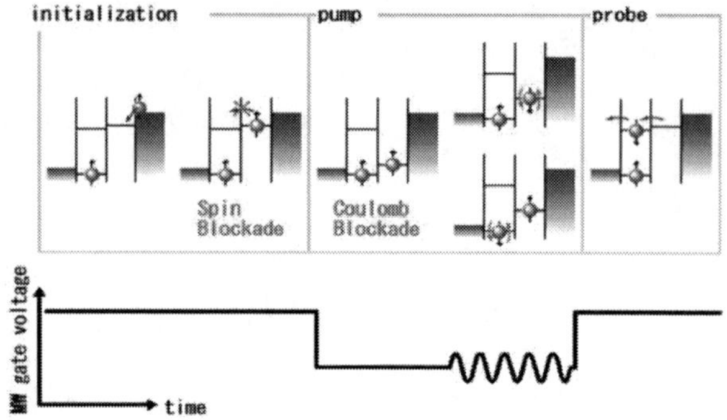

Figure 3. Pump and probe scheme for time-resolved EDSR measurement (upper) and pulsed gating with a MW burst in the probe stage (lower).

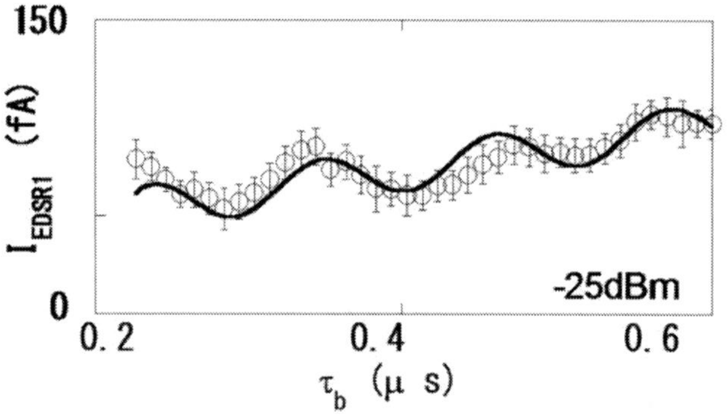

Figure 4. Rabi oscillation for dot 1 measured using a pump and probe technique (see Fig. 3).

3. Influence of hyperfine coupling

Coupling between electron spin and nuclear spin is represented by a hyperfine interaction Hamiltonian:

$$H_{HF} = A|\psi(\mathbf{x})|^2\left(\frac{I_- S_- + I_- S_-}{2} + I_Z S_Z\right) \quad ,$$

where A is the hyperfine constant, $\psi(x)$ is the electron wavefunction and S (I) is the electron (nuclear) spin operator. The first term in the bracket indicates a flip-flop interaction, while the second term accounts for the Overhauser effect. Nuclear spins are dynamically polarized through the flip-flop interaction to build up a Zeeman field on the electron spin (DNP). This causes an Overhauser shift of the EDSR and at the same time works to sustain the EDSR condition through a positive feedback in the case of up-sweep of external field B_0 or down-sweep of MW frequency $f_{a.c.}$. We studied the DNP effect for a device similar to that in Fig.1a but using a technique of charge sensing instead of transport measurement. The transconductance G = dI_{QPC}/dV_p of a quantum point contact (QPC) placed just outside the double dot is measured for the charge sensing. Here, I_{QPC} is the current through the QPC and V_p is the voltage applied to the plunger gate of Dot 1, which is one of the gate electrodes attached to Dot 1 (not shown) .

Figure 5 shows the QPC transconductance measured for sweeping up B_0 from 0.98 to 1.08 T with MW power P_{MW} as a parameter and $f_{a.c.}$ fixed at 5.6 GHz. Two peaks in dark blue or red representing EDSR for two dots are initially observed at 1.0, and 0.99 T, respectively, for the lowest P_{MW}. Both peaks become broader and larger, suffering from DNP as the P_{MW} increases. The peak height starts to increase for P_{MW} exceeding - 30 dBm, and more or less simultaneously the influence of DNP appears pronounced. The peak increase directly reflects the increase of B_{MW} or Rabi frequency compared to the dephasing rate due to statistical fluctuation of nuclear spin. From the threshold of the MW power we calculate the corresponding B_{MW} of a few mT comparable to the fluctuating nuclear field [10, 23]. This is reasonable because the influence of DNP should be visible when B_{MW} is larger than the fluctuating nuclear field. The peak broadening, and peak shift with increasing P_{MW} is due to the feedback effect, and the Overhauser effect, respectively, as described before. On the other hand, as the P_{MW} is further increased exceeding -25 dBm, the peak height starts to decrease. This is probably because of multiple-photon assisted tunneling through the outer tunnel barriers or electron heating to degrade the EDSR signal. Note when sweeping down B_0, no significant DNP effect is observed.

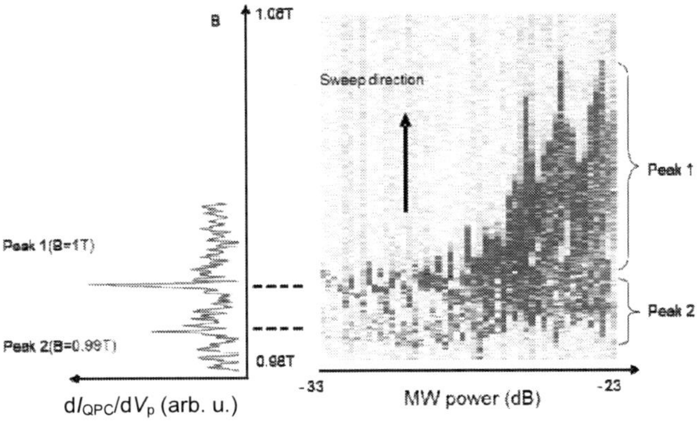

Figure 5. Two EDSR peaks measured for various microwave powers.

4. Conclusion

We have used a micro-magnet technique to realize CW and time-resolved electric dipole spin resonance of two individual electrons in a double quantum dot and study the influence of hyperfine coupling on the spin resonance. The spin resonance occurs when an electron oscillates inside the dot driven by the microwaves in the presence of a micro-magnet induced stray field across the dot with the microwave frequency equivalent to the Zeeman energy. We have observed two spin resonance peaks separated by the inhomogeneous Zeeman energy between the two dots which is also imposed by the micro-magnet. We have applied a pump and probe technique for the spin resonance to observe Rabi oscillations for one of the two dots. The Rabi oscillation frequency is proportional to the microwave field, consistent with theory. We have observed the influence of hyperfine coupling on the two CW spin resonance peaks through the dynamical nuclear polarization, which is stronger for the larger of the two spin resonance peaks and in general becomes more pronounced as the microwave power increases, both reflecting that the electron-nuclei interaction is more efficient for the stronger spin resonance.

Acknowledgments
This work was supported financially by Grants-in-Aid for Scientific Research S (No.19104007), B (No. 18340081), and Special Coordination Funds for Promoting Science and Technology. S. T. acknowledges support from the QuEST grant (HR0011-09-1-0007). S. T. and Y. T. acknowledge support from MEXT Grant-in-Aid for Scientific Research on Innovative Areas, Contract No. 21102003, Funding Program for World-Leading Innovative R&D on Science and Technology (FIRST) and ERATO-JST (080300000477). M.P.-L. acknowledges supports from the Japan Society for the Promotion of Science and the Canadian Institute for Advanced Research.

References

[1] Loss D and DiVincenzo D P 1998 *Phys. Rev.* A **57** 120
[2] Tarucha S, Austing D G, Honda T, van der Hage R J and Kouwenhoven L P 1996 *Phys. Rev. Lett.* **77** 3613
[3] Ciorga M, Sachrajda A S, Hawrylak P, Gould C, Zawadzki P, Jullian S, Feng Y and Wasilewski Z 2000 *Phys. Rev. B* **61** R16315
[4] For review see, for example van der Wiel, De Francesch S, Elzerman J M, Fujisawa T, Tarucha S and Kouwenhoven L P 2003 *Reviews of Modern Physics* **75**, 1-22
[5] Fujisawa T, Austing D G, Tokura Y, Hirayama Y and Tarucha S 2002 *Nature* **419** 278
[6] Elzerman J M, Hanson R, Willems van Beveren L H, Witkamp B, Vandersypen L M K and Kouwenhovenet L P 2004 *Nature* **430**, 431
[7] Amasha S, MacLean K, Radu I P, Zumbühl D M, Kastner M A, Hanson M P and Gossard A C 2008 *Phys. Rev. Lett.* **100**, 046803
[8] Koppens F H L, Buizert C, Tielrooij K J, Vink I T, Nowack K C, Meunier T, Kouwenhoven L P and Vandersypen L M K. 2006 *Nature* **442** 766
[9] Nowack K C, Koppens F H L, Nazarov Yu V and L.M.K. Vandersypen L M K 2007 *Science* **318** 1430
[10] Pioro-Ladriere M, Obata T, Tokura Y, Shin Y -S, Kubo T, Yoshida K, Taniyama T and Tarucha S 2008 *Nature Physics* **4** 776
[11] Obata T, Pioro-Ladri`ere M, Tokura Y, Shin Y -S, Kubo T, Yoshida K, Taniyama T and Tarucha S 2010 *Phys. Rev.* B **81** 085317
[12] Petta J R, Johnson A C, Taylor J M, Laird E A, Yacoby A, Lukin M D, Marcus C M, Hanson M P and Gossard A C 2005 *Science* **309** 2180
[13] Foletti S, Bluhm H, Mahalu D, Umansky V, and Amir Yacoby A, 2009 *Nature Physics* **5** 903.
[14] Golovach V N, Borhani M, and Loss D 2006 *Phys. Rev. B* **74** 165319
[15] Rashba E I and Efros Al L 2003 *Phys. Rev. Lett.* **91** 126405
[16] Laird E, Barthel A C, Rashba E I, Marcus C M, Hanson M P and Gossard A C 2007 *Phys. Rev. Lett.* **99** 246601
[17] Tokura Y, van der Wiel W G, Obata T and Tarucha S 2006 *Phys. Rev. Lett.* **96** 047202
[18] Ono K and Tarucha S 2004 *Phys. Rev. Lett.* **92** 256803
[19] Kodera T, Ono K, Kitamura Y, Tokura Y, Arakawa Y and Tarucha S 2009 *Phys. Rev. Lett.* **102** 146802
[20] Koppens F H L, Folk J A, Elzerman J M, Hanson R, Willems van Beveren L H. Vink I T, Tranitz H P, Wegscheider W, Kouwenhoven L P and vandersypen L M K 2005 *Science* **309** 1346
[21] Ono K D.G. Austing, Y. Tokura, and S. Tarucha 2002 *Science* **297** 1313.
[22] Obata T, Pioro-Ladriere M, Tokura Y, Brunner R, Shin Y -S, Kubo T, Yoshida K, Taniyama T, and Tarucha S 2009 *J. Phys.: Conf. Ser.* **193**, 012046 1-3
[23] Rashba E I 2008 *Phys. Rev.* B **78**, 195302

Journal of Physics: Conference Series **334** (2011) 012010 doi:10.1088/1742-6596/334/1/012010

Electronic energy levels, wavefunctions and potential landscape of nanostructures probed by magneto-tunnelling spectroscopy

O Makarovsky[1], O Thomas[1], A Patanè[1], R P Campion[1], C T Foxon[1], E E Vdovin[1], D K Maude[2], A G Balanov[3] and L Eaves[1]*

[1]School of Physics & Astronomy, University of Nottingham, NG7 2RD, UK
[2]Grenoble High Magnetic Field Laboratory, CNRS, F-38042 Grenoble, France
[3]Department of Physics, Loughborough University, LE11 3TU, UK

*E-mail: laurence.eaves@nottingham.ac.uk

Abstract. We create electrostatically induced quantum dots by thermal diffusion of interstitial Mn out of a p-type (GaMn)As layer into the vicinity of a GaAs quantum well. This leads to the formation of deep, approximately circular and strongly confined dot-like potential minima in a large mesa diode structure. The minima are formed without need for advanced lithography or electrostatic gating. Using fields of up to 30 T, magnetotunnelling spectroscopy of an individual dot reveals the symmetry of the electronic eigenfunctions and, for the approximately circular dots, a rich spectrum of Fock-Darwin-like states with an orbital angular momentum component $|l_z|$ ranging from 0 up to 11. We find that a small fraction of the dots has elongated potential minima, giving rise to quenching of the orbital angular momentum of the electronic eigenstates. By developing a model to describe the diffusion of the Mn interstitial ions, we determine the electrostatic potential landscape in the quantum well and hence the distribution of dot shapes and sizes. This is in a good agreement with our experimental data.

1. Introduction and brief history of magneto-tunnelling spectroscopy

The combination of high magnetic fields and quantum tunnelling provides a powerful tool for elucidating the nature of the quantized states of both free and bound electrons and holes in semiconductor materials, device and nanostructures. In magnetotunnelling spectroscopy (MTS) experiments, an applied bias is used to resonantly tune the energy of a charged carrier which tunnels through a barrier into a quantum state. When a magnetic field is applied in the plane of the barrier, the tunnelling electron or hole acquires an in-plane momentum due to the action of the Lorentz force. Thus the combination of applied bias and magnetic field can be used to control both the energy and the in-plane wavevector of a carrier when it enters the quantum state. The concept of MTS was developed from early studies of magneto-electric skipping states at a tunnel barrier interface [1] and from measurements of the change in the current-voltage characteristics, *I(V)*, of a resonant tunnelling device due to a strong magnetic field applied in the plane of the quantum well [2]. It was then used to measure resonant band anticrossing effects in the valence [3,4] and conduction bands [5,6] of quantum wells, thus providing detailed plots of the energy-wavevector dispersion relations, $\varepsilon(\boldsymbol{k})$, of holes and electrons. In addition to the study of the extended states of free carriers, MTS has also been used to map out in k-space the probability distribution function $|\psi(\boldsymbol{k})|^2$ of the bound electronic states, including the ground and excited states of hydrogenic donors [7], quantum wires [8] and self-assembled quantum dots [9-11]. Since the wavefunction of a stationary state is real, it is therefore possible to to deduce $\psi(\boldsymbol{k})$ and hence determine the form of the real space wavefunction $\psi(\boldsymbol{r})$ in these types of quantum structure. Magnetocapacitance spectroscopy, a variant of MTS, has recently been used to probe the wavefunctions of many-electron states of InAs quantum dots [12]. The MTS technique is able to probe the eigenfunctions and eigenstates of nanostructures which are buried deep below the surface of a functional semiconductor device. For this reason it complements scanning tunnelling microscopy, which can only access atoms and eigenstates at or very close to the surface of a solid.

Here we describe how MTS can be used to study the electronic properties of electrostatically-induced quantum dots (QDs) [13,14]. We show how studies of the dependence of the tunnel current on magnetic field applied parallel and perpendicular to the barrier plane (xy) can be combined to provide detailed information about the electronic eigenstates. We make these dots by post-growth annealing of a *p-i-n* diode in which the p-layer is composed of ferromagnetic (GaMn)As with 5% Mn and the undoped intrinsic "*i*" layer contains an AlAs/GaAs double barrier resonant tunnelling structure. For a narrow range of annealing conditions, it is possible to diffuse doubly-ionised Mn interstitial (Mn_i^{2+}) donor ions out of the (GaMn)As surface layer into an undoped GaAs layer adjacent to the quantum well structure. The random clustering of small numbers (typically $\lesssim 10$) of the Mn_i^{2+} ions modifies the electrostatic potential in the well, each cluster giving rise to deep and approximately parabolic potential energy minima which can quantum confine tunnelling electrons on a length scale of ~10 nm [13,14].

MTS reveals the presence of dots with approximately circular symmetry, in which the electron eigenstates in high magnetic fields approximate to the Fock-Darwin model [15]. Sharp resonant tunnelling peaks with the z-component of angular momentum ranging from zero up to 11 are observed. In addition, we also find that some of the dots are elongated, and the orbital momentum is quenched. Wavefunction images of the dot eigenstates are measured by studying the variation of the tunnel current as a function of the in-plane magnetic field. We compare our experimental data with a theoretical model which describes the thermal diffusion process, the potential landscape created by the diffused Mn_i^{2+} ions and the nature of the electronic eigenstates of the dots.

Mn interstitials form in (GaMn)As when the Mn concentration exceeds ~ 2%. They are doubly ionised donors, $Mn_i^{2+}(3d^5)$, and inhibit ferromagnetism as they compensate the itinerant holes introduced by substitutional Mn_{Ga} acceptors [16,17]. Thermal annealing of (GaMn)As is an effective means of increasing its Curie temperature and electrical conductivity by out-diffusion of the compensating Mn_i donors to the surface of the heterostructure [17]. We use controlled annealing to create QD-like potential minima in *p-i-n* resonant tunnelling diodes (RTDs) containing a (GaMn)As top layer, grown by molecular beam epitaxy on (001) n^+-GaAs substrates.

2. Details of the heterostructure and the effect of annealing

We focus on high magnetic field studies of a structure with the following composition, in order of growth on a (100) n^+GaAs substrate: a 300 nm layer of n^+-GaAs Si-doped to 2×10^{18} cm^{-3}; 100 nm of n-GaAs doped to 2×10^{17} cm^{-3}; an undoped central intrinsic region comprising a 20 nm GaAs layer, a 5 nm AlAs tunnel barrier, a 6 nm GaAs QW, a 5 nm AlAs tunnel barrier and a 10 nm GaAs layer; and, finally, a 50nm capping layer of p^+-(GaMn)As with 3% Mn. The (GaMn)As layer was grown at 250°C, the others at 600°C. The epilayers were processed into 200 μm diameter mesa diodes, shown schematically in Figure 1(a). These devices emit bright electroluminescence from the QW in forward bias above the flat band conditions [18]. A schematic band diagram of the device at zero bias is shown in Figure 1(b).

Prior to annealing, the forward bias current flowing through the device increases rapidly around the flat band condition, $V \approx V_{FB}$=1.41 V, rising to a peak at 1.8 V, which corresponds to resonant electron tunnelling through the lowest ($E1$) subband of the two-dimensional (2-D) conduction band QW. No current is detectable above the background noise level (< 0.3 pA) in the as-grown mesas for V < 1.3 V. However, following annealing at 150°C for 3 hours, sharp peaks appear in $I(V)$ below 1.3 V with peak heights in the range ~10-100 pA. We attribute these peaks to tunnelling into discrete quantum states arising from the formation of deep potential minima as shown schematically in Figure 1(c). Following this type of low temperature annealing, small clusters of diffused Mn_i ions form in the undoped GaAs region between the (GaMn)As layer and the upper AlAs barrier. Since each ion is doubly charged ($+2e$), this clustering gives rise to deep minima in the potential energy profile at the central plane of the QW. The deepest minima are formed by clusters containing typically 5-10 closely-separated Mn_i ions (all within ~10nm). The measured thermal broadening of the low voltage threshold of each current peak between 0.3 and 20 K confirms that the resonances are due to resonant tunnelling into a single quantised energy level by electrons close the Fermi level, E_F, of a thermalised degenerate electron gas in the n-GaAs emitter. These measurements also determine the fraction of applied voltage, f = 0.65±0.05, dropped between E_F and the level [19]. This f-value indicates that the states through which electrons tunnel are spatially confined in the QW between the two AlAs barriers; the energy of a particular state is given by $efV + a$, when a is a constant.

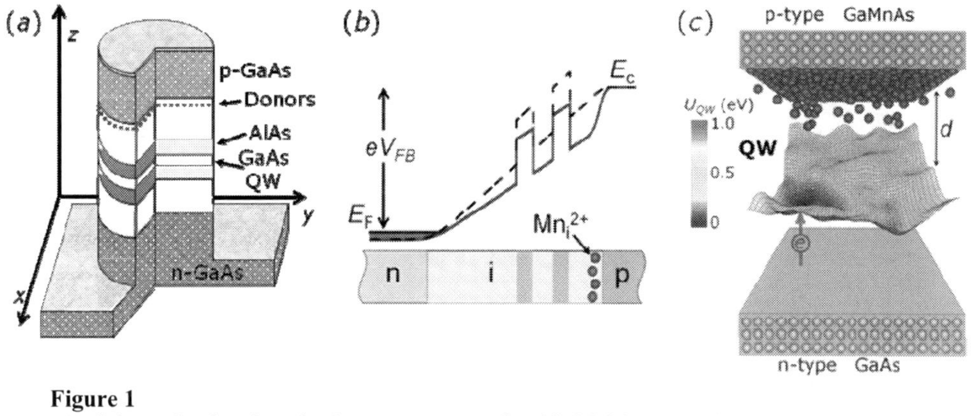

Figure 1

a) Schematic showing the layer structure of a (GaMn)As *p-i-n* diode and a layer of diffused Mn_i donors.

b) Calculated conduction band diagram of our device before (black) and after (blue) annealing.

c) Potential energy profile, $U_{QW}(x,y)$, in the central QW plane, and schematic of an electron tunnelling from the *n*-GaAs layer into quantised states of a minimum of $U_{QW}(x,y)$.

Higher temperature (up to 200°C) or longer annealing (>3 hours) leads to merging of the sharp peaks into an increasingly broad and strong shoulder in $I(V)$. We attribute this to the formation of additional current pathways arising from diffusion of increasing numbers of Mn_i ions from the (GaMn)As towards the QW. This conclusion is supported by measurements on control samples, in which a Be- or C-doped p-GaAs capping layer replaces the (GaMn)As: their *I*(*V*) are unaffected by annealing for 24 hours at 250°C, due to the much smaller diffusion coefficient of these dopants compared to Mn interstitials.

3. Experimental results: magnetotunnelling spectra in fields up to 30T

To understand the nature of the quantum states associated with the sharp tunnelling resonances, we first examine the effect of a magnetic field, B_z, applied perpendicular to the QW plane. Figure 2(a) shows a grey-scale plot of differential conductance, d*I*/d*V*, versus B_z for a sample annealed at 150°C for 3 hours (a darker shade corresponds to higher conductance). The lowest-energy peak does not split, but shifts to higher bias, consistent with the diamagnetic shift of a 1*s*-like orbital ground state. In contrast, the higher bias peaks reveal a rich orbital Zeeman splitting pattern. Figure 2(a) also plots the calculated B_z-dependence of the Fock-Darwin spectrum for a 2-D parabolic potential [19]. This provides a quantitative fit to our data with only one adjustable parameter, $\hbar\omega_0 = 22\pm2$ meV, the characteristic energy spacing of the Fock-Darwin levels at $B_z = 0$. The other parameters are the electron effective mass in GaAs and leverage factor *f*.

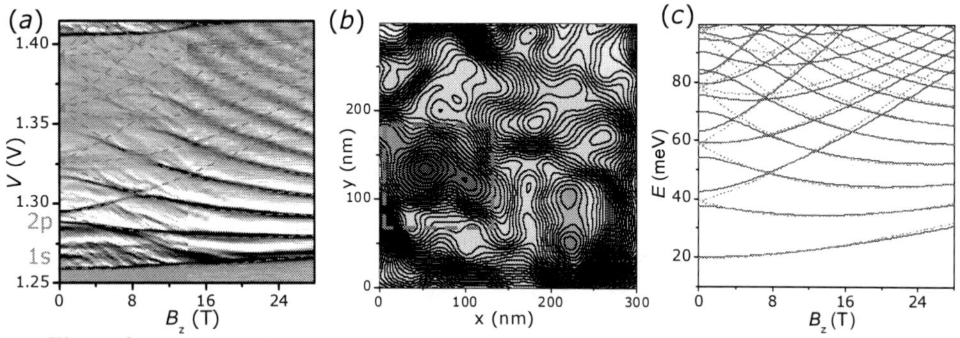

Figure 2

a) Grey-scale plot of differential conductance d*I*/d*V* at $T = 2$K versus B_z up to 28T for a relatively symmetric nanoelectrostatic QD. Red lines represent the fit to the Fock-Darwin spectrum.

b) Numerical simulation of the electrostatic potential created by randomly distributed nanoclusters of diffused Mn_i ions for $d = 18$nm.

c) Blue solid curves: B_z-dependences of the energy spectrum calculated for the deep potential minimum in the area marked by the red square in plot ***(b)***. Red dashed lines: Fock-Darwin spectrum.

The emergence, at high field ($B_z > 16$ T), of a series of equally-spaced peaks, which shift to lower energy and become more closely spaced with increasing B_z, is a characteristic feature of FD states. At high B_z, these evolve into nondegenerate states of the lowest Landau level, which skip around equipotential contours of the confining potential with different quantised angular momenta, l_z; we observe $|l_z|$ values from 0 to 11 (Fig. 2a). In the limit of high B_z, these states correspond to states of the lowst Landau level, modified by the presence of the near-parabolic confining potential. A comparison of our data with the FD model reveals that the confining potential does not have exact

Journal of Physics: Conference Series **334** (2011) 012010 doi:10.1088/1742-6596/334/1/012010

circular symmetry: the orbital degeneracy of the first two excited states, $2p_{\pm}$, is lifted even at $B_z = 0$ (see the split peaks at $V \sim 1.29$ V). This lowered symmetry also explains the lifting of the three-fold orbital degeneracy of the next set of excited states around $V = 1.325$ V and the level anti-crossings, e.g. that at $V = 1.31$ V and $B_z = 8$ T.

From our measurements, we deduce that the electrons are tunnelling through the bound states of a deep and approximately parabolic circular potential well, see Figure 1(c). To understand how such a potential can be produced by low temperature annealing, we use Fick's law to calculate the density distribution of Mn_i ions that have diffused out of the p-(GaMn)As, assuming a constant surface concentration of Mn_i at the (GaMn)As/GaAs interface.

The band bending along z resulting from the diffused Mn_i^{2+} ions is calculated from Poisson's equation and shown in Figure 1(b). Although the conduction band edge E_c in the GaAs spacer layer is lowered relative to the unannealed case, it remains well above the Fermi energy E_F, so that the Mn_i^{2+} donors remain ionised. Despite the electric field in the intrinsic region of the diode, the barrier height, Q, for hopping of Mn ions between interstitial sites is only slightly reduced (<5%) by electrostatic effects. This is supported by experiment: we observe no change in the diffusion rate of Mn_i for annealing at 150°C in the presence of applied voltages of ±1 V.

We simulate numerically the distribution of Mn_i ions neglecting Coulomb interactions between them. This is a reasonable approximation since the Mn_i density in the GaAs spacer is relatively low, changing from $\sim 10^{19}$ cm^{-3} at the (GaMn)As interface to effectively zero at the AlAs interface (corresponding to a mean Mn_i separation >5nm and a Coulomb interaction energy much less than Q). Our simulated positions of individual Mn_i in the xy-plane are defined by random numbers corresponding to "white noise" of uniform distribution, with the ion density along the growth axis z given by the Fick law. The simulations use 3×10^6 Mn_i^{2+} ions occupying a volume of $5 \times 5 \times d$ μm^3, where d is the distance from the (GaMn)As/GaAs interface to the central QW plane. Figures 1(c) and 2(b) show the resulting electronic potential energy $U_{QW}(x,y)$ in the middle of the QW, for a central area of 0.3×0.3 μm^2, over which edge effects can be neglected. The potential energy contours in Figures 2(b) reveal a complex landscape. This arises from random spatial variations in the density profile of diffused Mn_i with nanoscale regions ("clusters") extending over ~10 nm, where the local density of Mn_i considerably exceeds the mean value. The "clusters" giving rise to the deepest potential minima typically contain up to ~10 randomly-placed Mn_i ions and form in the GaAs layer just below the (GaMn)As layer. A deep minimum in $U_{QW}(x,y)$ directly below such a cluster can be seen to left of center in Fig. 2(b). The lowest energy quantised states of this potential minimum are ~ 0.3 eV below E_c. Hence they provide resonant channels for electron tunnelling at bias voltages far below the flat band condition and give rise to the sharp peaks in $I(V)$. By solving numerically the 2-D Schrödinger equation of an electron in this minimum as a function of B_z, we obtain a calculated magnetospectrum shown in Figure 2(c). This is in good qualitative agreement with the measured spectrum in Figure 2(a).

The spatial forms of the QD wavefunctions are related directly to the potential landscape by Schrödinger's equation. By measuring a tunnel current through a particular QD as a function of $\boldsymbol{B}_{xy}$ aligned along different directions in the xy-plane, we obtain a spatial map of the probability density $|\psi(\boldsymbol{k})|^2$ of an eigenstate in the k_x - k_y plane of momentum space. It is reasonable to assume that the z-motion is fully quantised into the ground state of motion along this direction and that the wavefunction can be separated into z- and xy-parts. The Lorentz force on the tunnelling electrons allows us to tune their in-plane momentum, $\hbar\boldsymbol{k}$, according to the relation $\boldsymbol{k} = -e\boldsymbol{s} \times \boldsymbol{B}/\hbar$, where $s = 10 \pm 1$nm is the length of the tunnelling path from the emitter into the bound states of the QD. Whereas the ground state wavefunction has approximately circular symmetry, the wavefunction maps of the first two excited states have the lobed character of $2p_x$ and $2p_y$ orbitals rather than the toroidal shape of degenerate $2p_{\pm}$ orbitals. This is consistent with the observed zero field energy splitting of the $2p$-like states in Figure

2(a) and provides additional evidence that the symmetry of the confining potential is lower than that of the FD model.

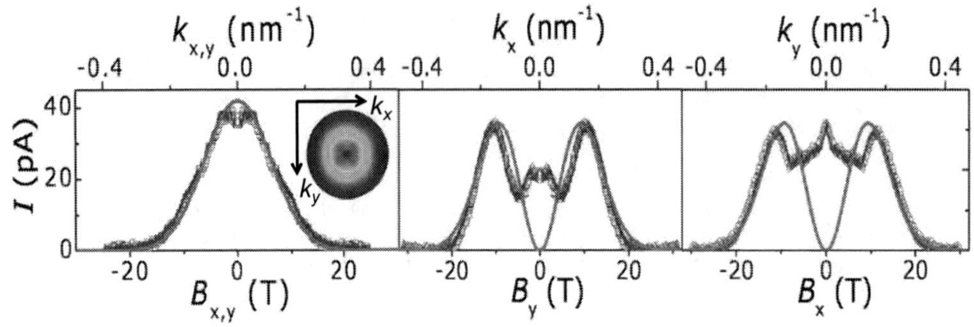

Figure 3 Plots along the symmetry axes k_x and k_y of $|\psi(\boldsymbol{k})|^2$ at B_{xy} up to 30 T for the 1*s*-like and $2p_{x,y}$ eigenstates of a deep potential minimum. Blue circles: data points; red curves: fits to simple harmonic oscillator potential with $\hbar\omega_0 = 22$ meV. The inset shows the circular symmetry of the 1*s*-like ground state wavefunction in the *xy*-plane.

Figure 3 probes the $|\psi(\boldsymbol{k})|^2$ probability density functions of the almost circular symmetric 1s state (see inset Fig. 3) and of the non-degenerate $2p_{x,y}$ orbitals along their respective symmetry axes using fields up to 30T. A fit to the simple harmonic oscillator (SHO) approximation gives $\hbar\omega_0 = 24\pm2$ meV, in agreement with the value 22 meV obtained independently from the data in Figure 2(a). However, the measured $|\psi(\boldsymbol{k})|^2$ of the $2p_{x,y}$ states also have a significant amplitude at $\boldsymbol{k}_{xy} \sim \boldsymbol{B}_{xy} = 0$, in contrast to the form expected for single particle SHO-states. We attribute this feature to the finite spread of *k*-states in the degenerate Fermi sphere of the emitter and to the electron-electron interactions between the tunnelling electron and nearby electrons in the emitter Fermi sea [20-22].

4. Observation of elongated quantum dots with quenched angular momenta

Many of the deep quantum dots that we have created and studied using our annealing method approximate quite well to Fock-Darwin states (Fig. 2a), indicating that they have approximately circular symmetry. However, for some of the dots the absence of orbital Zeeman splitting in the MTS spectra with $B \| z$ indicates that their orbital angular momentum is quenched even in fields of 30T and hence that the confining potential is far from circular. The presence of strongly elongated deep potential minima is evident from one of our simulations shown in Figure 4(a). Close to the minimum of the potential, the equipotentials are far from circular. Figure 4(b) shows the corresponding calculated energy spectra for the lowest energy levels of this confining potential. The energies of the lowest eigenstates are equally spaced to a reasonable approximation, indicating a harmonic potential. However, they have only weak magnetic field dependences indicating that orbital angular momentum of the 2p-like levels etc. is quenched even at fields of ~20 T. Figure 4(c) shows a grey scale conductance plot of the measured magnetotunnelling spectrum for such an elongated dot.

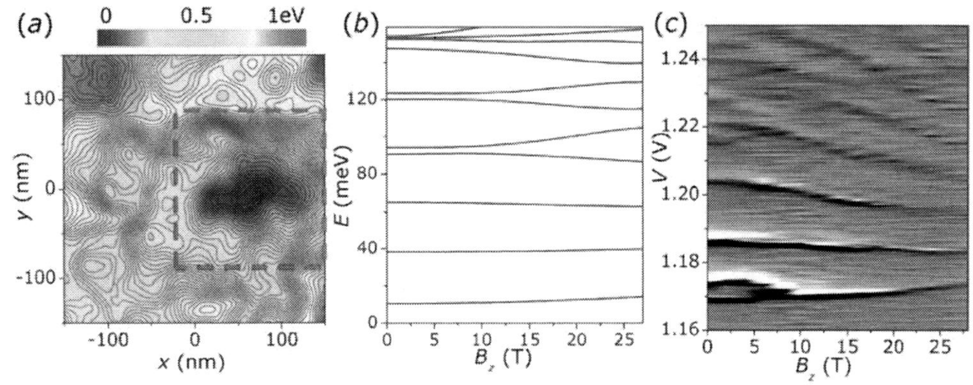

Figure 4
a) An elongated electrostatic potential minimum created by randomly distributed nanoclusters of Mn_i ions, as obtained from our numerical simulations.
b) B_z-dependences of the energy spectrum calculated in the area marked by the red square in plot ***(a)***. ***c)*** Grey-scale plot of dI/dV at $T = 2K$ versus B_z up to 28T for an elongated nanoelectrostatic QD.

Figure 5 is assembled from our numerical simulations by analysing the characteristics of 80 deep potential minima such as that shown in Figure 4(a). We use these to plot the probability of finding a dot with a particular value of the parameter ω_2/ω_1. The potential minima approximate well to $m_e^*(\omega_1^2x^2+\omega_2^2y^2)/2$, where ω_1 and ω_2 characterise an elliptical contour. It can be seen from the Figure 5 that only a small fraction of the dots have anisotropy ratio $\omega_1/\omega_2 < 0.6$. We measured in detail the magnetotunnelling spectra of 20 deep quantum dots. Most of these dots have energy spectra similar to the Fock-Darwin model, thus indicating an approximately circular symmetric potential. Only two of the measured samples have energy spectra that indicate large deviations from circular symmetry, as for the case shown in Figure 4, in good qualitative agreement with our model calculations.

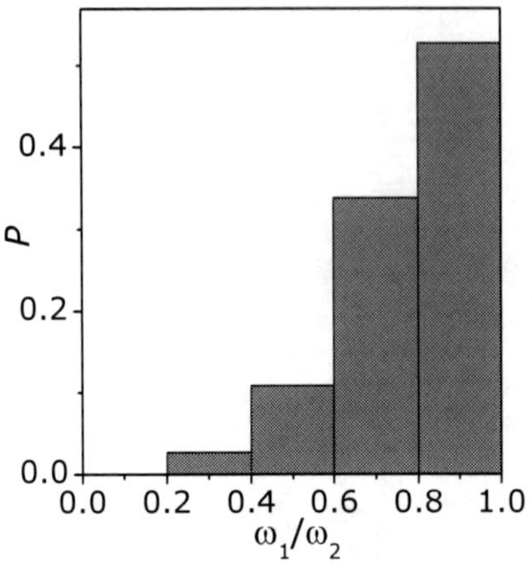

Figure 5
Probability of finding a deep QD potential minimum whose cross-section in the (x,y) plane is approximated by an ellipse with axis ratio ω_2/ω_1 ($\omega_2 \geq \omega_1$). The histogram is constructed from our simulations by examining the characteristics of 80 deep potential minima.

Finally, we address the question of what happens to the electrons after tunnelling from the n-type emitter into one of the deep potential minima. The typical current at a resonant peak (~10-100 pA) is consistent with electrons tunnelling out of the minima across the band gap and into the empty states at the top of the valence band in the GaMnAs layer. Future tunnelling experiments could help elucidate the nature of these states near the valence band edge of GaMnAs [18,24].

5. Conclusions

We have used high magnetic fields combined with resonant-tunnelling spectroscopy to study electronic properties formed by the controlled diffusion of doubly charged Mn_i donor ions in *p-i-n* resonant-tunnelling-diodes in which the capping *p*-layer is ferromagnetic (GaMn)As. Random clusters of these ions give rise to deep potential minima in the plane of the quantum well of the device. We observe a Fock-Darwin-like spectrum for the eigenstates of the minima with approximately circular symmetry, but we also observe elongated potential minima in which the orbital angular momentum of the bound electron states is quenched. These experiments provided insights into these "electrostatically formed" quantum dots and also into the nature of the diffusion process of Mn interstitial donors, which in turn has led on to our development of a new method of making submicron-size light-emitting diodes by means of laser assisted diffusion of Mn_i in *p-i-n* diodes [24].

References

[1] Snell B R, Chan K S, Sheard F W, Eaves L, Toombs G A, Maude D K, Portal J C, Bass S J, Claxton P, Hill G and Pate M A 1987 *Phys. Rev. Lett.* **59** 2806

[2] Leadbeater M L, Eaves L, Simmonds P E, Toombs G A, Sheard F W, Claxton P A, Hill G and Pate M A 1988 *Solid-State Electron.* **31** 707

[3] Hayden R K, Maude D K, Eaves L, Valadares E C, Henini M, Sheard F W, Hughes O H, Portal J C and Cury L 1991 *Phys. Rev. Lett.* **66** 1749

[4] Gennser U, Kesan V P, Syphers D A, Smith III T P, Iyer S S and Yang E S 1993 *Appl. Phys. Lett.* **63** 545

[5] Endicott J, Patanè A, Ibañez J, Eaves L and Bissiri M, 2003 *Phys. Rev. Lett.* **91** 126802

[6] Patanè A, Endicott J, Ibañez J, Brunkov P N, Eaves L, Healy S B, Lindsay A, O'Reilly E P and Hopkinson, M 2005 *Phys. Rev. B* **71** 195307

[7] Sakai J-W, Fromhold T M, Beton P H, Eaves L, Henini M, Main P C, Sheard F W and Hill G 1993 *Phys. Rev. B* **48** 5664

[8] Beton P H, Wang J, Mori N, Eaves L, Main P C, Foster T J and Henini M 1995 *Phys. Rev. Lett.*

75 1996
[9] Vdovin E E, Levin A, Patanè A, Eaves L, Main P C, Khanin Y N, Dubrovskii Y V, Henini M and Hill G 2000 *Science* **290** 122
[10] Patanè A, Hill R J A, Eaves L, Main P C, Henini M, Zambrano M L, Levin A, Mori N, Hamaguchi C, Dubrovskii Y V, Vdovin E E, Austing D G, Tarucha S and Hill G 2002 *Phys. Rev. B* **65** 165308
[11] Bester G, Reuter D, He Lixin, Zunger A, Kailuweit P, Wieck A D, Zeitler U, Maan J C, Wibbelhoff O and Lorke A 2007 *Phys. Rev. B* **76** 075338
[12] Wibbelhoff O S, Lorke A, Reuter D and Wieck A D 2005 *Appl. Phys. Lett.* **86** 092104
[13] Makarovsky O, Thomas O, Balanov A G, Eaves L, Patanè A, Campion R P, Foxon C T, Vdovin E E, Maude D K, Kiesslich G and Airey R J 2008 *Phys. Rev. Lett.* **101** 226807
[14] Makarovsky O, Balanov A G, Eaves L, Patanè A, Campion R P, Foxon C T and Airey R J 2010 *Phys. Rev. B* **81** 035323
[15] Johnson B L and Kirczenow G 2000 *Europhys. Lett.* **51** 367
[16] Jungwirth T, Wang K Y, Masek J, Edmonds K W, Konig J, Sinova J, Polini M, Goncharuk N A, MacDonald A H, Sawicki M, Rushforth A W, Campion R P, Zhao L X, Foxon C T and Gallagher B L 2005 *Phys. Rev. B* **72** 165204
[17] Edmonds K W, Boguslawski P, Wang K Y, Campion R P, Novikov S N, Farley N R S, Gallagher B L, Foxon C T, Sawicki M, Dietl T, Nardelli M B and Bernholc J 2004 *Phys. Rev. Lett.* **92** 037201
[18] Thomas O, Makarovsky O, Patanè A, Eaves L, Campion R P, Edmonds K W, Foxon C T and Gallagher B L 2007 *Appl. Phys. Lett.* **90** 082106
[19] Itskevich I E, Ihn T, Thornton A, Henini M, Foster T J, Moriarty P, Nogaret A, Beton P H, Eaves L and Main P C 1996 *Phys. Rev. B* **54** 16401
[20] Geim A K, Main P C, La Scala N, Eaves L, Foster T J, Beton P H, Sakai J W, Sheard F W, Henini M, Hill G and Pate M A 1994 *Phys. Rev. Lett.* **72** 2061
[21] Hapke-Wurst I, Zeitler U, Frahm H, Jansen A G M, Haug R J and Pierz K 2000 *Phys. Rev. B* **62** 12621
[22] Vdovin E E, Khanin Yu N, Makarovsky O, Dubrovskii Yu V, Patanè A, Eaves L, Henini M, Mellor C J, Benedict K A and Airey R 2007 *Phys. Rev. B* **75** 115315
[23] Jungwirth T, Sinova S, MacDonald A H, Gallagher B L, Novák V, Edmonds K W, Rushforth A W, Campion R P, Foxon C T, Eaves L, Olejník E, Masek J, Yang S-R Eric, Wunderlich J, Gould C, Molenkamp L W, Dietl T and Ohno H 2007 *Phys. Rev. B* **76** 125206
[24] Makarovsky O, Kumar S, Rastelli A, Patanè A, Eaves L, Balanov A G, Schmidt O G, Campion R and Foxon C T 2010 *Advanced Materials* **22** 3176-3180

Journal of Physics: Conference Series **334** (2011) 012011 doi:10.1088/1742-6596/334/1/012011

Magneto-optical cavity quantum electrodynamics effects in quantum dot - micropillar systems

S. Reitzenstein[1], S. Münch[1], P. Gold[1], P. Franeck[1], A. Rahimi-Iman[1], A. Löffler[1], S. Höfling[1], L. Worschech[1], I. V. Ponomarev[2], T. L. Reinecke[2] and A. Forchel[1]

[1]Technische Physik, University of Würzburg, Am Hubland, D-97074 Würzburg, Germany
[2]Naval Research Laboratory, Washington, DC 20375, USA

E-mail: `stephan.reitzenstein@physik.uni-wuerzburg.de`

Abstract. We report on magneto-optical studies of strongly coupled quantum dot – micropillar cavity systems. Large $In_{0.3}Ga_{0.7}As$ quantum dots (QDs) in the active layer of the micropillar facilitate the observation of strong coupling. In addition, they exhibit a particular large diamagnetic response which is exploited to demonstrate magneto-optical resonance tuning in the strong coupling regime. The magnetic field employed in Faraday configuration induces a transition from strong coupling towards the critical coupling regime which is explained in terms of a magnetic field dependent oscillator strength of the $In_{0.3}Ga_{0.7}As$ QDs. We further study the coherent interaction between spin resolved states of the QDs and microcavity photon modes. A detailed oscillator model is used to extract the associated coupling parameters of the individual spin and cavity modes and reveals an effective coupling between photon modes that is mediated by the exciton spin states.

1. Introduction

Following the first demonstration of strongly coupled quantum dot (QD)-microcavity systems enormous effort has been devoted to the field of cavity quantum electrodynamics (cQED) [1, 2]. For instance cQED experiments in semiconductor nanostructures have been motivated by an interest in demonstrating quantum optics effects such as the coherent exchange of energy between quantum emitters and resonant modes of optical cavities on a solid state platform, which can open opportunities for large scale integration in technologies such as quantum information[3, 4]. To date most experimental studies involving strong coupling in QD-microcavity systems have relied on temperature tuning [1, 2] or electro-optical resonance tuning based on the quantum confined Stark effect [5].

Here we demonstrate that an external magnetic field provides an additional degree of freedom to more fully explore the potential of coherently coupled QD-microcavities. A magnetic field induced diamagnetic shift and Zeeman splitting as well as magnetic localization of carriers present a means of controlling the transition energy [6, 7], the spin-state and the oscillator strength [8], respectively, of a QD interacting with a microcavity. In this sense the magnetic field not only acts as a tuning parameter but also opens a way of modifying in situ the coupling strength of the interacting system. A prerequisite for this in situ manipulation is the presence of $In_{0.3}Ga_{0.7}As$ QDs that are extended laterally in the active layer of the micropillar. Their large diamagnetic coefficient of up to 30 $\mu eV/T^2$ facilitates the exploitation of the diamagnetic shift as a tuning

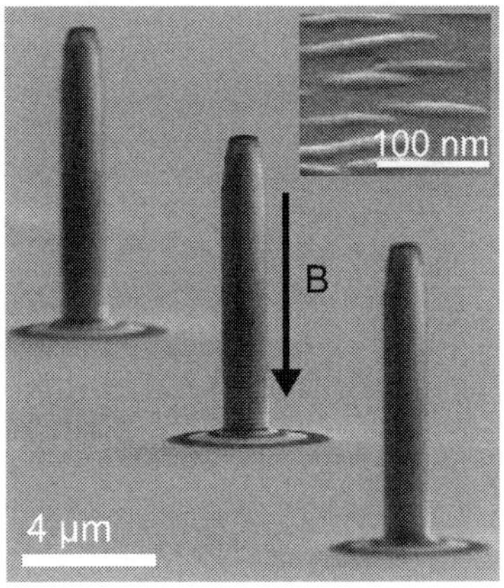

Figure 1. Scanning electron microscope (SEM) image showing micropillar cavities with a diameter of 1.3 μm. Inset: SEM image of laterally extended $In_{0.3}Ga_{0.7}As$ QDs.

parameter. The additional confinement introduced by the external magnetic field controls the oscillator strength of the large QDs from values close to 50 down to about 25 for the highest applied magnetic field (5 T) which induces a transition from the strong coupling regime towards the critical coupling regime when the magnetic field is increased. In addition, we address spin related cQED effects in a coherently coupled QD-micropillar system. The spin degree of freedom is accessed when a non-zero magnetic field is applied so that the spin degeneracy of the QD exciton is lifted. This allows us to study the coupling of individual spin states to cavity photon modes.

2. Technology

We have investigated high quality (Q) factor micropillar cavities based on planar GaAs/AlAs microcavities grown by molecular beam epitaxy. The planar distributed Bragg reflector (DBR) structure consists of 23 and 27 $\lambda/4$ thick GaAs/AlAs mirror pairs on the surface and the substrate side of an one-λ thick GaAs cavity layer, respectively. At the center of the GaAs cavity, a single layer of $In_{0.3}Ga_{0.7}As$ QDs is introduced during the epitaxial growth. The growth of the QDs was optimized in order to realize large QDs (in-plane extension of about 40 nm), providing increased dipole moments and oscillator strengths, which facilitate the observation of the strong coupling regime. Subsequent to the growth of the planar microcavities, high Q-factor micropillar cavities were realized by means of high resolution electron beam lithography and plasma etching in an electron cyclotron resonance system. We have realized micropillar cavities with diameters ranging from 1.0 μm to 2.0 μm, providing Q factors $Q = E_c/\gamma_c$ up to 15,000, where E_c denotes the mode energy and γ_c the full width at half-maximum (FWHM) of the mode. We refer to Refs. [9, 10] for details on the fabrication of the micropillar cavities.

Magneto-optical studies were performed at low temperature using a high resolution microphotoluminescence (μPL) setup. The sample was mounted to the cold-finger of a He flow cryostat providing magnetic fields up to 5 T in Faraday configuration. This means that the magnetic field was applied collinearly to the symmetry axis of the micropillars as sketched in Fig. 1. The sample was excited by a frequency doubled Nd:Yag laser emitting at 532 nm. The PL was detected using a nitrogen cooled Si charge coupled device camera.

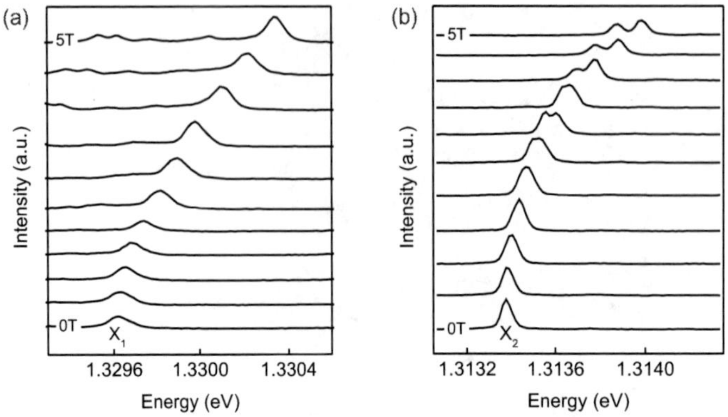

Figure 2. Magnetic field dependent μPL spectra of $In_{0.3}Ga_{0.7}As$ quantum dots. (a) Magnetic response of a quantum dot exciton line X_1. The QD line exhibits a large diamagnetic constant of 29 $\mu eV/T^2$ and a Landé g-factor close to zero. (b) Magnetic response of another QD exciton X_2 showing a slightly smaller diamagnetic constant (22 $\mu eV/T^2$) if compared to the exciton lines shown in panel (a). The QD exciton line splits into two Zeeman split components at large magnetic fields. From the splitting a Landé g-Factor $g_L = 0.46$ is determined.

3. Experimental

3.1. Magnetic Response of Large $In_{0.3}Ga_{0.7}As$ Quantum Dots

Magneto-optical studies allow one to probe the diamagnetic shift as well as the Zeeman-splitting of QDs, where the magnitude of the diamagnetic shift depends on the lateral extension of the electron and hole wavefunction. In fact, provided that the cyclotron energy $\hbar\omega_c$ is much smaller than the exciton binding energy, which is the case in the present experiment, the diamagnetic shift of QD excitons ΔE_{dia} can be expressed via [11]

$$\Delta E_{dia} = \kappa_x B^2 = \frac{e^2}{8}\left(\frac{\langle\rho_e^2\rangle}{m_e} + \frac{\langle\rho_h^2\rangle}{m_h}\right) B^2, \tag{1}$$

where $\langle\rho_{e,h}^2\rangle$ denotes the average quadratic extension of the electron (heavy hole) wavefunction and $m_{e,h}$ represents the in-plane mass of the electron (hole). In Fig. 2 we present magnetic field dependent μPL spectra of large $In_{0.3}Ga_{0.7}As$ QDs. The QD exciton line in panel (a) shows a pronounced blue shift of emission when the magnetic field increases from 0 to 5 T. From an lineshape analysis of the emission lines we determine a diamagnetic constant $\kappa_{X1} = 29$ μeV. The large diamagnetic shift of these type of QDs associated with diamagnetic constants between about 20 and 30 $\mu eV/T^2$ if compared to standard $In_{0.6}Ga_{0.4}As$ QDs exhibiting diamgnetic shifts close to 10 $\mu eV/T^2$ reflects clearly their enhanced lateral extension. While no significant Zeeman splitting can be observed of the exciton lines presented in Fig. 2(a), a clear splitting is observed for the exciton line investigated in Fig. 2(b) showing a slightly smaller diamagnetic constant $\kappa_{X2} = 22$ μeV. From the Zeemann splitting ΔE_Z we determine a Landé g-factor $g_X = 0.46$ according to $\Delta E_Z = g_X \mu_B B$. Comparing these diamagnetic constant and Landé g-factors of these QD lines suggests that QDs with a smaller diamagnetic shift exhibit larger Landé g-factors. Indeed, an statistical evaluation of the magnetic response of $In_{0.3}Ga_{0.7}As$ QDs revealed an anti-correlation between the Landé g-factor and the diamagnetic shift.

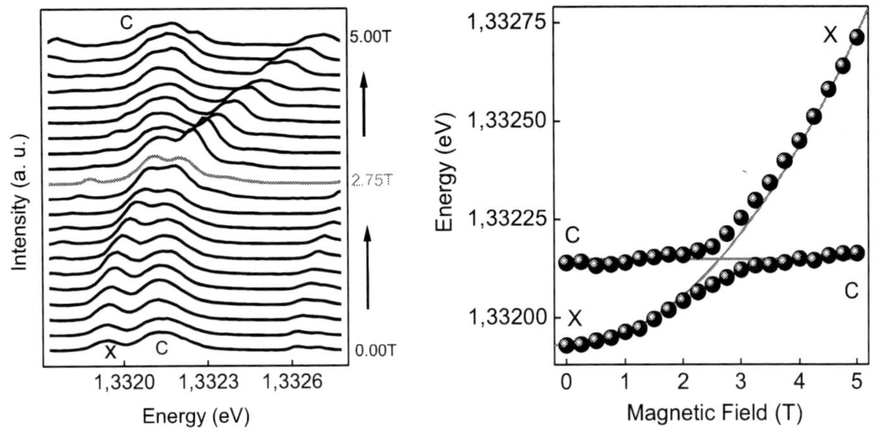

Figure 3. Magneto-optical tuning in the strong coupling regime. (a) Magnetic field dependent μPL spectra of a QD - micropillar with a diameter of 1.6 μm and a Q factor of about 10000 at 10 K. The coherent interaction between the QD exciton X and the photon mode C is reflected in a normal mode splitting on resonance at B = 2.75 T. (b) Energy dispersion of the strongly coupled QD - micropillar system. On resonance a vacuum Rabi-splitting of 110 μeV is observed.

3.2. Magneto-Optical Resonance Tuning

In this section we address magneto-optical resonance tuning in the strong coupling regime. The large QDs are very suitable for this study as they provide large light-matter coupling constant and also feature a large κ_x which allows for a sufficiently large tuning range of almost 1 meV for magnetic fields between 0 ant 5 T. This tuning range is similar to those obtained typically for temperature or electro-optical resonance tuning [1, 2, 5].

Magneto-optical tuning is presented in Fig. 3(a) which shows magnetic field dependent μPL spectra of a 1.6 μm diameter micropillar at 10 K. At B = 0 the exciton is located at the low energy side of the cavity mode. With increasing magnetic field the exciton experiences a diamagnetic blue shift and resonance with the cavity mode is achieved at $B = 2.75$ T. The avoided crossing of the excitonic and the photonic mode is a clear signature of the strong coulping regime. The avoided crossing of the interacting modes can also be seen in Fig. 3(b) which shows the corresponding energy dispersions. The energy of the cavity mode is not affected by the magnetic field. In contrast, the exciton line shows a blue shift of 780 μeV between 0 and 5 T which corresponds to $\kappa_x = 31\ \mu eV/T^2$. The coherent interaction of the exciton and photon mode is reflected in a vacuum Rabi-splitting of $\Delta E_r = 110\ \mu eV$ at $B = 2.75$ T.

3.3. Magnetic Field Control of the Coupling Strength

The magnetic field introduces an additional confinement of the carriers in the elongated QDs which is expected to change to dipole moment of the QD excitons and, thus, the light-matter coupling strength. In order to investigate the light-matter coupling strength as a function of magnetic field we recorded PL emission spectra at resonance for magnetic fields between 0 and 5 T. The corresponding spectra are depicted in Fig. 4. The mode splitting decreases with increasing magnetic field and for magnetic field exceeding 3 T a single, broad emission peak evolves at the resonance energy E_{res}. Lorentzian lineshape fitting the experimental allows us to

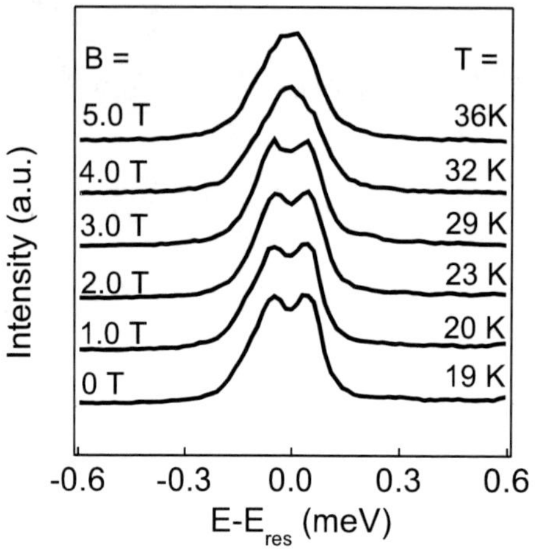

Figure 4. Normalized μPL spectra at resonance for magnetic fields between 0 and 5 T.

extract the vacuum Rabi splitting ΔE_R from which the light-matter coupling constant g can be determined via the relation $\Delta E_R = 2\sqrt{\left(g^2 - \frac{(\gamma_c-\gamma_x)^2}{16}\right)}$, where γ_c (γ_x) denote the cavity (exciton) mode linewidth. For small magnetic fields ΔE_R the QD exciton is confined within the extended potential of the QD and the coupling with the cavity mode results in a vacuum Rabi splitting of about 100 μeV and a coupling constant of about 60 μeV. For magnetic field exceeding 3 T the exciton wavefunction becomes localized and squeezed by the additional magnetic confinement when the magnetic length falls below the characteristic size of the QD potential [12]. This leads to a reduction of the diplole moment and the oscillator strength of the QD exciton which in turn explains the significant decrease of the coupling constant and the vacuum Rabi splitting to values of $g = 44$ μeV and $E_R = 65$ μeV, respectively. The reduction of the coupling strength is related to a decrease of the QD oscillator strength from about 45 at 0 T down to about 25 at 5 T which was confirmed independently by time resolved μPL measurements [12]. This results show that the interaction strength of the strongly coupled QD-microcavity system can be controlled in situ by an external magnetic field.

3.4. Exciton Mediated Photon-Photon Coupling

In this section we study the interaction of spin degree of freedom of excitons in InGaAs quantum dots with the photon cavity modes in pillar microcavities, and we address the question of how the polarization of the modes affects their coupling. The spin states are split with a magnetic field in Faraday configuration, and their interactions with individual cavity modes are identified

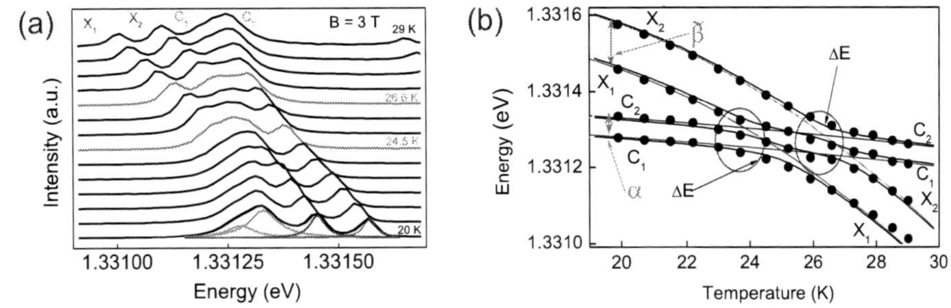

Figure 5. (a) Temperature-dependent μPL spectra of micropillar with a diameter of 1.6 μm 3 T. The zeeman split QD exciton lines X_1 and X_2 interact coherently at 24.5 K and 26.6 K, respectively, with the two linearly polarized components C_1 and C_2 of the fundamental cavity modes. (b) Corresponding energy dispersions obtained from Lorentzian lineshape fitting.

in magneto-photoluminescence experiments.

Spin resolved light-matter interaction is explored by temperature dependent μPL measurements at 3 T. Fig. 5(a) shows sets of μPL spectra of a 1.6 μm diameter micropillar at $B = 3$ T. The cavity peak has a distinctive left shoulder that can be deconvolved into two unequal Lorentzians (C_1 and C_2). Both mode components are split by $\alpha \approx 50$ μeV and have a Q factor of about 15000. The splitting of the fundamental cavity mode into two orthogonal linearly polarized mode components originates from the processing of the micropillars [9].

At $B = 0$ T the exciton shows a fine structure splitting of $\beta = 20$ μeV. Spin related effects come into play when a magnetic field splits the excitonic line into two components X_1 and X_2. The magnetic field also changes their polarizations, and a total Zeeman splitting of $\Delta E_{Zeem} =$ 117 μeV at $B = 3$ T can be seen in the lowest trace of Fig. 5(a). Magnetic field dependent measurements at constant temperature give a diamagnetic coefficient of $\kappa = 22$ μeV/T^{-2} for the QD exciton, which reflects again the extended electron-hole wavefunction of these large oscillator strength QDs [12].

The coupling of spin-related excitonic emission lines with photonic modes of a microcavity is studied at 3 T. Both exciton lines shift through resonance with the cavity modes when the sample temperature is increased from 20 K to 29 K (cf. Fig. 5(a)). Minima of the splitting in energy of the outermost emission lines are observed at about 24.5 K and 26.5 K, respectively. Significant broadening and splitting (marked by an arrows) of the associated emission lines occurs when the exciton components enter the interaction regime. The corresponding energy dispersions are plotted in Fig. 5(b). At both resonance temperatures anticrossings of all involved modes indicate unambiguously the coherent cQED coupling with the Zeeman split exciton components in the spin resolved cQED regime.

In order to analyze the interacting system we employ a detailed coupled oscillator model of the exciton and cavity modes described in Ref. [13]. The model considers a basis of cylindrically symmetric cavity modes and cylindrically symmetric QD excitonic states and experimental parameters α and β which determine the mixings of the two cavity modes and the two exciton

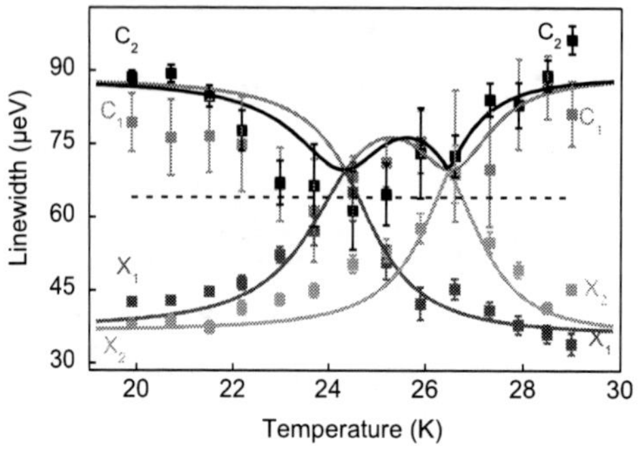

Figure 6. Emission mode linewidths of the experimental data presented in Fig. 5(a).

modes, respectively. The only free parameter is the coupling constant g between cavity and exciton states. It is interesting to note that a strong magnetic field adds an additional splitting to the fine structure splitting β and tends to restore the circular polarization. At intermediate fields excitons are neither linearly nor circularly polarized, with a polarization governed by ratio of parameters δ/β, where $\delta = g_L \mu_B B$ is the bare Zeeman splitting, and g_L is the Landé g-factor. In the present experiment $\delta/\beta = 6.25$ for $B = 3$ T. The total splitting of two excitons is determined by the parameter $\tilde{\beta} = \sqrt{\beta^2 + \delta^2}$.

The real and imaginary parts of the complex eigenenergies of the coupled modes as functions of T are calculated from the four oscillator model and are potted Fig. 5(b) and Fig. 6, respectively, along with the experimental data. The best agreement was obtained for $g = 38$ μeV for which the theory (solid lines) describes the experimental data (dots) very well. Additional insight into the physics of the coupled exciton spin-photon system can be obtained from the mode linewidths as a functions of temperature. The calculated imaginary parts of the energy-eigenvalues are shown in Fig. 6 together with the experimental results obtained fitting of the data in Fig. 5(a). Pronounced mixing of the lines occurs for temperatures between $T \approx 24.5$ K and $T \approx 26.5$ K where the exchange of linewidths between the X transitions and the C cavity modes indicates coherent interaction in the strong coupling regime.

Interestingly, the linewidths show considerable involvement of the C_2 (C_1) mode near the anticrossing at 24.5 K (26.5 K). Indeed, the crossing point of the X and the C linewidths is shifted upwards compared to the expected value $(\Gamma_C + \Gamma_X)/2 = 64$ μeV (cf. dashed line in Fig. 6) for strong coupling with only a single photonic mode. This effect is clearly observed for the experimental data the 26.5 K and means that the photonic character of the coupled modes is increased when two photon modes are involved in the spin related light-matter interaction. These observations demonstrate that there is an effective interaction between the two photon states mediated by the quantum dot exciton. The physical mechanism for this coupling is the photon-exciton interaction given by the parameter g.

4. Conclusions

In summary, we have presented magnetooptical QD-micropillar systems in the strong coupling regime. The large diamagnetic coefficient of large $In_{0.3}Ga_{0.7}As$ QDs with values up to 30 $\mu eV/T^2$ facilitate the exploitation of the diamagnetic shift as a tuning parameter. The additional confinement introduced by the external magnetic field allows us to control the oscillator strength of the large QDs from values close to 50 down to about 25 for $B = 3$ T which induces a transition from the strong coupling regime towards the weak coupling regime. Moreover, we have addressed spin related cQED effects in solid state by demonstrating strong (coherent) coupling between excitonic spin states and high quality factor photon modes of a pillar microcavity structure. We find a coupling of independent photonic modes via the excitonic transition in addition to the coupling between the spin states of the excitons and the cavity modes. Such a coupling is interesting from a fundamental point of view, and nonlinear interactions mediated by quantum dots are also of considerable interest in connection with quantum communications and quantum information.

5. Acknowledgments

The authors acknowledge financial support by the Deutsche Forschungsgemeinschaft via the Research Group Quantum Optics in Semiconductor Nanostructures,, the EU via the project "TREASURE", the BMBF through the project NanoQuit, the State of Bavaria and the U.S. Office of Naval Research. We thank M. Emmerling and A. Wolf for technical assistance.

6. References

[1] J. P. Reithmaier et al., Nature **432**, 197 (2004).
[2] T. Yoshie et al., Nature **432**, 200 (2004).
[3] I. Cirac et al., Phys. Rev. Lett. **78**, 3221 (1997).
[4] A. Imamoglu et al., Phys. Rev. Lett. **83**, 4204 (1999).
[5] C. Kistner et al., Optics Express **16**, 15006 (2008).
[6] J. Tignon et al., Phys. Rev. Lett. **74**, 3967 (1995).
[7] M.-F. Tsai et al., Phys. Rev. Lett. **101**, 267402 (2008).
[8] M. Lomascolo et al., Appl. Phys. Lett. **74**, 676 (1999).
[9] S. Reitzenstein et al., Appl. Phys. Lett. **90**, 251109 (2007).
[10] S. Reitzenstein, and A. Forchel, J. Phys. D: Appl. Phys. **43**, 033001 (2010).
[11] S. N. Walck, and T. L. Reinecke, Phys. Rev. B **57**, 9088 (1998).
[12] S. Reitzenstein et al., Phys. Rev. Lett. **103**, 127401 (2009).
[13] S. Reitzenstein et al., Phys. Rev. B **82**, 121306 (R) (2010).

Additional Evidence for the Surface Origin of the Peculiar Angular-Dependent Magnetoresistance Oscillations Discovered in a Topological Insulator $Bi_{1-x}Sb_x$

A A Taskin, Kouji Segawa and Yoichi Ando[1]

Institute of Scientific and Industrial Research, Osaka University
Ibaraki, Osaka 567-0047, Japan

E-mail: y_ando@sanken.osaka-u.ac.jp

Abstract. We present detailed data on the unusual angular-dependent magnetoresistance oscillation phenomenon recently discovered in a topological insulator $Bi_{0.91}Sb_{0.09}$. Direct comparison of the data taken before and after etching the sample surface gives compelling evidence that this phenomenon is essentially originating from a surface state. The symmetry of the oscillations suggests that it probably comes from the (111) plane, and obviously a new mechanism, such as a coupling between the surface and the bulk states, is responsible for this intriguing phenomenon in topological insulators.

1. Introduction

Topological insulators [1-3] are an emerging class of materials that host a new quantum-mechanical state of matter [4-7] where an insulating bulk state supports an intrinsically metallic surface state that is "topologically protected", meaning that it is stable against any disorder that does not break time-reversal symmetry. Recently, angle-resolved photoemission spectroscopy (ARPES) studies on a cleaved trigonal surface of $Bi_{1-x}Sb_x$ have revealed that the energy dispersions of its surface states possess the distinctive character to qualify this material as a topological insulator [8-10]. However, to directly probe the unique properties of the surface states and to elucidate whether they could really be exploited on a macroscopic level, transport and magnetic studies of high-quality single crystals are indispensable. Unfortunately, in "real-life" samples of topological insulators available today, there is always some bulk conductivity due to residual carriers, and separating the contributions from two-dimensional (2D) and three-dimensional (3D) states turns out to be challenging [11-13].

In this context, we have recently succeeded in observing both the de Haas-van Alphen (dHvA) oscillations [14] and the Shubnikov-de Haas (SdH) oscillations [15] in high-quality bulk single crystals of $Bi_{1-x}Sb_x$ alloy in the "insulating" regime ($0.07 \leq x \leq 0.22$), which is the first material to be known as a 3D topological insulator. These observations became possible by growing highly pure and homogeneous single crystals of this alloy and achieving the bulk carrier density of the order of 10^{16} cm^{-3}. The dHvA and SdH oscillations signified a previously-unknown Fermi surface (FS) with a clear 2D character that coexists with a 3D bulk FS [14,15]. Since $Bi_{1-x}Sb_x$ is a 3D material, the observed 2D FS is naturally assigned to the "surface", which could be internal surfaces such as twin boundaries.

[1] To whom any correspondence should be addressed.

Furthermore, we have extended our measurements to the angular dependence of the magnetoresistance (MR), which was successfully applied to the studies of quasi-2D organic conductors in the 1980s and lead to the discovery of the celebrated angular-dependent magnetoresistance oscillations (AMRO) [16-18]. Intriguingly, in our angular-dependent MR studies, we found oscillatory angular dependences in both the resistivity ρ_{xx} and the Hall resistivity ρ_{yx} [15]. The oscillations observed at lower fields were elucidated to be essentially a manifestation of the SdH oscillations, whereas the ones observed at higher fields were obviously of novel origin [15]. We proposed that the latter originates from the topological surface state on the cleaved (111) surface of $Bi_{1-x}Sb_x$, where the coupling between the surface and bulk states are probably playing a key role in the peculiar MR oscillation phenomenon [15].

In this paper, we present additional evidence to demonstrate that the "high-field" oscillations are most likely due to the exposed surface, by directly comparing the oscillations before and after the sample surface was chemically etched.

2. Experimental

High-quality $Bi_{0.91}Sb_{0.09}$ single crystals were grown from a stoichiometric mixture of high-purity (6N) Bi and Sb elements by a zone melting method. The resistivity was measured by a standard four-probe method on a rectangular sample, where the current was directed along the C_1 axis. Continuous rotations of the sample in constant magnetic fields were used to measure the angular dependence of the MR within the trigonal-binary (C_3-C_2) crystallographic plane. Magnetic fields up to 16 T were applied using a dc superconducting magnet in a ^{4}He cryostat. All the data shown here were taken at the lowest temperature of 1.5 K. To obtain a fresh surface on the $Bi_{1-x}Sb_x$ single crystals, we applied the following chemical etching procedure:

1) Keep for several minutes in the 1:4:9 mixture of HNO_3, CH_3COOH, and H_2O.
2) Rinse in distilled water.
3) Keep for several minutes in the diluted (50 mol-%) HCl.
4) Rinse in distilled water and dry.

3. Results

3.1. Angular-Dependent MR Oscillations

Figures 1(a) and 1(b) show the angular dependences of the transverse MR and the Hall resistivity,

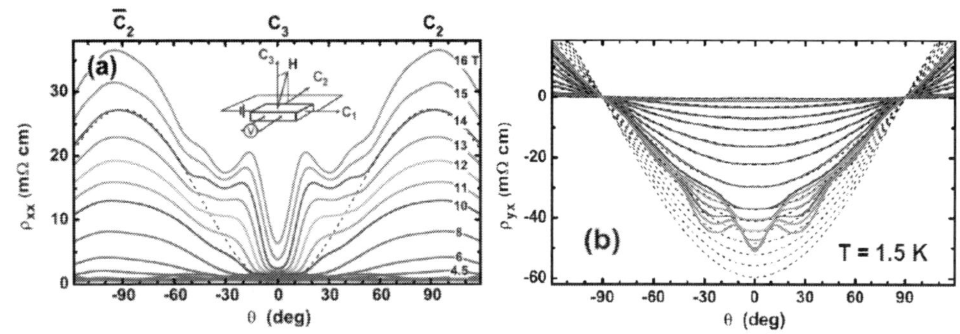

Figure 1. (a) Angular dependences of ρ_{xx} measured in the trigonal-binary (C_3-C_2) plane in constant magnetic fields. The dashed line is the MR background ~ (1-cos$^2\theta$) for B = 14 T. Inset depicts the measurement geometry, where θ = 0° corresponds to the C_3 axis. (b) Angular dependences of ρ_{yx} measured in the same conditions as in (a). The dashed lines are the expected angular dependences of the Hall effect plotted for all magnetic fields.

respectively, measured in magnetic fields rotated within the trigonal-binary (C_3-C_2) plane perpendicular to the current. The magnetic-field strength was kept constant during each rotation. As can be seen in Figs. 1(a) and 1(b), both ρ_{xx} and ρ_{yx} present an oscillatory behavior as a function of the rotation angle θ. As we have elaborated in our previous paper [15], two different types of oscillations can be distinguished: The first type [though not very clearly seen in Fig. 1] consists of oscillations appearing at lower fields, while the second one becomes prominent at higher fields ($B \geq 10$ T). The former "low-field" oscillations were demonstrated [15] to be essentially due to the SdH oscillations originating from the 2D state residing on the (2,-1,-1) plane, which was unambiguously observed in the $\rho_{xx}(B)$ data for magnetic-field sweeps in fixed magnetic-field directions.

The focus of the present paper is the "high-field" oscillations, which develop on a smooth field-dependent background coming from the anisotropy of $\rho_{xx}(B)$ along the different axes. An example of the fitting of the background for $B = 14$ T is shown by the dashed line in Fig. 1(a). Because of a large MR background in strong magnetic fields, only largest peaks in $\rho_{xx}(\theta)$ can be clearly seen in the raw data [Fig. 1(a)].

The $\rho_{yx}(\theta)$ data shown in Fig. 1(b) also present pronounced angular-dependent oscillations at high fields. The "background" for $\rho_{yx}(\theta)$ is simply the angular dependence of the Hall effect, $R_H B\cos\theta$, where R_H is the Hall coefficient. As can be clearly seen in Fig. 1(b), low-filed $\rho_{yx}(\theta)$ data follow this expected angular dependence very closely (we use $R_H = -37$ cm^3/C, obtained from the Hall measurements), and the large deviation from this simple behavior is observed only in magnetic fields above 10 T. Figure 2 shows "pure" oscillations in $\Delta\rho_{yx}(\theta)$ after subtracting the $R_H B\cos\theta$ contribution from $\rho_{yx}(\theta)$. One can clearly see a set of peaks, which are marked by short vertical ticks in Fig. 2. They are symmetric with respect to the C_3 axis and show a rather complicated magnetic-field dependence.

3.2. *Surface-Condition Dependence*

The data shown in Figs. 1 and 2 were taken on a refreshed surface immediately after chemically etching the sample. To investigate the effect of etching, we had conducted the same angular-dependent MR measurements on the same sample before etching. Incidentally, this sample was characterized 7 months before those measurements, and it was kept in a desiccator, exposing its surface to dry air for 7 months. Our intention was to see how such an "old" sample behaves. Intriguingly, the angular-dependent MR oscillations were smeared, as shown in Fig. 3 for ρ_{xx}. Figure 4 presents a direct comparison of the oscillations in ρ_{xx} and ρ_{yx} before and after etching. It is evident that the refreshed surface yields more pronounced oscillations, which gives direct evidence that the peculiar angular-

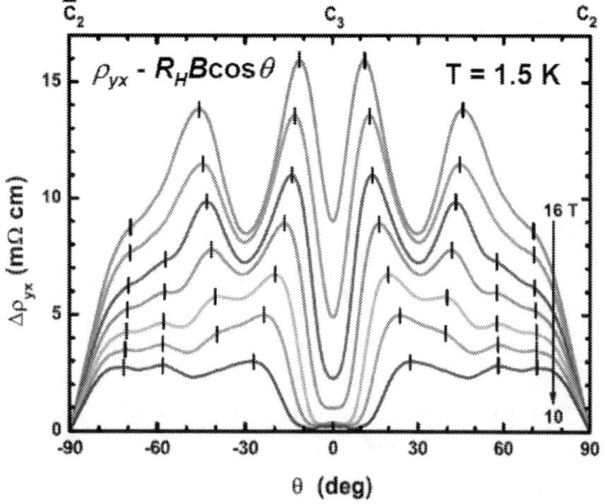

Figure 2. "High-field" angular-dependent oscillations in $\Delta\rho_{yx}(\theta)$ obtained by subtracting the expected $R_H B\cos\theta$ behavior from the ρ_{yx} data shown in Fig. 1(b). Ticks mark the positions of distinguishable peaks.

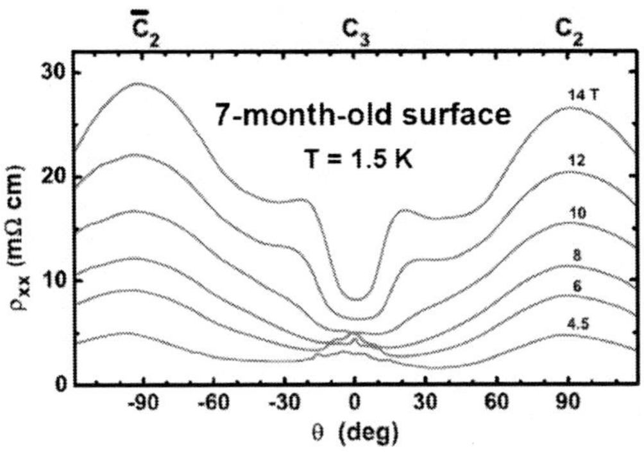

Figure 3. Angular dependences of ρ_{xx} measured in the trigonal-binary (C_3-C_2) plane on the 7-month-old surface of the same $Bi_{0.91}Sb_{0.09}$ sample before etching. Note that the angular-dependent MR oscillations are smeared compared to those shown in Fig. 1(a).

dependent oscillations are essentially originating from a surface state.

4. Discussions

In the "high-field" oscillations, an important feature is that the amplitude of the peaks weakens as the magnetic field is rotated away from the C_3 axis, which is somewhat reminiscent of the behavior of the ordinary AMRO in quasi-2D systems if the conduction planes lie perpendicular to the C_3 axis [16-18]. Thus, it is probable that the "high-field" oscillations are coming from the (111) plane (which is perpendicular to the C_3 axis), where surface states are seen in photoemission [8-10] and tunnelling [19] experiments. Another distinguishable feature of the "high-field" oscillations is that they survive up to rather high temperatures [15]. For example, even at 40 K there are still visible traces of oscillations while the SdH oscillations are already gone at this temperature [15], which is reminiscent of the behaviour of the ordinary AMRO in quasi-2D systems. In spite of these similarities to the quasi-2D AMRO, the peak positions of the "high-field" oscillations apparently shift with the magnetic field, which is not expected for the ordinary AMRO. Moreover, the existence of a finite coupling between

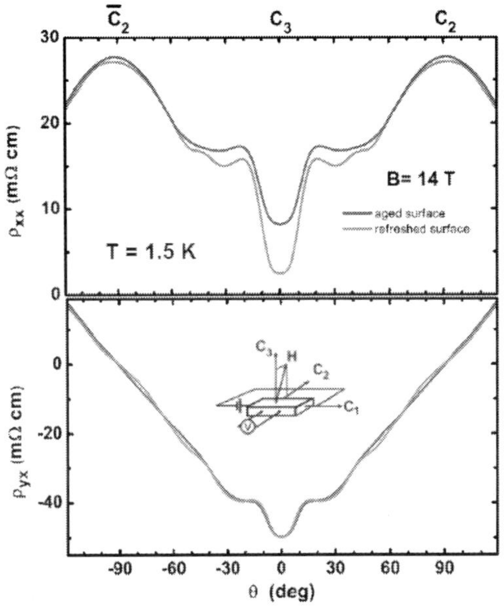

Figure 4. Direct comparison of the oscillations in ρ_{xx} and ρ_{yx} before and after etching the sample surface, measured on the same $Bi_{0.91}Sb_{0.09}$ sample. It is evident that the refreshed surface yields more pronounced oscillations, which gives evidence for the surface origin of these oscillations.

conduction planes is essential for the quasi-2D AMRO [20], but there is no such inter-plane coupling for the surface states as long as the crystal is thick enough. Therefore, the observed "high-field" angular oscillations are a new phenomenon apparently specific to topological insulators.

5. Conclusion

We present detailed data on the angular-dependent MR oscillation phenomenon which was recently discovered in a topological insulator $Bi_{0.91}Sb_{0.09}$. Direct comparison of the data taken before and after etching the sample surface gives compelling evidence that this novel phenomenon is essentially originating from a surface state. The symmetry of the oscillations suggests that it probably comes from the (111) plane. In the surface state of a topological insulator, there is no "quasi two-dimensionality" that introduces a finite warping to the 2D cylindrical Fermi surface, whereas the existence of such a warping is essential for the ordinary AMRO to occur; therefore, it is likely that a new mechanism, such as a coupling between the surface and the bulk states, is responsible for this intriguing phenomenon in topological insulators.

Acknowledgments

This work was supported by JSPS (KAKENHI Grant No. 19674002), MEXT (KAKENHI Grant No. 22103004), and AFOSR (AOARD Grant No. 10-4103).

References

[1] Qi X L and Zhang S C 2010 *Physics Today* **63** 33.
[2] Moore J E 2010 *Nature* **464** 194.
[3] Hasan M Z and Kane C L 2010 *arXiv:*1002.2895 (to be published in *Rev. Mod. Phys.*).
[4] Fu L. Kane C L and Mele E J 2007 *Phys. Rev. Lett.* **98** 106803.
[5] Fu L and Kane C L 2007 *Phys. Rev. B* **76** 045302.
[6] Moore J E and Balents L 2007 *Phys. Rev. B* **75** 121306(R).
[7] Qi X L, Hughes T L and Zhang S C 2008 *Phys. Rev. B* **78** 195424.
[8] Hsieh D *et al.* 2008 *Nature* **452** 970.
[9] Hsieh D *et al.* 2009 *Science* **323** 919.
[10] Nishide A *et al.* 2010 *Phys. Rev. B* **81** 041309(R).
[11] Eto K, Ren Z, Taskin A A, Segawa K and Ando Y 2010 *Phys. Rev. B* **81** 195309.
[12] Peng H *et al.* 2010 *Nature Materials* **9** 225.
[13] Checkelsky J G *et al.* 2009 *Phys. Rev. Lett.* **103** 246601.
[14] Taskin A A and Ando Y 2009 *Phys. Rev. B* **80** 085303.
[15] Taskin A A, Segawa K and Ando Y 2010 *Phys. Rev. B* **82** 121302(R).
[16] Kartsovnik M V *et al.* 1988 *JETP Lett.* **48** 541.
[17] Kajita K et al. 1989 *Solid St. Comm.* **70** 1189.
[18] Yamaji K 1989 *J. Phys. Soc. Jpn.* **58** 1520.
[19] Roushan P, Seo J, Parker C V, Hor Y S, Hsieh D, Qian D, Richardella A, Hasan M Z, Cava R J and Yazdani A 2009 *Nature* **460** 1106.
[20] Cooper B K and Yakovenko V M 2006 *Phys. Rev. Lett. 96* 037001.

Thermoelectric transport of perfectly conducting channels in two- and three-dimensional topological insulators

Shuichi Murakami[1,2], Ryuji Takahashi[1], O. A. Tretiakov[3], Ar. Abanov[3], Jairo Sinova[3]

[1] Department of Physics, Tokyo Institute of Technology, 2-12-1 Ookayama, Meguro-ku, Tokyo 152-8551, Japan

[2] PRESTO, Japan Science and Technology Agency (JST), Kawaguchi, Saitama 332-0012, Japan

[3] Department of Physics, Texas A&M University, College Station, Texas 77843, USA

E-mail: murakami@stat.phys.titech.ac.jp

Abstract. Topological insulators have gapless edge/surface states with novel transport properties. Among these, there are two classes of perfectly conducting channels which are free from backscattering: the edge states of two-dimensional topological insulators and the one-dimensional states localized on dislocations of certain three-dimensional topological insulators. We show how these novel states affect thermoelectric properties of the systems and discuss possibilities to improve the thermoelectric figure of merit using these materials with perfectly conducting channels.

1. Introduction

Topological insulators (TI) [1, 2, 3, 4, 5] have attracted a lot of attention in the recent years. In these materials the excitations in the bulk are gapped just like in the normal insulators, but due to specific topological properties of the band structure these insulators possess gapless modes at the edges (on the surface). These modes are topologically protected and are stable against any perturbations that do not break the time-reversal symmetry. Various novel properties of TIs have been theoretically proposed, and some of them have been experimentally observed. The advent of this class of materials has brought us renewed interest in nonmagnetic insulators which have rarely been considered as an interesting subject for studying electronic properties. The transport properties of the edge/surface excitations in topological insulators are very distinct from those of the low-energy excitations of conventional metals. In two-dimensional (2D) topological insulators, the edge states are free from elastic scattering by nonmagnetic disorder. In three-dimensional (3D) topological insulators [6, 7], elastic scattering for the surface states is at least partly suppressed. It is also proposed that in certain classes of 3D topological insulators, the lattice dislocations have gapless states, which are robust against elastic backscattering by nonmagnetic disorder [8].

Thus, we have two cases of topologically protected one-dimensional (1D) gapless states in topological insulators; (i) the edge states in 2D topological insulators, and (ii) the states bound to dislocations in certain 3D topological insulators. These states have similar physical properties, that they are perfectly conducting, and they carry spin current. Hence these 1D gapless states are expected to have large contributions to transport. Having in mind that some of the topological insulators are good thermoelectric materials, we studied the thermoelectric properties of these 1D states, i.e. for edge states [9] and for states bound to dislocations [10]. In the present paper, we review the results of these articles and discuss them in detail. The paper is organized as follows. In Section 2 we study thermoelectric transport by the edge states in 2D topological insulators, following [9]. Section 3 is devoted for thermoelectric transport by the states bound to dislocations in 3D topological insulators [10]. The results are summarized in Section 4.

Thermoelectric properties of the system are characterized by the thermoelectric figure of merit ZT defined by

$$ZT = \frac{\sigma S^2 T}{\kappa}, \tag{1}$$

where σ is the charge conductivity, S is the Seebeck coefficient, κ is the thermal conductivity, and T is the temperature. The larger ZT means better efficiency for thermoelectric conversion. There are various approaches toward improving the ZT. In many materials, the main obstacle for better ZT is a large phonon heat transport. One approach to impove ZT is called phonon-glass electron-crystal (PGEC) [11], i.e. a good conductor for electrons but a poor conductor for phonons. The phonon conduction is suppressed by varying the lattice structure, and thus making the system glassy for phonons. Nevertheless the electron conduction should remain good for a good thermoelectric transport. Another approach for improving ZT is low-dimensionality [12]. This idea is to control the electronic band structure, so that the density of states has a peaked behaviour [13]. Despite these attempts the maximum ZT remains of order of unity, awaiting for a breakthrough.

2. Thermoelectric transport by edge states of 2D topological insulators

Because the 2D topological insulators have gapless edge states which are free from elastic scattering, they might have a possibility for novel and enhanced transport properties. For charge transport it has been measured experimentally in HgTe quantum well. In multiterminal measurement, the edge channels are shown to be perfectly conducting [16, 17]. While the edge states form perfectly conducting channels, the number of edge states is typically much smaller than that of the bulk, and the overall transport is mainly contributed by the bulk. In charge transport, we can neglect the bulk transport at lower temperatures, if the chemical potential is within the bulk gap. Nevertheless, because we are interested in the heat transport, we should consider finite temperatures, and bulk carriers cannot be neglected. Hence, in order to make the edge states more prominent compared with the bulk, we consider the system in narrow ribbon geometry.

The charge and heat current in a solid is described by the following linear response,

$$\begin{pmatrix} j/q \\ w \end{pmatrix} = \begin{pmatrix} L_0 & L_1 \\ L_1 & L_2 \end{pmatrix} \begin{pmatrix} -\nabla\mu \\ -T^{-1}\nabla T \end{pmatrix}, \tag{2}$$

where j is the current density, q is the electron charge, w is the heat current density, and μ is the chemical potential. L_i are the transport coefficients, using which the various physical quantities are expressed as

$$\sigma = q^2 L_0, \quad S = \frac{L_1}{qTL_0}, \quad \pi = \frac{L_1}{qL_0}, \quad \kappa = \kappa_{\rm ph} + \kappa_{\rm e}, \quad \kappa_e = \frac{1}{T}\frac{L_0 L_2 - L_1^2}{L_0} \tag{3}$$

where $\kappa_{\rm ph}$ and $\kappa_{\rm e}$ are the thermal conductivity by the phonons and that by the electrons, respectively.

2.1. Thermoelectric transport by edge states only

There are three carriers which contribute to ZT: the edge electrons, bulk electrons, and phonons. We first consider an ideal situation of only edge states, while transport by the bulk electrons and phonons is neglected. The density of states is schematically described as in Fig. 1(a). We set the origin of the energy as the conduction band edge. We apply the Landauer-Büttiker formula

$$L_\nu = \frac{1}{h}\int \sum_i \mathcal{T}_i(E)(E-\mu)^\nu \left(-\frac{\partial f}{\partial E}\right) \mathrm{d}E, \tag{4}$$

where i is the channel index, $\mathcal{T}_i(E)$ is the transmission probability for the i-th channel at energy E, and $f(E) = 1/(e^{(E-\mu)/k_\mathrm{B}T}+1)$ is the Fermi distribution function [14]. In the present case, the transmission probability is equal to unity for the states with energy E at $-\Delta < E < 0$, and zero otherwise, with Δ being the size of the bulk gap. We then obtain

$$L_\nu^\mathrm{e} = \frac{2\ell}{sh}\int_{-\Delta}^{0} (E-\mu)^\nu \left(-\frac{\partial f}{\partial E}\right) \mathrm{d}E = c_e F_\mathrm{e}(\nu,\mu)(k_BT)^\nu, \tag{5}$$

$$c_\mathrm{e} = \frac{2\ell}{sh}, \quad F_\mathrm{e}(\nu,\mu) = \int_{-\bar{\Delta}-\bar{\mu}}^{-\bar{\mu}} x^\nu \frac{\mathrm{e}^x}{(\mathrm{e}^x+1)^2}\mathrm{d}x, \tag{6}$$

where s is the cross section of the ribbon, ℓ is an effective length of the ribbon, $\bar{\Delta} = \Delta/k_BT$, and $\bar{\mu} = \mu/k_BT$. The superscript "e" denotes the edge transport. The effective length ℓ is taken as the inelastic scattering length of the edge states ℓ_inel. If we neglect the phonon transport, the figure of merit ZT is

$$ZT = \frac{L_1^2}{L_0L_2 - L_1^2} \tag{7}$$

The resulting ZT is shown in Fig. 1(b). We can see that as the chemical potential becomes larger than zero and goes into the bulk band, the ZT increases rapidly. Compared with the ZT ever achieved in experiments which is around $ZT \sim 2$, the resulting ZT is extremely large. It is because ZT reflects the average energy per carrier in the unit of $k_\mathrm{B}T$; if the chemical potential is in the bulk band, the edge carriers have larger energy in average, giving large ZT.

In this regime of μ, the edge carriers are thermally excited holes. Therefore the carrier numbers are exponentially small, so are the transport coefficients L_i^e. Hence, if we include bulk carriers or phonons, they will easily become dominant over the edge contributions, suppressing the large ZT.

2.2. Competition between edge and bulk transport

In addition to edge transport, we consider the bulk transport. The schematic figure is shown in Fig. 2. For simplicity, we set the bulk gap to be sufficiently larger than the temperature, and the chemical potential μ to be near the conduction band edge. Under these assumptions the bulk valence band is neglected. We also note that we are considering very narrow ribbons, for which the transverse motion is quantized due to the confinement. For ribbons as narrow as W = 10nm wide or so, we can assume that only the first subband is involved.

Bulk transport is diffusive and calculated by the Boltzmann transport theory. For simplicity we put the lifetime τ to be constant.

$$L_\nu^\mathrm{b} = \frac{1}{s}\int_{-\infty}^{\infty} \frac{\mathrm{d}k}{2\pi}\, c\tau v^2(\epsilon(k)-\mu)^\nu \left(-\frac{\partial f(\epsilon(k))}{\partial \epsilon}\right) = \frac{4\sqrt{2m}\mu^* c}{esh}\int_0^\infty \mathrm{d}\epsilon\, \sqrt{\epsilon}(\epsilon-\mu)^\nu \left(-\frac{\partial f}{\partial \epsilon}\right)$$

$$= c_b F_\mathrm{b}(\nu,\mu)(k_BT)^\nu, \tag{8}$$

$$c_\mathrm{b} = \frac{4\sqrt{2mk_BT}\mu^* c}{esh}, \quad F_\mathrm{b}(\nu,\mu) = \int_{-\bar{\mu}}^{\infty} \sqrt{x+\bar{\mu}}\, x^\nu \frac{\mathrm{e}^x}{(\mathrm{e}^x+1)^2}\mathrm{d}x, \tag{9}$$

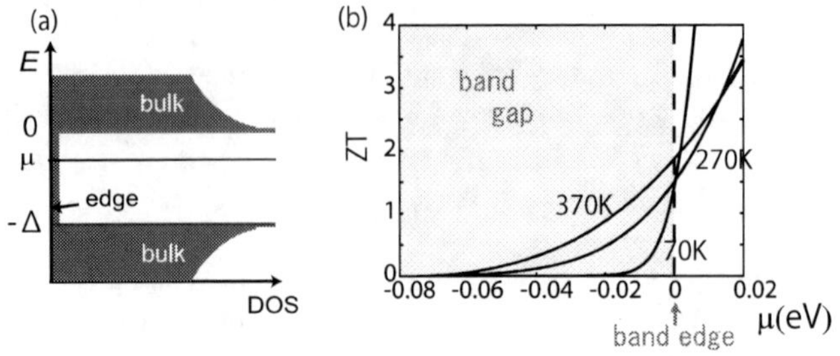

Figure 1. (a) Schematic density of states (DOS). (b) Figure of merit ZT as a function of chemical potential μ, calculated from edge transport only. Bulk carriers and phonons are neglected. Δ is taken to be $\Delta = 0.15$eV.

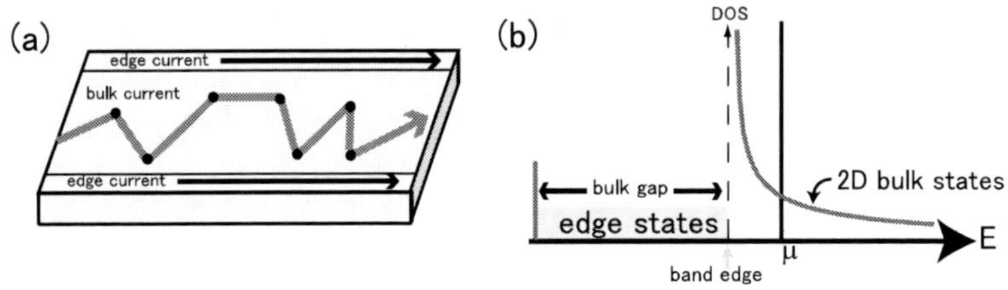

Figure 2. (a) Schematic picture for the two-dimensional ribbon. (b) Density of states including edge states and bulk states at the first subband.

where $\epsilon(k) = k^2/2m$, c is the number of carrier pockets, μ^* is the mobility, m is the effective mass, and the superscript "b" denotes the bulk transport. The prefactors c_e and c_b characterize the size of the transport for the edge and the bulk, respectively, and the total transport coefficient is given by their sum:

$$L_\nu = L_\nu^{\rm e} + L_\nu^{\rm b}. \tag{10}$$

The results for the transport coefficients at $T = 1.8$K are shown in Fig. 3 (a)-(d). We took Bi_2Te_3 as an example [28], and adopt the parameters as $c = 6$, $\ell_{\rm inel} = 1\mu$m, $m^* = 0.02m_e$. The cross section of the ribbon is set to be $s = 10$nm$\times$ 0.5 nm. The mobility is $\mu^* = 2000cm^2{\rm V}^{-1}{\rm s}^{-1}$, taken from the bulk mobility at 80K [18]. For the phonon thermal conductivity we assumed $\kappa_L = 0.1{\rm Wm}^{-1}{\rm K}^{-1}$ estimated from the bulk values [19, 20].

In Fig. 3, we observe interesting interplay between the edge and bulk transport. It is reasonable that in the charge conductivity (Fig. 3(a)) and the thermal conductivity (Fig. 3(c)), when the chemical potential is in the bulk gap ($\bar{\mu} \ll 0$), the transport is dominated by edge carriers. On the other hand, when the chemical potential is in the bulk band ($\bar{\mu} \gg 0$), the transport is dominated by bulk carriers. The behavior of the Seebeck coefficient S (Fig. 3(b)) is interesting. First, the edge and bulk contributions have opposite signs. It is because the sign of the Seebeck coefficient is the same as the sign of the carriers involved. When the chemical potential is near the bulk band edge, the carriers in the bulk conduction band are electrons, whereas the carriers in the edge states are holes. Thus they have opposite contributions for the Seebeck coefficients. It is also noted that the Seebeck coefficient from the edge states is an increasing function of μ, while that from the bulk states is a decreasing function of μ. They are both asymptotically linear functions.

As a whole, the figure of merit ZT (Fig. 3(d)) has a maximum around $\mu = 0$, the bulk band edge. We note that although the ZT from the edge is very large when μ is large, it is reduced drastically by the bulk carriers and phonons. It is because in this regime of μ, the carrier concentration of edge states is exponentially small, and it can be easily overcome by the bulk carriers and the phonons.

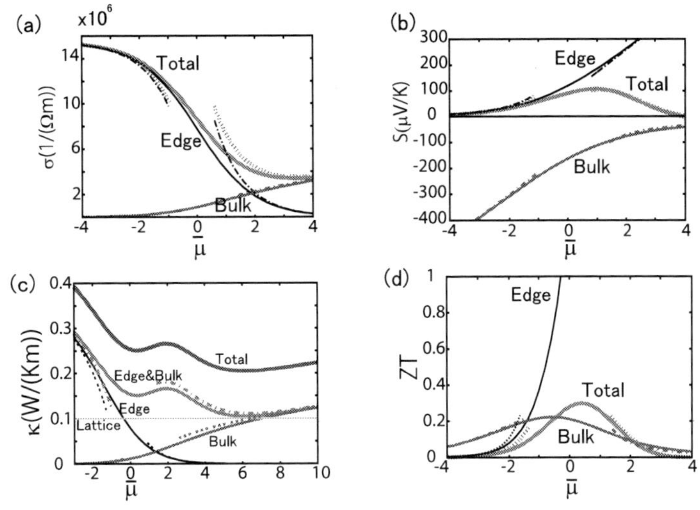

Figure 3. (a) Charge conductivity σ, (b) Seebeck coefficient S, (c) thermal conductivity κ, (d) thermoelectric figure of merit ZT. The horizontal axis represents $\bar{\mu} = \mu/k_BT$, and $\bar{\mu} = 0$ corresponds to the band edge.

We can evaluate the transport coefficients when $\bar{\mu}$ is far from the band edge, i.e. $|\bar{\mu}| \gg 1$. Here we also neglect the valence band and focus only on the conduction band and edge states, by setting $\bar{\Delta} = \Delta/(k_{\rm B}T) \to \infty$. For the edge states, when $\bar{\mu} \gg 1$ (chemical potential is deep inside the bulk conduction band), the asymptotic form of $F_{\rm e}(\nu, \mu)$ is

$$F_{\rm e}(0,\mu) \sim {\rm e}^{-\bar{\mu}}, \quad F_{\rm e}(1,\mu) \sim -(1+\bar{\mu}){\rm e}^{-\bar{\mu}}, \quad F_{\rm e}(2,\mu) \sim (2+2\bar{\mu}+\bar{\mu}^2){\rm e}^{-\bar{\mu}}, \tag{11}$$

and when $\bar{\mu} \ll -1$ (chemical potential is deep inside the bulk gap),

$$F_{\rm e}(0,\mu) \sim 1-{\rm e}^{\bar{\mu}}, \quad F_{\rm e}(1,\mu) \sim (\bar{\mu}-1){\rm e}^{\bar{\mu}}, \quad F_{\rm e}(2,\mu) \sim \frac{\pi^2}{3} + (-\bar{\mu}^2+2\bar{\mu}-2){\rm e}^{\bar{\mu}}. \tag{12}$$

On the other hand, for the bulk states, when $\bar{\mu} \gg 1$ (chemical potential is deep inside the bulk conduction band), the asymptotic form of $F_{\rm b}(\nu, \mu)$ is

$$F_{\rm b}(0,\mu) \sim \sqrt{\bar{\mu}}, \quad F_{\rm b}(1,\mu) \sim \frac{\pi^2}{6\sqrt{\bar{\mu}}}, \quad F_{\rm b}(2,\mu) \sim \frac{\pi^2}{3}\sqrt{\bar{\mu}} \tag{13}$$

and when $\bar{\mu} \ll -1$ (chemical potential is deep inside the bulk gap),

$$F_{\rm b}(0,\mu) \sim \frac{\sqrt{\pi}}{2}{\rm e}^{\bar{\mu}}, \quad F_{\rm b}(1,\mu) \sim \frac{\sqrt{\pi}}{2}{\rm e}^{\bar{\mu}}(\frac{3}{2}-\bar{\mu}), \quad F_{\rm b}(2,\mu) \sim \frac{\sqrt{\pi}}{2}{\rm e}^{\bar{\mu}}(\frac{15}{4}-3\bar{\mu}+\bar{\mu}^2). \tag{14}$$

From these asymptotic forms, we evaluate the asymptotics for the transport properties, shown in Fig. 3 as broken lines. In particular, the ZT from the edge state only has the asymptotic form

$$ZT_{\rm e} \sim (1+\bar{\mu})^2 : \bar{\mu} \gg 1, \tag{15}$$

$$ZT_{\rm e} \sim \frac{3}{\pi^2}(1-\bar{\mu})^2 e^{\bar{\mu}} : \bar{\mu} \ll -1, \tag{16}$$

and the ZT from the bulk state only has the asymptotic form

$$ZT_{\rm b} \sim \frac{\pi^2}{12\bar{\mu}^2} : \bar{\mu} \gg 1, \tag{17}$$

$$ZT_{\rm b} \sim \frac{2}{3}\left(\frac{3}{2}-\bar{\mu}\right)^2 : \bar{\mu} \ll -1. \tag{18}$$

2.3. Universal behavior described by dimensionless parameters

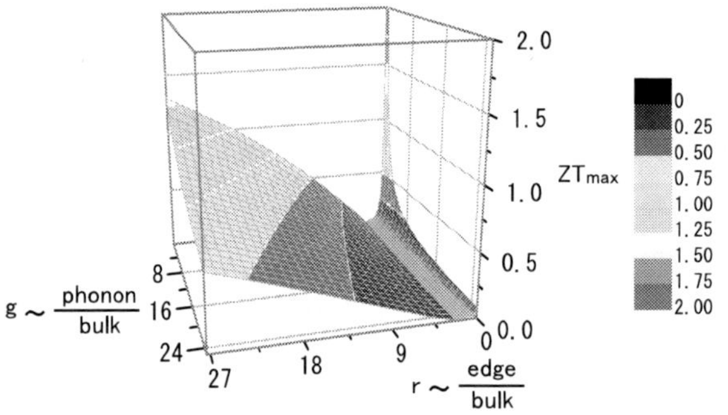

Figure 4. $ZT_{\rm max}$ as a function of r and g. $\bar{\mu}$ is set to maximize ZT.

We have calculated various thermoelectric properties using specific values of parameters. There is a wide range of choices for various parameters, such as sample geometry, material choice, temperature, and so forth. It is therefore important to develop a theory which is applicable for various choices of parameters. This can be achieved by introducing dimensionless parameters r and g defined by

$$r = c_{\rm e}/c_{\rm b} = \frac{e\ell}{2\sqrt{2mk_BT}\mu^* c}, \quad g = \left(\frac{\kappa_L}{k_B^2 T}\right)/c_{\rm b} = \frac{\kappa_L esh}{4\sqrt{2mk_B^5 T^3}\mu^* c}. \tag{19}$$

The parameter r characterizes the ratio of edge transport to bulk transport, and g characterizes the phonon transport to bulk electronic transport. The figure of merit ZT is then expressed as a function of r, g and $\bar{\mu}$. We then maximize ZT as a function of $\bar{\mu}$. In Fig. 4 we show the maximum value of ZT for each value of r and g. maximum. One can see from Fig. 4 that ZT has a minimum at $r \sim 2.3$. As we increase r, the edge states become more and more dominant, and bulk-to-edge crossover occurs. Because the edge and bulk tend to cancel each other, this crossover is accompanied by a minimum of ZT. In general the behavior of r and g as functions of the temperature T is complex. Nevertheless, the inelastic scattering length ℓ is a rapidly decreasing function of T, as is seen from the case of quantum Hall effect with $\ell_{\rm inel} \sim T^{-1.5}$ [21], and we can estimate $r \sim T^{-2}$. Thus we conclude that by lowering temperature, edge states becomes dominant, and there occurs a bulk-to-edge crossover. The ZT thereby has a minimum and then increases again as a function of temperature. For 10nm-wide ribbon, the crossover is expected to be around 5K-10K.

2.4. Discussion

In the above estimate, the value of ZT is sensitive to the edge-state inelastic scattering length ℓ. If ℓ becomes longer, it is more favorable for edge-state transport since it is free from scattering up to a longer distance. It is difficult to qualitatively estimate ℓ in topological insulators because it is dominated by electron-electron scattering. We can draw analogy with quantum Hall systems, where ℓ is estimated to be 1μm at 1K [21]. On the other hand, in HgTe quantum well, known as a 2D topological insulator, perfectly conducting channels for the system size 1μm are observed at 1.8K, meaning that $\ell > 1\mu$m [16]. From these data we used $\ell = 1\mu$m in our calculation.

Because r is inversely proportional to the ribbon width, narrower ribbons are more favorable for edge-dominated transport, in agreement with our expectations. On the other hand, there is a minimum width for which the ribbon supports the perfectly conducting edge channels. It is determined by a penetration depth λ of edge states. If the ribbon width is narrower than the penetration depth of edge states, there occurs a backscattering between the edge states on the opposite sides of the ribbon, and the edge state transport is suppressed. In HgTe quantum well the penetration depth is 6nm-50nm [22, 24]. On the other hand, in some systems such as bismuth(111) 1-bilayer films, λ is predicted to be as short as a lattice constant [23, 24], and the ribbon can be made very narrow with keeping the perfectly conducting edge channels.

3. Thermoelectric transport by dislocations in 3D topological insulators

Compared with 2D topological insulators where the edge states are free from scattering, the surface states of 3D topological insulators undergo some scattering. In a case of Dirac cone as seen in Bi_2Se_3, the perfect backscattering by 180 degrees is prohibited, but other scattering processes can happen. Therefore, the surface states do not show perfect conduction.

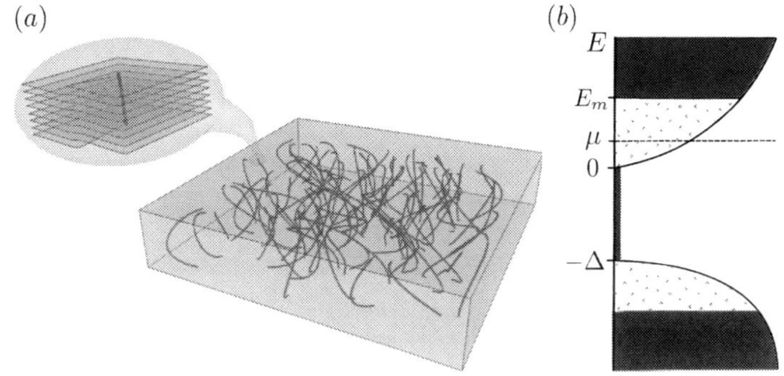

Figure 5. (a) Dislocations in 3D topological insulator. The inset shows one such dislocation of the screw type. (b) Schematic picture of the band structure for the calculation.

Nevertheless, it is theoretically predicted that in some of the 3D topological insulators there occurs a Kramers pair of gapless states on a lattice dislocation [8]. This state carries a pure spin current and is topologically protected. Thus these dislocations form perfectly conducting channels, as is similar to the edge states in 2D topological insulators. It is then expected that if we increase the number of dislocations, as shown in Fig. 5(a), the transport by these 1D channels becomes dominant over the bulk transport and phonon transport. Increasing the number of dislocations in the crystal is also helpful for other reasons: the mean free path of phonons is reduced by disorder, and the bulk carriers localize below the mobility edge formed at the bottom of the conduction band. Thus more dislocations enhance the 1D channel transport while at the same time suppress the bulk electron transport and phonon transport.

The framework of the ZT calculation is similar to the 2D ribbon, and we do not provide it here. The details can be found in [10]. All the calculations are performed for room temperature, T =300 K. We assume that the phonon thermal conductivity is reduced due to disorder from the bulk value of $\kappa_{ph} = 1\text{Wm}^{-1}\text{K}^{-1}$ to $\kappa_{ph} = 0.01\text{Wm}^{-1}\text{K}^{-1}$ for the average distances between dislocations $\sim$ 3 nm. The gap between the valence and conduction bands is taken to be $\Delta = 0.15$ eV. We also set the mobility edge to be $E_m = 0.05$ eV, measured from the bottom of the conduction band, see Fig. 5(b). The resulting ZT for the 1D channel and the 3D bulk carriers are given in Fig. 6(a) and (b), respectively. As a function of the concentration of dislocations the figure of merit ZT is calculated in Fig. 6(c). The highest value ZT reaches when the chemical potential μ, which can be changed by an external gate, is near the mobility edge. Remarkably the maximum value of ZT is over 6 at room temperature.

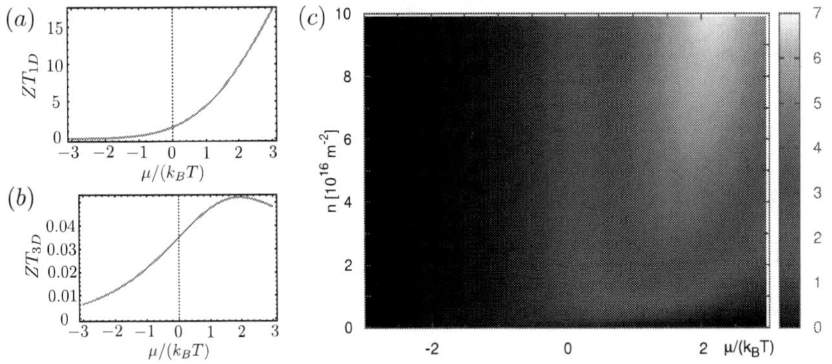

Figure 6. (a) Figure of merit ZT_{1D} from 1D channels bound to dislocations as a function of chemical potential μ. (b) Figure of merit ZT_{3D} for the bulk of the 3D topological insulator. The 1D channels are neglected here. (c) Contour plot of the total ZT as a function of the dislocations density n and the chemical potential μ.

It is important to mention here that, if the impurities are non-magnetic, the transition to a bulk Anderson insulator should not destroy the 1D gapless states, because the time reversal invariance is not broken. Note also that by increasing the density of dislocations and/or non-magnetic impurities one can reach even greater values of ZT. Nevertheless, these densities are less likely to be achieved and can as well result in the tunneling of electrons between the channels of neighboring dislocations. This tunneling could lead to the gap opening and localization of the protected 1D states. This is the reason for our choice of an upper limit of the dislocation density of $n \sim 10^{17}\text{m}^{-2}$ which corresponds to an average dislocation spacing of around 3 nm.

We note that not every topological insulator has dislocations carrying 1D gapless protected modes. The condition for the 1D gapless protected modes is expressed as $\mathbf{B} \cdot \mathbf{M}_\nu = \pi \ (\text{mod} 2\pi)$, where the vector $\mathbf{M}_\nu$ (so called time-reversal-invariant momentum) is a part of the Z_2 topological number for the topological insulator considered and $\mathbf{B}$ is the Burgers vector of the dislocation [8]. Therefore, only the materials which Z_2 topological numbers satisfy the above criterion can carry 1D gapless modes through the dislocations. In this sense, the dislocations in Bi_2Te_3 [25, 26, 27, 28] and Bi_2Se_3 do not carry 1D gapless modes, but the dislocations in $Bi_{1-x}Sb_x$ [29, 30] ($0.07 < x < 0.22$) possess 1D gapless modes.

4. Conclusion

In the present paper, we consider two types of topologically protected 1D gapless states in 2D and 3D topological insulators, as current-carrying paths for thermoelectric transport. In both cases, the 1D modes alone give large thermoelectric figure of merit ZT, whereas the bulk states and phonons suppress the ZT. Thus in order to have good ZT, it is necessary to have high density of 1D modes. It is accomplished by making narrow ribbons for 2D topological insulators, and by making dislocations densely distributed in 3D topological insulators.

Acknowledgments

This work is supported partly by Grant-in-Aid for Specially Promoted Research (21000004), by Grant-in-Aid for Scientific Research (C) (22540327), by Grant-in-Aid for Global COE Program "Nanoscience and Quantum Physics" from the Ministry of Education, Culture, Sports, Science and Technology of Japan, by NSF under Grant Nos. DMR-0547875 and 0757992, by the Research Corporation Cottrell Scholar Award, and by the Welch Foundation (A-1678).

References

[1] Kane C L and Mele E J 2005 *Phys. Rev. Lett.* **95** 226801
[2] Kane C L and Mele E J 2005 *Phys. Rev. Lett.* **95** 146802
[3] Bernevig B A and Zhang S C 2006 *Phys. Rev. Lett.* **96** 106802
[4] Wu C J, Bernevig B A and Zhang S C 2006 *Phys. Rev. Lett.* **96** 106401
[5] Xu C and Moore J E 2006 *Phys. Rev.* B **73** 045322
[6] Fu L, Kane C L and Mele E J 2007 *Phys. Rev. Lett.* **98** 106803
[7] Moore E J and Balents L 2007 *Phys. Rev.* B **75** 121306
[8] Ran Y, Zhang Y and Vishwanath A 2009 *Nature Phys.* **5** 298
[9] Takahashi R and Murakami S 2010 *Phys. Rev.* B **81** 161302(R)
[10] Tretiakov O A, Abanov Ar, Murakami S and Sinova J 2010 *Appl. Phys. Lett.* **97** 073108
[11] Slack G A , in "*CRC Handbook of Thermoelectrics*" (Ed. D.M. Rowe) pp. 407-440, (CRC Press, Boca Raton, 1995).
[12] Hicks L D and Dresselhaus M S 1993 *Phys. Rev.* B **47** 12727
[13] Mahan G D and Sofo J O 1996 *Proc. Natl. Acad. Sci. U.S.A.* **93** 7436
[14] Sivan U and Imry Y 1985 *Phys. Rev.* B **33** 551
[15] Bernevig B A, Hughes T L and Zhang S C 2006 *Science* **314** 1757
[16] König M, Wiedmann S, Brune C, Roth A, Buhmann H, Molenkamp L W, Qi X L and Zhang S C 2007 *Science* **318** 766
[17] Roth A, Brüne C, Buhmann H, Molenkamp L W, Maciejko J, Qi X -L and Zhang S -C 2009 Science **325** 294
[18] Champness C H and Kipling A L 1966 *Can. J. Phys.* **44** 769
[19] MacDonald D K C *et al.* 1959 *Philos. Mag.* **4** 443
[20] Zou J and Balandin A 2000 *J. Appl. Phys.* **89** 2932
[21] Machida T, Hirai H, Komiyama S and Shiraki Y 1996 *Phys. Rev.* B **54** 16860
[22] Zhou B, Lu H -Z, Chu R -L, Shen S -Q and Niu Q 2008 *Phys. Rev. Lett.* **101** 246807
[23] Murakami S 2006 *Phys. Rev. Lett.* **97** 236805
[24] Wada M, Murakami S, Freimuth F, Bihlmayer G, arXiv:1005.3912.
[25] Hsieh D, Xia Y, Qian D, Wray L, Meier F, Dil J H, Osterwalder J, Patthey L, Fedorov A V, Lin H, Bansil A, Grauer D, Hor Y S, Cava R J and Hasan M Z 2009 *Phys. Rev. Lett.* **103** 146401
[26] Hsieh D, Xia Y, Qian D, Wray L, Dil J H, Meier F, Osterwalder J, Patthey L, Checkelsky L G, Ong N P, Fedorov A V, Lin H, Bansil A, Grauer D, Hor Y S, Cava R J and Hasan M Z 2009 *Nature* **460** 1101
[27] Zhang H J, Liu C X, Qi X L, Dai X, Fang Z and Zhang S C 2009 *Nature Phys.* **5** 438
[28] Chen Y L, Analytis J G, Chu J H, Liu Z K, Mo S K, Qi X L, Zhang H J, Lu D H, Dai X, Fang Z, Zhang S C, Fisher I R, Hussain Z, Shen Z X 2010 *Science* **325** 178
[29] Hsieh D, Qian D, Wray L, Xia Y, Hor Y S, Cava R J and Hasan M Z 2008 *Nature* **452** 970
[30] Nishide A, Taskin A A, Takeichi Y, Okuda T, Kakizaki A, Hirahara T, Nakatsuji K, Komori F, Ando Y and Matsuda I 2010 *Phys. Rev.* B **81** 041309

Journal of Physics: Conference Series **334** (2011) 012014 doi:10.1088/1742-6596/334/1/012014

Zero-resistance states in bilayer electron systems induced by microwave irradiation

S. Wiedmann[1,2,3*], G. M. Gusev[4], O. E. Raichev[5], A. K. Bakarov[6], and J. C. Portal[2,3]

[1]Radboud University Nijmegen, Institute for Molecules and Materials and High Field Magnet Laboratory, Toernooiveld 7, 6525 ED Nijmegen, The Netherlands
[2]LNCMI, UPR 3228, CNRS-INSA-UJF-UPS, BP 166, 38042 Grenoble Cedex 9, France
[3]INSA Toulouse, 31077 Toulouse Cedex 4, France
[4]Instituto de Física da Universidade de São Paulo, CP 66318 CEP 05315-970, São Paulo, SP, Brazil
[5]Institute of Semiconductor Physics, NAS of Ukraine, Prospekt Nauki 41, 03028, Kiev, Ukraine
[6]Institute of Semiconductor Physics, Novosibirsk 630090, Russia

E-mail: *s.wiedmann@science.ru.nl

Abstract. We experimentally demonstrate the existence of microwave-induced zero-resistance states (ZRS) in bilayer electron systems created in wide quantum wells. In contrast to single-layer two-dimensional electron systems, ZRS are developed from the strongest magneto-intersubband oscillation peaks inverted by microwave radiation. Our experimental work is discussed within the theories of microscopic mechanisms of photoresistance, and the calculations reveal that the condition for absolute negative resistivity correlates with the appearance of ZRS.

1. Introduction

Exposing a two-dimensional electron system (2DES) to microwave (MW) irradiation leads to the appearance of zero dissipative resistance (zero-resistance states, ZRS) for samples with high electron mobility [1]. This phenomenon is closely related to microwave-induced resistance oscillations (MIROs) [2]. Both MIROs and ZRS occur at low magnetic fields before Shubnikov-de Haas (SdH) oscillations are developed, and are governed by the ratio of the radiation frequency ω to the cyclotron frequency ω_c. Since their discovery, properties of ZRS and MIROs have been experimentally investigated in numerous works including dependence on temperature, MW power (see Refs. [1, 2] as well as Ref. [3] and references therein), and MW field polarization [4].

MW-induced vanishing resistance has attracted attention of theorists who suggested several mechanisms responsible for both MIROs and ZRS. The so-called displacement mechanism which accounts for a spatial displacement of electrons along the applied dc field under scattering-assisted microwave absorption has been predicted by Ryzhii in the 70s [5] and found again by Durst *et al.* [6]. Later on, the so-called inelastic mechanism, based on inelastic-scattering-controlled oscillatory contribution to the electron distribution function, has been developed by Dmitriev *et al.* [7]. Both these mechanisms can lead to absolute negative resistivity, which causes the instability of homogeneous current flow and spontaneous breaking of the sample into domains [8] giving rise to vanishing dissipative resistance R_{xx}.

Until now, all studies of MIROs and ZRS have been carried out in high-mobility 2D systems with one populated subband. Recently, the observation of MIROs in 2D systems with more than one populated subband, formed by two or more coupled quantum well layers, has demonstrated that MW-induced oscillation phenomena are not restricted to single-layer systems [9, 10, 11] but there is no evidence for MW-induced ZRS neither in bilayer nor in other systems. This raises the question whether vanishing dissipative resistance appear only in a one-subband system or might occur in multilayer or in quasi-three-dimensional (3D) electron systems.

2. Experimental observation

Our samples are wide GaAs quantum wells (w=45 nm) with a high electron density of $n_s \simeq 9.1\times10^{11}$ cm^{-2} and a mobility of $\mu \simeq 1.9\times10^6$ cm^2/V s. We have studied samples in both Hall bar (length $L\times$ witdh W=250 μm $\times$ 50 μm) and van der Pauw (3 mm $\times$ 3 mm) geometry. Our system forms a bilayer configuration due to charge redistribution, when two quantum wells near the interfaces are separated by an electrostatic potential barrier, illustrated in the inset of Figure 1(a). The two lowest subbands are separated by the energy Δ_{SAS} = 1.4 meV, extracted from the periodicity of magneto-intersubband (MIS) oscillations [12] which appear as a result of a consecutive alignment of Landau levels of different subbands as the magnetic field increases. The measurements of the resistance have been performed in a cryostat with a variable temperature insert, and we used special probes to deliver MW radiation down to the sample and to control the orientation of linear MW polarization in the temperature range from 1.4 to 6.5 K.

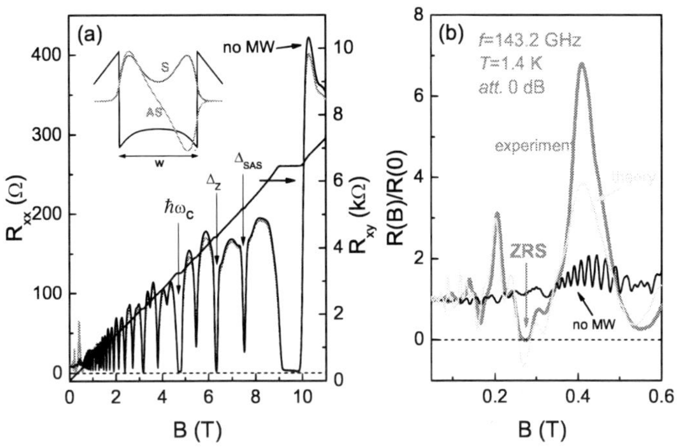

Figure 1. (Color online) (a) Longitudinal resistance and quantum Hall effect in a high-mobility bilayer system, with and without MW excitation (energy gaps are labeled). For MW frequency of 143.2 GHz (red trace) at a temperature of 1.4 K, zero resistance appears in the region of the inverted MIS oscillation peak at $B \simeq 0.27$ T. Inset: wide quantum well with symmetric (S) and antisymmetric (AS) wave functions. (b) Expanded view of the low-field data showing $R_{xx} \to 0$ in the vicinity of 0.27 T.

In Figure 1(a), we show measurements of longitudinal and Hall resistance with and without MW irradiation (no MW). For high magnetic field we have labeled corresponding energy gaps $\hbar\omega_c$, Δ_Z (Zeeman gap), and Δ_{SAS}. We observe MIS oscillations for $B < 1$ T thereby confirming bilayer nature of our system. Under a MW irradiation of 143.2 GHz, we find enhanced and inverted MIS oscillations for $B < 1$ T whereas in high magnetic fields MWs only damp SdH oscillations due to heating of the electron gas [9]. The Hall resistance in Figure 1(a) remains essentially unaffected. In the region of small magnetic fields, see Figure 1(b), we observe a profound ZRS for $B \simeq 0.27$ T developed from an inverted MIS oscillation peak. This behavior is distinct from the one in single-layer systems where ZRS occur at $\omega/\omega_c = j + 1/4$ (j is an integer).

3. Experimental analysis and theoretical model

A detailed study in the whole frequency range to our disposal (35 to 170 GHz) reveals that, in total, four inverted MIS oscillation peaks evolve into ZRS, as presented in Figure 2. Corresponding ZRS are marked with roman numbers from (I) to (IV).

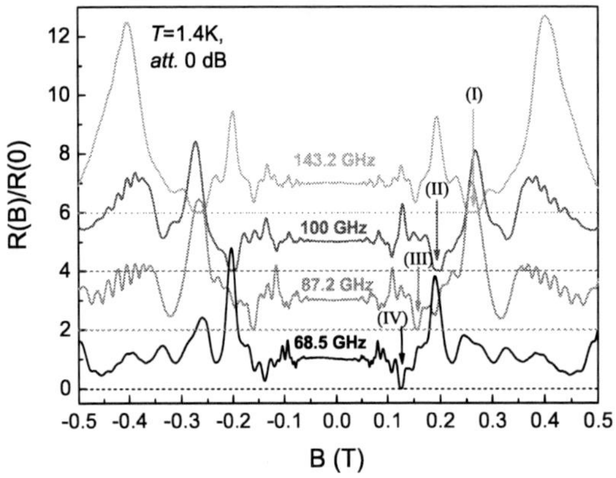

Figure 2. (Color online) Normalized magnetoresistance for positive and negative magnetic field exposed to MW irradiation from 68.5 to 143.2 GHz for highest intensity (attenuation 0 dB). ZRS can be observed for four inverted MIS oscillations, denoted with peak (I) to peak (IV). Traces are shifted up for clarity except the one for 68.5 GHz. Dashed lines mark R_{xx}=0.

A decrease in MW frequency leads to the appearance of ZRS at lower magnetic fields, because the conditions for inversion of MIS oscillations are determined by MW frequency [9]. For each frequency, only the strongest inverted MIS peak, closest to the cyclotron resonance $\omega = \omega_c$, evolves into a ZRS; the other inverted MIS peaks do not reach zero resistance. In contrast to single-layer systems, the regions where ZRS occur are narrow and strongly sensitive to MW frequency. An overview of ZRS is presented in Figure 3 showing the frequency regions where each of the inverted MIS peaks evolves into a ZRS. The width of ZRS correlates with the MIS peak width, which implies that the regions of ZRS are narrower for lower frequencies.

We focus now on the MIS oscillation peak (I) which exhibits a profound ZRS at $B = 0.27$ T. In Figure 4(a) the evolution of magnetoresistance exposed to a MW irradiation of 143 GHz for several chosen temperatures from 1.4 to 6.5 K is presented. As we approach ZRS by lowering temperature, the amplitude of $R(B)/R(0)$ rapidly decreases. A fit to exponential dependence $R \propto \exp(-E_{ZRS}/T)$, see the Arrhenius plot shown in Figure 4(b), gives us E_{ZRS}=7 K, which is of the same order of magnitude as in single-layer systems [1].

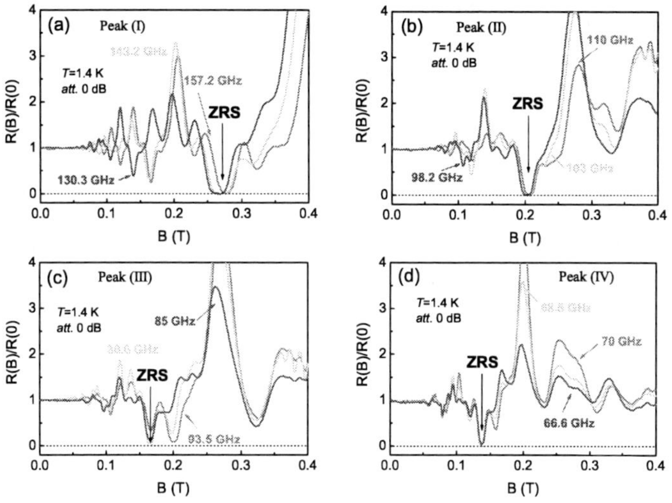

Figure 3. (Color online) Regions of ZRS in bilayer electron systems: (a) for peak (I), (b) for peak (II), (c) for peak (III), and (d) for peak (IV). With decreasing MW frequency, the frequency region for existence of ZRS becomes smaller.

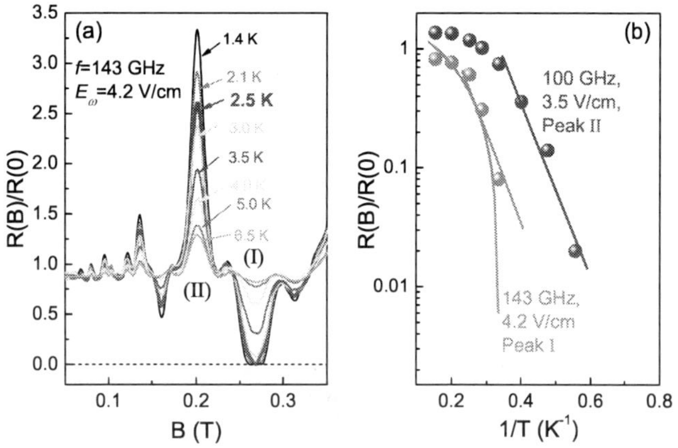

Figure 4. (Color online) (a) Temperature dependence of longitudinal resistance from 1.4 to 6.5 K. The ZRS first narrows at 2.1 K and vanishes for $T > 2.5$ K. (b) $R(B)/R(0)$ for 100 GHz [ZRS, peak (II)] and 143 GHz [ZRS, peak (I)]. The solid lines are fits to $R \propto \exp(-E_{ZRS}/T)$. The red trace is a theoretical calculation.

We analyze our experimental results within the models of displacement and inelastic mechanisms of photoresistance generalized (see Refs. [9, 10]) to the two-subband case. The inelastic mechanism is found to be the dominant contribution under conditions of our experiment, while the displacement mechanism becomes essential only at higher MW power or at $T > 4$ K [13]. The effective electron temperature, which depends on MW power, frequency, and magnetic field, has been determined by assuming energy relaxation of electrons due to their interaction with acoustic phonons. By analyzing dark magnetoresistance data at different temperatures, we have extracted the quantum lifetime and transport scattering time of electrons in our sample [13]. The inelastic relaxation time, which characterizes the strength of the inelastic mechanism, is determined from theoretical estimates [7]. The MW electric field strength at 0 dB attenuation, E_ω=4.2 V/cm, is found by fitting the theory to experimental traces. As a result, the calculated magnetoresistance at 143.2 GHz in Figure 1(b) reproduces experimentally observed features. In particular, the negative resistivity around $B = 0.27$ T can be the reason of inhomogeneous current pattern (domains) which is responsible for ZRS in this region of magnetic fields.

4. Conclusion

We have demonstrated the evidence of MW-induced ZRS for 2DES which are different from conventional single-layer electron systems. Theoretical calculations have shown that this phenomenon can be explained by the absolute negative resistivity under MW irradiation governed by the inelastic mechanism of photoresistance. Our results are the first step to generalization of ZRS phenomenon for the systems beyond the single-subband case, since we show that the additional intersubband scattering leading to MIS oscillations is not a hindrance for observation of ZRS in bilayers. These results also open the possibility for experimental and theoretical studies of ZRS in quasi-3D systems.

The authors acknowledge financial support from CNPq and FAPESP (Brazilian agencies). For MW measurements we have used MW sources from the French project ANR MICONANO.

References

[1] R. G. Mani, J. H. Smet, K. von Klitzing, V. Narayanamurti, W. B. Johnson, and V. Umansky, *Nature* **420**, 646 (2002); M. A. Zudov, R. R. Du, L. N. Pfeiffer, and K. W. West, *Phys. Rev. Lett.* **90**, 046807 (2003); R. L. Willett, L. N. Pfeiffer, and K. W. West, *Phys. Rev. Lett.* **93**, 026804 (2004).
[2] M. A. Zudov, R. R. Du, J. A. Simmons, and J. L. Reno, *Phys. Rev. B* **64**, 201311(R) (2001); P. D. Ye, L. W. Engel, D. C. Tsui, J. A. Simmons, J. R. Wendt, G. A. Vawter, J. L. Reno, *Appl. Phys. Lett.* **79**, 2193 (2001).
[3] A. T. Hatke, M. A. Zudov, L. N. Pfeiffer, and K. W. West, *Phys. Rev. Lett.* **102**, 066804 (2009).
[4] J. H. Smet, B. Gorshunov, C. Jiang, L. Pfeiffer, K. West, V. Umansky, M. Dressel, R. Meisels, F. Kuchar, and K. von Klitzing, *Phys. Rev. Lett.* **95**, 116804 (2005).
[5] V. I. Ryzhii, *Sov. Phys. Solid State* **11**, 2078 (1970); V. I. Ryzhii, R. A. Suris, and B. S. Shchamkhalova, *Sov. Phys. Semicond.* **20**, 1299 (1986).
[6] A. C. Durst, S. Sachdev, N. Read, and S. M. Girvin, *Phys. Rev. Lett* **91**, 086803 (2003).
[7] I. A. Dmitriev, M. G. Vavilov, I. L. Aleiner, A. D. Mirlin, and D. G. Polyakov, *Phys. Rev. B* **71**, 115316 (2005).
[8] A. V. Andreev, I. L. Aleiner, and A. J. Millis, *Phys. Rev. Lett* **91**, 056803 (2003).
[9] S. Wiedmann, G. M. Gusev, O.E. Raichev, T. E. Lamas, A. K. Bakarov, and J. C. Portal *Phys. Rev. B* **78**, 121301(R) (2008); O. E. Raichev, *Phys. Rev. B* **78**, 125304 (2008).
[10] S. Wiedmann, G. M. Gusev, O.E. Raichev, T. E. Lamas, A. K. Bakarov, and J. C. Portal *Phys. Rev. B* **81**, 085311 (2010).
[11] S. Wiedmann, N. C. Mamani, G. M. Gusev, O. E. Raichev, A. K. Bakarov, and J. C. Portal, *Phys. Rev. B* **80**, 245306 (2009).
[12] N. C. Mamani, G. M. Gusev, T. E. Lamas, A. K. Bakarov, and O. E. Raichev, *Phys. Rev. B* **77**, 205327 (2008).
[13] S. Wiedmann, G. M. Gusev, O. E. Raichev, A. K. Bakarov, and J. C. Portal, *Phys. Rev. Lett* **105**, 026804 (2010).

Nonequilibrium magnetooscillations in spatially non-uniform quantum Hall systems

Ivan A. Dmitriev
Institut für Nanotechnologie, Karlsruher Institut für Technologie, 76021 Karlsruhe, Germany;
A.F. Ioffe Physical Technical Institute, 194021 St. Petersburg, Russia
E-mail: ivan.dmitriev@kit.edu

Abstract. A theory of nonequilibrium magnetotransport in high Landau levels of inhomogeneous quantum Hall systems is presented. A nonlinear current is calculated in response to given local gradients of the chemical and electrostatic potential in the presence of moderately strong microwave radiation. In the regime of high temperature, theory generalizes previously obtained results describing microwave- and Hall-induced oscillations of magnetoresistivity for the case of spatially non-uniform systems. Additionally, the regime of low temperatures is studied, where strong modification of Shubnikov-de Haas oscillations by the ac and dc fields is demonstrated.

1. Introduction

Over the past decade, a family of fascinating nonequilibrium phenomena has been discovered in 2D electron gas (2DEG) including the microwave-induced (MIRO) [1–18], Hall-induced (HIRO) [19–32], and phonon-induced (PIRO) [33–39] resistance oscillations. Particularly interesting are the zero-resistance states (ZRS) [40–56] which form in the minima of MIRO at sufficiently high microwave power. The magnetoresistivity in these structures is governed by spectral and spatial resonances set by the cyclotron frequency $\omega_c = eB/mc$ and the cyclotron diameter $R_c = v_F/\omega_c$ in various combinations with the parameters of the external fields. All these effects were observed in moderately strong transverse magnetic fields $\mathbf{B}\|\mathbf{e}_z$ where the dynamics of electrons remains essentially quasiclassical. At the same time, Landau quantization of the orbital motion can be pronounced and leads to the emergence of novel spectral and spatial resonances. This paper extends existing theories of these phenomena [54–92] to include the case of spatial inhomogeneity as well as effects of strong dc and ac fields on Shubnikov-de Haas oscillations (SdHO).

2. Nonlinear magnetotransport in irradiated inhomogeneous 2DEG

For a 2DEG in a classically strong magnetic field ($\omega_c\tau_{\rm tr} \gg 1$, where $\tau_{\rm tr}$ is the transport scattering time), and in high Landau levels (chemical potential $\mu(\mathbf{r}) \gg \omega_c$), the local dissipative current is calculated in response to a moderately strong microwave illumination as well as to local gradients of the chemical potential and electrostatic potential, $e\mathbf{E}_\mu = -\nabla\mu$ and $\mathbf{E} = -\nabla\phi(\mathbf{r})$. Local gradient of the electrochemical potential $\eta(\mathbf{r}) = \mu(\mathbf{r}) + e\phi(\mathbf{r})$, given by $\mathbf{E}_\eta = -\nabla\eta/e = \mathbf{E}_\mu + \mathbf{E}$, is zero in equilibrium, in the absence of microwaves and external bias. In a given nonequilibrium situation, the solution of electrostatic problem which provides the relation between the local electron density $n(\mathbf{r})$ and electrostatic potential should be found self-consistently using setup-specific boundary conditions.

In the classical limit $\omega_c\tau_q \ll 1$, where τ_q is the quantum scattering time, the density of electronic states (DoS) in a 2D parabolic band is constant, $\nu = \nu_0 = m/2\pi$ (we use $\hbar = 1$). Provided interaction and memory effects [89] can be ignored, the local direct current $\mathbf{j}(\mathbf{r}) = en(\mathbf{r})\mathbf{v}_H + \sigma_D\mathbf{E}_\eta$. Here the Hall velocity $\mathbf{v}_H = c\mathbf{E}\times\mathbf{B}/B^2$, and Drude conductivity $\sigma_D = e^2n(\mathbf{r})/m{\omega_c}^2\tau_{tr}$ is insensitive to the strength of dc fields, to the microwave irradiation, and to the details of the energy distribution of electrons [68].

We will see that even weak modulation of the DoS due to Landau quantization,

$$\tilde{\nu}(\varepsilon) \equiv \nu(\varepsilon)/\nu_0 = 1 - 2\lambda\cos(2\pi\varepsilon/\omega_c), \quad \lambda = \exp(-\pi/\omega_c\tau_q) \ll 1 \tag{1}$$

drastically changes this picture and makes the transport properties of 2DEG very sensitive both to external perturbations and to internal inhomogeneities.

Being expressed in terms of migration of the guiding center $\mathbf{R}$ of the cyclotron orbits, the current in translationally invariant system with given gradients of the chemical and electrostatic potentials reads

$$\mathbf{j}(\mathbf{r}) = en(\mathbf{r})\mathbf{v}_H + 2e\nu_0\int d\varepsilon\Big\langle\Delta\mathbf{R}_{\varphi\varphi'}\Pi_{\varphi\varphi'}(\varepsilon)\Big\rangle_{\varphi\varphi'}, \tag{2}$$

where φ and φ' denote the angle of momentum before and after the scattering, $\Delta\mathbf{R}_{\varphi\varphi'} = R_c\mathbf{e}_z\times(\mathbf{n}_{\varphi'} - \mathbf{n}_\varphi)$ with $\mathbf{n}_\varphi = (\cos\varphi, \sin\varphi)$ is the corresponding shift of the guiding center, $\Pi_{\varphi\varphi'}(\varepsilon)$ stands for the probability of such scattering event for electron with initial kinetic energy ε, and angular brackets denote the angular average. In the presence of disorder, characterized by the scattering rate $\tau_\theta^{-1} = \sum \tau_n^{-1}e^{in\theta}$, and microwave radiation of (cyclic) frequency ω, $\Pi_{\varphi\varphi'} = \Pi^{(el)}_{\varphi\varphi'} + \sum_\pm \Pi^{(\pm\omega)}_{\varphi\varphi'}$, where

$$\Pi^{(el)}_{\varphi\varphi'}(\varepsilon) = \tau^{-1}_{\varphi-\varphi'}[1 - \mathcal{P}_\omega\sin^2(\varphi/2 - \varphi'/2)]\mathcal{M}(\varepsilon, e\mathbf{E}\Delta\mathbf{R}_{\varphi\varphi'}, e\mathbf{E}_\eta\Delta\mathbf{R}_{\varphi\varphi'}) \tag{3}$$

accounts for elastic scattering modified by the microwave radiation, and

$$\Pi^{(\pm\omega)}_{\varphi\varphi'}(\varepsilon) = [\mathcal{P}_\omega/2\tau_{\varphi-\varphi'}]\sin^2(\varphi/2 - \varphi'/2)]\mathcal{M}(\varepsilon, e\mathbf{E}\Delta\mathbf{R}_{\varphi\varphi'} \pm \omega, e\mathbf{E}_\eta\Delta\mathbf{R}_{\varphi\varphi'} \pm \omega) \tag{4}$$

gives the rate of microwave photon absorption and emission. In equations above we introduced the dimensionless microwave power, $\mathcal{P}_\omega = [ev_FE_\omega/\omega(\omega\pm\omega_c)]^2$, where E_ω is the (screened [13; 84]) microwave field and + (−) sign corresponds to the passive (active) circular polarization [93].

$$\mathcal{M}(x, y) = \tilde{\nu}(\varepsilon)\tilde{\nu}(\varepsilon + x)f(\varepsilon, \mu)[1 - f(\varepsilon + x, \mu + x - y)] \tag{5}$$

contains the product of initial and final DoS, and occupation of initial and final states given by the the local energy distribution function $f(\varepsilon, \mu)$. In equilibrium, the distribution function $f^{(T)}(\varepsilon, \mu) = 1/(\exp[\varepsilon/T - \mu/T] + 1)$ is position-independent for a given total energy of electron $\varepsilon^{(tot)} = \varepsilon + e\phi$, since it is a function of the difference $\varepsilon^{(tot)} - \eta$ with $\eta = \text{const}(\mathbf{r})$, giving rise to vanishing dissipative current.

In the presence of external electric field, $\mathbf{E}_\eta \neq 0$, and/or microwave radiation, the stationary distribution function obeys the kinetic equation

$$-\tau_{\text{in}}^{-1}\tilde{\nu}(\varepsilon)[f(\varepsilon, \mu) - f^{(T)}(\varepsilon, \mu)] = \Big\langle\Pi_{\varphi\varphi'}(\varepsilon) - \underline{\Pi}_{\varphi\varphi'}(\varepsilon)\Big\rangle_{\varphi\varphi'} \tag{6}$$

which balances the inelastic relaxation (described by the inelastic scattering rate ${\tau_{\text{in}}}^{-1}$) and redistribution of electrons due to spectral diffusion and absorption/emission of the microwave quanta ($\underline{\Pi}$ is the probability of the inverse process with the final state $\{\varepsilon, \varphi\}$). In the classical limit $\lambda \to 0$, Eq. (6) has a solution of the Fermi-Dirac form with temperature T increased due to heating by the dc field and microwave radiation. Much more interesting is the oscillatory correction due to Landau quantization, which at linear order $\mathcal{O}(\lambda)$ has the form

$$f(\varepsilon, \mu) = f^{(T)}(\varepsilon, \mu) + \pi^{-1}\omega_c \lambda A \sin(2\pi\varepsilon/\omega_c)\partial_\varepsilon f^{(T)}(\varepsilon, \mu), \tag{7}$$

$$A = \frac{\mathcal{P}_\omega \tilde{\omega} \sin\tilde{\omega}\bar{\gamma}(\zeta) + \zeta_\eta \cos(\widehat{\mathbf{E}_\eta \mathbf{E}})\partial_\zeta \left[2\mathcal{P}_\omega \sin^2(\tilde{\omega}/2)\bar{\gamma}(\zeta) - \gamma(\zeta)\right]}{\tau_{\rm in}^{-1} + \tau_0^{-1} + 2\mathcal{P}_\omega \sin^2(\tilde{\omega}/2)\bar{\gamma}(\zeta) - \gamma(\zeta)}. \tag{8}$$

In terms of the Bessel function $J_n(x)$ and $\bar{\tau}_n^{-1} = \tau_n^{-1} - \tau_{n-1}^{-1}/2 - \tau_{n+1}^{-1}/2$, we used here

$$\tilde{\omega} = \frac{2\pi\omega}{\omega_c}; \ \{\zeta, \zeta_\eta, \zeta_\mu\} = \frac{2\pi e R_c}{\omega_c}\{E, E_\eta, E_\mu\}; \ \gamma(x) = \sum \tau_n^{-1} J_n^2(x); \ \bar{\gamma}(x) = \sum \bar{\tau}_n^{-1} J_n^2(x). \tag{9}$$

In the homogeneous case $\mathbf{E} = \mathbf{E}_\eta$, $\mathbf{E}_\mu = 0$, Eqs. (7), (8) reproduce Eq. (4.5) of Ref. [90].

Calculation of the current, Eqs. (2)-(5), with known distribution function, Eqs. (7), (8), is straightforward. At order $\mathcal{O}(\lambda)$ the result reads

$$\begin{aligned} \mathbf{j}(\mathbf{r}) = en(\mathbf{r})\mathbf{v}_H + \sigma_D \mathbf{E}_\eta - \sigma_D S_{\rm SdH}[\mathcal{K}(\mathbf{E}_\eta) + \mathcal{K}(\mathbf{E}) + \mathcal{K}(\mathbf{E}_\mu)] \\ + 2\sigma_D S_{\rm SdH}\mathcal{P}_\omega \sin^2(\tilde{\omega}/2)[\bar{\mathcal{K}}(\mathbf{E}_\eta) + \bar{\mathcal{K}}(\mathbf{E})] + \sigma_D S_{\rm SdH} A\, \mathcal{K}(\mathbf{E}_\mu), \end{aligned} \tag{10}$$

where $\mathcal{K}(\mathbf{E}) = \dfrac{\mathbf{E}\tau_{\rm tr}}{2\zeta}\dfrac{\partial\gamma(\zeta)}{\partial\zeta}$, $\bar{\mathcal{K}}(\mathbf{E}) = \dfrac{\mathbf{E}\tau_{\rm tr}}{2\zeta}\dfrac{\partial\bar{\gamma}(\zeta)}{\partial\zeta}$ [in $\mathcal{K}(\mathbf{E}_\eta)$ one should make the replacement $\mathbf{E} \to \mathbf{E}_\eta$ and $\zeta \to \zeta_\eta$ etc.], and $S_{\rm SdH} = -4\lambda\cos[2\pi\mu(\mathbf{r})/\omega_c]X_T/\sinh X_T$ with $X_T = 2\pi^2 T/\omega_c$.

It is interesting to mention that the inelastic correction to SdHO, represented by the last term in Eq. (10), vanishes in the homogeneous case $E_\mu = 0$. In the limit $\zeta, \zeta_\eta \to 0$ of weak electric fields, Eq. (10) gives ordinary SdHO with first-order microwave-induced corrections

$$\begin{aligned} \mathbf{j}(\mathbf{r}) = en(\mathbf{r})\mathbf{v}_H + \sigma_D \mathbf{E}_\eta + \sigma_D S_{\rm SdH}\mathbf{E}_\eta[1 - (\tau_{\rm tr}/\tau_*)\mathcal{P}_\omega \sin^2(\tilde{\omega}/2)] \\ - \frac{1}{2}\sigma_D S_{\rm SdH}\mathbf{E}_\mu \mathcal{P}_\omega \left[\frac{\tau_{\rm tr}}{\tau_*}\sin^2\frac{\tilde{\omega}}{2} + \frac{\tilde{\omega}\sin\tilde{\omega}}{\tau_{\rm tr}/\tau_{\rm in} + 2\mathcal{P}_\omega \sin^2(\tilde{\omega}/2)}\right], \end{aligned} \tag{11}$$

where $\tau_*^{-1} = 3/\tau_0 - 4/\tau_1 + 1/\tau_2$ (in terms of τ_n, $\tau_q = \tau_0$ and $1/\tau_{\rm tr} = 1/\tau_0 - 1/\tau_1$). Another interesting case is the limit of strong electric field $\zeta_\eta \gg 1$ in homogeneous 2DEG, $\mathbf{E} = \mathbf{E}_\eta$, where

$$\mathbf{j}(\mathbf{r}) = en(\mathbf{r})\mathbf{v}_H + \sigma_D \mathbf{E} - \sigma_D S_{\rm SdH}\mathbf{E}\frac{2\tau_{\rm tr}}{\pi\tau_\pi}\frac{\cos 2\zeta}{\zeta^2}\left[1 - 4\mathcal{P}_\omega \sin^2\frac{\tilde{\omega}}{2}\right]. \tag{12}$$

Here $\tau_\pi^{-1} = \sum \tau_n^{-1}\cos\pi n$ is the backscattering rate. This expression shows periodic change of the phase of SdHO (modified by the microwave radiation) with increasing dc bias.

At temperature higher than the Dingle temperature $T_D = (2\pi\tau_q)^{-1}$, first-order corrections $\propto S_{\rm SdH}$ to the Drude result become exponentially suppressed. Neglecting the first-order terms and collecting second-order corrections which do not oscillate with μ/ω_c and, therefore, survive high $T > T_D$, one obtains

$$\begin{aligned} \mathbf{j}(\mathbf{r}) = en(\mathbf{r})\mathbf{v}_H + 8\lambda^2\sigma_D A\, \mathcal{K}(\mathbf{E}) + 4\lambda^2\sigma_D\tilde{\omega}\sin\tilde{\omega}\mathcal{P}_\omega\bar{\mathcal{K}}(\mathbf{E}) \\ + 2\lambda^2\sigma_D\tau_{\rm tr}\Big[\frac{\mathbf{E}(\mathbf{E}_\eta\cdot\mathbf{E})}{E^2}\frac{\partial^2}{\partial\zeta^2} + \frac{\mathbf{v}_H(\mathbf{E}_\eta\cdot\mathbf{v}_H)}{v_H^2}\frac{\partial}{\zeta\partial\zeta}\Big]\Big\{2\sin^2(\tilde{\omega}/2)\mathcal{P}_\omega\bar{\gamma}(\zeta) - \gamma(\zeta)\Big\}. \end{aligned} \tag{13}$$

In different limits, Eq. (13) reproduces known results for MIRO and HIRO obtained in Refs. [65; 67; 71; 73; 79; 81; 87; 90] for various models of disorder potential, as well as results

[58; 83] for photovoltage and photocurrent oscillations which arise in inhomogeneous system due to nonequilibrium violation of the Einstein relation. Furthermore, Eqs. (10) and (13) predict new effects in the nonlinear response of irradiated inhomogeneous systems, which deserve further investigation.

Discussions with A. A. Bykov, S. A. Dorozhkin, M. Khodas, A. D. Mirlin, D. G. Polyakov, S. A. Studenikin, M. G. Vavilov, S. A. Vitkalov, M. A. Zudov and support from the DFG, the DFG-CFN, Rosnauka Grant no. 02.740.11.5072, and the RFBR are gratefully acknowledged.

References

[1] Zudov M A, Du R R, Simmons J A and Reno J L 2001 *Phys. Rev. B* **64** 201311(R)

[2] Ye P D, Engel L W, Tsui D C, Simmons J A, Wendt J R, Vawter G A and Reno J L 2001 *Appl. Phys. Lett.* **79** 2193

[3] Dorozhkin S I 2003 *JETP Lett.* **77** 577

[4] Mani R G, Smet J H, von Klitzing K, Narayanamurti V, Johnson W B and Umansky V 2004 *Phys. Rev. Lett.* **92** 146801

[5] Dorozhkin S I, Smet J H, Umansky V and von Klitzing K 2005 *Phys. Rev. B* **71** 201306(R)

[6] Studenikin S A, Potemski M, Sachrajda A, Hilke M, Pfeiffer L N and West K W 2005 *Phys. Rev. B* **71** 245313

[7] Yuan Z Q, Yang C L, Du R R, Pfeiffer L N and West K W 2006 *Phys. Rev. B* **74** 075313

[8] Dorozhkin S I, Smet J H, von Klitzing K, Pfeiffer L N and West K W 2007 *JETP Lett.* **86** 543

[9] Dorozhkin S I, Bykov A A and and AK Bakarov I P 2007 *JETP Lett.* **85** 543

[10] Bykov A A 2008 *JETP Lett.* **87** 233

[11] Hatke A T, Zudov M A, Pfeiffer L N and West K W 2009 *Phys. Rev. Lett.* **102** 066804

[12] Tung L C, Yang C, Smirnov D, Pfeiffer L, West K, Du R and Wang Y J 2009 *Solid State Commun.* **149** 1531

[13] Fedorych O M, Potemski M, Studenikin S A, Gupta J A, Wasilewski Z R and Dmitriev I A 2010 *Phys. Rev. B* **81** 201302(R)

[14] Bykov A A 2009 *JETP Lett.* **89** 575

[15] Mani R G, Johnson W B, Umansky V, Narayanamurti V and Ploog K 2009 *Phys. Rev. B* **79** 205320

[16] Wiedmann S, Gusev G M, Raichev O E, Bakarov A K and Portal J C 2010 *Phys. Rev. B* **81** 085311

[17] Dai Y, Du R R, Pfeiffer L N and West K W 2010 *Phys. Rev. Lett.* **105** 246802

[18] Hatke A T, Zudov M A, Pfeiffer L N and West K W 2011 *Phys. Rev. B* **83** 121301(R)

[19] Yang C L, Zhang J, Du R R, Simmons J A and Reno J L 2002 *Phys. Rev. Lett.* **89** 076801

[20] Zhang W, Chiang H S, Zudov M A, Pfeiffer L N and West K W 2007 *Phys. Rev. B* **75** 041304(R)

[21] Hatke A T, Zudov M A, Pfeiffer L N and West K W 2009 *Phys. Rev. B* **79** 161308(R)

[22] Zhang W, Zudov M A, Pfeiffer L N and West K W 2007 *Phys. Rev. Lett.* **98** 106804

[23] Hatke A T, Chiang H S, Zudov M A, Pfeiffer L N and West K W 2008 *Phys. Rev. B* **77** 201304(R)

[24] Zhang J Q, Vitkalov S, Bykov A A, Kalagin A K and Bakarov A K 2007 *Phys. Rev. B* **75** 081305(R)

[25] Bykov A A, Zhang J Q, Vitkalov S, Kalagin A K and Bakarov A K 2007 *Phys. Rev. Lett.* **99** 116801
[26] Romero N, McHugh S, Sarachik M P, Vitkalov S A and Bykov A A 2008 *Phys. Rev. B* **78** 153311
[27] Vitkalov S 2009 *Int. J. of Mod. Phys. B* **23** 4727
[28] Hatke A T, Chiang H S, Zudov M A, Pfeiffer L N and West K W 2010 *Phys. Rev. B* **82** 041304(R)
[29] Zhang J Q, Vitkalov S and Bykov A A 2009 *Phys. Rev. B* **80** 045310
[30] Bykov A A, Zhang J, Vitkalov S, Kalagin A K and Bakarov A K 2005 *Phys. Rev. B* **72** 245307
[31] Hatke A T, Zudov M A, Pfeiffer L N and West K W 2011 *Phys. Rev. B* **83** 081301(R)
[32] Hatke A T, Chiang H S, Zudov M A, Pfeiffer L N and West K W 2008 *Phys. Rev. Lett.* **101** 246811
[33] Zudov M A, Ponomarev I V, Efros A L, Du R R, Simmons J A and Reno J L 2001 *Phys. Rev. Lett.* **86** 3614
[34] Bykov A A, Kalagin A K and Bakarov A K 2005 *JETP Lett.* **81** 523
[35] Zhang W, Zudov M A, Pfeiffer L N and West K W 2008 *Phys. Rev. Lett.* **100** 036805
[36] Hatke A T, Zudov M A, Pfeiffer L N and West K W 2009 *Phys. Rev. Lett.* **102** 086808
[37] Bykov A A and Goran A V 2009 *JETP Lett.* **90** 578
[38] Zudov M A, Chiang H S, Hatke A T, Zhang W, Pfeiffer L N and West K W 2009 *Int. J. of Mod. Phys. B* **23** 2684
[39] Bykov A A, Goran A V and Vitkalov S A 2010 *Phys. Rev. B* **81** 155322
[40] Mani R G, Smet J H, von Klitzing K, Narayanamurti V, Johnson W B and Umansky V 2002 *Nature* **420** 646
[41] Zudov M A, Du R R, Pfeiffer L N and West K W 2003 *Phys. Rev. Lett.* **90** 046807
[42] Yang C L, Zudov M A, Knuuttila T A, Du R R, Pfeiffer L N and West K W 2003 *Phys. Rev. Lett.* **91** 096803
[43] Du R R, Zudov M A, Yang C L, Yuan Z Q, Pfeiffer L N and West K W 2004 *Int. J. of Mod. Phys. B* **18** 3465
[44] Du R R, Zudov M A, Yang C L, Pfeiffer L N and West K W 2004 *Physica E* **22** 7
[45] Zudov M A 2004 *Phys. Rev. B* **69** 041304(R)
[46] Yang C L, Du R R, Pfeiffer L N and West K W 2006 *Phys. Rev. B* **74** 045315
[47] Bykov A A 2007 *JETP Lett.* **86** 608
[48] Mani R G 2005 *Phys. Rev. B* **72** 075327
[49] Zudov M A, Du R R, Pfeiffer L N and West K W 2006 *Phys. Rev. B* **73** 041303(R)
[50] Zudov M A, Du R R, Pfeiffer L N and West K W 2006 *Phys. Rev. Lett.* **96** 236804
[51] Smet J H, Gorshunov B, Jiang C, Pfeiffer L, West K, Umansky V, Dressel M, Meisels R, Kuchar F and von Klitzing K 2005 *Phys. Rev. Lett.* **95** 116804
[52] Bykov A A, Bakarov A K, Islamov D R and Toropov A I 2006 *JETP Lett.* **84** 391
[53] Bykov A A 2010 *JETP Lett.* **91** 361
[54] Andreev A V, Aleiner I L and Millis A J 2003 *Phys. Rev. Lett.* **91** 056803
[55] Auerbach A, Finkler I, Halperin B I and Yacoby A 2005 *Phys. Rev. Lett.* **94** 196801
[56] Finkler I G and Halperin B I 2009 *Phys. Rev. B* **79** 085315

[57] Dmitriev I A, Gellmann R and Vavilov M G 2010 *Phys. Rev. B* **82** 201311
[58] Dmitriev I A, Dorozhkin S I and Mirlin A D 2010 *Physica E* **42** 1159
[59] Raichev O E 2008 *Phys. Rev. B* **78** 125304
[60] Ryzhii V I 1970 *Sov. Phys. Solid State* **11** 2078
[61] Ryzhii V I, Suris R A and Shchamkhalova B S 1986 *Sov. Phys. Semicond.* **20** 1299
[62] Durst A C, Sachdev S, Read N and Girvin S M 2003 *Phys. Rev. Lett.* **91** 086803
[63] Shi J and Xie X C 2003 *Phys. Rev. Lett.* **91** 086801
[64] Lei X L and Liu S Y 2003 *Phys. Rev. Lett.* **91** 226805
[65] Dmitriev I A, Mirlin A D and Polyakov D G 2003 *Phys. Rev. Lett.* **91** 226802
[66] Ryzhii V, Chaplik A and Suris R 2004 *JETP Lett.* **80** 363
[67] Vavilov M G and Aleiner I L 2004 *Phys. Rev. B* **69** 035303
[68] Dmitriev I A, Mirlin A D and Polyakov D G 2004 *Phys. Rev. B* **70** 165305
[69] Dietel J, Glazman L I, Hekking F W J and von Oppen F 2005 *Phys. Rev. B* **71** 045329
[70] Torres M and Kunold A 2005 *Phys. Rev. B* **71** 115313
[71] Dmitriev I A, Vavilov M G, Aleiner I L, Mirlin A D and Polyakov D G 2004 *Physica E* **25** 205
[72] Alicea J, Balents L, Fisher M P A, Paramekanti A and Radzihovsky L 2005 *Phys. Rev. B* **71** 235322
[73] Dmitriev I A, Vavilov M G, Aleiner I L, Mirlin A D and Polyakov D G 2005 *Phys. Rev. B* **71** 115316
[74] Kashuba A 2006 *Phys. Rev. B* **73** 125340
[75] Dmitriev I A, Mirlin A D and Polyakov D G 2007 *Phys. Rev. Lett.* **99** 206805
[76] Dietel J 2006 *Phys. Rev. B* **73** 125350
[77] Park K 2004 *Phys. Rev. B* **69** 201301
[78] Volkov V A and Takhtamirov E E 2007 *JETP* **104** 1063
[79] Dmitriev I A, Mirlin A D and Polyakov D G 2007 *Phys. Rev. B* **75** 245320
[80] Lei X L 2009 *Phys. Rev. B* **79** 115308
[81] Dmitriev I A, Khodas M, Mirlin A D, Polyakov D G and Vavilov M G 2009 *Phys. Rev. B* **80** 165327
[82] Raichev O E 2010 *Phys. Rev. B* **81** 165319
[83] Dmitriev I A, Dorozhkin S I and Mirlin A D 2009 *Phys. Rev. B* **80** 125418
[84] Pechenezhskii I V, Dorozhkin S I and Dmitriev I A 2007 *JETP Lett.* **85** 86
[85] Lei X L 2008 *Phys. Rev. B* **77** 205309
[86] Dmitriev I A, Mirlin A D and Polyakov D G 2008 *Physica E* **40** 1332
[87] Vavilov M G, Aleiner I L and Glazman L I 2007 *Phys. Rev. B* **76** 115331
[88] Lei X L 2007 *Appl. Phys. Lett.* **90** 132119
[89] Dmitriev I A, Evers F, Gornyi I V, Mirlin A D, Polyakov D G and Wölfle P 2008 *Phys. St. Sol. B* **245** 239
[90] Khodas M and Vavilov M G 2008 *Phys. Rev. B* **78** 245319
[91] Lei X L 2007 *Appl. Phys. Lett.* **91** 112104
[92] Khodas M, Chiang H S, Hatke A T, Zudov M A, Vavilov M G, Pfeiffer L N and West K W 2010 *Phys. Rev. Lett.* **104** 206801
[93] In the general case of elliptic polarization, the results usually (see, e.g., Ref. [81]) change only slightly but the analytic expressions are much more complex.

Symmetry breaking as the origin of zero-differential resistance states of a 2DEG in strong magnetic fields.

Manuel Torres [1] **and Alejandro Kunold** [2]

[1] Instituto de Física, Universidad Nacional Autónoma de México, 01000, México
[2] Departamento de Ciencias Básicas, Universidad Autónoma Metropolitana-Azcapotzalco, 02200, México

E-mail: `torres@fisica.unam.mx`

Abstract.
Zero resistance differential states have been observed in two-dimensional electron gases (2DEG) subject to a magnetic field and a strong dc current. In a recent work we presented a model to describe the nonlinear transport regime of this phenomenon. From the analysis of the differential resistivity and the longitudinal voltage we predicted the formation of negative differential resistivity states, although these states are known to be unstable. Based on our model, we derive an analytical approximated expression for the Voltage-Current characteristics, that captures the main elements of the problem. The result allow us to construct an energy functional for the system. In the zero temperature limit, the system presents a quantum phase transition, with the control parameter given by the magnetic field. It is noted that above a threshold value ($B > B_{th}$), the symmetry is spontaneously broken. At sufficiently high magnetic field and low temperature the model predicts a phase with a non-vanishing permanent current; this is a novel phase that has not been observed so far.

1. Introduction

The study of non-equilibrium magnetotransport in high mobility two-dimensional electron gases (2DEG) has acquired great experimental and theoretical interest. Strong magnetoresistance oscillations (MIRO) and zero resistance states (ZRS) have been observed in high mobility $GaAs/Al_xGa_{1-x}As$ heterostructures subject to a magnetic field and to microwave radiation [1, 2]. Our current understanding of this phenomenon rests upon models that predict the existence of negative-resistance states (NRS) yielding an instability that rapidly drives the system into a ZRS [3]. Two distinct mechanisms for the generation of NRS are known, one is based in the microwave-induced impurity scattering [4, 5, 6, 7, 8, 9], while the second is linked to inelastic processes leading to a non-trivial electron distribution function [10, 11, 12].

More recently, Hall field-induced resistance oscillations (HIRO) and zero differential resistance states (ZDRS) have been observed in response to strong dc electric current excitations [13, 14, 15]. The effect of a direct dc current on electron transport can be quite dramatic leading to zero differential resistance states (ZDRS)[16]. At low temperature and above a threshold bias current the differential resistance vanishes and the longitudinal dc voltage becomes constant. Bykov et al.[16] discussed their experimental results following an approach similar to that of Andreev et al. [3]. In terms of the differential longitudinal resistivity ρ_{xx} the stability condition reads

$$\rho_{xx} = \frac{\partial E_x}{\partial J_x} \geq 0\,. \tag{1}$$

Hence, according to the condition in Eq. (1) the 2DES is unstable at negative differential resistance. The presence of the ZDRS can be attributed to the formation of negative differential resistance states (NDRS) that yields an instability that drives the system into a ZDRS.

2. Model and results

In a recent work we presented a model [17, 18] to describe the nonlinear response to a direct dc current applied to a 2DEG in a strong magnetic field. The model incorporates the exact dynamics of two-dimensional damped electrons in the presence of arbitrarily strong magnetic and dc electric fields, while the effects of randomly distributed impurities are perturbatively added. From the analysis of the differential resistivity and the longitudinal voltage we observe the formation of negative differential resistivity states. Both the effects of elastic impurity scattering as well as those related to inelastic processes play an important role. The theoretical predictions correctly reproduce the main experimental features provided that the inelastic scattering rate obeys a T^2 temperature dependence, consistent with electron-electron interaction effects.

The more relevant results are presented here. After the current density is worked out, it splits into a Drude and an impurity induced contribution: $\mathbf{J} = \mathbf{J}^D + \mathbf{J}^{imp}$. The Drude contribution is given by $\mathbf{J}^D = n_e\, e\, \mathbf{v}_d$, where n_e is the electron density and the drift velocity is given as

$$\mathbf{v}_d = \frac{e\tau}{m^*}\frac{\mathbf{E} - \omega_c\tau\, \mathbf{e}_z \times \mathbf{E}}{1+\omega_c^2\tau_{tr}^2}\,. \tag{2}$$

Here, e_z is the unit vector normal to the plane of the system, ω_c is the cyclotron frequency, m^* is the effective mass of the charge carrier, and τ_{tr} is the transport scattering time. It is important to point out that the Drude term is not added by hand; instead it explicitly appears in our formalism because the exact solution of the Schrödinger equation in the presence of magnetic an electric fields is obtained from a unitary transformation [18] that incorporates the solution of the classical equation of motion.

In the presence of elastic scattering an electron exchanges a momentum $\mathbf{q} = \mathbf{k} - \mathbf{k}'$ with the impurity scatterers. Hence, the components of the impurity induced density current can be expressed as

$$J_i^{imp} = \frac{e\,\omega_c\, n_{imp}}{\hbar^2}\sum_{\mu\mu'}\int\frac{d^2q}{2\pi}\left[f\left(\mathcal{E}_{\mu,\frac{q}{2}}\right) - f\left(\mathcal{E}_{\mu',-\frac{q}{2}}\right)\right]G^i_{\mu\mu'}(q)\,, \tag{3}$$

where $i = x, y$; n_{imp} is the impurity density, and f is the Fermi distribution function evaluated at the energies $\mathcal{E}_{\mu,q/2}$ and $\mathcal{E}_{\mu',-q/2}$ respectively. The Landau levels are tilted by the electric field according to $\mathcal{E}_{\mu,k} = \hbar\omega_c\left(\mu+\frac{1}{2}\right) + \hbar\omega_k$; with $\omega_k = \mathbf{k}\cdot\mathbf{v}_d$. Notice that the x component of the exchanged momentum $q_x = k_x - k_x'$ is equivalent to a hopping (shifting of the guiding center) in the y direction $\Delta y = l_B^2 q_x$; with $l_B = \sqrt{\hbar/eB}$ the magnetic length. The function $G^i_{\mu\mu'}$ in Eq. (3) is given by

$$G^i_{\mu\mu'} = |V(q)|^2\,|D_{\mu\mu'}(z_q)|^2 \times \frac{q_i\Delta_{\mu\mu'} + 2\,|\epsilon_{ij}|\, q_j\omega_c\Gamma}{\Delta^2_{\mu\mu'} + 4\omega_c^2\Gamma^2}, \tag{4}$$

in the previous equation $\epsilon_{i,j}$ is the 2D antisymmetric tensor, $\Delta_{\mu\mu'} = [\omega_q + \omega_c(\mu-\mu')]^2 - \omega_c^2 + \Gamma^2$, $V(q)$ is the Fourier component of the scattering potential, and Γ is the Landau level broadening factor. In order to take into account the known fact that the width of LLs depends on the magnetic field [19], we consider $\Gamma^2 = \beta(2\omega_c/\pi\tau_{tr})$ where β is a phenomenological parameter that takes into account the difference between the transport scattering time τ_{tr} and the quantum scattering time τ_q. In the case of short range neutral impurities $\beta = 1$. The matrix elements $D_{\mu,\nu}$ are given in terms of the associated Laguerre polynomial, see [18].

In recent work is has been reported that the temperature dependence of nonlinear oscillatory magnetoresistance in 2DEG subject to a strong dc electric field can be explained if the quantum scattering time τ_q incorporates a temperature dependence that is attributed to electron-electron interactions [20]. The quantum scattering rate is written as

$$\frac{1}{\tau_q} = \frac{1}{\tau_{imp}} + \frac{1}{\tau_{ee}}, \qquad \frac{1}{\tau_{ee}} = \frac{\lambda\,(k_B T)^2}{\hbar\, E_F}, \tag{5}$$

the parameter λ has to be experimentally determined, but it is a constant of the order of the unity. For short range neutral scatterers the impurity rate can be estimated as $1/\tau_{imp} = 4\pi^2\hbar\alpha n_{imp}/m^*$, where α is related to the scattering length [8].

Let us now consider the nonlinear transport regime. In a current controlled scheme: the longitudinal density current is fixed to a constant value J_{dc} and there is no transverse current, J_y. This leads to a set of two implicit equations for the density current

$$\begin{aligned} J_x\,(E_x, E_y) &= J_{dc}\,, \\ J_y\,(E_x, Ey) &= 0\,. \end{aligned} \tag{6}$$

They represent two implicit equation for the unknown E_x and E_y; the equations can be solved following a self-consistent iteration [18]. However, it is verified that for the conditions that apply in experiments and in the separated LL regions ($\omega_c\tau_{tr} \gg 1$), the solution of the previous equations simplify because the following conditions $E_x \ll E_y$ and $J_x^{imp} \ll en_eE_y/B$ are simultaneously satisfied. Hence, from the first equation in (6) it follows that the leading contribution to the Hall electric field is given by the classical result $E_y = BJ_{dc}/en_e$. Whereas the second equation in (6) yields a nonlinear $E_x - J_{dc}$ relation

$$E_x = \frac{m^*}{e^2\, n_e\tau_{tr}}\, J_{dc} + \frac{B}{en_e}\, J_y^{imp}\left(E_x \sim 0, E_y \sim \frac{BJ_{dc}}{en_e}\right). \tag{7}$$

The expression for the impurity induced currents J_y^{imp} is given in Eq. (3). In what follows we set $J_{dc} \equiv J_x$.

We had previously presented numerical results in very good agreement with the experimental ones, based in Eq. (7) [17], and in the self consistent solution of equations in (6) [18]. However in this work we point out that an approximated analytical $E_x - J_x$ relation can be obtained. As we are interested in the separated LL region the dominant effect is provided by the intra-Landau transitions. We include the following assumptions: (1) only the LL nearest to the Fermi level contribute. (2) The potential is taken as a constant $V(q) \equiv 2\pi\hbar^2\alpha/m^*$, corresponding to a short-range impurity scattering, and (3) the integrand in Eq. (3) has a dominant peak at the value $q_{max} \approx \sqrt{8\pi\, n_e}$, this allow us to carry out explicitly the integral over the Laguerre polynomials. After the angular integral are evaluated, a simply $E_x - J_x$ relation is obtained

$$E_x = C_1\,\frac{J_x}{\tau_{tr}} + C_2\,\frac{B^{3/2}}{\tau_{tr}\,J_x}\left[1 - \frac{1}{\sqrt{1 + \frac{C_3\,\tau_{tr}\,J_x^2}{B}}}\right], \tag{8}$$

the parameters in this equation are given as

$$C_1 = \frac{m^*}{e^2 n_e}, \qquad C_2 = \hbar n_{imp}\,\alpha^2\left(\frac{2\pi e}{m^*}\right)^{3/2}, \qquad C_3 = \frac{2\pi^2 m^*}{e^3 n_e}. \tag{9}$$

Eq. (8) depends on the a temperature through the T-dependence of τ_{tr}. In the simplest approximation, short range impurity scatterers, we can consider that τ_{tr} is proportional to τ_q, consequently the T dependence is dictated by Eq. (5). Utilizing this assumption and the simple equation in (5) we carried out the comparison with the experimental results, observing a very good agreement. In particular the transition from positive to zero differential resistance states shows the correct dependence on the temperature [21].

In this work we shall concentrate in a simple but interesting limit, we take $T = 0$ and consider the analysis of the stability of the NDRS. Therefore the transport scattering rate coincides with the impurity scattering rate: $1/\tau_{tr} = 1/\tau_{imp} = 4\pi^2\hbar\alpha n_{imp}/m^*$. In the region of positive differential resistivity $\rho_{xx} = \frac{\partial E_x}{\partial J_x} > 0$, the relation between between E_x and J_x is single-valued and an homogeneous uniform electron density throughout the system is stable. On the other hand, in the region in which ρ_{xx} is negative corresponding to a NDRS, the system becomes unstable. An approximated scheme (Maxwell construction) replaces the negative slope by a horizontal line, *i.e.* a ZDRS. The real reason for the appearance of a negative slope in the $E_x - J_x$ plot is the implicit restraint of uniform electron density throughout the system. However in these regions configurations with a non uniform current distribution turn out to be the equilibrium configuration of the system. The simplest possible pattern is a domain wall: two parts of the sample carry stable density currents J_{x1} and J_{x2} with the same value of E_x. Hence we expect a phase transition defined by a critical iso-B line, that passes through a critical point that is a point of inflection of this iso-B line; so both $\frac{\partial E_x}{\partial J_x}$ and $\frac{\partial^2 E_x}{\partial J_x^2}$ vanish at this point. The critical values: $E_x = E_c$, $J_x = J_c$ and $B = B_c$; at which these conditions are satisfied are readily determined as

$$J_c = \xi_1 \frac{C_1}{\tau_{imp}^2\, C_2\, C_3^{3/2}}, \qquad E_c = \xi_2 \frac{C_1^2}{\tau_{imp}^3\, C_2\, C_3^{3/2}}, \qquad B_c^{1/2} = \xi_3 \frac{C_1}{\tau_{imp}^{3/2}\, C_2\, C_3}. \tag{10}$$

with the values of the constants ξ_i given by: $\xi_1 \approx 40.7\,,\ \ \xi_2 \approx 161.9\,,\ \ and\ \ \xi_3 \approx 20.85$. Notice that the method follows similar steps those used in the analysis of the critical point of the Van der Waals equation [22].

Using Eqs. (9), and (10) we can express J_c, E_c and B_c is terms of the parameters of the system. In order to get an estimation we consider: $n_e = 8 \times 10^{15} m^{-2}$, $\mu = 100 m^2 V/s$, $\alpha = 0.1$, $n_{imp} = 4.5 \times 10^{11} m^{-2}$. So we obtain:

$$J_c \sim 0.4\, A/m\,, \qquad E_c \sim 6.5\, V/m\,, \qquad B_c \sim 0.12\, T\,. \tag{11}$$

These values correspond to zero temperature, so are not directly comparable with the experimental ones, however the order of magnitude show a reasonable agreement [16].

We now define reduced variables

$$J_r = \frac{J_x}{J_C}\,, \qquad E_r = \frac{E_x}{E_C}\,, \qquad B_r = \frac{B}{B_C}\,. \tag{12}$$

Hence the critical point is localized at $J_r = E_r = B_r = 1$. Using (8) and (12), we readily obtain the reduced voltage-current characteristic

$$E_r = 0.25\, J_r + \frac{1.38}{J_r} B_r^{3/2} \left(1 - \left[1 + 3.8\frac{J_r^2}{B_r}\right]^{-1/2}\right)\,, \tag{13}$$

which is "universal", in the sense that it does not explicitly contains any parameter of the system.

Fig. 1 display a series of plots of reduced electric field E_r as a function of the reduced longitudinal density current J_r ; for various values of the reduced magnetic field. We clearly observe a phase transition: below the threshold $B_r < 1$ the slope of the curve is always positive

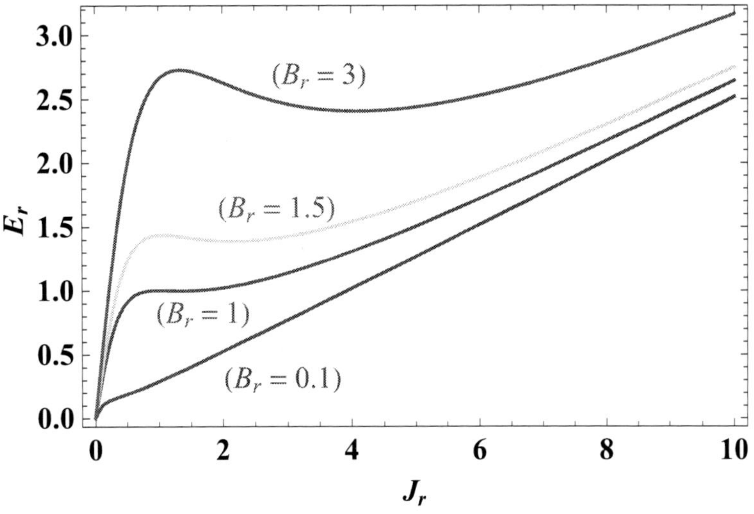

Figure 1. (Color on line). Reduced longitudinal electric field E_r as function of the reduced longitudinal current J_r, the values of B_r are displayed near to each line.

and the system is stable. On the other hand if $B_r > 1$ and above the threshold bias current ($J_r > 1$) the differential resistivity becomes negative.

We can define a scalar Lyapunov functional $W[J_r] = \int d^2r\, w\,[J_r]$, where

$$w[J_r] = \int^{E_r} J_r\,[E'_r]\; dE'_r \;=\; \int^{J_r} J'\,\frac{dE_r}{dJ'_r}\,dJ'_r\,. \tag{14}$$

Using Eq. (13) a simple calculations yields:

$$w[J_r] = 0.125\,J_r^2 + 1.38\,B_r^{3/2}\left[1 - \log\left\{2\left(1+\sqrt{1+3.8\,\frac{J_r^2}{B_r}}\right)\right\} - \left[1+3.8\,\frac{J_r^2}{B_r}\right]^{-1/2}\right]. \tag{15}$$

One can show that $W[J_r]$ is a Lyapunov functional, *i.e.* a non-increasing function of time, so its minima are stable steady states. To the expression for $w[J_r]$ in Eq. (15) one should add a field (current) derivative contribution of the form $\lambda|\nabla\cdot\mathbf{E}|^2$, where λ introduces a domain wall thickness scale that produces a positive contribution to W. Fig. 2 display a series of plots of w as a function of J_r for various values of the reduced magnetic field. If $B_r < 1$ the functional $W[J_r]$ has a single minima at $J_r = 0$, however when $B_r > 1$ new minima appear at $J_r \neq 0$. Furthermore we observe that for sufficiently strong magnetic field the minima at $J_r = 0$ is not longer the ground state. In general the system is expected to relax to one of the local minimum of W, not necessarily the ground state. However in the presence of noise the system is expected to escape from the high-lying minima. This suggest that at sufficiently high magnetic field and low temperature the model predicts a phase with a permanent non-vanishing current J_m, where J_m is the minima of W. This is a novel phase that has not been observed so far, hence it deserves further studies. In future work we will use the present results to analyze in detail the formation and structure of the domain walls and also to include the finite temperature effects.

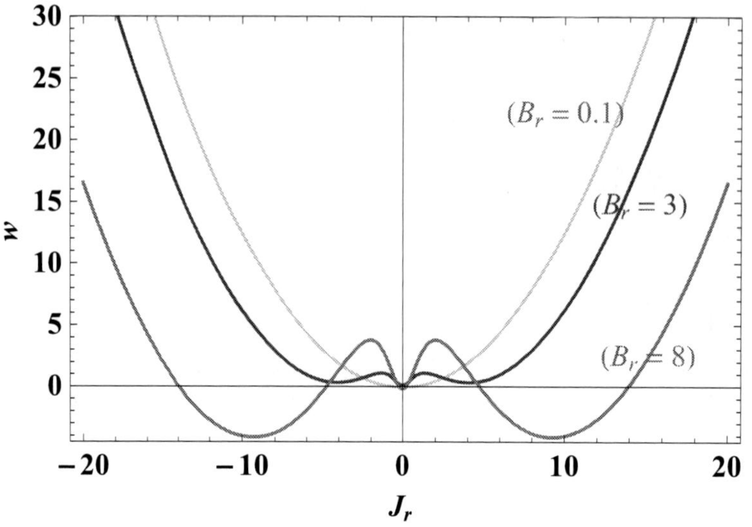

Figure 2. (Color on line). A plot of w vs J_r, for values of $B_r = 0.1,\ 3,\ and 8$.

2.1. Acknowledgments
We acknowledge support from UNAM project PAPIIT-IN118610.

References

[1] M. A. Zudov, R. R. Du, J. A. Simmons, and J. L. Reno, Phys. Rev. B **64**, 201311(R) (2001).
[2] R. G. Mani, J. H. Smet, K. von Klitzing, V. Narayanamurti, W. B. Johnson, and V. Umansky, Nature **420**, 646 (2002).
[3] A. V. Andreev, I. L. Aleiner, and A. J. Millis, Phys. Rev. Lett. **91**, 056803 (2003).
[4] V. I. Ryzhii, Sov. Phys. Solid State **11**, 2078 (1970).
[5] A. C. Durst, S. Sachdev, N. Read, and S. M. Girvin, Phys. Rev. Lett. **91**, 086803 (2003).
[6] X. L. Lei and S. Y. Liu, Phys. Rev. Lett. **91**, 226805 (2003).
[7] M. G. Vavilov and I. L. Aleiner, Phys. Rev. B **69**, 035303 (2004).
[8] M. Torres and A. Kunold, Phys. Rev. B **71**, 115313 (2005).
[9] J. Inarrea and G. Platero, Phys. Rev. B **76**, 073311 (2007).
[10] I. A. Dmitriev, A. D. Mirlin, and D. G. Polyakov, Phys. Rev. Lett. **91**, 226802 (2003).
[11] I. A. Dmitriev, M. G. Vavilov, I. L. Aleiner, A. D. Mirlin, and D. G. Polyakov, Phys. Rev. B **71**, 115316 (2005).
[12] J. P. Robinson, M. P. Kennett, N. R. Cooper, and V. I. Fal'ko, Phys. Rev. Lett. **93**, 036804 (2004).
[13] A. A. Bykov, J. Q. Zhang, S. Vitkalov, A. K. Kalagin, and A. K. Bakarov, Phys. Rev. B **72**, 245307 (2005).
[14] J. Q. Zhang, S. Vitkalov, A. A. Bykov, A. K. Kalagin, and A. K. Bakarov, Phys. Rev. B **75**, 081305(R) (2007).
[15] W. Zhang, H.-S. Chiang, M. A. Zudov, L. N. Pfeiffer, and K. W. West, Phys. Rev. B **75**, 041304(R) (2007).
[16] A. A. Bykov, J. Q. Zhang, S. Vitkalov, A. K. Kalagin, and A. K. Bakarov, Phys. Rev. Lett. **99**, 116801 (2007).
[17] A. Kunold and M. Torres, Physica B **403**, 3803 (2008).
[18] A. Kunold and M. Torres, Phys. Rev. B **80**, 205314 (2009).
[19] T. Ando, A. B. Fowler, and F. Stern, Rev. Mod. Phys. **54**, 437 (1982).
[20] A. T. Hatke, M. A. Zudov, L. N. Pfeiffer, and K. W. West, Phys. Rev. B **79**, 161308(R) (2009b).
[21] A. Kunold and M. Torres, To be published.
[22] R. K. Pathria, *Statistical Mechanics* (Butterworth-Heinemann, Oxford., 1996).

Investigation of Serial Quantum Hall arrays as a Quantum Resistance Standard

J. Könemann, C. Leicht, F.-J. Ahlers, E. Pesel, K. Pierz, and H. W. Schumacher

Physikalisch-Technische Bundesanstalt, Bundesallee 100, 38116 Braunschweig, Germany

E-mail: Jens.Koenemann@ptb.de

Abstract. In this work, we test the feasibility of a resistance standard based on a serial array of ten lithographically interconnected quantum Hall bars based on an Al(Ga)As heterostructure. Our investigation involves a novel multiple layer process for the fabrication of our samples which allows to realize a thin SiO_2 insulations by means of a lift-off process. Additionally, we have performed preliminary precision measurements of the i=2 Hall plateau in order to characterize the metrological properties of our samples.

1. Introduction

In electrical quantum metrology the electrical units are referred to natural constants like the electronic charge *e* and the Planck constant *h*. Namely, the quantum Hall effect (QHE) links the unit Ohm to $R=h/e^2$ [1]. For this quantum Hall resistance the von Klitzing constant $R_K=h/e^2$ was introduced and its value was recommended in the year 1990 as $R_{K\text{-}90}$=25812.807 Ω. For realizing such quantum resistance standards, QHE devices with a two-dimensional electron gas (2DEG) in an Al(Ga)As heterostructure are widely used. In such samples, metrological calibrations are carried out at filling factor $i = 2$, meaning that he corresponding quantized resistance has a value of $R_K/2 \sim 12.9$ kΩ [2,3]. Combining quantum Hall devices in series or in parallel with the help of the multiple link technique [4] allows for tailoring of quantum standards with almost arbitrary resistance values and also over a broad resistance range, i.e. the realization of decadic quantum resistors [5]-[8]. This would increase the number of well known resistance ratios for the validation of resistance bridges and comparator measurement systems. Moreover, quantum Hall array resistance standards (QHARS) might simplify the calibration of standard platinum resistors by shortening the calibration chain. On the other hand, QHE arrays could act as a stable high-precision traveling standard [9]. And, finally, serial QHE-arrays could benefit from their noise behavior which is for resistance values larger than 100 kΩ, expected to be much lower than for equivalent wire resistors. Such low noise, low temperature coefficient, high resistance devices of several MΩ would improve the signal-to-noise ratio in a low current experiment, as, i.e. a direct realization of the quantum metrological triangle experiment [10].

2. Experiment and results

A sketch of the interconnection scheme for a series array of two Hall devices is shown in Fig. 1. The elements of the array are elongated Hall bars connected in series with respect to the current path *I* (between terminals 2 and 3) which is indicated in the figure by the thickened wiring line. Furthermore, two connections of the voltage terminals are present: One direct connection and one cross connection. The direct connection serves to add up the Hall voltages U_{xy} of the two elements measured between terminal 1 and 4 and thus to realize a serial Hall resistor. The additional cross connection serves to suppress the influence of the non negligible contact resistance of the interconnect on the serial Hall voltage. Without the cross connection a voltage measurement between the voltage terminals 1 and 4 would yield a value $U_{1,4} = 2U_{xy} + IR_c$ where U_{xy} is the Hall voltage at filling factor i and R_c is the resistance of the current interconnect. The cross connections serve to reduce the contribution of the contact resistance to $U_{1,2}$: Each additional connection carries a current which is smaller than the current in the preceding connection by a factor $R_C/(R_K/i)$ which is in the order of 10^{-4} for typical R_C values of 1 Ω. Hence, high precision realization of the serial Hall resistance $R_{1,4} = U_{1,4}/I=2(R_K/i)$ is possible. Note however, that this effect of the cross connections only occurs if both the current and voltage interconnections between the two Hall bars terminate on the same equipotential line. Under quantized Hall conditions the two-terminal voltage drop across the Hall bar occurs at two opposing corners of the current terminals, the so called hot spots. The Hall bar edges between these two hot-spots are on the same potential as indicated by the full black and dashed red lines.

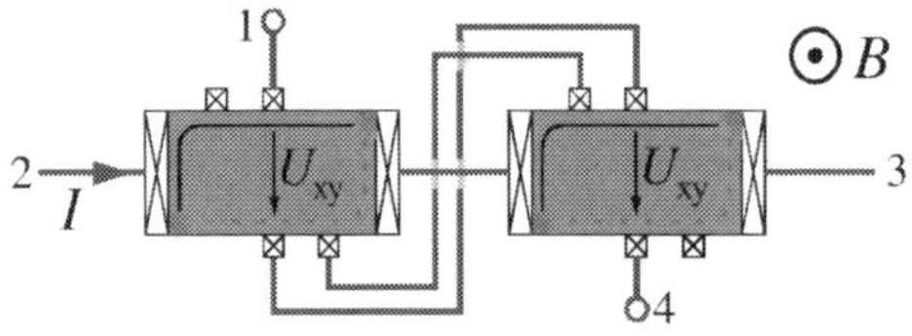

Figure 1. Series connections of two Hall elements with additional cross-connections, see text.

Our samples were grown at a growth temperature of 620°C by solid source MBE on 2" GaAs wafers of a set of 10 nominally identical heterostructures for complete-wafer processing. The heterostructures with a deep two-dimensional electron gas 72 nm below the surface had an electron density of $n \sim 5 \cdot 10^{11}$ cm^{-2} with an electron mobility of $\mu \sim 3.94 \cdot 10^5$ cm^2/Vs. These values were extracted from magnetotransport measurements at 2.2 K of 4 mm × 4 mm pieces in van-der Paauw (vdP)-geometry cleaved from the center of the wafers. From these heterostructures we processed lithographic serial arrays consisting of ten rectangular Hall bars having linear dimensions of 1.05 mm x 0.31mm. The design was directly written by a laser lithographic pattern generator Heidelberg instruments DWL66fs and the sample fabrication involved a multiple layer process.

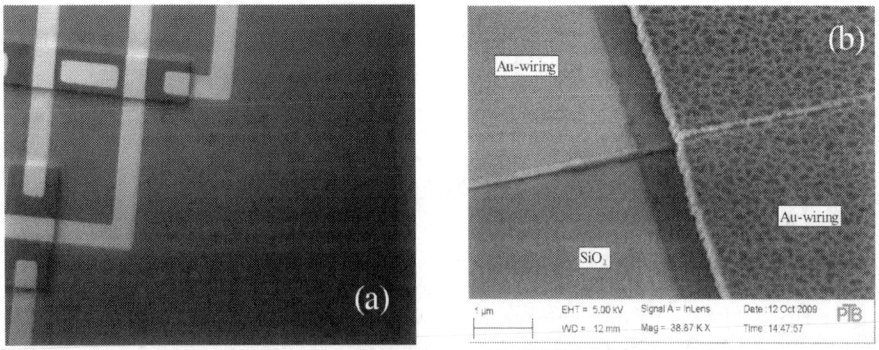

Figure 2. (a) Zoom-in of the interconnection area with SiO2 layer on top of the Au wiring. (b) Electron micrograph of the interconnections.

Journal of Physics: Conference Series **334** (2011) 012017 doi:10.1088/1742-6596/334/1/012017

In a first step, the Hall bars and the alignment markers were defined by a 100 nm deep slow wet etch process. Afterwards standard AuGeNi ohmic contacts with eutectic AuGe and Ni were thermally evaporated and annealed at 450°C for 55 seconds. Then the wiring layer was produced by evaporating 10 nm Ti and 70 nm Au with a connection line width of 10 µm. As discussed before, the serial array needed interconnections for compensating the contact resistances. Hence, we have realized thereafter an insulation layer for the interconnections of the Au wiring with a layer of 100 nm SiO_2 deposited by a PECVD in a lift-off process. The PECVD process was performed in three steps with a rotation of the sample after each turn to avoid bad insulation due to pinholes. The interconnection area is shown in Fig. 2. Two crossing gold wirings separated by SiO2 (not visible) are shown. The electron micrograph well visualizes the steepness of the insulator edge after lift-off. Nevertheless, by realizing the upper wire by a 200 nm Au layer a good bridge connection over the insulator edge can be ensured.

The resulting magnetotransport data for one of our samples at $T = 0.3$ K is presented in Fig. 3. Here an array of 10 Hall bars connected in series according to the scheme of Fig. 1 is investigated. Within the resolution of our pre-characterization, which was better than 60 mΩ, we found for the $i = 2$ plateau vanishing longitudinal resistivities over the corresponding magnetic field range. This indicated the absence of parasitic bulk conductivity and demonstrates the good quantization of the 2DEGs. Comparing the R_{xx}-curves of the two Hall bars at the ends of the array, they fitted quite well to each other within a deviation of less than 2 % in the oscillation period demonstrating the homogeneity of the serial array. The Hall curves showed a flat i=2 resistance plateau with a value of 129.07 kΩ ≈ $10(R_K/2)$ as expected and a width of 1.3 T centered around 9.55 T. Interestingly, reversing the magnetic field direction (B_+) reduced the quantized value of the resistance plateau by exactly the contribution of two Hall bars in the series connection. The reason for this behavior is the shift of the quantum Hall hot spots upon field reversal and the related redistribution of the edge state current paths which results in a partial exchange of the role of the voltage and current terminals. As a consequence the last two Hall bars do not contribute to the total serial voltage. A more detailed discussion of this effect can be found in reference [11,12]. High-precision measurements of R_H were performed at $T = 80$ mK and a current of $I = 1$ µA by means of a cryogenic current comparator (CCC) bridge [13,14] with our well-known QHE standard P137-18.

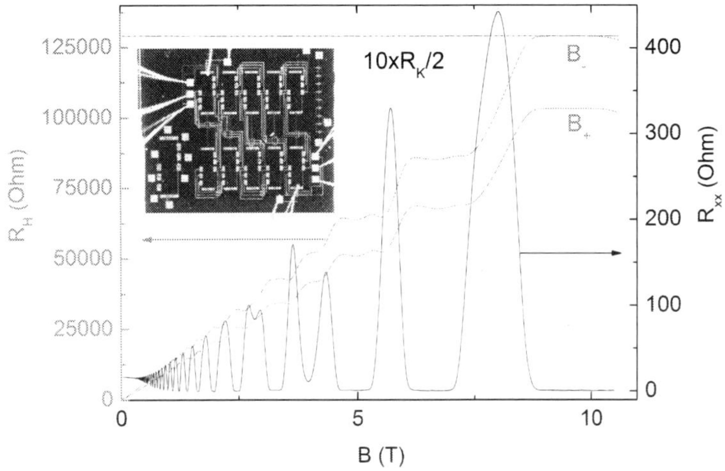

Figure 3. Experimental value of the transverse and longitudinal resistance of lithographic serial 10-Hall bar array for both field configurations, T=0.3 K. Inset shows a photograph of the sample mounted in a TO8 socket suitable for precision measurements.

The quantization of the *i*=2 resistance was measured at the center of the plateau yielding a relative deviation of 1.5×10^{-7} from the expected value $10(R_K/2)$.. The reasons for this deviation are subject of further investigation. Possible reasons could be e.g. an imperfect insulation of the interconnection region or a too high Ohmic contact resistance in some of the Hall bars of the array.

3. Conclusion

In summary, we have investigated a serial array of ten quantum Hall bars as a Quantum Hall Array Resistance standard. We were able to process samples showing a relative deviation of 1.5×10^{-7} from the expected resistance value of $10(R_K/2)$. For future applications in electrical quantum metrology a deviation of 10^{-8} must be obtained. Such values seem feasible upon optimization of fabrication parameters.

C. Leicht has been supported by International Graduate School of Metrology IGSM, Braunschweig.

References

[1] K. von Klitzing, G. Dorda, and M. Pepper," New Method for High-Accuracy Determination of the Fine-Structure Constant Based on Quantized Hall Resistance", *Phys. Rev. Lett.* vol. 45, 494, 1980.
[2] H. Bachmair, E.O. Göbel, G. Hein, J. Melcher, B. Schumacher, J. Schurr, L. Schweitzer, P. Warnecke, "The von Klitzing resistance standard", *Physica* E 20, vol. 14, 2003.
[3] F. Delahaye and B. Jeckelmann, "Revised technical guidelines for reliable dc measurements of the quantized Hall resistance", *Metrologia* vol. 40, 217, 2003.
[4] F. Delahaye, "Series and Parallel Connection of Multiterminal Quantum Hall-Effect devices," in *J. Appl. Phys.*, vol. 74, 7914 (1993).
[5] W. Poirier, A. Bounouh, K. Hayashi, H. Fhirna, F. Piquemal, G. Genevès, and J.P. André, "$R_K/100$ and $R_K/200$ quantum Hall array resistance standards". *J. Appl. Phys.* Vol. 92, 2844, 2002.
[6] W. Poirier, A. Bounouh, F. Piquemal and J.P. André, A new generation of QHARS: discussion about the technical criteria for quantization", *Metrologia*, vol. 41, pp. 285-294, 2004.
[7] G. Hein, B. Schumacher, and F. J. Ahlers, Preparation of Quantum Hall Effect Device Arrays, *CPEM 2004 Conf. Digest*, London, UK, 273, (2004).
[8] N. Kaneko, C. Urano, T. Itatani, and S. Kiryu, Development of a 10 kΩ Quantum Hall Array Resistance Standard, *CPEM 2006 Conf. Digest*, Torino, Italy, pp. 512, (2006.).
[9] R. Goebel, F. Delahaye, B. Jeckelmann, F. Schopfer, and W. Poirier, "Preliminary Investigation of the use of Quantum Hall Array Resistance Standards as Traveling Standards", *CPEM 2006 Conf. Digest*, Torino, Italy, pp. 514, (2006).
[10] K. K. Likharev and A.B. Zorin, "Theory of Bloch oscillations in small Josephson junctions", *J. Low Temp. Phys.*, **59**, 347 (1985).
[11] F. Ahlers, G. Hein, H. W. Schumacher, and K. Pierz,"Quantum Hall Series Arrays under Magnetic Field Reversal", *CPEM Conf. Digest*, Torino, pp. 508 – 509, (2006).
[12] J. Könemann, F. Ahlers, E. Pesel, K. Pierz, H. W. Schumacher, „Serial Arrays of Square Quantum Hall Devices for Resistance Standards", in *Conf. Precision Electromagn. Meas. Dig.*, 2010, pp. 631-632.
[13] E. Pesel, B. Schumacher and P. Warnecke, "Resistance Scaling up to 1 MΩ at PTB with a Cryogenic Current Comparator", *IEEE Trans. Instrum. Meas.*, vol. 44, vol. 2, pp. 273-275, 1995.
[14] E. Pesel, B. Schumacher and P. Warnecke, IEEE Trans. Instrum. Meas., vol. 44, vol. 2, pp. 273-275, 1995.

Spatial mapping of potential fluctuation in GaAs/AlGaAs and graphene by a scanning nanoelectrometer

Yukio Kawano and Koji Ishibashi

RIKEN, 2-1, Hirosawa, Wako, Saitama 351-0198, Japan

ykawano@riken.jp

Abstract. We present observations of potential fluctuation distributions in two-dimensional systems of a GaAs/AlGaAs interface and of a graphene surface. We have exploited scanning electrometers based on gate effects for the sensor via capacitive coupling with a sample. The experimental data for the GaAs/AlGaAs sample show that the temporal potential fluctuation (noise voltage) largely occurs in the transition region between two adjacent quantum-Hall plateaus and that it is concentrated on a sample boundary. This fluctuation can be understood as arising from unstable electron relaxation in the non-equilibrium edge state. For the graphene sample, a Hall potential profile was imaged and a potential fluctuation with the period of 50-100nm was observed.

1. Introduction

Electric potential fluctuations in space and/or time domains play important roles in quantum electron transport of meso- and nano-scopic devices, such as electron localization and delocalization. Interest in such phenomena has attracted much interest, and several scanning probe techniques for obtaining potential images have been developed, such as electro-optic effect imaging [1,2], a Kelvin force microscope (KFM) [3], and a scanning single-electron transistor (SET) [4,5]. Although the KFM and the SET are useful methods for obtaining high-resolution images, one can not rule out the possibility that the application of high voltage between a KFM tip and a sample may disturb the actual potential distribution. The SET is only used in a dilution or ^{3}He refrigerator, because observations of clear coulomb oscillations are required. This condition restricts the range of measurements that are feasible with this method.

We have developed two types of scanning electrometers; a two-dimensional electron gas (2DEG) sensor in GaAs/AlGaAs heterostructure [6] and a cantilever-integrated 2DEG sensor [7]. The sensing mechanism is that local potential is detected through gate effects for the 2DEG sensor via capacitive coupling with a sample. Using this technique, we have demonstrated mapping of the potential distribution in other 2D systems of a GaAs/AlGaAs interface and of a graphene surface. We have imaged spatial distribution of the temporal potential fluctuation (noise voltage) in the GaAs-based 2DEG sample. The experimental data revealed linear distribution, which results from unstable relaxation of edge-state electrons. For the graphene sample, the potential amplitude was shown to fluctuate with the period of 50-100nm. This feature has not been obtained for the 2DEG in the GaAs/AlGaAs heterostructure. We discuss relevant mechanisms of charge dynamics.

2. Experimental setup

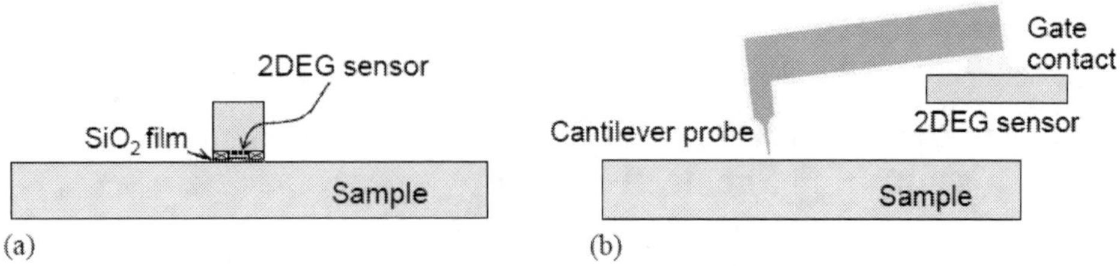

Figure 1. Sketch of the scanning electrometers; (a) 2DEG sensor and (b) cantilever-integrated 2DEG sensor.

Figures 1(a) and (b) schematically depict the two types of potential imaging systems. In Fig. 1(a), a sensor (electrometer) is in contact with a sample. In order to prevent the two devices from being damaged, the surface of the sensor is coated with a 50 nm-thick SiO_2 film. A magnetic field B was applied perpendicular to the 2DEG planes. In this scheme, since the distance between the 2DEG sensor and the sample is very short, they are capacitively coupled. It follows that when a source-drain current I_{sam} is passed through the sample, excess charges ΔQ are induced into the 2DEG, accordingly to $CV(x,y)=\Delta Q$. Here, $V(x,y)$ is a local voltage in the sample right below the sensor, and C is the capacitance between the sensor and the sample. The generation of ΔQ results in the change, ΔR_{sen}, in the sensor resistance, R_{sen}. By moving the sensor over the sample surface and simultaneously measuring $\Delta R_{sen}\times I_{sen}$, a two-dimensional distribution of $V(x,y)$ in the sample is imaged (I_{sen} is the current for the sensor).

Based on the above mechanism, we carried out spatial mapping of the potential fluctuation (noise voltage). Using a lock-in amplifier, we measured the mean-square voltage, S_V, of the voltage fluctuations ΔV, which is expressed as $S_V=\overline{|\Delta V(f)|^2}/\Delta f$. Here, f is a central frequency for the noise detection and Δf is a frequency bandwidth. In this experiment, we set f=1kHz and Δf=10Hz in the lock-in amplifier. The S_V value of the pre-amplifier was about $4\times10^{-18}V^2Hz^{-1}$ at f=1kHz, a level which is low enough to measure only sample noise. In order to remove the noise of the sensor itself, we subtracted S_V at a sample current $I_{sam}=0$ from that at $I_{sam}\neq0$.

The spatial resolution of the above system was about 2μm. In this scheme, the resolution is restricted by the size of the 2DEG sensing area. In order to resolve this problem and improve the resolution, we constructed the second type of electrometer system, as illustrated in Fig. 1(b). Here, a metal-coated cantilever probe with a tip diameter of 20 nm is integrated with the gate contact of the 2DEG sensor. In this system, local potential is detected by the following processes: (i) The metallic probe tip senses local potential through capacitive coupling with the sample; (ii) As a result, the tip region of the probe is electrically charged, leading to electrical polarization of the gate contact of the 2DEG sensor according to charge conservation law; (iii) The polarization of the gate contact results in generation of excess charges, ΔQ, in the 2DEG of the sensor, and thereby the sensor resistance changes. With this system, we were able to obtain a much higher resolution (24nm), which was primarily determined by the tip diameter (20nm).

The sensors were fabricated from GaAs/AlGaAs heterostructure wafers, where the 2DEG layers are located 0.1μm beneath the crystal surface. As samples, we used two types of 2D systems; GaAs/AlGaAs and graphene. We used an exfoliated graphene on a SiO_2/Si substrate. The GaAs/AlGaAs (graphene) sample is a rectangular Hall bar with a length of 200 (10) μm and a width of 50 (0.9) μm.

For potential imaging measurements, the samples were moved by an X-Y translation stage, while the sensors were stationary. The whole system was immersed in a ^{4}He cryostat, and all the measurements were performed at 4.2K.

3. Experimental results

3.1. GaAs/AlGaAs

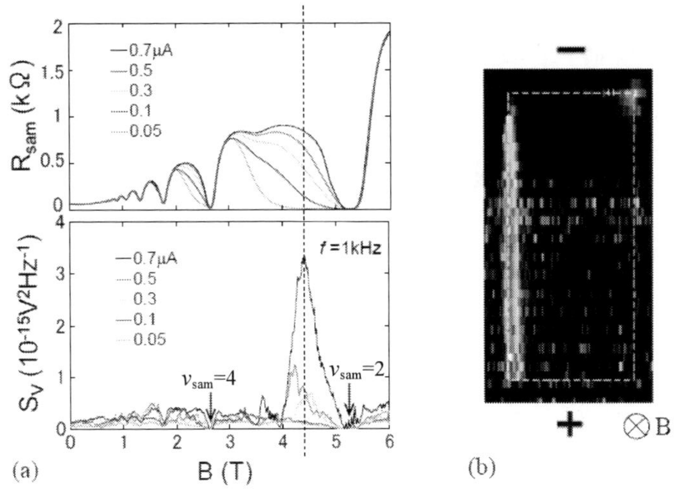

Figure 2. (a) The longitudinal resistance, R_{sam}, of the GaAs/AlGaAs sample (upper panel) and the mean-square voltage, S_V, of the temporal potential fluctuations (lower panel) as a function of magnetic field B. (b) Two-dimensional S_V map in the GaAs/AlGaAs sample at B=4.41T (ν_{sam}=2.38) and I_{sam}=0.7μA. The value of B=4.41T corresponds to the peak position of S_V vs B, as indicated by the broken line in (a). The length and width of the Hall bar sample are 200μm and 50μm, respectively.

First, we studied the basic features of the sample noise with conventional four-terminal transport measurements. Figure 2 (a) displays a longitudinal resistance, R_{sam}, of the sample, and a corresponding noise voltage, S_V, with the sweep of B at I_{sam}=0.05, 0.1, 0.3, 0.5, and 0.7μA. The R_{sam} spectra between ν_{sam}=4- and ν_{sam}=2-quantum Hall plateau s in the upper panel of Fig. 2(a) show that, when I_{sam}<0.5μA, the amplitude of the oscillations normally observed in R_{sam} vs. B is strongly reduced, whereas as the current increases, the amplitude of R_{sam} recovers. (ν_{sam} is Landau-level filling factor of the sample.) These are well-known phenomena, which originate from the non-equilibrium edge transport decoupled from the bulk transport and its quenching for the current above 0.5~1μA [8,9]. The data of S_V vs. B show that, with increasing the current beyond I_{sam}~0.5μA, S_V rapidly grows in the region of 2<ν_{sam}<4. Comparing the B dependence of S_V with that of R_{sam} suggests that the S_V enhancement is related to the unstable electron transfer taking place when the non-equilibrium edge state equilibrates, as the current increases, with the bulk state.

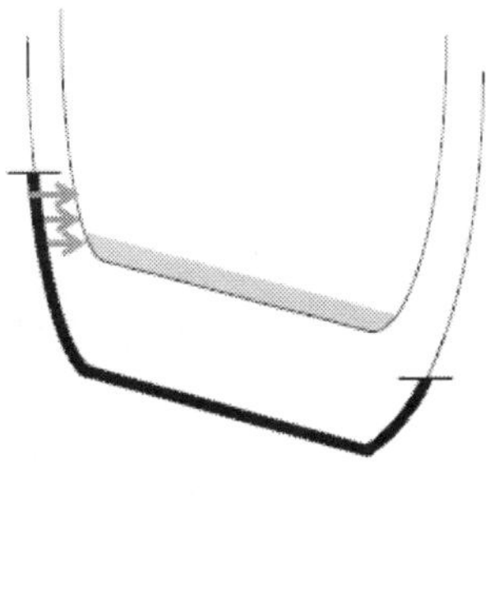

Figure 3. Schematic representation of spatial profile of Landau levels and electron dynamics in the region of 2<ν_{sam}<4.

Figure 2(b) displays S_V mapping for the same sample as that used in the measurements shown in Fig. 2(a) at I_{sam}=0.7μA and B=4.41T (ν_{sam}=2.38). The value of B=4.41T corresponds to the maximum peak position of S_V with the sweep of B at I_{sam}=0.7μA, shown in the lower panel of Fig. 2(a). The S_V map reveals that the most of the noise is concentrated in a higher-potential (lower Hall voltage) edge region of the Hall bar. This result provides clear proof that the S_V enhancement in the region of $2<\nu_{sam}<4$ results from the equilibration process of the electrons from the non-equilibration edge state to the bulk state, as schematically represented in Fig. 3.

This data was taken at f=1kHz and the dynamics with the time constant of 1ms was investigated. The S_V distribution should strongly depend on the time constant characteristic of the charge dynamics and hence on the f value. The S_V mapping in the frequency domain is an interesting next step.

3.2. Graphene

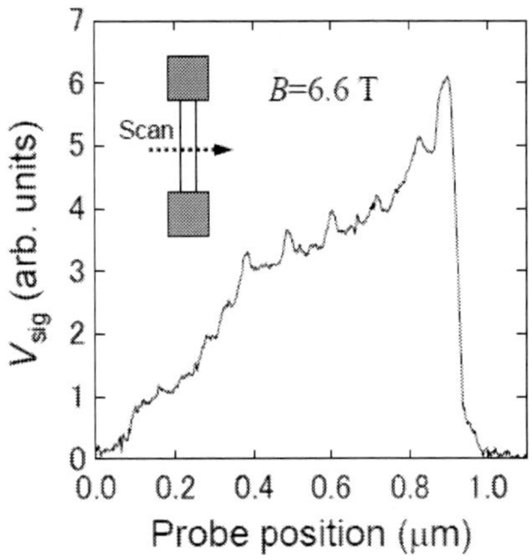

Figure 4. A line-scan plot of the potential profile along the channel-width direction of the graphene sample at B=6.6 T and I_{sam}=0.4μA.

Using the second electrometer system with much higher resolution (24nm), we performed the potential imaging in the graphene Hall bar device. Figure 4 shows a line-scan plot of the Hall potential profile along the channel-width direction at I_{sam}=0.4μA. The graphene device has a channel length of 10μm and a channel width of 0.9μm. The data of Fig. 4 revealed that the potential amplitude fluctuates with the period of 50-100nm. We observed that such potential fluctuation in space was also seen for different positions of the graphene device. This feature has not been obtained on the 2DEG in the GaAs/AlGaAs heterostructure. At present we interpret that the potential fluctuation may arise from topological variation in the graphene surface and resulting change in Hall effect or from trapped charges in the SiO_2 substrate. The elucidation of this feature requires further investigation, including a comparison with potential imaging for epitaxial graphene.

4. Remarks and Summary

We would like to make two remarks on the potential imaging. First, we note future improvement in the detection sensitivity and the spatial resolution of our system. It is known that a quantum point contact (QPC) device works as an ultra-sensitive electrometer [10]. We anticipate that the use of the QPC electrometer would result in the large enhancement of the sensitivity. In our second type of electrometer system, the resolution is determined by the tip diameter. Based on this fact, using a carbon nanotube probe with a smaller tip diameter is expected to lead to more improvement in the resolution. Second, the potential variation in graphene and its effect on transport and optical properties are still questions to be resolved. Miller *et al.* have reported that the peak position of Landau level obtained with scanning tunneling spectroscopy varies with the tip location relative to the graphene surface [11]. We plan to investigate the screening of the potential and the charge redistribution with our scanning electrometer.

In summary, we have studied the potential fluctuations in the 2D systems of GaAs/AlGaAs and of graphene by using two types of scanning electrometers. We have imaged the spatial distribution of the temporal potential fluctuation (noise voltage) in the GaAs-based 2DEG sample. The experimental data revealed linear distribution resulting from unstable relaxation of non-equilibrium edge-state electrons. For the graphene sample, we observed that the potential amplitude fluctuates with the period of 50-100nm. Studying the potential fluctuation and its effect on macroscopic transport properties is our next target.

5. Acknowledgements

This work was supported in part by Precursory Research for Embryonic Science and Technology (PRESTO) of Japan Science and Technology Agency (JST) and by a Grant-in-Aid from the Ministry of Education, Culture, Sports, Science and Technology.

References

[1] Fontein P F, Kleinen J A, Hendriks P, F. Blom A P, Wolter J H, Lochs H G M, Driessen F A J M, Giling L J and Beenakker C W J 1991 *Phys. Rev. B* **43** 12090.
[2] Dohi M, Yonamine R, Oto K and Muro K 2007 *Int. J. Mod. Phys. B* **21** 1414
[3] Yoo M J, Fulton T A, Hess H F, Willett R L, Dunkleberger L N, Chichester R J, Pfeiffer L N, and West K W 1997 *Science* **276** 579
[4] Yacoby A, Hess H F, Fulton T A, Pfeiffer L N and West K W 1999 *Solid State Commun.* **111** 1
[5] McCormick K L, Woodside M T, Huang M, Wu M, McEuen P L, Duruoz C, and Harris J S Jr 1999 *Phys. Rev. B* **59** 4654.
[6] Kawano Y and Okamoto T 2004 *Appl. Phys. Lett.* **84** 1111
[7] Kawano Y and Ishibashi K 2010 *Appl. Phys. Lett.* **96** 142109
[8] McEuen P L, Szafer A, Richter C A, Alphenaar B W, Jain J K, Stone A D, Wheeler R G and Sacks R N 1990 *Phys. Rev. Lett.* **64** 2062
[9] Komiyama S and Nii H 1993 *Physica B* **184** 7
[10] Vandersypen L M K, Elzerman J M, Schouten R N, Beveren L H Willems van, Hanson R and Kouwenhoven L P 2004 *Appl. Phys. Lett.* **85** 4394
[11] Miller D L, Kubista K D, Rutter G. M, Ruan M, Heer W A de, First P N and Stroscio J A 2009 *Science* **324** 924

Hall-potential distribution in AC quantum Hall effect

Hiroshi Akera

Division of Applied Physics, Faculty of Engineering, Hokkaido University, Sapporo, Hokkaido, 060-8628, Japan

E-mail: akera@eng.hokudai.ac.jp

Abstract. The distribution of the Hall voltage induced by low-frequency AC current is studied theoretically in the incoherent linear transport of quantum Hall systems with slowly-varying con ning potential. It is shown that in the low-frequency limit the Hall potential has a steeper slope where the diagonal conductivity is smaller, while at higher frequencies it exhibits a peak or a dip where the Hall conductivity has a slope.

1. Introduction

In the quantum Hall e ect [1, 2] observed in two-dimensional electron systems (2DES) in strong magnetic elds, the Hall voltage V_H divided by the current I is quantized as $V_H/I = h/(ie^2)$ with i an integer. The distribution of this quantized Hall voltage along the width of the 2DES has been studied theoretically [3, 4, 5, 6, 7, 8, 9, 10, 11, 12, 13] and experimentally [14], but has problems that remain to be solved. One of the problems is the distribution in the AC current. In this paper we study theoretically the Hall-voltage distribution in low-frequency AC current with the angular frequency ω in the incoherent linear transport in a system with slowly-varying con ning potential. The Hall-voltage distribution in a uniform 2DES with sharp edges has been studied elsewhere [15].

2. Model and Equations

In this paper we consider the 2DES in the xy plane with two boundaries which are both parallel to the x axis. We assume that the conductivity and the current are uniform along x. From the equation of charge conservation, the Hall charge density $\rho(y,\omega)$ is given by

$$i\omega\rho(y,\omega) = \quad \nabla_y j_y(y,\omega). \tag{1}$$

We assume that the phase coherence length is shorter than the length scale of variations of the electric eld induced by the current, and then we obtain the local response of the current density to the electric eld:

$$j_y(y,\omega) = \quad _{yx}(y,\omega)E_x(\omega) + \quad _{yy}(y,\omega)E_y(y,\omega). \tag{2}$$

Here we have also assumed that E_x has no dependence on y since $\nabla_y E_x \quad \nabla_x E_y \quad 0$ when ω is small.

We restrict our discussion to the low-frequency region, where $\hbar\omega$ is much smaller than relevant energy scales. We expand $\quad _{\beta}(\omega)$ ($, \beta = x, y$) in a power series of ω. Since the real and

Published under licence by IOP Publishing Ltd

imaginary parts of $_{\beta}(\omega) = {}'_{\beta}(\omega) + i\ {}''_{\beta}(\omega)$ satisfy the following relation: $'_{\beta}(\ \omega) = {}'_{\beta}(\omega)$ and $''_{\beta}(\ \omega) = \ {}''_{\beta}(\omega)$, we can write $_{\beta}(\omega)$, by retaining terms up to the rst order of ω, as

$$_{\beta}(\omega) = {}^{0}{}_{\beta} + i\omega\ {}^{0}{}_{\beta}, \tag{3}$$

where $^{0}{}_{\beta}$ is the DC conductivity and $^{0}{}_{\beta}$ is the DC susceptibility.

We assume that the phase coherence length is shorter than the length scale of variations of the con ning potential and de ne the local equilibrium chemical potential $_{\text{eq}}$ as a function of y, which decreases as approaching the boundary. We describe this y dependence by a simple function:

$$\begin{aligned} {}_{\text{eq}}(y) &= {}_{\text{b}}, \qquad (0 < y < W/2) \\ {}_{\text{eq}}(y) &= {}_{\text{b}}\left[1 \quad (y \quad W/2)^2/W_{\text{e}}^2\right], \qquad (y > W/2) \end{aligned} \tag{4}$$

and $_{\text{eq}}(\ y) = {}_{\text{eq}}(y)$.

We assume that the y dependence of $^{0}_{yx}(y)$, $^{0}_{yy}(y)$, $^{0}_{yx}(y)$ and $^{0}_{yy}(y)$ originates from the y dependence of $_{\text{eq}}(y)$, that is $^{0}_{yx}(y) = {}^{0}_{yx}(\ {}_{\text{eq}}(y))$ and so on. As for the $_{\text{eq}}$ dependence of $^{0}_{yx}$ and $^{0}_{yy}$, we employ a model [16] which retains the observed features of $^{0}_{yx}(B)$ and $^{0}_{yy}(B)$:

$$^{0}_{yx}(\ {}_{\text{eq}}) = \frac{2e^2}{h}\sum_{N} f_N, \tag{5}$$

$$^{0}_{yy}(\ {}_{\text{eq}}) = \frac{2e^2 D_0}{k_{\text{B}}T}\sum_{N}(2N+1) f_N\,(1 \quad f_N)\,, \tag{6}$$

$$f_N = \{1 + \exp\left[(\varepsilon_N \quad {}_{\text{eq}})/(k_{\text{B}}T)\right]\}^{\ 1} \tag{7}$$

with $\varepsilon_N = \hbar\omega_{\text{c}}(N + 1/2)$, $N = 0, 1, 2,$ and D_0 is a constant. As for $^{0}_{yx}(\ {}_{\text{eq}})$ and $^{0}_{yy}(\ {}_{\text{eq}})$ we use simple formulas

$$^{0}_{yx}(\ {}_{\text{eq}}) = 0, \qquad {}^{0}_{yy}(\ {}_{\text{eq}}) = e^2\nu/(\hbar\omega_{\text{c}}), \tag{8}$$

which are obtained by neglecting the Landau-level mixings and by neglecting the Landau-level broadening compared to $\hbar\omega_{\text{c}}$. In calculating the lling factor ν, we use the following density of states:

$$\begin{aligned} D(\varepsilon) &= 1/(2\ l^2\Gamma), \qquad (|\varepsilon \quad \varepsilon_N| < \Gamma) \\ D(\varepsilon) &= 0, \qquad (\text{otherwise}) \end{aligned} \tag{9}$$

Figure 1 shows the $_{\text{eq}}$ dependence of $^{0}_{yx}$, $^{0}_{yy}$ and $^{0}_{yy}$ given by eqs.(5), (6) and (8). Figure 2 presents the y dependence of $_{\text{eq}}$, $^{0}_{yx}$, $^{0}_{yy}$ and $^{0}_{yy}$ for which the Hall potential is calculated in the following.

The electrostatic potential due to the Hall charge, (y), in a 2DES uniform along x is given by

$$(y) = \int_{\ \infty}^{\infty} dy' K(y \quad y')\rho(y'). \tag{10}$$

where $K(y \quad y')$ is the potential due to the unit line charge at a distance $|y \quad y'|$. Here and in the following the ω dependence of the variables and coefficients is not shown explicitly. We consider the two models of the electrostatic interaction. One is the long-range interaction with

$$K(y \quad y') = \ \frac{2}{\varepsilon}\ln|y \quad y'|. \tag{11}$$

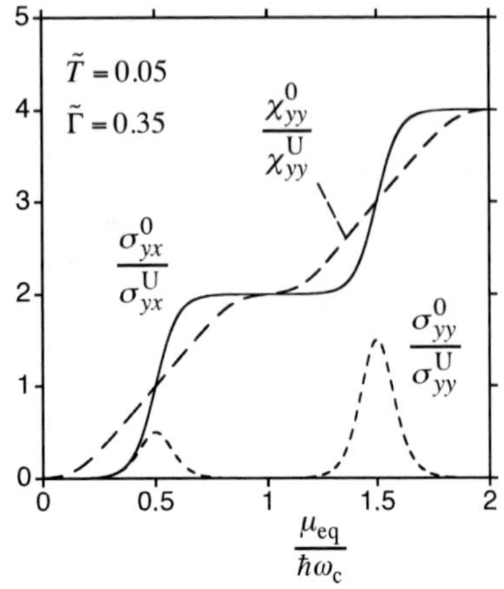

Figure 1. $_{\rm eq}$ dependence of $^0_{yx}$, $^0_{yy}$ and $^0_{yy}$. $^{\rm U}_{yx} = e^2/h$, $^{\rm U}_{yy} = e^2 D_0/(k_{\rm B}T)$, $^{\rm U}_{yy} = e^2/(h\omega_{\rm c})$, $\tilde{T} = k_{\rm B}T/(\hbar\omega_{\rm c})$, and $\tilde{\Gamma} = \Gamma/(\hbar\omega_{\rm c})$.

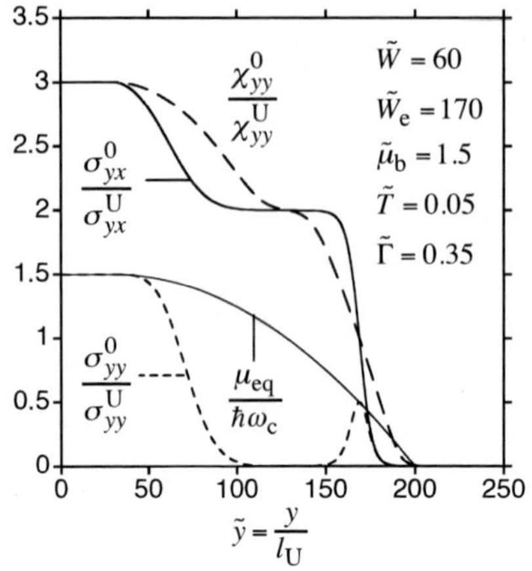

Figure 2. y dependence of $^0_{yx}$, $^0_{yy}$ and $^0_{yy}$ as well as $_{\rm eq}$. $\tilde{}_{\rm b} = {}_{\rm b}/(\hbar\omega_{\rm c})$. $\tilde{W} = W/l_{\rm U}$ and $\tilde{W}_{\rm e} = W_{\rm e}/l_{\rm U}$ with $l_{\rm U}$ de ned in eq.(16).

This is the potential in a dielectric material with the dielectric constant ε. The other is the short-range interaction with

$$K(y \quad y') = r_K \ (y \quad y'). \tag{12}$$

In this model

$$(y) = r_K \rho(y). \tag{13}$$

This model is valid when the range of $K(y \quad y')$ is much shorter than the length scale of variation of $\rho(y)$, L_ρ. If we consider a 2DES with a gate electrode, this condition becomes $d \ll L_\rho$ where d is the distance between the 2DES and the gate. In such system

$$r_K = 4 \ d/\varepsilon. \tag{14}$$

We calculate the Hall potential (y) by solving eqs.(1) and (10). We consider a periodic array of in nitely-long 2DES strips. The centerline of the nth strip is at $y = nW_{\rm p}$ where n is the integer and $W_{\rm p}$ is the periodicity. The Hall potential satis es $(\ y) = \ (y)$ and $(y + W_{\rm p}) = \ (y)$, which lead to $(y) = 0$ at $y = \ W_{\rm p}/2$. Then we expand (y) in the Fourier series and solve for the Fourier coefficients. We have con rmed that (y) within the 2DES has little dependence on $W_{\rm p}$ if the gap between 2DES strips is wide enough.

3. Calculated Results

We use the following dimensionless variable:

$$\tilde{y} = y/l_{\rm U}, \tag{15}$$

with a unit

$$l_{\rm U} = 2 \ ^{\rm 0b}_{yy}/\varepsilon = \nu_{\rm b} l^2/(\ a_{\rm B}), \tag{16}$$

where $^{\rm 0b}_{yy}$ is the value of $^0_{yy}$ in the bulk region $W/2 < y < W/2$ and $a_{\rm B} = \hbar^2\varepsilon/me^2$ is the e ective Bohr radius. When we use $B = 5$T, $\nu_{\rm b} = 4$ and the value of m and ε of GaAs, we have

Journal of Physics: Conference Series **334** (2011) 012019 doi:10.1088/1742-6596/334/1/012019

$l_U \sim l$, while $l_U(\propto B^{\ 2})$ becomes larger at smaller B. We introduce a normalized Hall potential

$$\tilde{\ } = {}^{0b}_{yy}/(E_x\ {}^{0b}_{yx} l_U), \tag{17}$$

and a normalized angular frequency

$$\tilde{\omega} = \omega\ {}^{0b}_{yy}/\ {}^{0b}_{yy}. \tag{18}$$

Figures 3 and 4 present the Hall potential (y) as a function of y in the long-range interaction eq.(11). In the low-frequency limit, (y) shows a steeper slope where ${}^{0}_{yy}$ is smaller and a plateau where ${}^{0}_{yy}$ is larger. At higher frequencies, on the other hand, (y) exhibits a peak (or a dip) where ${}^{0}_{yx}$ has a slope. These features are also seen in (y) calculated in the short-range interaction eq.(12) as shown in Figs.5 and 6.

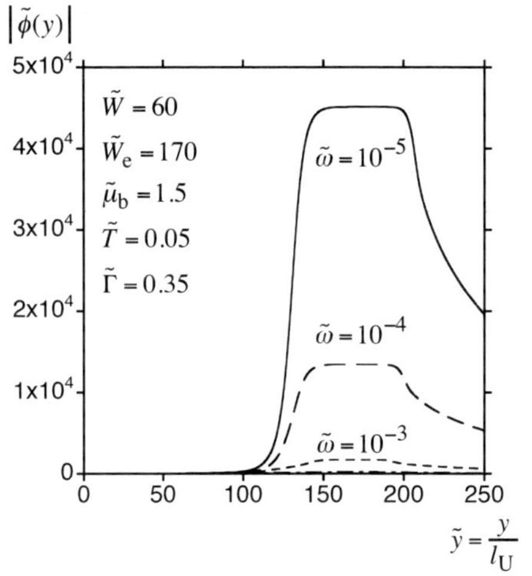

Figure 3. Hall potential (y) in the long-range interaction eq.(11) at lower frequencies. $\tilde{W}_p = W_p/l_U = 800$.

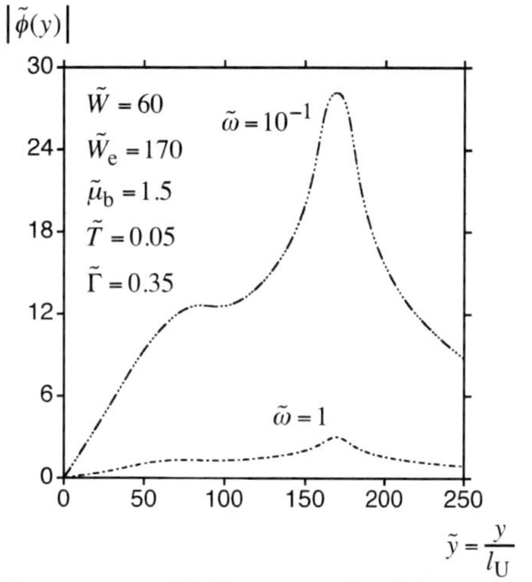

Figure 4. Hall potential (y) in the long-range interaction eq.(11) at higher frequencies. $\tilde{W}_p = W_p/l_U = 800$.

Mechanisms of such features are understood from the equation for (y) in the short-range interaction. Equations (1) and (13) give $i\omega\ (y) = \nabla_y D(y)\nabla_y\ (y) + r_K\nabla_y i\omega\ {}^{0}_{yy}(y)\nabla_y\ (y)$ $r_K(\nabla_y\ {}^{0}_{yx}(y))E_x$, where $D(y) = r_K\ {}^{0}_{yy}(y)$ is a di usion coefficient, the second term $r_K\nabla_y i\omega\ {}^{0}_{yy}(y)\nabla_y\ (y)$ represents a screening due to the polarization charge, and the third term $r_K(\nabla_y\ {}^{0}_{yx}(y))E_x$ is the source term. When $\tilde{\omega} \ll 1$, the second term is negligible compared to the rst term. Because of this rst term, (y) is at where $D(y)$ is larger, while it shows a slope where $D(y)$ is smaller. When $\tilde{\omega} \gg 1$, on the other hand, the rst term is negligible. In this case, the third term, or the slope of ${}^{0}_{yx}(y)$, determines the spatial dependence of (y). In the limit of $\tilde{\omega} \to \infty$, $(y)\omega/(iE_x)$ approaches a real value independent of ω, which is the case also in the long-range interaction.

4. Conclusions and Discussion

We have studied the Hall potential (y) as a function of y (in the width direction) in quantum Hall systems in the case of low-ω AC current in the incoherent linear transport where the dynamics of the local Hall-charge density is determined by the complex diagonal conductivity

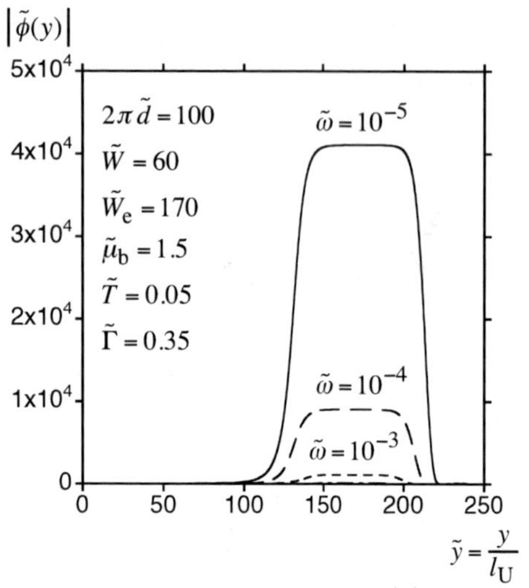

Figure 5. Hall potential (y) in the short-range interaction eq.(12) at lower frequencies. $\tilde{W}_p = W_p/l_U = 800$. $\tilde{d} = d/l_U$.

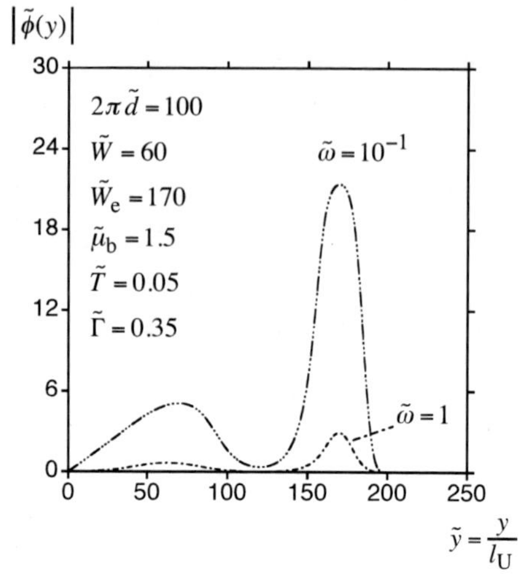

Figure 6. Hall potential (y) in the short-range interaction eq.(12) at higher frequencies. $\tilde{W}_p = W_p/l_U = 800$. $\tilde{d} = d/l_U$.

$_{yy} = {}^0_{yy} + i\omega\ {}^0_{yy}$ as well as the Hall conductivity ${}^0_{yx}$. We have considered the case of slowly-varying con ning potential, in which ${}^0_{yy}(y)$, ${}^0_{yy}(y)$ and ${}^0_{yx}(y)$ vary with y. We have found the following conclusions both in the long-range interaction and in the short-range interaction. In the low-frequency limit, in the region of smaller ${}^0_{yy}(y)$, the Hall potential (y) shows a steeper slope and the current is concentrated. This has been shown before in the DC current [11]. At higher frequencies, on the other hand, (y) exhibits a peak or a dip where ${}^0_{yx}(y)$ has a slope.

We have also calculated the frequency dependence of the magnetoresistance $R_{xx}(\omega) = V_x(\omega)/I_x(\omega)$. With increasing ω, $|R_{xx}|$ exhibits a stepwise increase and the phase delay of I_x relative to V_x increases from 0 to $/2$.

References

[1] K. von Klitzing, G. Dorda and M. Pepper: Phys. Rev. Lett. **45** (1980) 494.
[2] S. Kawaji and J. Wakabayashi: in *Physics in High Magnetic Fields*, edited by S. Chikazumi and N. Miura (Springer, Berlin, 1981) p. 284.
[3] A.H. MacDonald, T.M. Rice and W.F. Brinkman: Phys. Rev. B **28** (1983) 3648.
[4] D.J. Thouless: J. Phys. C **18** (1985) 6211.
[5] O. Heinonen and P.L. Taylor: Phys. Rev. B **32** (1985) 633.
[6] D. Pfannkuche and J. Hajdu: Phys. Rev. B **46** (1992) 7032.
[7] D.J. Thouless: Phys. Rev. Lett. **71** (1993) 1879.
[8] C. Wexler and D.J. Thouless: Phys. Rev. B **49** (1994) 4815.
[9] H. Hirai and S. Komiyama: Phys. Rev. B **49** (1994) 14012.
[10] J.J. Palacios and A.H. MacDonald: Phys. Rev. B **57** (1998) 7119.
[11] K. Güven and R. R. Gerhardts: Phys. Rev. B **67** (2003) 115327.
[12] A. Siddiki and R. R. Gerhardts: Phys. Rev. B **70** (2004) 195335.
[13] S. Kanamaru, H. Suzuura and H. Akera: J. Phys. Soc. Jpn. **75** (2006) 064701.
[14] P.F. Fontein, J.A. Kleinen, P. Hendriks, F.A.P. Blom, J.H. Wolter, H.G.M. Lochs, F.A.J.M. Driessen, L.J. Giling and C.W.J. Beenakker: Phys. Rev. B **43** (1991) 12090.
[15] H. Akera: to be published in Physica E (published online at: http://dx.doi.org/10.1016/j.physe.2011.02.008).
[16] H. Akera and H. Suzuura: J. Phys. Soc. Jpn. **74** (2005) 997.

A Numerical Study of Non-local Magneto Transport Effects in Quantum Hall Device Structures

Josef Oswald, Christoph Uiberacker, Christian Stecher
Institute for Physics, University of Leoben, Franz Josef Str. 18, A-8700 Leoben, Austria
E-mail: josef.oswald@unileoben.ac.at

Abstract. We developed a non-equilibrium network model for magneto transport in 2D electronic systems in the quantum Hall effect regime, which is able to capture the real sample geometry, including contacts, leads and in-homogeneities like introduced by gate electrodes. In this paper we investigate non-local effects in magneto transport for a number of contact configurations of a quantum Hall system, including the effect of having unused metallic contacts between the active region of classical current flow and the remote region where the non-local signals are obtained.

1. Introduction

There are an increasing number of devices like applications of quantum Hall systems for various purposes. One is spintronics in context with injection and detection of spin polarized electrons, but there are also other device concepts like electron interferometers or terahertz emitters, which make use of the unique properties of edge states[1]. While theoretical approaches mostly rely on the equilibrium situation, experiments drive the system out of equilibrium so that in real experiments one always measures non-equilibrium currents and potentials. Especially non-local effects are driven by the presence of a non-equilibrium between bulk and edge channels (EC). An appropriate transport model should take care of this and has also to provide a link between the microscopic physics and the experimental answer of the finally macroscopic sample. We developed a non-equilibrium network model for magneto transport in 2D electronic systems in the quantum Hall effect regime, which is able to capture the real sample geometry, including contacts, leads and in-homogeneities like introduced by gate electrodes [2, 3, 4, 5, 6]. In this paper we are going to demonstrate, that also non-local effects can be addressed successfully.

2. The non-equilibrium network model

The intention of our network approach to magneto transport is to address directly the non-equilibrium current flow. This means that we have to deal with the situation of a non-uniform lateral distribution of the chemical potentials. We use a network of directed quantum channels, which interconnect a 2-dimensional arrangement of nodes, which physically correspond to the saddles of the potential fluctuations in real systems. From this point of view, the basic structure of the network appears similar to the Chalker-Coddington model [7], but the handling of the nodes is substantially different. The directed channels keep their chemical potentials between the nodes, while the nodes transmit and modify the chemical potentials from the incoming to

Published under licence by IOP Publishing Ltd

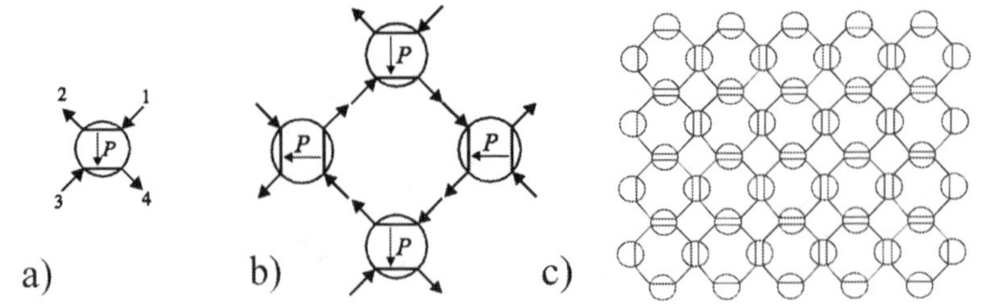

Figure 1. a) Node of the network with two incoming and two outgoing channels. The channels $1 \to 2$ and $3 \to 4$ are treated like ECs with back scattering, where $P = R/T$ corresponds to the relation between reflection and transmission coefficients according to the Landauer- Büttiker formalism. b) Arrangement of the nodes for building the minimal physical element of a network, which is the closed loop of a so called magnetic bound state. c) The complete network is composed by putting together a sufficient number of such adjacent loops.

the outgoing channels. The main facts are given below and for further details refer to the cited papers.[8, 9].

Fig.1a shows a single node of our network, which transmits potentials from the incoming to the outgoing channels, Figs.1b-c demonstrate, how the network is composed. The potentials transmitted by one node are calculated by using $u_2 = (u_1 + P \cdot u_3)/(1 + P)$ and $u_4 = (u_3 + P \cdot u_1)/(1 + P)$. We attribute the non-equilibrium currents to channel pairs, like e.g. the current from the right to the left $I = (e^2/h) \cdot (u_1 - u_4)$ (see Fig.1a). It is important, that in this way our nodes provide a handle to both, the injected currents and the potentials. P results from tunneling across saddle points of the potential landscape in the bulk[8]: $P = \exp\left[-\frac{L^2 E_F}{e\tilde{V}} \frac{eB}{h}\right]$, E_F is the Fermi energy relative to the saddle energy which corresponds to the center of the LL, eB/h is the number of LL states, L is the period and $\tilde{V}$ the amplitude of a representative two-dimensional Cosine-potential modulation, which has the same Taylor expansion like the actual saddle potential. Therefore the ratio $L^2/\tilde{V}$ can be understood as a measure of the "smoothness" of the potential modulation near the saddle. The design of the sample is done by shaping the lateral confining bare potential and using a self consistent Hartree type approximation to calculate the Fermi energy and the lateral carrier distribution. On this basis a gate electrode can be easily modeled by biasing the bare potential of the designated gate region (for further details see cited papers).

3. Results and Discusssion

For the numerical study we used a Hall bar with contact leads and also a more simple rectangular Hall conductor with point contacts at the edges, both with a bulk carrier density of $4 \cdot 10^{11} cm^{-2}$. Fig.2 shows the simulations for the Hall bar structure. The current injection is done in an unusual fashion like indicated in the insert of Fig.2 on the left: One current contact at one end of the Hall bar is used as usual, but the other current contact is applied across the bulk at about half length of the Hall bar. The width of that contact stripe is chosen that it just touches the edge region, where the side depletion starts to develop. For geometry reasons there should be no signal at the remote (Hall) contact pair. As can be seen on the left of Fig.2, already at low fields there comes up a small non-local resistance which forms a kind of maximum at about B=0.5 Tesla, before Schubnikow-de-Haas oscillations set-in. In the high magnetic field regime on the right, a strong non-local signal develops, which appears like the longitudinal resistance in the QHE regime. The main reason for that is the introduced non-equilibrium between edge and

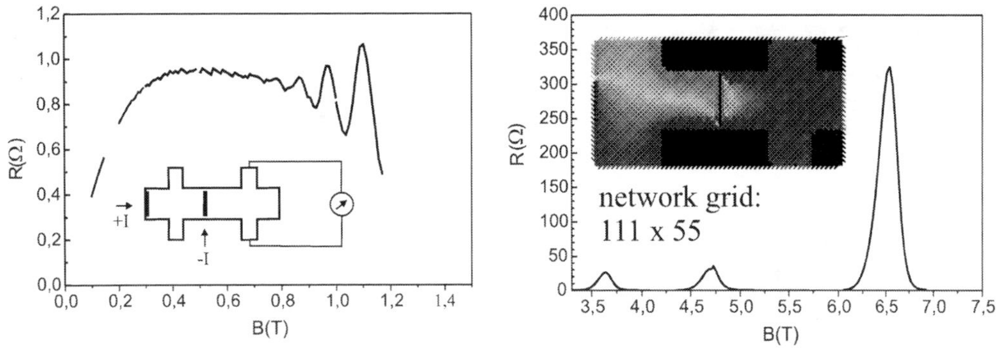

Figure 2. Left: Low field non-local resistance of a Hall bar. Insert: Contact scheme. Right: High field non-local resistance according to the contact scheme on the left. Insert: Screen shot of network layout. The gray scale inside the Hall bar represents the lateral distribution of the excitation potential (not discussed).

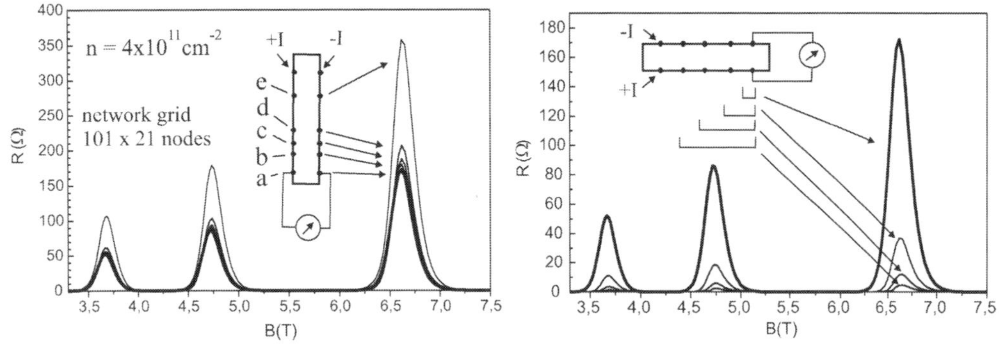

Figure 3. Left: Non-local resistance at different distance from current injection. Insert: Contact scheme of the point contact pairs with a-e marking the different positions for taking the signal. In this case only one voltage contact pair is present at a time. The contacts for current injection are marked with +I and -I. Right: Non-local resistance at fixed position, but with different number of additional unused contact pairs attached at the same time, like indicated in the insert (indicated by the square brackets)

bulk, while the symmetry of the sample is destroyed by the opposite directed ECs, which pass the interior current injection contact.

For obtaining a better understanding of the non-local signal, we change now to the rectangular geometry with point contacts at the edges, but no contact leads. That provides more flexibility in changing conduct configurations without changing the Hall conductor as a whole and thus allowing us to compare different contact configurations more easily. The non-equilibrium is now created with a current contact pair at opposite edges near one end of the conductor and the non-local resistance is investigated for different contact configurations. On the left of Fig. 3 the dependence of the non-local resistance on the distance from current injection is demonstrated and one can see that the non-local resistance goes down with distance. On the right of Fig. 3

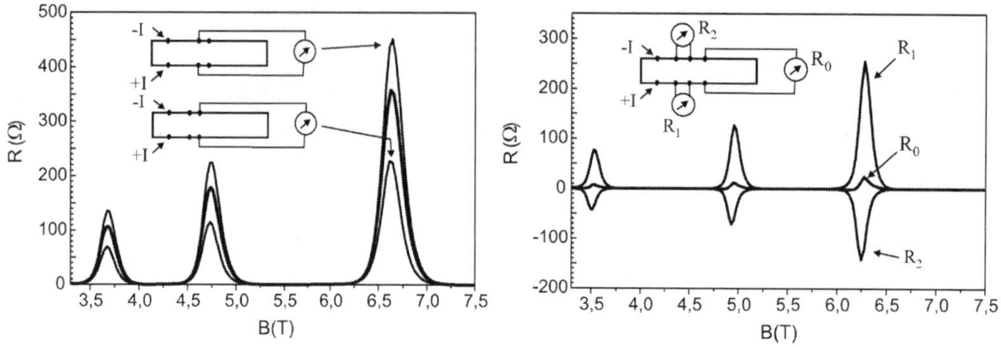

Figure 4. Left: Non-local resistance of a QH conductor taken at fixed voltage contact pair, without and with an unused contact pair, but introduced once before and once after the voltage probes. Right: Non-local resistance at 3 voltage contacts pairs, one across the Hall bar (R_0), one at the lower (R_1) and one at upper edge (R_2) like indicated in the insert.

we show the behavior of the non-local resistance at fixed distance but with a different number of unused contact pairs between current injection and remote voltage probes. The non-local resistance goes down with the number of unused contacts, because any additional contacts introduce additional equilibration between edge and bulk, which removes a part of the non-equilibrium already before arriving at the voltage probes. Since in our study the non-local signal is observed as a potential difference between opposite edges, this at the same time indicates an edge current that is delivered to the remote region. However, since there are no further current carrying contacts in the remote region, this edge current has to be balanced by a counter flowing bulk current for current conservation reasons.

The left side of Fig.4 shows the non-local resistance at fixed position, but comparing the case without additional contacts and the cases with an additional contact pair, but once in front and once behind the voltage probes (see insert on the left). One can clearly see, that the additional contact pair in front of the voltage probes reduces the non-local resistance as compared without additional contacts, but if the additional contacts are on the other (right) side of the voltage probes, the non-local resistance increases. This can be understood quite easily in the following way: The direction of the ECs is counter clockwise in this simulations. Therefore the lower ECs pick up the positive potential of the lower current contact and carry that all around the sample to the negative current contact. However, the ECs lose successively some of the potential due to the ongoing equilibration all along the edge and at possibly present additional contacts. If there are additional contacts on the right side, they appear on the way from the voltage contact on the lower edge to the voltage contact at the upper edge, which enhances the voltage drop between these non-local voltage contacts. This increases also the non-local edge current to be delivered to the remote side, which has to pass the voltage probes. Therefore the resulting additional non-local voltage can alternatively also be roughly understood in terms of a Hall voltage of that remote edge current caused by the equilibration processes in the remote region.

In order to clarify the discussion so far, Fig.4 on the right shows the non-local resistance of the remote voltage probes like before, but now there are 2 additional longitudinal voltage contact pairs at the upper and lower edge in order to measure also the longitudinal non-local resistance. As one can see, R_0 gets almost suppressed because of the additional contact pairs, but one can also see, that in contrast to R_1, R_2 at the upper edge is negative. This means that the voltage drop along the upper edge is opposite to that of the lower edge. This is consistent with the above discussion concerning the remote edge current, which comes up because of the equilibration process in the remote region. Since the conducting bulk channels, which are needed for balancing the remote edge current, exist only in the transition regime between quantum Hall plateaus, such non-local signals appear only in the plateau transition regime, just like the local longitudinal resistance.

4. Summary

Using a non-equilibrium network model we have studied different non-local configurations of a quantum Hall system. We have shown that any additional voltage probes affect the results at all other contacts by introducing an additional edge-bulk equilibration. We have also demonstrated, that edge-bulk equilibration in the remote region causes a remote edge current, which has to be balanced by a bulk current. This remote edge current is mainly responsible for the non-local signal, which can be roughly understood as a type of Hall voltage of that remote edge current.

5. Acknowledgments

Financial support by the Austrian Science Foundation (FWF) Project Nr. P19353.

References

[1] K. Ikushima, et al, Physica E **42** (4), 1034 (2010)
[2] J. Oswald, International J. Mod. Phys. B**21**, 1424 (2007)
[3] M. Oswald, Oswald, R.G. Mani, Phys. Rev. B72(3), 35334 (2005)
[4] M. Oswald, Oswald, Phys. Rev. B74(15), 153315 (2006)
[5] C. Uiberacker, C. Stecher, J. Oswald, Phys. Rev. B 80(23), 235331 (2009)
[6] J. Oswald, et al, JOURNAL OF LOW TEMPERATURE PHYSICS 159, 180 (2010)
[7] J.T. Chalker and P.D. Coddington, J. Phys. C 21, 2665 (1988)
[8] J. Oswald and M. Oswald, J. Phys. Cond. Mat. **18**, R101 (2006)
[9] C. Sohrmann, J. Oswald, and R.A. Römer, Lecture Notes in Physics **762**, 163 (2009)

High-sensitive optical measurement of spin polarization in a quantum Hall system

H Ito[1], T Yamazaki[1], D Fukuoka[1], K Oto[1], K Muro[1], Y Hirayama[2,4] and N Kumada[3]

[1]Graduate School of Science, Chiba University, Chiba, Japan
[2]Graduate School of Science, Tohoku University, Sendai, Japan
[3]NTT Basic Research Laboratories, NTT Corporation
[4]ERATO Nuclear Spin Electronics Project, Sendai, Japan

E-mail: 07sm2102_ito@graduate.chiba-u.jp

Abstract. Spin polarization measurement is important for the study of a variety of spin states in quantum Hall system. Kerr rotation is proportional to spin polarization, so we developed a high sensitive measurement of Kerr rotation by using homodyne detection and a variety of modulation techniques. Furthermore, we developed Kerr rotation spectra measurement system base on the multi-channel homodyne detection, which enables the assignment of the optical transitions and semi-quantitative estimation of spin polarization by integrating the spectra over the specific optical transition. The spin polarization presents a rapid spin depolarization on both sides of $\nu=1$ due to Skyrmionic excitation. However the top of spin polarization presents a narrow flat region. Furthermore the spin polarization around $\nu=3$ also shows a rapid spin depolarization which suggest the existence of Skyrmion at higher odd filling factor.

1. Introduction

In the quantum Hall system, the freedom of kinetic motion is quenched due to Landau quantization and the electron-electron interaction brings a variety of condensed states, like fractional quantum Hall states. Especially for the electrons in GaAs/AlGaAs quantum well where the Zeeman energy is much smaller than the electron-electron interaction, a variety of spin states appears in the quantum Hall system. The existence of quantum Hall ferromagnet and Skyrmion were first demonstrated by the spin polarization measurement due to Knight shift in NMR experiment [1]. However, because of the low sensitivity, the NMR measurement needs a special sample of multi quantum wells. In order to measure the spin polarization in a single quantum well, a variety of spin polarization measurement methods were developed such as polarization analysis of photo luminescence [2] and optical absorption [3]. However, luminescent measurement is not a direct method, and absorption measurement also needs a special sample and the sensitivity is not high enough. So, we still have unsolved problems, like instability of ferromagnetic state at $\nu=1$ [3] and the existence of Skyrmionic excitation at higher odd filling factor [4, 5]. In order to clarify these problems, a more sensitive and reliable measurement method of spin polarization in a quantum well layer is highly desired.

2. High-sensitive measurement of Kerr rotation by homodyne detection

Kerr rotation is the phenomenon that polarization of the reflected beam is rotated from that of the incident beam, and the angle is proportional to the magnetization along the optical axis. In the non-magnetic material, such as GaAs/AlGaAs, the Kerr rotation angle is proportional to the difference of the electron spin population. In the quantum Hall system, the angle is in the order of mill-radian. So we developed a high sensitive Kerr rotation measurement method by homodyne detection. The sample was single side δ–doped GaAs/AlGaAs quantum well. The g-factor was measured to be 0.23 by TRKR experiment for the 15nm quantum well [6]. The electron density can be changed in the range of $1.45\sim3.2\times10^{11}cm^{-2}$ by a back gate voltage and the mobility is about 1×10^{6} cm^2/Vs.

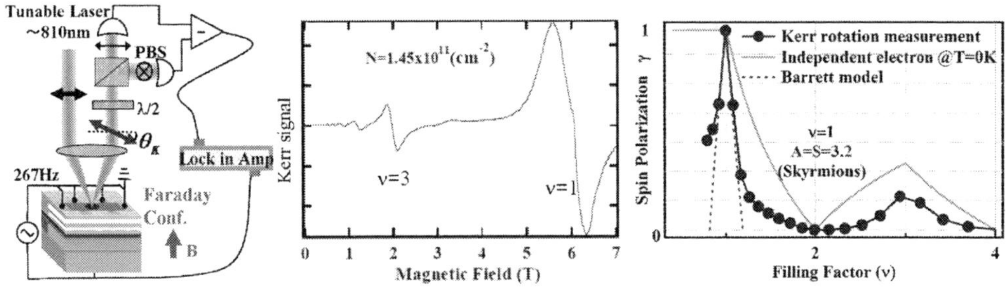

(a) (b) (c)

Figure 1. Kerr rotation measurement by gate modulation: (a) Experimental setup, (b) Gate modulation signal, (c) Estimated spin polarization vs. filling factor

At first, we introduced a gate voltage modulation technique for the Kerr rotation measurement. Figure 1 shows the experimental setup (a), the modulation signal as a function of magnetic field (b) and the estimated spin polarization as a function of filling factor (c). Kerr rotation for the 810nm radiation from a tuneable laser was observed in Faraday configuration at 1.5K by using a balanced detector and lock-in detection. The gate modulation signal shows a dispersive profile around odd integer filling factor, where the quantum ferromagnetism is expected.

The spin polarization against filling factor is derived from the gate modulated Kerr signal in the following procedure. We define the spin polarization function per particle as $\gamma(\nu,N)=S_z(\nu,N)/N$. Assuming the spin polarization function per particle is same for N and $N+\Delta N$, i.e. $\gamma(\nu, N+\Delta N)=\gamma(\nu, N)=\gamma(\nu)$, the gate modulated Kerr signal is expressed as,

$$\Delta\theta_K \approx \alpha[\gamma(\nu+\Delta\nu)\times(N+\Delta N)-\gamma(\nu)\times N] \approx \alpha\left[\frac{d\gamma(\nu)}{d\nu}\Delta\nu\times N+\gamma(\nu)\times\Delta N\right] \quad (1)$$

,where α is a proportionality constant. Gate modulated Kerr signal contains the components induced by the change of $\Delta\nu$ and ΔN as seen in Eq.(1). $\gamma(\nu)$ satisfies the following integral equation,

$$\gamma(\nu)=\int\frac{d\gamma(\nu)}{d\nu}d\nu \approx \int\left(\frac{1}{\alpha N}\frac{\Delta\theta_K}{\Delta\nu}-\frac{\gamma(\nu)}{\nu}\right)d\nu \quad (2)$$

So we derive $\gamma(\nu)$ self-consistently by an iteration method with the starting function $\gamma(\nu)=0$ at each even filling factor.

Spin polarization presents a rapid spin depolarization on both side of ν=1, reflecting existence of Skyrmion. However in this method, gate voltage modulation disturbs the electron system and precise determination of the filling factor dependence of spin polarization is difficult because the spin polarization is averaged over the modulation depth.

Journal of Physics: Conference Series **334** (2011) 012021 doi:10.1088/1742-6596/334/1/012021

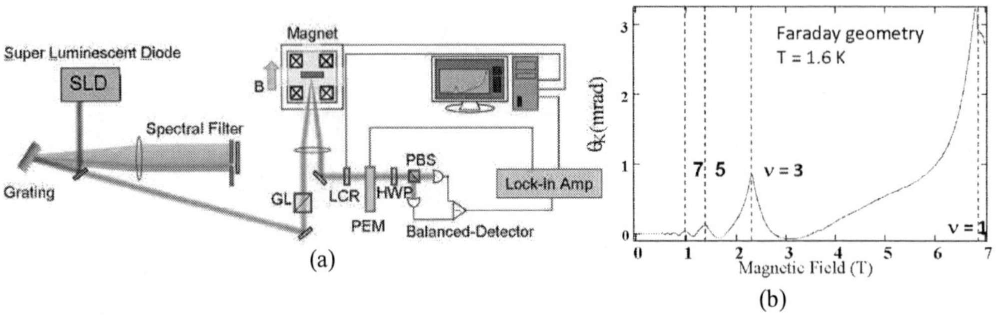

Figure 2. Kerr rotation measurement by optical phase modulation: (a) Experimental setup, (b) Kerr rotation as a function of magnetic field

So, we employed a light phase modulation technique in the Kerr rotation measurement as shown in Fig.2 (a). Filtered radiation from super luminescent diode was used for this experiment. Rapid phase modulation by a photo elastic modulator enables a very high-sensitive measurement of Kerr rotation. A sharp peak at ν=3, and some fine structure around ν=1 are clearly seen in Fig.2 (b).
Kerr rotation measurement for the radiation around a specific wavelength enables a high sensitive detection of spin polarization. However the sensitivity depends on a magnetic field and wavelength of the light. So, for the quantitative measurement of spin polarization and detailed analysis of spin states, we need whole Kerr rotation spectrum.

3. Kerr rotation spectrum measurement by multi-channel homodyne detection

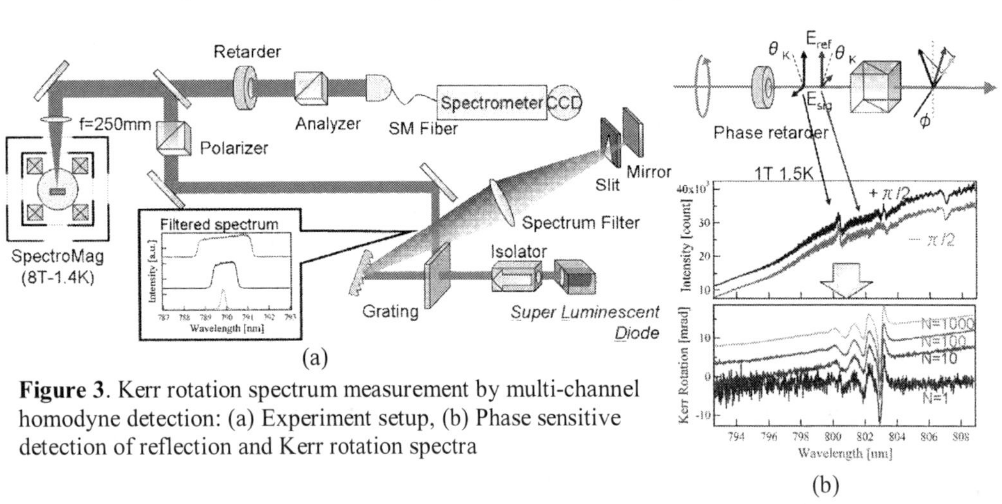

Figure 3. Kerr rotation spectrum measurement by multi-channel homodyne detection: (a) Experiment setup, (b) Phase sensitive detection of reflection and Kerr rotation spectra

Figure 3 (a) shows the experimental setup for Kerr rotation spectrum measurement. Broad band light from super-luminescent diode (SLD) was introduced to the sample in Faraday configuration. The incident light was polarized by Glan laser prism. Optical pass of the reflected light was separated from the incident pass by non-polarizing beam splitter and analyzed by another Glan laser prism after passing through the liquid crystal retarder. Finally, the beam is introduced to the spectrometer with CCD multi channel detector. Unnecessary radiation from SLD was eliminated by spectrum filter and the intensity of incident beam was kept below 1μW for 50μmϕ spot to minimize the optical disturbance to the quantum Hall system.

Figure 3 (b) shows the phase sensitive measurement of reflection spectra by liquid crystal retarder and the Kerr rotation spectra. Two spectra were measured by changing the relative phase between signal and reference axis by π. Kerr rotation spectra are derived from the spectra with the following equation

$$\theta_K(\lambda) = \frac{\cos\phi}{2\sin\phi} \times \frac{I_{+\pi/2}(\lambda) - I_{-\pi/2}(\lambda)}{I_{+\pi/2}(\lambda) + I_{-\pi/2}(\lambda)},$$

where ϕ is the angle between the analyzer axis and the signal axis. The measurement was repeated and Kerr rotation spectra are averaged in order to decrease shot noise. We can choose the Kerr rotation spectra and magnetic dichroic spectra by adjusting the center phase of phase retarder. Hereafter, we focus on the magnetic dichroic (MD) spectra.

Figure 4 (a) shows magnetic dichroic spectra observed at ν=1 and 1.5K. We also measured optical absorption spectra for σ^+ and σ^- polarized light by photo luminescent excitation method to check the correspondence between the MD spectra and the absorption spectra. In Fig.4 (b), upper trace shows the absorption for σ^- polarization and lower trace show the absorption for σ^+ polarization. MD spectra correspond to the difference between σ^- and σ^+ absorption. Figure 4 (c) shows the assignment of the optical transition according to Fromer [7].

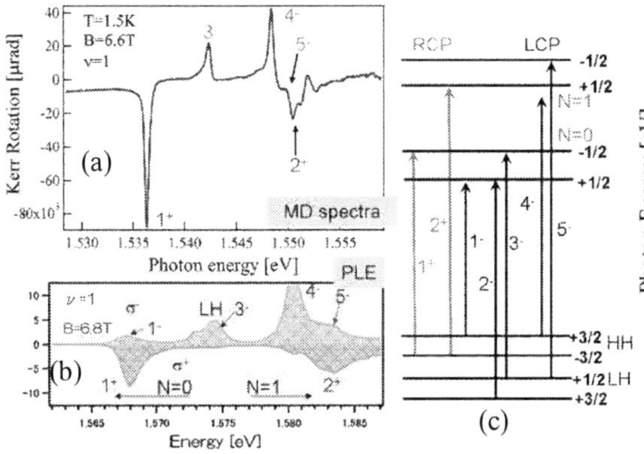

Figure 4. MD spectra and optical transitions observed at ν=1 and 1.5K: (a) MD spectra, (b) Absorption spectra, (c) Assignment of optical transition

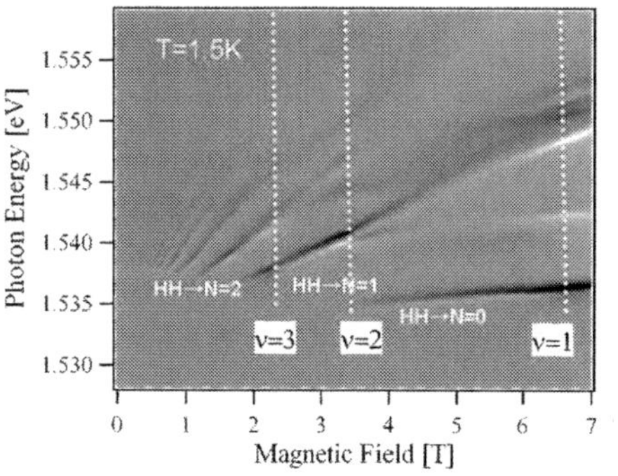

Figure 5. Gray scale plot of MD intensity observed at 1.5K as a function of magnetic field and photon energy

Figure 5 is 2D plot of MD intensity as a function of photon energy and magnetic field. We can see the magnetic field dependence of the optical transitions. Lowest energy transition is heavy hole state to N=0 electron Landau level, and the next lower energy transition is heavy hole state to N=1 electron Landau level. The Landau fan is strongly deviated from simple Landau fan. Furthermore we can see the optical transitions become sharper at integer filling factor, and absorption line presents splitting at integer filling factor. However, these are not the subject of this paper, so move to the evaluation of spin polarization.

4. Evaluation of spin polarization from magnetic dichroic spectra

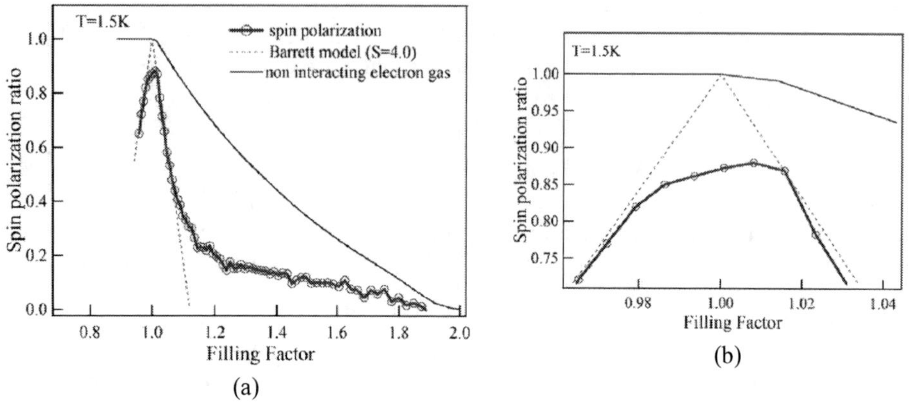

Figure 6. Filling factor dependence of spin polarization estimated from MD spectra: (a) Spin polarization in the range of ν=0.95~1.9, (b) At the top region of ν=1

Intensity of σ^- transition is proportional to the density of unoccupied spin up states and intensity of σ^+ transition is proportional to that of unoccupied spin down states. So, we can derive the spin polarization by integrating the MD spectrum intensity across the optical transitions from the heavy hole state to the lowest unoccupied electron Landau level.

Figure 6 (a) shows the filling factor dependence of the spin polarization ratio in the range of ν=0.95~1.9, which is obtained by integrating the MD spectra intensity across of the transition to N=0 Landau level. The solid line is the spin polarization calculated based on the non interacting electron gas model at T=0. Dotted line is the expected spin polarization due to Skyrmionic excitation according to Barrett model where the Skyrmion size is assumed to be 4.0[1]. Circles show the spin polarization ratio obtained in the present experiment, whose vertical scale is adjusted to the Barrett model. Spin polarization measurement by Kerr rotation reconfirmed the Skyrmionic excitation around ν=1. However, spin polarization has a flat region at the top, which means the existence of non polarized spins at ν=1 as shown in the Fig 6 (b). Similar result was recently reported by P.Plochocka at 1.6K [3]. These unusual filling factor dependence suggest the existence of an intriguing spin state at ν=1, where a stable quantum ferromagnetic state is expected.

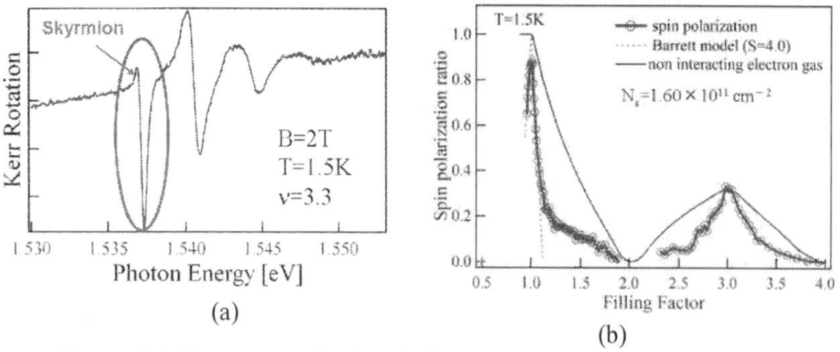

Figure 7. MD spectra and spin polarization in the range of ν=2~4
(a) MD spectra observed at ν=3.3 (b) Spin polarization around ν=3

Figure7 (a) shows the MD spectrum observed at ν=3.3. In this magnetic field, the lowest absorption line is the transition from the heavy hole state to lowest unoccupied Landau level, N=1. On the low energy side of this transition, a peak corresponding to σ^- absorption was observed. This indicates the unoccupied spin up state, which reflects the existence of Skyrmion around ν=3.3. Figure7 (b) shows the spin polarization obtained around ν=3 (ν=1) for the electron density 1.6×10^{11} cm^{-2}. The rapid spin depolarization was also observed on both sides of ν=3. The existence of Skyrmion at higher odd filling factor is theoretically predicted for the finite well width [4, 5]. The obtained result seems to support the theoretical prediction, however, there is some significant discrepancies between them. To confirm the existence of Skyrmion at higher integer odd filling factor, we need a more systematic study of spin polarization at lower temperature.

We are now developing a fiber based Kerr rotation spectra measurement system which is applicable to the dilution refrigerator. Kerr rotation spectra at mill-Kelvin may solve the unsettled problems and reveal the new spin condensed states in the quantum Hall states.

This work was in part supported by Global COE program at ChibaUniversity (G-03, MEXT).

5. References

[1] S E Barrett, G Dabbagh, L N Pfeiffer, K W West and R Tycko 1995 *Phys. Rev. Lett.* **74** 5112

[2] I V Kukushkin, K v Klitzing and K Eberl 1999 *Phys. Rev. Lett.* **82** 3665

[3] P.Plochocka, J M Schneiber, D K Maude and M Potemski 2009 *Phys. Rev.Lett.* **102** 126806

[4] N. R. Cooper 1997 *Phys.Rev.B* **55** R1934

[5] H.A.Fertig, Luis Brey, R Cote, A H MacDonald, A Karlhede and S L Sondhi 1997 *Phys.Rev.B* **55** 10671

[6] D. Fukuoka, T Yamazaki, N Tanaka, K Oto, K Muro, Y Hirayama, N Kumada and H Yamaguchi 2008 *Phys.Rev.B* **78**, 041304(R)

[7] N.A.Fromer, C Schuller, C W Lai and D S chemia 2002 *Phys.Rev.B* **66** 205314

Journal of Physics: Conference Series **334** (2011) 012022 doi:10.1088/1742-6596/334/1/012022

Optical probing of quantum Hall effect of composite fermions and of the liquid-insulator transition

F Rossella[1], V Bellani[1,#], F Dionigi[1,£], M Amado[2], E Diez[2], K Kowalik[3,§], G Biasiol[4] and L Sorba[5]

[1]Dipartimento di Fisica "A. Volta", Università di Pavia, Pavia, Italy
[2]Laboratorio de Bajas Temperaturas, Universidad de Salamanca, Salamanca, Spain
[3]Laboratoire National des Champs Magnétiques Intenses, Grenoble, France
[4]Istituto Officina dei Materiali CNR-TASC, Trieste, Italy
[5]NEST, Istituto Nanoscienze-CNR and Scuola Normale Superiore, Pisa, Italy

Present addresses: [£]Institut for Fysik, Danmarks Tekniske Universitet, Lyngby, Denmark; [§]Department für Physik, Ludwig Maximilians Universität, München Germany

E-mail: [#]bellani@unipv.it

Abstract. In the photoluminescence spectra of a two-dimensional electron gas in the fractional quantum Hall regime we observe the states at filling factors $\nu = 4/5$, $5/7$, $4/11$ and $3/8$ as clear minima in the intensity or area emission peak. The first three states are described as interacting composite fermions in fractional quantum Hall regime. The minimum in the intensity at $\nu = 3/8$, which is not explained within this picture, can be an evidence of a suppression of the screening of the Coulomb interaction among the effective quasi-particles involved in this intriguing state. The magnetic field energy dispersion at very low temperatures is also discussed. At low field the emission follows a Landau dispersion with a screened magneto-Coulomb contribution. At intermediate fields the hidden symmetry manifests. At high field above $\nu = 1/3$ the electrons correlate into an insulating phase, and the optical emission behaviour at the liquid-insulator transition is coherent with a charge ordering driven by Coulomb correlations.

1. Introduction

The two dimensional electron gas (2DEG) created in a modulation-doped semiconductor quantum well (QW) has been widely investigated using magneto-transport and magneto-optical techniques [1-4]. In particular, in magneto-photoluminescence experiments realized in an asymmetric modulation doped QW, the photo-excited holes are able to probe the plasma of interacting electrons, since the holes weakly affect the 2DEG when the separation between electrons and holes is large [5]. A magnetic field normal to the QW enhances many-body effects and gives access to the observation of quantum phenomena such as the integer and fractional quantum Hall effect (IQHE and FQHE, respectively).These phenomena have been explained within theories which invoke the many body interactions that rule the arrangement of 2D electrons placed in high magnetic field, leading them to form highly correlated states. In particular, a solid theoretical scenario [6] which envisages the formation of Landau levels (LLs) and quasi-particles named composite fermions (CFs), has allowed the interpretation of several experimental results. However, FQHE at fractional filling factors that cannot be explained within the CFs theory has been observed in transport experiments [2] and very recently in optical experiments [7], and not all the features observed in the PL experiments are well understood but they keep on producing sources of research and discussion [8]. Besides, the presence of residual disorder is a fundamental ingredient for the observation of these quantum phenomena, and in samples with intermediate electron density (~ 10^{11} cm^{-2}) and sufficiently high mobility (~ 2×10^6 cm^2/Vs) the interplay between disorder and Coulomb interactions can reveal interesting interaction-related effects. Among these, the two-dimensional metal-insulator transition represents an intriguing example, whose nature in still an open question. When the magnetic field exceeds a critical value B_c, the system from a quantum Hall liquid exhibiting the quantum Hall plateau becomes insulating, characterized by a diverging longitudinal resistance [9]. This transition to the insulating state has been studied by electrical transport in several semiconductor heterostructures [9, 10] and recently in graphene [11], and both the disorder-induced localization (Anderson) and the correlation-induced localization (Mott) have been invoked for the explanation of the phenomenon [12-13]. In particular in [12] the authors illustrate a scenario where the metal-insulator transition in the 2DEG can be attributed to strong correlations that are effectively enhanced even far away from integer filling, due to incipient charge ordering driven by non-local Coulomb interactions; optical results supporting this scenario have been obtained recently [14].

In this work we illustrate the results of our recent magneto-photoluminescence experiments, trying to enlighten the role of many body effects, strong correlations among the electrons and interaction-related effects. Analyzing the intensity and area of the 2DEG PL emission as a function of the magnetic field B at temperature down to 600 mK, we observe several structures in the spectra at integer and fractional filling factors ν, which we interpret within the CFs picture and taking into account the coulomb screening of the 2DEG [15]. At temperature down to 40 mK, studying the magnetic field dispersion of the emission energy and monitoring simultaneously the longitudinal and Hall resistances, we evidence a significant change in the PL in correspondence of the liquid-to-insulator transition of the 2DEG.

2. Experimental

The experiments were performed on a 20 nm-thick asymmetric modulation-doped GaAs/AlGaAs single QW with carrier density n = 1.8 × 10^{11} cm^{-2} and mobility μ = 1.6 × 10^6 cm^2 V^{-1} s^{-1}. Indium contacts were placed with a Hall bar geometry on a rectangular sample of millimetre size, allowing simultaneous transport and optical experiments. The sample was excited with the light from a Ti:sapphire laser at the energy of 1.748 eV, with an excitation power smaller than 300 μW. The experiments were performed in the temperature range 40 mK - 1.5 K.

3. Results and discussion

In Fig. 1 (top) is reported the low temperature PL emission deriving from the recombination of the 2DEG electrons in their first Landau level L_0 with photo-created valence holes in their first LL. The intensity of the L_0 emission varies markedly and in particular it exhibits clear minima at specific values of the magnetic field B, as can be seen in Fig. 1 (bottom) which shows the intensity of the PL peak as a function of B. Strong suppression of the PL intensity is observed at ν = 1 (B = 7.6 T), while a strong step is present at ν = 2 at B ~ 3.8 T; a weak minimum and a small step are also observed at ν = 3 at ν = 4, respectively. These behaviours, and similar ones reported by other authors [3,8], can be traced back to the influence by the electrons in the higher LLs on the emission from L_0. At higher fields when ν < 1, the emission is suppressed at values of the magnetic field that can be differentiated in two groups: i) the ones corresponding to fractional filling at ν = 2/3, 3/5, 2/5, 1/3, 2/7 and at ν on both side of 1/2, and ii) the ones corresponding to the filling factors ν = 4/5, 5/7, 4/11 and 3/8 (for the last two see in particular Fig. 2). Fractions in group i) agree with the ones expected by the theory of CFs [6,16], according to which the FQHE for the electrons is seen as the IQHE of CFs, quasi-particles originating from the mutual interaction of the electrons in the magnetic field and formed by the bounding of an electron with an even number of magnetic flux quanta. The net result is the passage from a system of interacting electrons in a magnetic field B to another of vanishing interacting CFs in a residual magnetic field B* which leads to the formation of LLs for the CFs. This allows to rescale the known picture for the IQHE of electrons to the case of the quasi-particles described by J. K. Jain [6]. The analysis of the PL intensity minima, with simultaneous monitoring the areas under the PL curves, allow to give a physical interpretation of the results, basing on the argument that the overlap between the wave-functions of the 2D electrons and photo-created holes depends on the Coulomb screening efficiency of the 2DEG [15]. The fact that the CFs form LLs, in analogy to what happens for electrons, makes possible to extend this interpretation for the behavior of the electrons in the IQH regime [4,15] to the case of FQHE PL intensity minima such as the ones observed in group i).

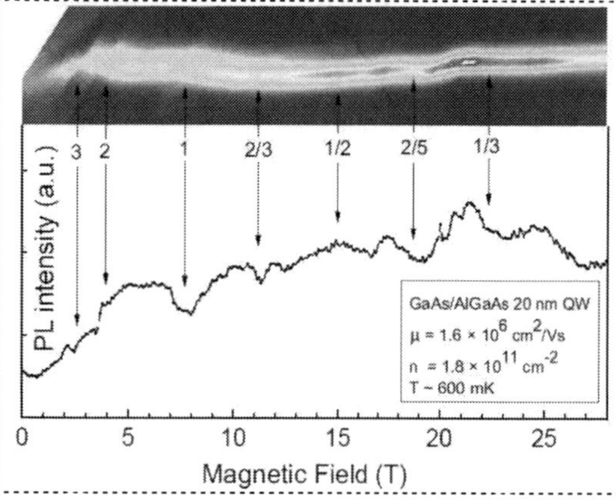

Figure 1. (top) contour plot of the PL emission from the first Landau level ~ 600 mK; **(bottom)** PL intensity vs magnetic field.

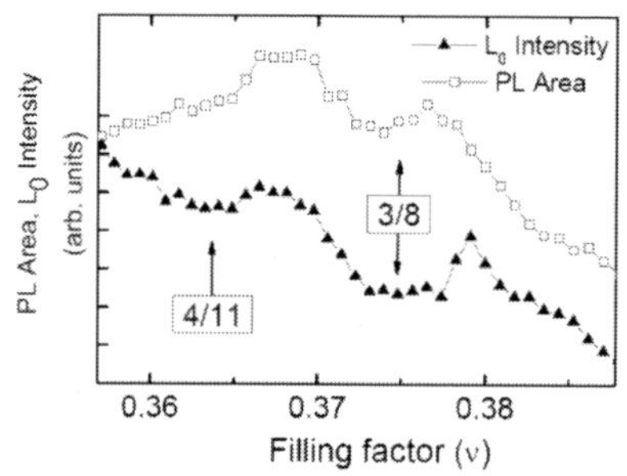

Figure 2. PL intensity and area vs B in the magnetic field range between 2/5 and 1/3. Clear depression of the optical emission is observed at 4/11 and 3/8.

The states belonging to group ii) do not belong to the IQHE of CF series, but at least the first three can be explained in the frame of the theory of J. K. Jain [6], imagining the corresponding quantum states as FQHE of CFs. This requires to assume that the CF-CF interaction is not vanishing, but the residual interaction manifests itself as FQHE, which implies composing CFs of CFs (see for example the experimental work of Gallais et al. [17]). On the contrary, the origin of $\nu = 3/8$ can neither be explained as IQHE nor as FQHE of CFs. Indeed, the enigmatic origin of this state, which has been observed only in few recent magneto-transport experiments as a small minimum in R_{xx} [2], is still under debate. For its explanation, a model invokes an instability in the Fermi sea with opening of a gap involving a p-wave pairing of CFs [18], with similarities to the pairing proposed to explain the minimum in R_{xx} observed at $\nu = 5/2$ due to the occurrence of a BCS-like "Pfaffian" state [19-21]. A different approach proposes the grouping of quasi-particles as a possible explanation [18]. Another picture suggests the clustering of composite bosons that carrying flux positive as well as negative quanta [22]. In the framework of all these theories describing the occurrence of a minimum in R_{xx} at $\nu = 3/8$, the optical observation of the 3/8 state in the optical emission represents a cue for a better understandings of its nature.

We also investigated the magnetic field dispersion of the emission energy. We observe that the strongest L_0 emission is accompanied by the ones from higher LLs (at higher energy) and the one from a shake-up (SU1) process [23] (at lower energy). All these emissions approximately follow a linear dispersion, but the accurate analysis of the L_0 energy vs B at very low temperature shows several interesting features [14]. At low field L_0 follows quite close the LL dispersion, while above $B_{\nu=2}$ it deviates and stays well below the linear behavior $1/2(\hbar\omega_{c\text{-}e} + \hbar\omega_{c\text{-}h})$, also exhibiting two small inflections or shifts at $\nu = 2/3$ and between $\nu = 2/5$ and $\nu = 1/3$ (also observed by I. Bar-Joseph et al. [23], and by M. Byszewski et al. [8]).

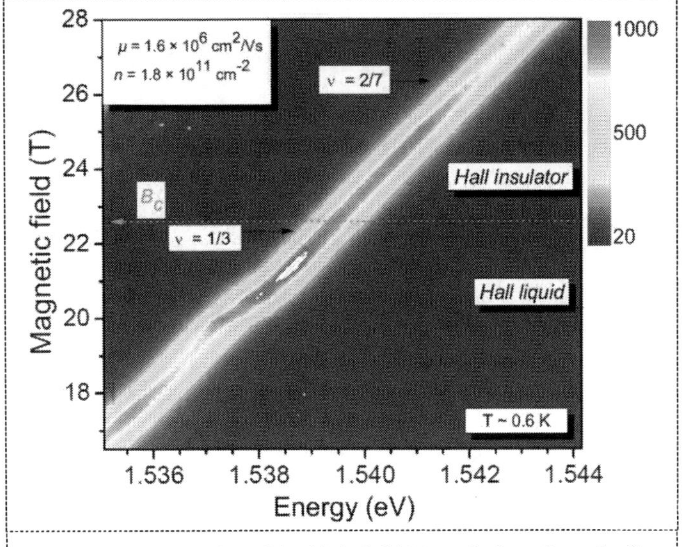

Figure 3. Contour plot of the high field PL emissions from the first Landau level L_0. Transport results and the analysis of the emission energy indicate that the 2DEG undergoes the liquid-insulator transition beyond the critical magnetic field B_c.

In this range we observe another change in the overall behavior of the energy dispersion: it becomes quadratic up to ~19 T, as typically observed for electron-hole bound states like neutral or charged excitons [24]. Above ~ 21 T the emission follows another dispersion law that we find out to be linear, suggesting that the system is evolving from a liquid phase to an insulating one [14]. Fig. 3 shows the contour plot of the emission at magnetic fields, below and above the critical field B_c of the liquid-to-insulator transition, obtaining B_c from the transport measurements, because at B_c the R_{xy} crosses the values h/e^2 [14]. In high mobility GaAs-AlGaAs samples the transition takes place beyond the $\nu = 1/3$ state (as in our case) or the $\nu = 1/5$ state [9]. The frame for the interpretation of our results, obtained in simultaneous optical and transport experiments also at ~ 40 mK [14], is provided by recent theoretical findings concerning the strong correlation effects in 2DEG [25], showing that the insulating phase results from the incipient charge ordering driven by Coulomb interactions, and also by experimental works showing that at intermediate density and close to criticality the electrons exhibit an increased effective mass [26] also due to the strong correlations.

4. Conclusions

In conclusion, the quantum states at several fractional ν have been observed in optical experiments as suppression of the PL emission and interpreted using a model based on the CFs picture and the Coulomb screening of the 2DEG ($\nu = 4/5, 5/7, 4/11$), or attributed to the weakening of the screening of the Coulomb interaction among the quasi-particles which compose the state ($\nu = 3/8$). The correlation effects in our 2DEG have also been investigated by simultaneous optical and transport experiments, observing a drastic change in the emission energy when the system undergoes the metal-insulator transition.

Acknowledgments
We are grateful to M. Hilke, W. Pan, G. Pastori Parravicini and M. Potemski and for enlightening discussions. This work has been supported by the Cariplo Foundation *QUANTDEV*, the Spanish Ministry of Science and Innovation (FIS2009-07880, PPT310000-2009-6, PCT310000-2009-3), the Junta de Castilla y León (SA049A10), and by EuroMagNET, EU contract n° 228043.

References

[1] Stormer H L 1998 *Rev. Mod. Phys.* **71**, 4875
[2] Pan W, Stormer H L, Tsui D C, Pfeiffer L N, Baldwin K W and West KW 2003 *Phys. Rev. Lett.* **90**, 016801
[3] Goldberg B B, Heiman D, Pinczuk A, Pfeiffer L and West K W 1990 *Phys. Rev. Lett.* **65**, 641
[4] Turberfield A J, Haynes S R, Wright P A, Ford R A, Clark R G, Ryan J F, Harris J J and Foxon C T 1990 *Phys. Rev. Lett.* **65**, 637
[5] Yusa G, Shtrikman H and Bar-Joseph I 2002 *Physica E* **12**, 49
[6] Jain J K 2000 *Physics Today* **53**, 39
[7] Bellani V, Dionigi F, Rossella F, Amado M, Diez E, Biasol G and Sorba L 2010 *Phys. Rev. B* **81**, 155316
[8] Byszewski M, Chwalisz B, Maude D K, Sadowski M L, Potemski M, Saku T, Hirayama Y, Studenikin S, Austing D G, Sachrajda A S and Hawrylak P 2006 *Nature Physics* **2**, 239
[9] Hilke M, Shahar D, Song S H, Tsui D C, Shayegan M, Xie Y H 1999 *Ann. Phys.* **8**, 603
[10] Hilke M, Shahar D, Song S H, Tsui D C, Xie Y H and Monroe D 1998 *Nature* **395**, 675
[11] Amado M, Diez E, Lopez-Romero D, Rossella F, Caridad J M, Dionigi F, Bellani V and Maude D K 2010 *New J. Phys.* **12** 053004
[12] Camjayi A, Haule K, Dobrosavljevic V, Kotliar G 2008 *Nature Phys.* **4**, 932
[13] Aguiar M C O, Dobrosavljevic V, Abrahams E, Kotliar G 2009 *Phys. Rev. Lett.* **102**, 156402
[14] Dionigi F, Rossella F, Bellani V, Amado M, Diez E, Kowalik K, Biasol G and Sorba L 2010 submitted to *Phys. Rev. Lett.*, arXiv:1007.2936v1
[15] Dahl M, Heiman D, Pinczuk A, Goldberg B B, Pfeiffer L N and West K W 1992 *Phys. Rev. B.* **45**, 6957
[16] Jain J K 2003 *Physica E* **20**, 79
[17] Gallais Y, Kirschenmann T H, Dujovne I, Hirjibehedin C F, Pinczuk A, Dennis B S, Pfeiffer L N and West K W 2006 *Phys. Rev. Lett.* **97**, 036804
[18] Wójs A, Yi K-S and Quinn J J 2004 *Phys. Rev. B.* **69**, 205322
[19] Pan W, Xia J S, Shvarts V, Adams D E, Stormer H L, Tsui D C, Pfeiffer L N, Baldwin K W and West K W 1999 *Phys. Rev. Lett.* **83**, 3530
[20] Peterson M R, Jolicoeur Th and Das Sarma S 2008 *Phys. Rev. Lett.* **101**, 016807
[21] Pan W, Xia J S, Stormer H L, Tsui D C, Vicente C, Adams E D, Sullivan N S, Pfeiffer L N, Baldwin K L and West K W 2008 *Phys. Rev. B.* **77**, 075307
[22] Jolicoeur Th 2007 *Phys. Rev. Lett.* **99**, 036805
[23] Bar-Joseph I, Yusa G and Shtrikman H 2003 *Sol. State Comm.* **127,** 765; Bar-Joseph I 2005 *Semicond. Sci. Technol.* **20,** R29; Finkelstein G, Shtrikman H and Bar-Joseph I 1998 *Physica B* **575,** 249
[24] Yoon H W, Sturge M D and Pfeiffer L N 1997 *Sol. State Comm.* **5**, 287
[25] Kravchenko S V and Sarachik M P 2004 *Rep. Prog. Phys.* **67**, 1
[26] Anissimova S, Venkatesan A, Shashkin A A, Sakr M R, Kravchenko S V T and Klapwijk M 2006 *Phys. Rev. Lett.* **96**, 046409

Journal of Physics: Conference Series **334** (2011) 012023 doi:10.1088/1742-6596/334/1/012023

Modulation induced stripe phase at fractional fillings

Akira Endo[1], **Naokazu Shibata**[2] **and Yasuhiro Iye**[1]

[1] Institute for Solid State Physics, University of Tokyo, Kashiwa, Chiba 277-8581, Japan

[2] Department of Physics, Tohoku University, Sendai, Miyagi 980-8578, Japan

E-mail: akrendo@issp.u-tokyo.ac.jp

Abstract. We have experimentally examined the state of a two-dimensional electron gas subjected to unidirectional periodic potential modulation in the vicinity of the filling ν=5/3 (equivalent to ν=1/3 by the particle-hole symmetry). In addition to a peak in the longitudinal resistivity ρ_{xx}, we find small amplitude oscillations roughly periodic in the magnetic field B in the Hall resistivity ρ_{yx}. The oscillations appear in a narrow range of the electron density n_e, and the period ΔB of the oscillations increases with the slight increase in n_e. The oscillations are interpreted as resulting from the commensurability between the high harmonics of the modulation potential and the stripe state, which is predicted by the density matrix renormalization group calculations to be the ground state in our system.

1. Introduction

In recent studies [1, 2], the present authors have investigated the effect of unidirectional periodic potential modulation $V(x)$ on the fractional quantum Hall (FQH) states. We have experimentally found a peak that grows with decreasing temperature (see Fig. 3 (a)), instead of a minimum, in the resistivity ρ_{xx} in the vicinity of the filling ν=5/3 when the amplitude of the modulation is made sufficiently large [2]. In search of the origin of the peak, we performed calculations by the density matrix renormalization group (DMRG) method at ν=1/3 (equivalent to ν=5/3 by the particle-hole symmetry), which revealed phase transition from the FQH state to the stripe state having a period $\sim 4l$ (with $l=\sqrt{\hbar/eB}$ the magnetic length) [2] (see the next section) reminiscent of the stripe phase at higher Landau levels (LLs) [3, 4, 5]. The result strongly suggests that the observed peak is originating from the stripe phase, representing the first experimental indication for the stripe phase in the lowest Landau level (LLL). Note that experimental evidence for the stripe phase in the LLL, in spite of the theoretical prediction [6, 7], has been elusive thus far. In the present paper, we study the state at $\nu\sim 5/3$ in more detail. The Hall resistivity ρ_{yx} is examined in addition to the longitudinal resistivity ρ_{xx}. We explore the behavior of the resistivities under the application of the dc current bias I_{dc}, which is expected to affect the pinning or sliding of the stripe state.

2. Result of DMRG calculations

In this section, we briefly review the result of the calculation by the DMRG method. We show the results for the rectangular as well as the sinusoidal modulation. Both types of the modulation exhibit qualitatively the same behavior. As shown in Fig. 1(a), the FQH gap $\Delta_{1/3}$ decreases with increasing modulation amplitude V_0, until it vanishes at the value of V_0 close to $\Delta_{1/3}$ at V_0=0 for sinusoidal $V(x)$, and at a smaller V_0 for rectangular $V(x)$. Charge density wave (CDW) with the

Published under licence by IOP Publishing Ltd

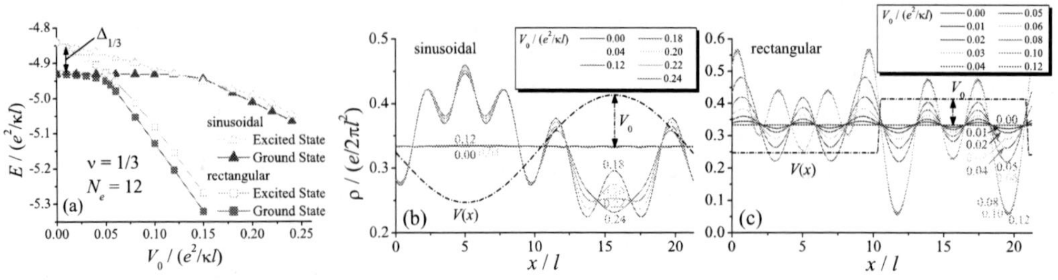

Figure 1. (a) Energy levels of the ground state (solid symbols) and the first excited state (open symbols) as a function of V_0. Triangles (squares) are for sinusoidal (rectangular) modulation. (b)(c) Charge density profile for various values of V_0; (b) sinusoidal modulation, (c) rectangular modulation. The profiles of the modulation potential $V(x)$ are plotted by dot-dashed lines. ν=1/3 and calculated for N_e=12 electrons.

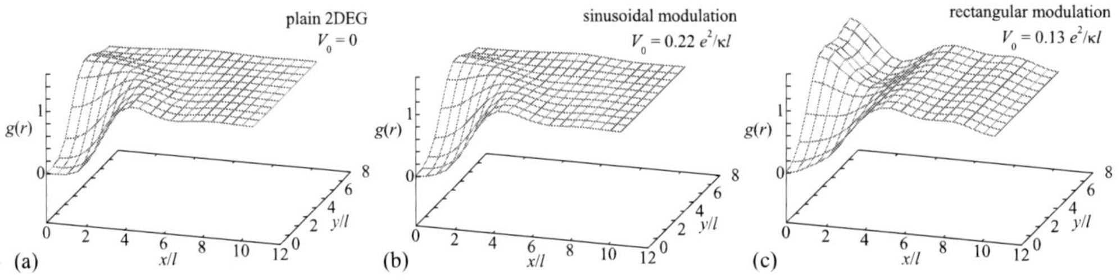

Figure 2. The pair correlation functions g(x, y) at ν=1/3 (N_e=12) for plain 2DEG (a), and for ULSL with sinusoidal modulation (b) and rectangular modulation (c).

distance between adjacent peaks $\sim 4l$ develops after $\Delta_{1/3}$ vanishes; the emergence of the CDW is abrupt for the sinusoidal $V(x)$, whereas gradual development is already seen while the value of $\Delta_{1/3}$ is still finite for rectangular $V(x)$ (Fig. 1(b)(c)). The pair correlation functions displayed in Fig. 2 reveal the correlation at $4l$ not only in the x-direction but also in the y-direction (along the stripe), resembling the stripe-II phase in [7]. Despite the prominent stripe-like charge density profile in Fig. 1 (b), however, the correlation is rather unclear for sinusoidal $V(x)$ (Fig. 2(b)) (for the reason not known at present).

3. Experimental details

The device of the unidirectional lateral superlattice (ULSL) in the Hall-bar geometry was fabricated from a GaAs/AlGaAs two-dimensional electron gas (2DEG) wafer with the electron density n_e=2.1×10^{15} m^{-2} and the mobility μ=77 m^2/Vs. Unidirectional periodic modulation of the electrostatic potential $V(x)$ having the period a=184 nm was introduced via strain-induced piezoelectric effect [8] by placing a grating of electron-beam resist on the surface [9], with the direction x parallel to the direction of the current (both I_{ac} for the lock-in measurement and the dc bias I_{dc}) in the Hall bar. To maximize the piezoelectric effect and hence the modulation amplitude, the direction x of the modulation was aligned to the [110] direction of the host crystal [8, 10]. The device also contains a section without the modulation (plain 2DEG) for reference adjacent to the ULSL section. The profile of the modulation $V(x)=\Sigma_n V_n \cos(2\pi n x/a + \phi_n)$ was determined by the analysis of the commensurability oscillations (CO) [11, 12]. Up to the

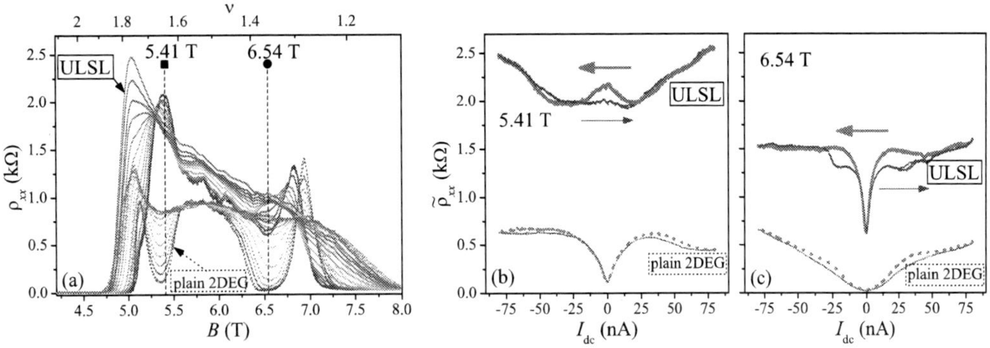

Figure 3. (a) Resistivity ρ_{xx} ($1<\nu<2$) for ULSL and the adjacent plain 2DEG at temperatures T (mK) = 15 (bottom, blue), 25, 36, 49, 70, 90, 117, 170, 218, 275, 323, 475, 434, 480 (top, red). (b)(c) Resistivity $\tilde{\rho}_{xx}$ subjected to dc current bias $I_{\rm dc}$, with down-sweeps (up-sweeps) of $I_{\rm dc}$ plotted with thick red lines (thin blue lines), measured at T=15 mK; (b) B=5.41 T (position of the peak in the ULSL) and (c) B=6.54 T (ν = 4/3 minimum). ULSL and plain 2DEG are plotted by solid and dotted lines, respectively.

fourth harmonics were detected with V_1=0.31 meV, V_2=0.1 meV, V_3=0.07 meV, V_4=0.05 meV (CO have no sensitivity to the phases ϕ_n). The profile is in between the two extreme cases, the sinusoidal and the rectangular modulation, but owing to the dominance of V_1, may be viewed as a sinusoidal modulation $V(x)=V_0\cos(2\pi x/a)$ with $V_0=V_1$ in the roughest approximation. We will show below, however, that higher harmonics may be playing an important role in the oscillations observed in ρ_{yx}. The value of V_1 is larger than the ν=5/3 FQH gap ~0.1 meV obtained by Lifshitz-Kosevich analysis [2, 13] in the plain 2DEG section, and therefore our sample is expected to be placed in the stripe phase according to Fig. 1(a). The measurement was performed in a dilution fridge at temperatures T=15–480 mK.

4. Experimental results and discussion

Traces of the longitudinal resistivity ρ_{xx} measured by the standard ac lock-in technique using an ac current $I_{\rm ac}$=0.5 nA (f=13 Hz) are shown in Fig. 3(a) for both the ULSL and the plain 2DEG sections. The ULSL section exhibits a peak that grows with decreasing temperature at $\nu\sim5/3$ below ~300 mK, while the FQH state survives at ν=4/3 albeit with rather complicated T- and B-dependences which may be signaling the possible competition between the FQH and the stripe states [2]. Assuming that the peak at $\nu\sim5/3$ results from the pinning of the stripe phase, the depinning caused by the application of a dc bias current $I_{\rm dc}$ is expected to reduce the peak height. In fact, the resistivity $\tilde{\rho}_{xx}$ measured by the ac lock-in technique with the current $I_{\rm ac}+I_{\rm dc}$ (corresponding to the differential resistivity $\propto dV/dI$) displays decrease with the increase of $|I_{\rm dc}|$ at smaller $|I_{\rm dc}|$ (Fig. 3(b)). However, the decrease is rather gradual and not abrupt as expected from the pinning-depinning transition. The decrease can rather be explained by the increase in the electron temperature by $I_{\rm dc}$ (due to the Joule heating), noting that the peak height reduces with increasing temperature at low temperatures. The electron-heating picture is consistent with the increase of $\tilde{\rho}_{xx}$ for larger $|I_{\rm dc}|$ and also with the increase of $\tilde{\rho}_{xx}$ in the plain 2DEG. Interesting behavior to be noticed in Fig. 3(b) is the hysteresis between the up-sweeps and down-sweeps of $I_{\rm dc}$, observed only in the ULSL section. Several traces, almost perfectly overlapping with each other, are plotted for both up- and down-sweeps (by thin blue and thick

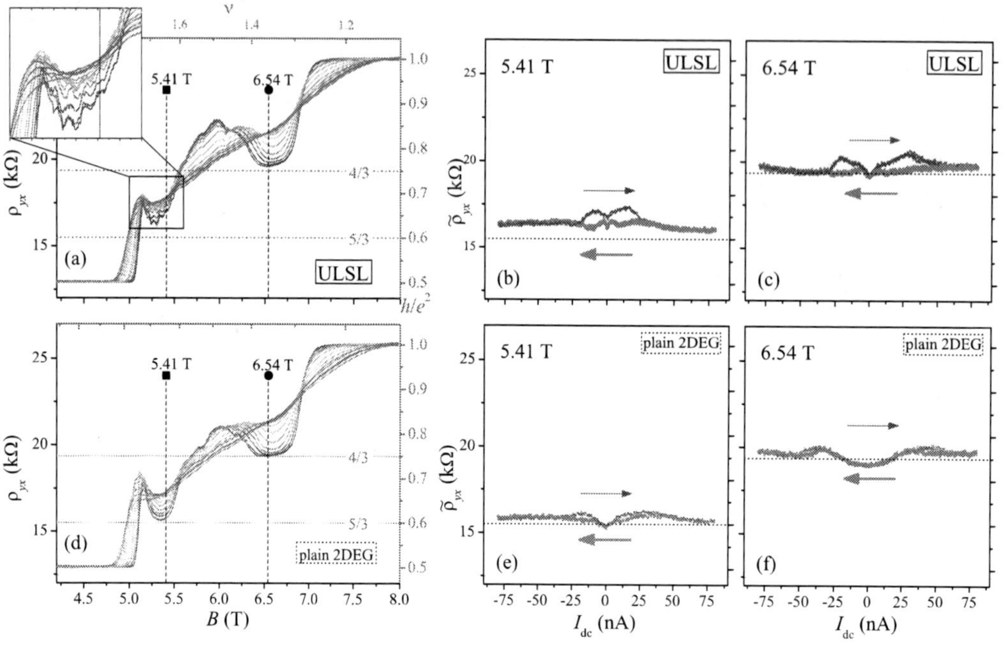

Figure 4. (a)(d) Hall resistivity ρ_{yx} ($1<\nu<2$) for ULSL and the adjacent plain 2DEG at temperatures T (mK) = 15 (bottom, blue), 25, 36, 49, 70, 90, 117, 170, 218, 275, 323, 475, 434, 480 (top, red). (b)(c)(e)(f) Hall resistivity $\tilde{\rho}_{yx}$ subjected to dc current bias $I_{\rm dc}$, with down-sweeps (up-sweeps) of $I_{\rm dc}$ plotted with thick red lines (thin blue lines), measured at T=15 mK; (b)(e) B=5.41 T (position of the peak in ρ_{xx} of the ULSL) and (c)(f) B=6.54 T (ν = 4/3). ULSL (a–c) and plain 2DEG (d–f) are plotted by solid and dotted lines, respectively.

red lines, respectively), certifying the reproducibility of the hysteresis. Although the origin of the hysteresis is not known at present, it is consistent with the presence of the collective phase. Slightly more complicated hysteretic behavior is observed also at ν=4/3, possibly related to the co-presence of the two different phases, the FQH state and the stripe phase.

Similar plots for the Hall resistivity ρ_{yx} are shown in Fig. 4 for ULSL (a–c) and plain 2DEG (d–f). Again, the ULSL section exhibits the hysteretic behavior. In the vicinity of ν=5/3, the Hall resistivity in ULSL shows small amplitude oscillations roughly periodic in B, as highlighted in the inset of Fig. 4(a). The oscillations reduce their amplitude with increasing temperature, and vanish at $T\sim$ 300 mK. The temperature range the oscillations are observed coincides with the range the peak develops in ρ_{xx}.

To further investigate the nature of the oscillations, we vary the electron density $n_{\rm e}$ by illumination, employing the persistent-photoconductivity effect. The variation in $n_{\rm e}$ leads to the shift in the magnetic field B at which ν=5/3 takes place, hence to the change in the magnetic length l or the strength of the Coulomb interaction $e^2/\kappa l$. The evolution of the oscillations in the Hall resistivity $\tilde{\rho}_{yx}$ (with the dc bias current $I_{\rm dc}$) is presented in Fig. 5. The period of the oscillations ΔB is seen to increase with increasing $n_{\rm e}$, and the oscillations disappear for $n_{\rm e}$ larger than $\sim 2.6\times 10^{15}$ m^{-2}. Again, the role of $I_{\rm dc}$ can be interpreted simply as heating of the electrons; the amplitude of the oscillations diminishes with increasing $I_{\rm dc}$ but the period and phase of the

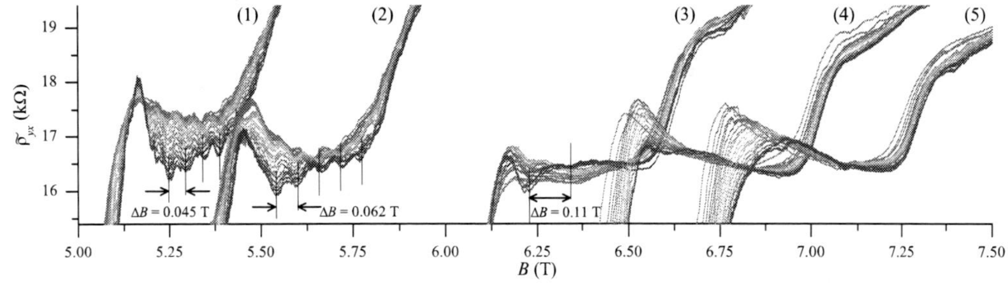

Figure 5. Hall resistivity $\tilde{\rho}_{yx}$ in the vicinity of ν=5/3 for different electron densities: n_e (in 10^{15} m^{-2}) are (1) 2.1, (2) 2.2, (3) 2.5, (4) 2.7 and (5) 2.8. Traces with different colors represent different values of dc current bias I_{dc} ranging from I_{dc}=0 nA (blue) down to either of −5 nA (1),(2) or −4 nA (3),(5) or −3 nA (4) (red).

oscillations remain unaltered. In the range of n_e the oscillations are observed ((1)–(3) in Fig. 5), the expected period of the stripe state $\sim 4l$ is close to the fourth harmonic of the modulation $a/4$=46 nm. Furthermore, the change in the Coulomb energy during the oscillation period ΔB is also close to the amplitude of the fourth harmonic potential V_4. We therefore speculate that the oscillations result from some commensurability effect between the stripe state and the fourth harmonic potential. Our scenario is that the stripe phase in itself is basically brought about by the large fundamental component V_1, resulting in the peak in ρ_{xx}, but the oscillations require the high harmonics. Further studies are necessary, however, to elucidate the exact mechanism through which the commensurability affects the resistivities.

Acknowledgments

This work was supported by Grant-in-Aid for Scientific Research (C) (18540312) and (A) (18204029) from the Ministry of Education, Culture, Sports, Science and Technology (MEXT).

References

[1] Endo A, Shibata N and Iye Y 2010 *Physica E* **42** 1042
[2] Endo A, Shibata N and Iye Y 2010 *J. Phys. Soc. Jpn.* **79** 103707
[3] Koulakov A A, Fogler M M and Shklovskii B I 1996 *Phys. Rev. Lett.* **76** 499
[4] Fogler M M, Koulakov A A and Shklovskii B I 1996 *Phys. Rev. B* **54** 1853
[5] Moessner R and Chalker J T 1996 *Phys. Rev. B* **54** 5006
[6] Shibata N and Yoshioka D 2003 *J. Phys. Soc. Jpn.* **72** 664
[7] Shibata N and Yoshioka D 2004 *J. Phys. Soc. Jpn.* **73** 1
[8] Skuras E, Long A R, Larkin I A, Davies J H and Holland M C 1997 *Appl. Phys. Lett.* **70** 871
[9] Endo A, Katsumoto S and Iye Y 2000 *Phys. Rev. B* **62** 16761
[10] The [110] and [1$\bar{1}$0] directions are equivalent for the piezoelectric effect [8]. Since, in our sample, the anisotropy artificially introduced by $V(x)$ is rather large and the mobility is not so high, our measurements are unlikely to be affected by the fragile built-in anisotropy between the two orientations observed in ultrahigh mobility 2DEGs [14, 15]. See also [16] for the comparison between the two crystallographic orientations.
[11] Weiss D, v Klitzing K, Ploog K and Weimann G 1989 *Europhys. Lett.* **8** 179
[12] Endo A and Iye Y 2005 *J. Phys. Soc. Jpn.* **74** 2797
[13] Lifshitz I M and Kosevich A M 1956 *Sov. Phys. JETP* **2** 636
[14] Lilly M P, Cooper K B, Eisenstein J P, Pfeiffer L N and West K W 1999 *Phys. Rev. Lett.* **82** 394
[15] Du R R, Tsui D C, Stormer H L, Pfeiffer L N, Baldwin K W and West K W 1999 *Solid State Commun.* **109** 389
[16] Endo A and Iye Y 2002 *Physics of Semiconductors 2002, Proc. 26th Int. Conf. Phys. Semicond.* ed Long A R and Davies J H (Bristol: IOP) p H132

HMF-19 IOP Publishing
Journal of Physics: Conference Series **334** (2011) 012024 doi:10.1088/1742-6596/334/1/012024

Critical tunnel currents and dissipation of Quantum-Hall bilayers in the excitonic condensate state

Y. Yoon[a], X. Huang, E. Yarar, W. Dietsche, L. Tiemann[b], S. Schmult and K. v.Klitzing

Max-Planck-Institut für Festkörperforschung, Heisenbergstraße 1, 70569 Stuttgart, Germany

E-mail: `w.dietsche@fkf.mpg.de`

Abstract.
Transport and tunneling is studied in the regime of the excitonic condensate at total filling factor one using the counterflow geometry. At small currents the coupling between the layers is large making the two layers virtually electrically inseparable. Above a critical current the tunneling becomes negligible. An onset of dissipation in the longitudinal transport is observed in the same current range.

1. Introduction

Two-layer quantum-Hall systems in which the lowest Landau level in each layer is approximately half-filled may form a superfluid state. This can be visualized by considering the filled and unfilled electron states in the Landau levels of the two layers as excitons which condense into a coherent ground state [1, 2, 3]. The first experimental evidence of an unusal state at the total (combined) filling factor one had already been reported early on as a distinct minimum in the longitudinal and a plateau in the Hall resistance [4]. This was later verified by several other groups. In counterflow experiments where equal but oppositely directed currents were passed through the two layers, a nearly complete vanishing of both Hall and longitudinal resistance was observed [5, 6, 7]. The coherence of the condensed excitons is thought to be responsible for the enhanced tunnel conductance at zero bias [8, 9, 10] and the Josephson-like tunneling with a large critical current which became observable in samples with sufficiently large tunneling probability and area [11, 12].

Theoretically this state has been described by theories resembling the BCS theory of superconductors with the Cooper pairs being replaced by the excitons. Condensation takes place at sufficient low temperatures if the magnetic length $l_B = \sqrt{\hbar/eB}$ becomes comparable to the layer separation d. The existence of an energy gap of the condensed state manifests itself by the observation of the phenomena of the Quantum-Hall effect at the total filling factor one.

In this contribution we report our results on the critical tunnel current in the counterflow geometry [13] and discuss its relation to the onset of dissipation in the in-plane transport.

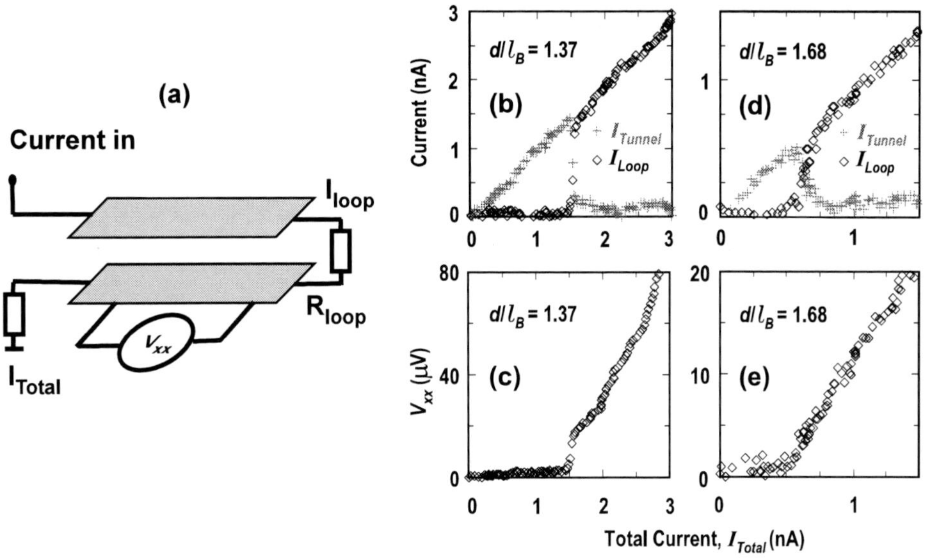

Figure 1. (a) A schematic of the measurement set-up. Current is first injected into the top layer and then redirected via a loop resistor into the bottom layer and then guided into ground. At d/l_B=1.37 (b and c) and d/l_B=1.68 (d and e) I_{Tunnel} & I_{Loop} and V_{xx} are plotted versus I_{Total}.

2. Experimental details

The double quantum well system used in this study have been grown by molecular beam epitaxy. Two 19 nm GaAs wells are separated by a 9.6 nm superlattice barrier layer consisting of alternating layers of AlAs (1.7 nm) and GaAs (0.28 nm). The quantum wells are populated by Si-doped sheets placed 300 nm below and 280 nm above the wells, yielding intrinsic densities of approximately 4.0×10^{10} cm^{-2} in the upper and 4.6×10^{10} cm^{-2} in the lower quantum well. Mobilities at these densities are 4.3×10^{5} cm^{2}V^{-1}s^{-1} and 4.1×10^{5} cm^{2}V^{-1}s^{-1}, respectively. A Hall bar device with a length of 880 μm and a width of 80 μm is prepared by optical lithography. Independent electrical contacts to the two layers are achieved by applying appropriate negative voltages to the pre-patterned buried back gates and the metallic front gates crossing the contact arms (selective depletion technique [14, 15]). With the intrinsic densities, the interlayer resistance at zero magnetic field and 4.2 K is around 380 kΩ. The densities in the two layers can be adjusted independently by additional front and back gates and are balanced in all data shown. The ratio (d/l_B) of the layer separation (d = 28.6 nm) to the magnetic length l_B is set to values between 1.37 and 1.68. All data were taken in a dilution refrigerator at a bath temperature of 20 - 30 mK.

Counterflow experiments are performed by passing two oppositely directed currents through the two layers as shown in Fig. 1(a) similar as in previous experiments [5, 6, 7]. We use a dc input current I which is generated by applying a voltage to a large resistor (100 MΩ, not shown) in series to the bilayer system. The current passes first through the top layer, is redirected via a loop resistor (10 kΩ) into the bottom layer in the opposite direction, and finally guided via another resistor (10 kΩ) into ground. The total I_{Total} and the loop I_{Loop} currents follow from the voltage drops across the 10 kΩ resistors. The interlayer tunneling current I_{Tunnel} is the difference I_{Total} - I_{Loop}.

3. Results

In Fig. 1(b) we plot I_{Tunnel} and I_{Loop} as function of the total (dc) input current for a d/l_B=1.37 at ν_T=1. Strikingly, at small total currents, we find that I_{Loop} is nearly zero. This is equivalent to almost all current tunneling before reaching the end of the Hall bar. With increasing total current, a critical current value I_C=1.5 nA is reached, where I_{Tunnel} suddenly becomes negligible and nearly all current flows through the loop resistor. At the same current the longitudinal voltage (V_{xx}) measured along one of the layers jumps from almost zero to a finite value and continues to increase rapidly with increasing I_{Total} (Fig. 1(c)). Below I_C the interpretation of the V_{xx} data is difficult because the current might tunnel before passing both voltage probes. At larger d/l_B values, Fig. 1(d) and (e), the transition at the critical current is less sharp but the general behavior is similar .

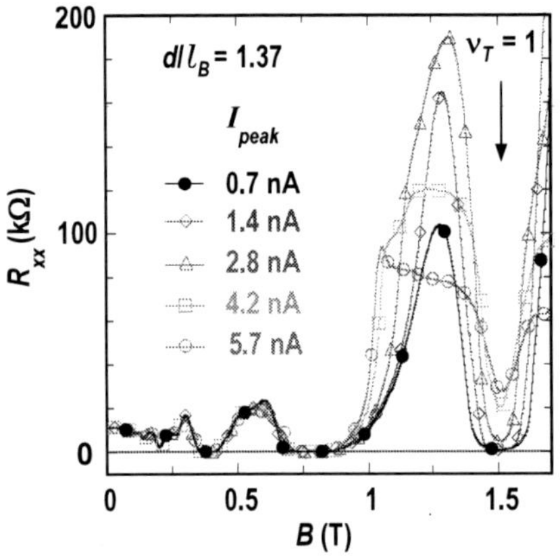

Figure 2. Longitudinal resistance of the sample of Fig. 1 at different ac-current amplitudes.

We estimate from the increase of the loop current at I_C that the effective interlayer resistance must be less than 300 Ω below I_C. The values of both I_C and the interlayer resistance are comparable to the ones reported in Ref.[11].

Naively, one would expect that reaching the maximum current for dissipationless tunneling signals the collapse of the condensed exciton state. This is, however, not the case as becomes obvious from data like the ones presented in Fig. 2 which shows several traces of R_{xx} vs the magnetic field B in the same counterflow geometry as shown in Fig. 1. For this set of data ac-current is used. The respective peak values of the sinusoidal excitation currents are given in the figure. The value of d/l_B was set to =1.37 where $I_C = 1.5\ nA$. Each trace corresponds to a B-sweep with a different I_{peak} value. It is striking that upon increasing I_{peak} above I_C the formerly deep and nearly complete minimum in R_{xx} weakens but does not disappear. Furthermore, although not shown here, R_{xy} remains small and nearly constant (at around 2 kΩ) as observed in the earlier counterflow experiments. This result corresponds also to earlier results [5, 6, 7] where clear signatures of the ν_T=1 were found in counterflow experiments while the observed tunneling conductance was much less than reported here.

In another, more recent experiment with a different sample we studied the current dependence of the longitudinal transport. This sample was grown with a thinner (7.6 nm) tunneling barrier

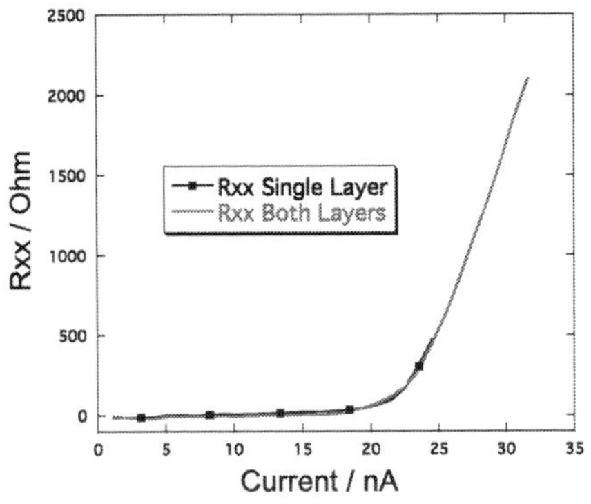

Figure 3. Longitudinal resistance as function of excitation current with only one layer being contacted and without contact separation ("Both Layers") at ν_T=1.

between the two layers. The density was adjusted for $d/l_B = 1.65$ at a magnetic field of 2.6 T. The longitudinal resistance was monitored as function of the ac current amplitude. Results are shown in Fig. 3. For the black curve with square markers the two layers were electrically separated by the gating technique. The red trace was obtained with the gates being turned off and the current being fed into both layers simultaneously. Strikingly, both traces are very similar. At small currents, the R_{xx} values are small and independent from current. Exceeding about 20 nA, both traces show a sharp upturn indicating a rapid increase in dissipation. The identity of the two traces is another sign that at small currents the two layers are virtually indistinguishable. The R_{xx}- values at the largest currents in the figure are still small compared with those at other (unquantized) filling factors. The critical tunnel currents have not yet been measured with this sample and a comparison with the critical current for the onset of dissipation seen in Fig. 3 is not yet possible. It is nevertheless intriguing to speculate that its value would be similar to that observed in longitudinal transport. The identity of the critical current for longitudinal transport and tunneling has actually been predicted theoretically [16] but has not yet been verified experimentally.

4. Conclusions

In conclusion, our findings show that exceeding the critical tunnel current does not cause the breakdown of the ν_T=1 quantum Hall state. It marks, however, the beginning of an increasing dissipation along the current-carrying layers. Recent theoretical suggestions attribute the dissipation in these exciton condensates to the unbinding and the current-induced motion of vortices connected to charged quasiparticles (merons) [17, 18, 19, 20] or to vortices parallel to the layers which move dissipatively if there is a bias voltage difference [21]. Below the critical current the bilayer system is indistinguishable from a single layer system.

5. Acknowledgements

The wafer used for the experiments of Fig. 1 and 2 had been grown by the group of W. Wegscheider. We would like to acknowledge discussions and correspondence with P. R. Eastham, D. V. Fil, M. Gilbert, T. Hyart, A. H. MacDonald, C. de Morais Smith and B. Rosenow, This project was supported by the BMBF (German Ministry of Education and Research) Grant No. 01BM456 and 01BM900.

[a]Present address: Helsinki University of Technology, Finland
[b]Present address: NTT Basic Research Laboratories, Atsugi, Japan

References

[1] Fertig H A 1989 *Phys. Rev. B* **40** 1087–1095
[2] MacDonald A H and Rezayi E H 1990 *Phys. Rev. B* **42** 3224–3227
[3] Eisenstein J P and MacDonald A H 2004 *Nature* **432** 691–694
[4] Eisenstein J P, SBoebinger G, NPfeiffer L, WWest K and He S 1992 *Phys. Rev. Lett.* **68** 1383–1386
[5] Kellogg M, Eisenstein J P, Pfeiffer L N and West K W 2004 *Phys. Rev. Lett.* **93** 036801
[6] Tutuc E, Shayegan M and Huse D A 2004 *Phys. Rev. Lett.* **93** 036802
[7] Wiersma R D, Lok J G S, Kraus S, Dietsche W, von Klitzing K, Schuh D, Bichler M, Tranitz H P and Wegscheider W 2004 *Phys. Rev. Lett.* **93** 266805
[8] Spielman I B, Eisenstein J P, Pfeiffer L N and West K W 2000 *Phys. Rev. Lett.* **84** 5808–5811
[9] Wiersma R, Lok J, Kraus S, Dietsche W, von Klitzing K, Schuh D, Bichler M, Tranitz H P and Wegscheider W 2006 *Physica E* **34** 16 – 21
[10] Finck A D K, Champagne A R, Eisenstein J P, Pfeiffer L N and West K W 2008 *Phys. Rev. B* **78** 075302
[11] Tiemann L, Dietsche W, Hauser M and von Klitzing K 2008 *New Journal of Physics* **10**
[12] Tiemann L, Yoon Y, Dietsche W, von Klitzing K and Wegscheider W 2009 *Phys. Rev. B* **80** 165120–165120–6
[13] Yoon Y, Tiemann L, Schmult S, Dietsche W, von Klitzing K and Wegscheider W 2010 *Physical Review Letters* **104** 116802
[14] Eisenstein J P, Pfeiffer L N and West K W 1990 *Appl. Phys. Lett.* **57** 2324–2326
[15] Rubel H, Fischer A, Dietsche W, von Klitzing K and Eberl K 1998 *Materials Science and Engineering B* **51** 207–211
[16] Ezawa Z and Iwazaki A 1994 *I.J.Mod.Phys.* **8** 2111–2155
[17] Fertig H A and Murthy G 2005 *Phys. Rev. Lett.* **95** 156802
[18] Huse D A 2005 *Phys. Rev. B* **72** 064514
[19] Roostaei B, Mullen K, Fertig H and Simon S 2008 *Phys. Rev. Lett* **101** 046804
[20] Eastham P R, Cooper N R and Lee D K K 2009 *Phys. Rev. B* **80** 045302
[21] Fil D V and Shevchenko S I 2009 *Journal of Physics: Condensed Matter* **21** 215701

Journal of Physics: Conference Series **334** (2011) 012025 doi:10.1088/1742-6596/334/1/012025

Density-Imbalance Stability Diagram of the $\nu_T = 1$ Bilayer Electron System at Full Spin Polarization

Keiko Takase and Koji Muraki

NTT Basic Research Laboratories, NTT Corporation, 3-1 Morinosato-Wakamiya, Atsugi 243-0198, Japan

E-mail: takase@will.brl.ntt.co.jp

Abstract.
We investigate the evolution of the total Landau level filling factor $\nu_T = 1$ bilayer quantum Hall (QH) state versus density imbalance at full spin polarization under a tilted magnetic field. When the system is well below the compressible-incompressible transition point at the balanced density, the $\nu_T = 1$ QH state extends widely versus density imbalance, continuously merging into the single-layer $\nu = 1$ QH state. In the vicinity of the transition point, the $\nu_T = 1$ QH state is only weakly developed at small imbalance but increases in strength toward $\nu_T = 1/3 + 2/3$, where it is clearly separated from the single-layer $\nu = 1$ QH state. These results suggest that the system at the imbalance of $\Delta\nu = 1/3$ undergoes a transition from the correlated $\nu_T = 1$ QH state to single-layer fractional QH states with increasing density.

1. Introduction

Closely spaced double layers of two-dimensional (2D) charge systems have provided a variety of fascinating quantum states arising from interlayer Coulomb interactions. Of particular interest is the bilayer $\nu_T = 1$ quantum Hall (QH) state in the limit of small tunneling, where all electrons reside in coherent superposition of the states in the upper and lower layers sharing the same phase. This collective state is characterized by broken symmetry with spontaneous interlayer phase coherence. In term of the pseudospin language encoding electrons in the upper (lower) layer as up (down) pseudospin, this state can be viewed as an easy-plane ferromagnet [1], where the spontaneous interlayer phase coherence is represented as a magnetization which develops in the x-y plane. As a consequence of the strong interlayer correlation, the $\nu_T = 1$ QH state exhibits various exotic phenomena such as Josephson-like interlayer tunneling [2], quantized Hall drag [3], and/or excitonic superfluidity [4]. The key parameter determining the stability of the state is the ratio between the intralayer and interlayer Coulomb interactions, which is parameterized by d/ℓ_B (d: distance between the layers, ℓ_B: magnetic length). It is now established that in the regime of small tunneling the $\nu_T = 1$ QH state exists below a critical value $d/\ell_B \sim 2$.

Most of previous studies on the $\nu_T = 1$ QH system have focused on the regime at balanced density, where both layers are at Landau level filling factor $\nu = 1/2$. The charge density imbalance between the layers, however, should also play a crucial role in determining the nature of the $\nu_T = 1$ QH state, since in the pseudospin space the density imbalance represents the z component of the magnetization induced out of the easy plane by a gate electric field. Several experiments have revealed that the $\nu_T = 1$ QH state is stabilized by a small density imbalance [5, 6, 7], which was discussed in relation with increased pseudospin stiffness in the collective mode under density imbalance [8]. Using tunneling spectroscopy as a probe of interlayer coherence, Champagne *et al.* have investigated in detail the phase boundary between the coherent $\nu_T = 1$ QH state and incoherent state as a function of the total density, or d/ℓ_B, and the density imbalance [7]. They found that the boundary depends on the two parameters in a rather nontrivial manner; in the vicinity of the phase boundary a small density imbalance stabilized or even restored the coherent state, whereas further increase in the imbalance destroyed it.

All these experiments have been carried out with the magnetic field applied perpendicular to the 2D plane. On the other hand, a recent experiment performed in the balanced density condition under tilted magnetic fields with various tilt angles found a notable role of spin in the phase diagram of the $\nu_T = 1$ system [9]. The experiment revealed that, different from the $\nu_T = 1$ QH state, under a perpendicular magnetic field the incoherent compressible state is not fully spin polarized near the phase boundary. Consequently, the phase boundary shifts to a much larger value of d/ℓ_B when the Zeeman energy is increased by adding an in-plane magnetic field. This has a profound implication that the reported phase diagram obtained under perpendicular magnetic fields does not represent the intrinsic property of the system and thus should be modified in the absence of the spin degree of freedom. It is further noted that the collective excitation mode, whose dispersion is expected to be Zeeman independent, does not account for the reported dependence of the phase boundary on density imbalance.

The aim of this work is to investigate the stability of the $\nu_T = 1$ QH state versus density imbalance at full spin polarization using a tilted magnetic field. We present data taken over a wide range of density imbalance from the balanced density to the single-layer limit, where all charges have been transferred to one layer.

2. Experimental

The sample was fabricated from an electron-doped GaAs double quantum well (DQW) containing 18-nm-thick GaAs wells separated by 10-nm-thick AlAs/GaAs (2.1 nm/0.56 nm) superlattice barrier with the corresponding interlayer distance (d) of 28 nm. The structure has a virtually negligible tunneling gap calculated to be $\Delta_{\rm SAS} = 150$ μK . The DQW is modulation-doped from both sides with the upper and lower setbacks of 180 and 200 nm, respectively. This enabled a nearly equal electron density of 0.8×10^{11} cm^{-2} to be produced in each layer without gate voltages. Transport measurements were carried out using standard low-frequency lock-in techniques with an excitation current of 10 nA or less. The sample geometry is a standard Hall bar with ohmic contacts connecting both layers in parallel. The low-temperature mobility was 9×10^5 cm^2/Vs at a total density of 1.1×10^{11} cm^{-2}. The electron densities of the individual layers were controlled through voltages $V_{\rm FG}$ and $V_{\rm BG}$ applied to the surface Schottky front gate and n^+-GaAs back gate, respectively. The whole measurement was performed at a temperature of around 50 mK. All data presented here were taken with the sample normal tilted by $\theta = 65°$ with respect to the applied magnetic field B.

3. Results and Discussion

Before investigating the effects of density imbalance, we first show the magnetotransport characteristics of our sample in the balanced density condition. Figure 1 depicts the longitudinal (R_{xx}) and Hall (R_{xy}) resistances taken at $\theta = 65°$, plotted as a function of the perpendicular component $B_\perp = B\cos\theta$. Here we show two representative results with different total densities (a) $n_T = 0.71 \times 10^{11}$ and (b) 1.0×10^{11} cm^{-2}, which correspond to (a) $d/\ell_B = 1.88$ and (b) 2.24, respectively. At both densities, we observe features relevant to the $\nu_T = 1$ QH state, characterized by a minimum in R_{xx} and a plateau in R_{xy}. At this high tilt angle, the large Zeeman energy forces the spin to be aligned in the competing compressible state, making it energetically unfavorable. This allows the QH phase to extend to a much larger value of $d/\ell_B \sim 2.3$ at

$\theta = 65°$ than $d/\ell_B \sim 1.9$ at $\theta = 0°$ [9]. The smaller electron density $n_T = 0.71 \times 10^{11}\text{cm}^{-2}$, with the corresponding $d/\ell_B = 1.88$ well below the critical value ~ 2.3 for this fully spin polarized system, implies that the system is deep in the QH phase. Accordingly, the $\nu_T = 1$ QH effect is well developed at this density, with a deep R_{xx} minimum and a wide quantized Hall plateau. On the other hand, at the higher electron density $n_T = 1.0 \times 10^{11}\text{cm}^{-2}$ with the corresponding $d/\ell_B = 2.24$ only slightly below the critical value, the system is close to the phase transition point. Since the energy gap is significantly reduced near the transition point [10], the R_{xx} and R_{xy} features signaling the $\nu_T = 1$ QH effect become much weaker at this density.

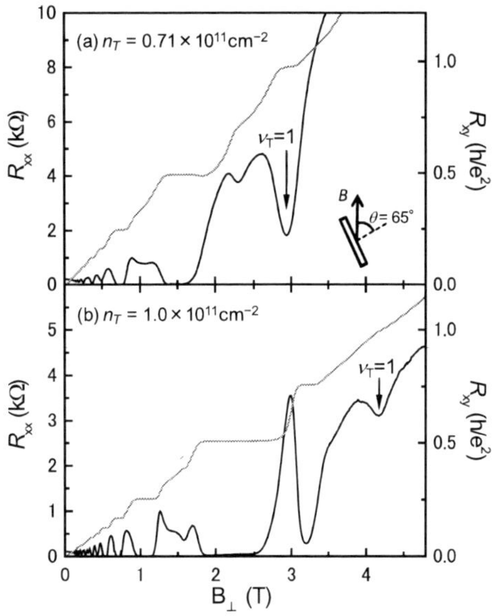

Figure 1. R_{xx} and R_{xy} versus perpendicular magnetic field at the total density of (a) $0.71 \times 10^{11}\text{cm}^{-2}$ corresponding to $d/\ell_B = 1.88$ and (b) $1.0 \times 10^{11}\text{cm}^{-2}$ corresponding to $d/\ell_B = 2.24$. The arrows indicate the $\nu_T = 1$ QH state. The inset shows a schematic illustration of the sample alignment with respect to the applied magnetic field.

To study the stability of the $\nu_T = 1$ QH state versus density imbalance, we recorded R_{xx} as a function of $V_{\rm FG}$ and $V_{\rm BG}$ at a fixed magnetic field. Figures 2(a) and 2(b) display color plot of R_{xx} measured at two different magnetic fields (a) $B = 7$ T ($d/\ell_B = 1.88$) and (b) 10 T ($d/\ell_B = 2.24$); these d/ℓ_B values correspond to those in Fig. 1. In the color plot, QH states appear as dark regions. The dash-dotted line in the diagonal direction represents the balanced density condition, while the dashed line in the orthogonal direction indicates the line of constant total filling factor $\nu_T = 1$, or the axis of density imbalance. At $d/\ell_B = 1.88$, the $\nu_T = 1$ QH regime is extended along the line of constant total filling factor over a wide range of imbalance, continuously merging into the single-layer $\nu = 1$ QH regime. Similar behavior has been reported for $\theta = 0°$ [11]. It is also seen in Fig. 2(a) that the $\nu_T = 1$ QH state becomes stronger at around a density imbalance of $\nu_T = 1/3 + 2/3$ and above. In addition to the $\nu_T = 1$ QH state, the data also reveal features running almost parallel to the $V_{\rm FG}$ or $V_{\rm BG}$ axes. These features, reflecting mostly the filling factor in each layer, are associated with the formation of $\nu = 2/3$ fractional QH states in individual layers. We note here that at density imbalance of $\nu_T = 1/3 + 2/3$ the $\nu_T = 1$ QH region is clearly separated from those regions with $\nu = 2/3$ fractional QH states formed in individual layers. At $d/\ell_B = 2.24$, the $\nu_T = 1$ QH state becomes much weaker and only barely discernible. Yet, it does extend over wide range of density imbalance and develops into a well-defined QH regime at $\nu_T = 1/3 + 2/3$, where it is separated from the single-layer $\nu = 1$ QH state. Turning to the $\nu = 2/3$ fractional QH regions in individual layers, we note that these regions are now connected more smoothly to the QH state at $\nu_T = 1/3 + 2/3$.

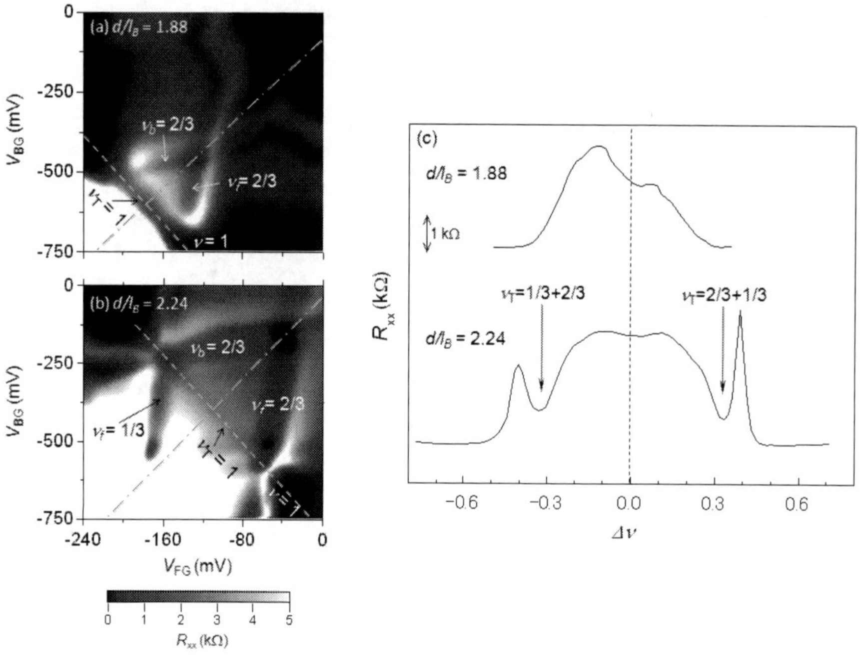

Figure 2. R_{xx} mapped as a function of $V_{\rm FG}$ and $V_{\rm BG}$ at (a) $d/\ell_B = 1.88$ ($B = 7$ T) and (b) $d/\ell_B = 2.24$ ($B = 10$ T) under tilted magnetic fields with $\theta = 65°$. (c) R_{xx} traces along $\nu_T = 1$ as a function of the filling factor difference between the layers at different d/ℓ_B.

The contrasting behavior of the $\nu_T = 1$ QH state at $d/\ell_B = 1.88$ and 2.24 becomes more apparent by plotting the line profile of R_{xx} along the total filling of $\nu_T = 1$. Figure 2 shows the R_{xx} traces along $\nu_T = 1$ as a function of the filling factor difference $\Delta\nu = \nu_f - \nu_b$, where ν_f (ν_b) denotes the filling factor in the front (back) layer. $\Delta\nu$ is determined by setting $\Delta\nu = 1/3$ ($-1/3$) at the point where the linear extrapolation of the $\nu = 2/3$ region in the front (back) layer intersects with the $\nu_T = 1$ line. At $d/\ell_B = 2.24$, there are pronounced peaks in R_{xx} which clearly separate the state at $|\Delta\nu| \leq 1/3$ from that at $|\Delta\nu| \gtrsim 0.4$, whereas at $d/\ell_B = 1.88$ such a peak is absent. This suggests that the nature of the $\nu_T = 1$ QH state at $|\Delta\nu| = 1/3$ has changed between these two situations. We remark that at $d/\ell_B = 2.24$ the R_{xx} mapping shows features associated with the $\nu = 1/3$ fractional QH state formed in individual layers. In addition, the $\nu_T = 1$ QH state increases in strength toward $|\Delta\nu| = 1/3$, i.e., the crossing point of the $\nu = 1/3$ and $\nu = 2/3$ fractional QH regions in the individual layers. From the features observed, we suppose that the system at $|\Delta\nu| = 1/3$ undergoes a transition from the interlayer correlated $\nu_T = 1$ QH state at $d/\ell_B = 1.88$ to independent single-layer fractional QH states at $d/\ell_B = 2.24$. By measuring the tunneling conductance between the layers together with the

Hall resistance in one layer, Champagne *et al.* have shown that under perpendicular magnetic fields such a transition occurs around $d/\ell_B = 1.80 \sim 1.84$ [7]. It is noteworthy that this value of d/ℓ_B coincides with the coherent-incoherent phase transition point at $\Delta\nu = 0$ in their sample. This suggests that the interlayer correlation is lost all together over the entire range of $|\Delta\nu|$. In marked contrast, in the present case, the $\nu_T = 1$ QH state does exist at $\Delta\nu = 0$ at the same d/ℓ_B where the two layers seem to be uncorrelated at $|\Delta\nu| = 1/3$. One clue may be that the $\nu = 2/3$ QH state has two ground states with different spin polarizations. We verified by a separate experiment that in our sample the single-layer $\nu = 2/3$ QH state is spin polarized at the magnetic fields used here. This is in contrast to the situation in Ref. [7], where the $\nu = 2/3$ QH state is likely to be spin unpolarized. At present, however, it remains unknown how the interlayer correlation is lost with the increase in $|\Delta\nu|$ and specifically how the spin polarization is concerned with this issue.

Another interesting feature is the camelback shape observed in the R_{xx} trace at $d/\ell_B = 1.88$ with its local minimum at $\Delta\nu = 0$. The asymmetry with respect to $\Delta\nu = 0$ is probably due to unintentional asymmetry in the sample. The camelback feature becomes less pronounced at $d/\ell_B = 2.24$. This shape indicates that a finite but small density imbalance makes the $\nu_T = 1$ QH state less stable. A similar feature has been reported for a bilayer hole system [11], where it was discussed in conjunction with a phase diagram proposed taking account of the ground state energies of two independent single-layer states. However, such behavior was not observed in previous activation gap measurements under a perpendicular magnetic field [5, 6], which showed a minimum in the gap at $\Delta\nu = 0$. We do not know the origin of the different behavior; a possible cause may be sample disorder or different current distribution between the two layers.

4. Conclusion

In summary, we have investigated the evolution of the bilayer $\nu_T = 1$ QH state versus density imbalance when the system is in full spin polarization under a tilted magnetic field. At d/ℓ_B much smaller than at the transition point, the $\nu_T = 1$ QH state evolves over a wide range of density imbalance and continuously develops into the single-layer $\nu = 1$ QH state. In the vicinity of the transition point, on the other hand, the $\nu_T = 1$ QH state becomes significantly weaker and merges into a well-developed QH regime at $\nu_T = 1/3 + 2/3$, where it is separated from the single-layer $\nu = 1$ QH regime. These observations suggest that the system at the imbalance of $\Delta\nu = 1/3$ undergoes a transition from the correlated $\nu_T = 1$ QH state to independent single-layer fractional QH states with increase in d/ℓ_B.

5. Reference

[1] K. Yang, K. Moon, L. Zheng, A. H. MacDonald, S. M. Girvin, D. Yoshioka, and S. C. Zhang, Phys. Rev. Lett. **72**, 732 (1994).
[2] I. B. Spielman, J. P. Eisenstein, L. N. Pfeiffer, and K. W. West, Phys. Rev. Lett. **84**, 5808 (2000).
[3] M. Kellogg, I. B. Spielman, J. P. Eisenstein, L. N. Pfeiffer, and K.W. West, Phys. Rev. Lett. **88**, 126804 (2002).
[4] J. P. Eisenstein, and A. H. MacDonald, Nature **432**, 691 (2004).
[5] E. Tutuc, S. Melinete, E. P. De Poortere, R. Pillarisetty, and M. Shayegan *Phys. Rev. Lett.* **91**, 076802 (2003)
[6] R. D. Wiersma et al., *Phys. Rev. Lett.* **93**, 266805 (2004).
[7] A. R. Champagne, A. D. K. Finck, J. P. Eisenstein, L. N. Pfeiffer, and K. W. West *Phys. Rev. B* **78**, 205310 (2008).
[8] Y. N. Joglekar and A. H. MacDonald, *Phys. Rev. B* **65**, 235319 (2002).
[9] P. Giudici, K. Muraki, N. Kumada, Y. Hirayama, and T. Fujisawa *Phys. Rev. Lett.* **100**, 106803 (2008).
[10] P. Giudici, K. Muraki, N. Kumada, and T. Fujisawa *Phys. Rev. Lett.* **104**, 056802 (2010).
[11] W. R. Clarke, A. P. Micolich, A. R. Hamilton, M. Y. Simmons, C. B. Hanna, J. R. Rodriguez, M. Pepper, and D. A. Ritchie *Phys. Rev. B* **71**, 081304(R) (2005).

Journal of Physics: Conference Series **334** (2011) 012026 doi:10.1088/1742-6596/334/1/012026

Emergent fractional quantum Hall effect at even denominator ν=3/2 in a triple quantum well in tilted magnetic fields

S. Wiedmann[1,2,3*], G. M. Gusev[4], A. K. Bakarov[5], and J. C. Portal[2,3]

[1]Radboud University Nijmegen, Institute for Molecules and Materials and High Field Magnet Laboratory, Toernooiveld 7, 6525 ED Nijmegen, The Netherlands
[2]LNCMI, UPR 3228, CNRS-INSA-UJF-UPS, BP 166, 38042 Grenoble Cedex 9, France
[3]INSA Toulouse, 31077 Toulouse Cedex 4, France
[4]Instituto de Física da Universidade de São Paulo, CP 66318 CEP 05315-970, São Paulo, SP, Brazil
[5]Institute of Semiconductor Physics, Novosibirsk 630090, Russia

E-mail: *`s.wiedmann@science.ru.nl`

Abstract. This work reports on the experimental observation of the even-denominator state ν=3/2 in a trilayer electron system in tilted magnetic fields. The ν=3/2 state demonstrates a strong minimum in longitudinal resistance and is accompanied by a plateau in Hall resistance in a narrow range of tilt angles.

1. Introduction

The observation of fractional quantum Hall (FQH) states at even denominators, in particular at filling factor ν=5/2, has attracted much attention, motivated by implications of fault-tolerant quantum computation [1]. After its discovery [2], many theoretical works indicate that the 5/2 FQH state, observed in a single quantum well, is an exotic non Abelian Pfaffian state. This state is a subject of ongoing discussions and recent experimental studies indicate implications for the Pfaffian State and can be of importance in clarifying Pfaffian versus anti-Pfaffian as the relevant ground state [3].

On the other hand, an even-denominator FQH state has been found in high mobility bilayer systems in the lowest Landau level (LL) at filling factor ν= 1/2 which has no counterpart in a single-layer 2D system [4, 5]. So far, it is assumed that the observed ν=1/2 state in bilayer systems is most likely described by the Abelian so-called Halperin {331} state. A recent theoretical study on the ν=1/2 FQH state as function of tunneling strength and layer separation discusses whether bilayers support both an Abelian Halperin {331} and a non-Abelian Pfaffian state [6]. This work also discusses if the transition between these two states might be observed experimentally in standard transport measurements of the FQH-effect [6].

A further advance in physics of FQH states with even denominators depends on experimental search for non-Abelian states, especially in systems which are distinct from bilayers. Multi-component quantum Hall (QH) systems, which consist of multiple quantum wells separated by thin barriers, exhibit new interesting FQH states due to many-body phenomena and might contribute to the understanding of FQH states [5]. In the present work we have carried out magnetotransport measurements in a coupled trilayer electron system, formed by a triple quantum well (TQW) in magnetic fields up to B=34 T and at a temperature of T=100 mK. If we apply a parallel component of the magnetic field, we observe an even-denominator state at total filling factor ν=3/2, which is to the best of our knowledge, the first experimental observation of an even-denominator state in a trilayer electron system. Note that the observed ν=3/2 FQH state is a hole conjugate of the ν=1/2 FQH state (ν=3/2=2-1/2).

2. Trilayer electron system

Our samples are symmetrically doped GaAs TQWs, separated by $Al_xGa_{1-x}As$ barriers, with a high total electron sheet density of $n_s = 6.9 \times 10^{11}$ cm^{-2} and a mobility of 8×10^5 cm^2/V s. The central well width is about 220 Å and both side wells have equal widths of 100 Å. The barrier thickness is d_b=20 Å. In order to populate the central, we increased its width. Corresponding energy gaps $\Delta_{jj'}$ between populated subbands are Δ_{12}=1.34 meV, Δ_{13}=3.65 meV, Δ_{23}=2.31 meV, extracted from a self-consistent Hartree-Fock calculation, and are in agreement with the periodicity of magneto-intersubband oscillations [7, 8]. The trilayer system with corresponding parameters is sketched in the inset of Figure 1. The energies in TQWs are described by the expression $\hbar\omega_c(N+1/2) \pm \Delta_Z/2 + E_j$, where $\hbar\omega_c$ is the cyclotron energy, Δ_Z the Zeeman energy, and E_j ($j = 1, 2, 3$) the energies of quantization in the TQW potential.

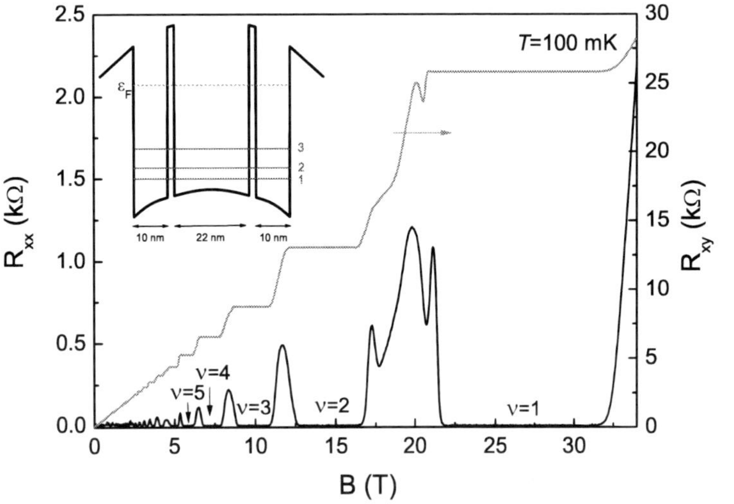

Figure 1. (Color online) Longitudinal resistance and quantum Hall effect in the trilayer electron system subjected to a perpendicular magnetic field at a temperature of 100 mK. Inset: Sketch of a TQW with corresponding subbands.

In contrast to previous studies in trilayer systems [9], the Landau fan diagram of this particular trilayer system investigated here deviates from the standard sequence of spin-split LLs separated by the subband gaps. Whereas in Ref. [9], we found FQH effect for the highest subband of each Landau level (FQH effect between, e.g. filling factors ν=2 and ν=3), we observe now FQH effect between integer filling factors ν=1 and ν=2, see Figure 1. The variation of energy gaps $\Delta_{jj'}$ with the magnetic field might be ascribed to charge transfer from the central well to the lateral wells.

3. Observation of the even-denominator state ν=3/2

Tilting the magnetic field leads to the observation of numerous minima in R_{xx} accompanied by plateaus in the Hall resistance R_{xy} at integer and fractional filling factors ν >3. For $\Theta > 37°$, integer filling factor ν=2 starts to collapse and a new FQH states are developed for ν <2. With increasing tilt angle, a new deep and broad minimum in R_{xx} appears at fractional filling factor ν= 3/2, see Figure 2(a). This minimum persists up to an angle of Θ=43.2° and then collapses. In Hall resistance, we observe a plateau quantized at $R_{xy} = 2h/3e^2$, which is shown in Figure 2(c). The even-denominator state is very sensitive to the component of the parallel magnetic field which is pointed out in Hall resistance in Figures 2(b) and (d).

A further increase of the tilt angle leads to a symmetric order in FQH states around even-denominator state ν=3/2 with denominators 5, 7, 9 etc. The best pronounced state is ν=7/3 which occurs at $B_\perp \simeq$14 T.

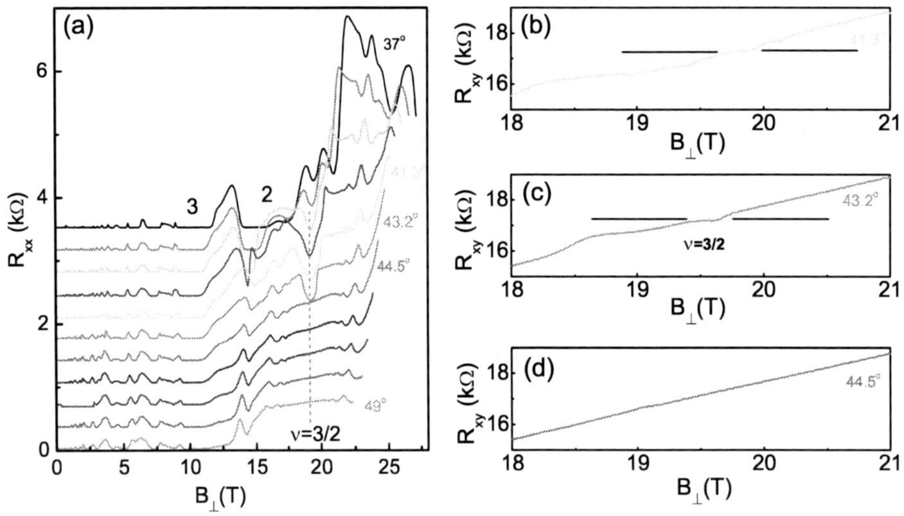

Figure 2. (Color online) (a) Longitudinal resistance R_{xx} as a function of the perpendicular component of the magnetic field for several chosen tilt angles from Θ=37° to Θ=49°. For $\Theta > 37°$ a minimum at even-denominator ν=3/2 is developed which persists until Θ=43.2°. (b)-(d) Corresponding Hall resistance exhibits a well developed plateau at ν=3/2 for Θ=43.2°.

Using the derivative of R_{xy}, we demonstrate in Figure 3(a)-(c) the evolution of magnetoresistance with increasing tilt angle for Θ=41.3°, Θ=43.2° and Θ=44.5°. Starting from the integer filling factor ν=3, we observe for Θ=41.3° a profound minima which first narrows for Θ=43.2° and is almost vanished for Θ=44.5°.

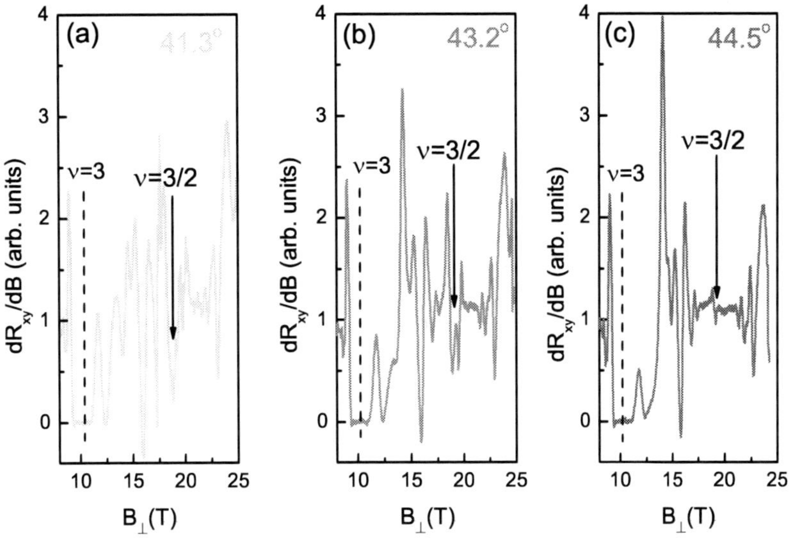

Figure 3. (Color online) (a)-(c) Evolution of magnetoresistance with increasing tilt angle. Derivative dR_{xy}/dB demonstrates a clear minimum for Θ=41.3° and Θ=43.2°. For Θ=44.5°, the even- denominator state at ν=3/2 disappears and we find several developed minima on the low- and high-field side with respect to ν=3/2.

4. Discussion

Multicomponent FQH states can be obtained by a generalization of the Laughlin state. Such incompressible states have been predicted for trilayer systems in Ref. [10]. The state at total filling factor ν=5/7 which is among the strongest states in a trilayer system, is a combination of filling factors 2/7 in the side layers and 1/7 in the central layer. This state has been found experimentally [11]. Recently, FQH states at even-denominator filling factors ν=1/2 and ν=1/4 in electron systems confined to a wide GaAs quantum well have been found with a significantly asymmetric charge distribution [12]. Those states disappear when the charge distribution is made symmetric and the subband splitting is lowered.

However, first we point out that the observation of the even-denominator state ν=3/2 in our trilayer electron system cannot be ascribed to a generalization of the Laughlin state. Second, we observe ν=3/2 in the presence of both perpendicular and parallel magnetic field and in a very narrow range $\Delta\Theta$. In general, the presence of an in-plane magnetic field adds an Aharonov-Bohm phase to the tunneling amplitude, which consequently leads to oscillations of the tunnel coupling between electronic states in the layers [13] and to a suppression of this coupling for low Landau levels. Oscillations in the tunnel coupling affect fractional quantized Hall phenomena in highly tilted samples as it has been demonstrated in Ref. [9].

Having a closer look to the sequence of fractional states around ν=3/2, one might suggest that this state in our trilayer electron system is an analog of the two-component ν=1/2 state in a wide quantum well [12]. We can further assume that the electron density is 1/3 of the total density which implies that our ν=3/2 is a ν=1/2+1 state, i.e., a ν=1/2 state with an inert background of a fully occupied LL. However, comparing our results to Ref. [12], we notice that the induced asymmetry by gating the wide quantum well gives rise to the appearance of a deep minimum in resistivity. In our TQW, we might have the situation of a slight asymmetry at high magnetic fields as well under certain conditions (tilt angles) but it could also be likely that oscillations of the tunneling gap create a favourite condition (charge distribution) for the observation of the even-denominator state ν=3/2.

5. Conclusion

In the present work we have found first experimental evidence of an even-denominator state at total fractional filling factor ν=3/2. This state is sensitive with respect to the component of the parallel magnetic field and is developed around $\Theta \simeq 42°$. We assume that our experimental finding challenges theory in order to understand the origin of this emergent ν=3/2 FQH state in a trilayer electron system in a tilted magnetic field.

Authors acknowledge financial support from CNPq and FAPESP (Brazilian agencies). Part of this work has been supported by EuroMagNET II under the EU contract number 228043.

References

[1] C. Nayak, S. H. Simon, A. Stern, M. Freedman, and S. Das Sarma, *Rev. Mod. Phys.* **80**, 1083 (2008).
[2] R. L. Willet, J. P. Eisenstein, H. L. Störmer, D. C. Tsui, A. C. Gossard, and J. H. English, *Phys. Rev. Lett.* **59**, 1776 (1987).
[3] Chi Zhang, T. Knuuttila, Yanhua Dai, R. R. Du, L. N. Pfeiffer, and K. W. West, *Phys. Rev. Lett.* **104**, 166801 (2010).
[4] J. P. Eisenstein, G. S. Boebinger, L. N. Pfeiffer, K. W. West, and Song He, Phys. Rev. Lett. 68, 1383 (1992).
[5] *Perspective in Quantum Hall Effects: Novel Quantum Liquids in Low-Dimensional Semiconductor Structures*, edited by S. Das Sarma and A. Pinczuk (Wiley and Sons, New York, 1998).
[6] Michael R. Peterson and S. Das Sarma, *Phys. Rev.* B **81**, 165304 (2010).
[7] S. Wiedmann, N. C. Mamani, G. M. Gusev, O. E. Raichev, A. K. Bakarov, and J. C. Portal, *Phys. Rev. B* **80**, 245306 (2009).
[8] S. Wiedmann, N. C. Mamani, G. M. Gusev, O. E. Raichev, A. K. Bakarov, and J. C. Portal, *Physica E* **42**, 1088 (2010).
[9] G. M. Gusev, S. Wiedmann, O. E. Raichev, A. K. Bakarov, and J. C. Portal, *Phys. Rev. B* **80**, 161302(R) (2009).
[10] A. H. MacDonald, *Surf. Sci.* **229**, 1 (1990).
[11] J. Jo, Y. W. Suen, L. W. Engel, M. B. Santos, M. Shayegan, *Phys.Rev. B* **46**, 9776 (1992).
[12] J. Shabani, T. Gokmen and M. Shayegan, *Phys. Rev. Lett.* **103**, 046805 (2009).
[13] J. Hu and A. H. MacDonald, *Phys. Rev. B* **46**, 12554 (1992).

Comparison of the Magnetoresistance Hysteresis between Bilayer $\nu = 4/3$ and Monolayer $\nu = 2/3$ Quantum Hall States

S. Tsuda[1], A. Fukuda[2], D. Terasawa[2], N. Minh Hai[1], Y. Ogasawara[1], Y. D. Zheng[3], and A. Sawada[3]

[1]Department of Physics, Graduate School of Science, Kyoto Univ., Kyoto 606-8502, Japan
[2]Department of Physics, Hyogo College of Medicine, Nishinomiya, 663-8501, Japan
[3]Research Center for Low Temperature and Materials Sciences, Kyoto Univ., Kyoto 606-8501, Japan

E-mail: shibun@scphys.kyoto-u.ac.jp

Abstract. We study the hysteresis of magnetoresistance in the bilayer $\nu = 4/3$ and monolayer $\nu = 2/3$ quantum Hall states (QHSs), where ν is the Landau level filling factor, in order to understand the mechanism of dynamic nuclear polarizations (DNPs) that affects transport properties of GaAs/AlAs two-dimensional electron gas system. Basically hysteresis around the bilayer $\nu = 4/3$ is resulted in the monolayer $\nu = 2/3$ in each layer. However, we also find hysteresis that appears around the bilayer $\nu = 1$ QHS region caused by the strong effects of DNPs as a consequence of the sweep over $\nu = 2/3$ QHS in one layer. These findings suggest the effects of nuclear spin polarizations in the bilayer QHSs over a wide range of filling factors.

1. Introduction

Much attention has been paid to the understanding of nuclear spins in semiconductors for the realization of the quantum computing because of their long decoherence time. One possibility to control nuclear spins is using the spin transitions in the quantum Hall states (QHSs). In the Landau level filling factor $\nu = 2/3$ monolayer QHSs, two ground states are degenerated owing to the competition between the Coulomb and the Zeeman interaction. Proper tuning of the Zeeman energy and the electron density enables us to change the electron spin polarizations, which changes the nuclear spin polarizations through the hyperfine interaction, accompanied by hysteresis of magnetoresistance [1, 2, 3, 4]. However, the mechanism of the dynamic nuclear spin polarizations (DNPs) at the spin transition point in the bilayer system has not yet been understood completely. In this report, we compare the hystereses of magnetoresistance at the monolayer $\nu = 2/3$ and the bilayer $\nu = 4/3$ QHSs in the GaAs/AlAs double quantum well structures by changing the electron density in one layer while keeping the electron density in another layer constant. We observe a large hysteresis for a wide range of filling factors, even in no QHSs or in the bilayer $\nu = 1$ QHS, when the filling of one layer is around 2/3.

2. Experiments

Samples employed in the experiments consist of two 20-nm-wide GaAs quantum wells separated by a 3-nm-wide AlAs barrier layer and processed into a 50-μm-wide Hall bar with the source-

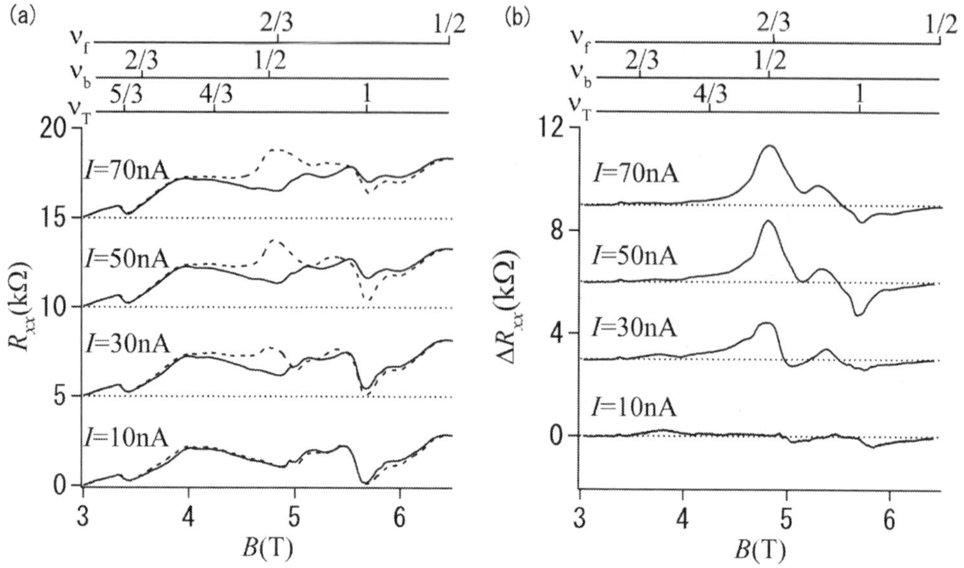

Figure 1. (a) R_{xx} versus B for several source-drain currents I =10, 30, 50, 70 nA around bilayer $\nu_T = 4/3$ at $\sigma = 0.15$, $n_T = 1.36 \times 10^{11}$ cm^{-2} and T = 70 mK. The solid (broken) lines denote that the magnetic field is swept upward (downward), respectively. Sweep rate is fixed at 0.1 T/min. Above the graph, ν_f, ν_b and ν_T are indicated by additional axes. The data are offset by 5 kΩ. (b)Difference between R_{xx} in up-sweep and one in down-sweep ΔR_{xx} versus B. The data are offset by 3 kΩ.

drain distance of 180 μm. The tunneling energy Δ_{SAS} is about 1 K. We can control electron density of the front layer n_f by changing front gate voltage V_f. The electron density of the back layer n_b is kept constant unless electrons in the front layer are depleted. We measure magnetoresistance R_{xx} and Hall resistance R_{xy} by using a standard low-frequency AC lock-in technique with a frequency of $f = 17.7$ Hz in a dilution refrigerator with a base temperature of T = 50 mK. Sweep rate of magnetic field is fixed at 0.1 T/min. Hereafter, we use n_f and n_b for representing electron densities in the front and back layers. We define the total electron density n_T as $n_T = n_f + n_b$ and the layer imbalance parameter σ as $\sigma = (n_f - n_b)/n_T$. We also use ν_f, ν_b and ν_T to represent the filling factor for n_f, n_b and n_T, respectively.

3. Results and Discussions

Figure 1 shows R_{xx} as a function of the magnetic field B at 70 mK for several source-drain currents I around bilayer $\nu_T = 4/3$ QHS with $n_T = 1.36 \times 10^{11}$ cm^{-2} and $\sigma = 0.15$. B is swept up to 6.5 T and then immediately swept down to 0 T. The solid (broken) lines denote R_{xx} when B is swept upward (downward). When I = 10 nA, there is a small hysteresis between the up and down sweeps. However, a large hysteresis appears at I = 30 nA and becomes larger with increasing I. Interestingly, the hysteresis does not disappear even in the $\nu_T = 1$ QHS region. The hysteresis in the $\nu_T = 1$ QHS is unexpected because both ν_f and ν_b are no longer 2/3 and there is no hysteresis at $\nu_T = 5/3$ while ν_b is close to 2/3, suggesting that the effect of

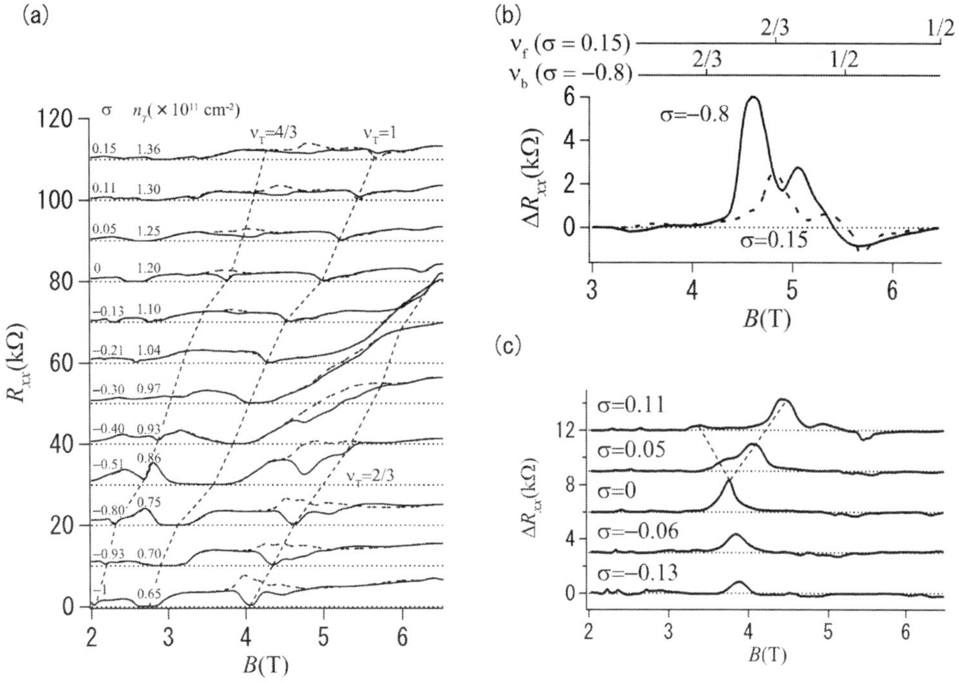

Figure 2. (a) R_{xx} versus B for different sweep direction for various σ. The sweep rate is 0.1 T/min and the current is 50 nA. The data are offset by 10 kΩ. (b) A comparison of ΔR_{xx} versus B between $\nu_b = 2/3$ ($\sigma = -0.8$) and $\nu_T = 4/3$ ($\sigma = 0.15$). The solid (dashed) line denotes $\nu_T = 2/3$ ($\nu_T = 4/3$). Above the graph, ν_f and ν_b are indicated by additional axes. (c) ΔR_{xx} versus B around $\nu_T = 4/3$ from $\sigma = -0.13$ to 0.11. The data are offset by 3 kΩ.

DNPs remains over the $\nu_T = 1$ bilayer QHS. In the monolayer $\nu = 1$ QHS, the ground state is considered as a complete spin ferromagnet. No hysteresis has been observed at this filling followed by the sweep over $\nu = 2/3$ QHS because fluctuations of electron spins in skyrmions, which is thought as a elementary excitation of spins, relax DNPs just around the off-fillings of the $\nu = 1$ QHS[3]. In the case of bilayer $\nu = 1$ QHS, spin excitations are considered to be less dominant than the pseudospin excitations, where pseudospin represents the layer degree of freedom. This difference may be the cause of remaining nuclear polarizations in the system. Figure 1 (b) shows the difference of R_{xx} between the up and down sweeps (ΔR_{xx}) against B. ΔR_{xx} becomes largest at $I = 50$ nA both around $\nu_T = 4/3$ and $\nu_T = 1$, and the linewidth of the peak is smaller than the one at $I = 70$ nA. They indicate that large currents ($I = 70$ nA) damp hysteresis, thus the following measurements are performed with $I = 50$ nA.

Figure 2 (a) shows R_{xx} as a function of B from $\sigma = -1.0$ to $\sigma = 0.15$ by increasing n_f while keeping n_b constant. Note that increasing n_f also changes both n_T and σ. At $\sigma = -1.0$, where all electrons are in the back layer, $\nu_b = 2/3$ monolayer QHS develops around $B = 4$ T with a large hysteresis. This hysteresis exists at $\sigma = -0.80$, where induced electrons are transferred to the back layer by the effect of the many-body exchange correlation[5], thus $\nu_T \simeq \nu_b$. When σ is increased above -0.80, the position of hysteresis stays at almost the same field of 4.5 T since n_b stays constant. Indeed, the hysteresis in $\sigma \leq -0.51$ is larger than the other hystereses due to the higher current density that leads to larger DNPs in the back layer. However, as σ

is increased above -0.13, the hysteresis almost disappears, alternated by the emergence of the coherent bilayer $\nu_T = 1$ QHS. On the other hand, in the lower field, we detect a small hysteresis at $\sigma = -0.13$. This newly developed hysteresis grows to the hysteresis at the bilayer $\nu_T = 4/3$ QHS as σ is increased to 0, and then the hysteresis occurs at $\nu_f = 2/3$ separated from $\nu_T = 4/3$. At $\sigma = 0.15$, the hysteresis finally appears in the bilayer $\nu_T = 1$ QHS.

In Fig. 2 (b), we make a comparison between the hysteresis of $\nu_T = \nu_b = 2/3$ at $\sigma = -0.8$ (monolayer $\nu = 2/3$) and $\nu_T = 4/3$ at $\sigma = 0.15$ (bilayer $\nu = 4/3$). The solid (dotted) line denotes ΔR_{xx} of $\nu_T = 2/3$ ($\nu_T = 4/3$). Both lines have two peaks separated within about 0.5 T. The shapes of the two lines are similar, which reflect the property of the two QHSs.

In Fig. 2 (c), we demonstrate ΔR_{xx} for σ from -0.13 to 0.11 around the bilayer $\nu_T = 4/3$ QHS. At $\sigma = -0.13$, a small peak is recognized at $B = 3.9$ T, which grows to a sharp peak at $\sigma = 0$. Above $\sigma = 0.05$, the peak separates into two peaks at $\nu_f = 2/3$ and $\nu_b = 2/3$. As a result, the overall behavior of hysteresis in R_{xx} is well described as a superposition of hystereses of $\nu_f = 2/3$ and $\nu_b = 2/3$.

These unusual hystereses found at no QHS and at $\nu_T = 1$ QHS are caused basically by the DNPs of the $\nu = 2/3$ QHS in each layer, probably because a small Δ_{SAS} works more prominently for the *intralayer* interaction than the *interlayer* interaction. However, hysteresis in the $\nu_T = 1$ implies that the effects of DNPs induced by the monolayer $\nu = 2/3$ in each layer cannot be erased by spin fluctuations by skyrmions near or on the $\nu = 1$ QHS. They agree with the fact that the dominant excitations in the bilayer $\nu = 1$ QHS is pseudospin excitations.

4. Conclusions

We observe hystereses in the bilayer $\nu_T = 4/3$ QHSs for a wide range of σ. The similar hysteresis is detected in the monolayer $\nu_b = 2/3$ QHSs and the behavior is well described as a superposition between $\nu_f = 2/3$ and $\nu_b = 2/3$. The hysteresis is also observed in no QHS regions and even at $\nu_T = 1$ QHS. These findings imply that the remaining nuclear spin polarization affects the transport in the bilayer QHSs over a wide range of filling factors.

5. Acknowledgments

We are grateful to N. Kumada and K. Muraki in NTT Basic Research Laboratories, and Y. Hirayama in Tohoku University, for providing us not only high-mobility samples but also fruitful discussions. We also thank to T. Saku for growing the heterostructures of the samples. This research was supported in part by Grants-in-Aid for the Scientific Research (Nos. 20029012, 20654028, 21340082 and 21740236).

References

[1] Kronmüller S, Dietsche W, Weis J, von Klitzing K, Wegscheider W and Bichler M 1998 *Phys. Rev. Lett.* **81** 2526
[2] Smet J H, Deutschmann R A, Wegscheider W, Abstreiter G and von Klitzing K 2001 *Phys. Rev. Lett.* **86** 2412
[3] Hashimoto K, Muraki K, Saku T and Hirayama Y 2002 *Phys. Rev. Lett.* **88** 176601
[4] Kumada N, Terasawa D, Morino M, Tagashira K, Sawada A, Ezawa Z F, Muraki K, Hirayama Y and Saku T 2004 *Phys. Rev.* B **69** 155319
[5] Ying X, Parihar S. R, Manoharan H. C and Shayegan M 1995 *Phys. Rev. B* **52** R11611.

HMF-19 IOP Publishing
Journal of Physics: Conference Series **334** (2011) 012028 doi:10.1088/1742-6596/334/1/012028

Dynamic nuclear plarization induced by the breakdown of fractional quantum Hall effect

Minoru Kawamura[1,2], Masashi Ono[3], Yoshiaki Hashimoto[4], Shingo Katsumoto[4,5], and Tomoki Machida[3,5]
[1]RIKEN Advanced Science Institute, 2-1 Wako Saitama, Japan
[2]PRESTO, Japan Science and Technology Agency, 4-1-8 Kawaguchi, Saitama 333-0012, Japan
[3]Institute of Industrial Science, University of Tokyo, 4-6-1 Komaba, Tokyo 153-8505, Japan
[4]Institute for Solid State Physics, University of Tokyo, 5-1-5 Kashiwa, Chiba 277-8581, Japan
[5]Institute for Nano Quantum Information Electronics, University of Tokyo, 4-6-1 Komaba, Tokyo 153-8505, Japan

E-mail: minoru@riken.jp

Abstract. Dynamic polarization of nuclear spins has been studied in the breakdown regime of fractional quantum Hall effect using a Corbino-disk device. We find that nuclear spins are polarized in the Corbino disk in the breakdown regime of fractional quantum Hall effect. Since edge channel is completely absent in the Corbino disk, the demonstration of dynamic nuclear polarization in the Corbino disk shows that nuclear spins are polarized and detected in the bulk channel of the quantum Hall conductor.

Breakdown phenomena of quantum Hall effect (QHE) has been a subject of research interest not only for developing a QHE resistance standard but also for understanding fundamental physics of electron transport in quantum Hall (QH) systems ever since the discovery of the QHE[1, 2]. The QHE breakdown has attracted a renewed interest because of the recent development of dynamic nuclear polarization technique in the breakdown regime of odd-integer QHE[3, 4]. It was demonstrated that nuclear spins at the interface of $GaAs/Al_{0.3}Ga_{0.7}As$ heterostructure was dynamically polarized via the hyperfine interaction when the two-dimensional electron system (2DES) located at the interface was driven to the breakdown regime of odd-integer QHE. The DNP technique makes it possible to detect nuclear magnetic resonance (NMR) signals by simply measuring the transport properties of the 2DES. The dynamic nuclear polarization (DNP) technique has been utilized for the initialization procedure of the nuclear spin quantum bits[5, 6, 7]. Because of its extremely high sensitivity, the resistively detected NMR has been also utilized to investigate electron spin properties of the QH conductors[8, 9, 10], which had not been explored otherwise.

In our earlier work[11], we found that DNP is also induced by the breakdown of fractional QHE. We found that the voltage-current characteristics strongly depend on current-sweep rates at Landau-level filling factors $\nu = 2/3$ and $1/3$, and that the value of V_{xx} decreased slowly over 600 s after a sudden increase in the bias current from 0 to 30 nA. The relevance of DNP in the sweep-rate dependent $V_{xx} - I$ characteristics and slow evolution of V_{xx} was confirmed by resistive detection of NMR signals under irradiation of rf-magnetic field. However, the demonstration of DNP in the breakdown regimes of fractional QHE was conducted using Hall-bar devices, in which both edge and bulk channels coexist. Therefore, it has not been clarified yet which

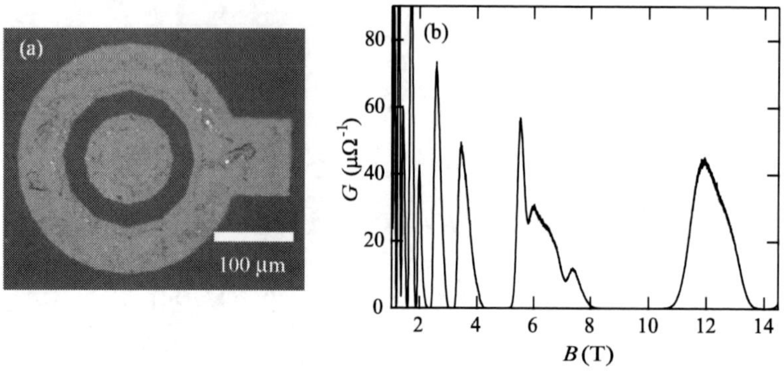

Figure 1. (a) Micrograph of the Corbino-disk device used in the present study. The inner and outer diameters of the Corbino disk are 120 and 180 μm, respectively. (b) Magnetic field dependence of the two-terminal conductance. The data were obtained using a standard lock-in technique with an ac excitation voltage 10 μV and a frequency 18 Hz.

channel is relevant to the DNP. In order to understand the mechanism of the DNP induced by the fractional QHE breakdown, it should be clarified experimentally which channel is relevant to the DNP.

In this paper, we report dynamic nuclear polarization in a Corbino-disk device in the breakdown regime of fractional QHE. We detected nuclear magnetic resonance signals by measuring the current flowing between the inner and outer electrodes of the Corbino disk in the breakdown regime of fractional QHE. Since edge channel is completely absent in the Corbino disk, the demonstration of the DNP in the Corbino-disk device shows that nuclear spins are polarized and detected in the bulk channel of the QH conductor.

Experiments were conducted using a device with Corbino disk geometry fabricated photolithographically from a wafer of $GaAs/Al_{0.3}Ga_{0.7}As$ single heterostructure with a 2DES at the interface. The mobility and carrier density of the 2DES at 4.2 K are 115 m^2/Vs and 2.27×10^{15} m^{-2}, respectively. The inner and outer diameters of the Corbino disk are 120 and 180 μm, respectively, as shown in Fig. 1(a). All the measurements were performed in a dilution refrigerator with a base temperature of 20 mK. An external magnetic field B was applied perpendicular to the 2DES plane by using a superconducting solenoid. The current flowing between the inner and outer electrodes was measured using standard dc method, unless otherwise mentioned. A single-turn coil around the device was used to apply radio-frequency (rf) magnetic fields for NMR measurements.

Figure 1(b) shows the magnetic field B dependence of the two-terminal conductance between the outer and inner electrodes of the Corbino disk. The data were obtained using a standard lock-in technique with an excitation voltage 10 μV and a frequency 18 Hz. The zero conductance plateaus are clearly seen at around Landau level filling factors $\nu = 4$ (B = 2.34 T), $\nu = 3$ (B = 3.14 T), $\nu = 2$ (B = 4.68 T), $\nu = 1$ (B = 9.36 T), and $\nu = 2/3$ (B = 14.05 T). The $\nu = 2/3$ fractional QH state observed in the present experiment belongs to the spin polarized phase because the applied magnetic field of B = 14.05 T is much higher than the typical magnetic field for the spin phase transition in $\nu = 2/3$[12].

Figure 2(a) shows the current (I) - voltage (V) characteristic curves obtained at $\nu = 2/3$ (B = 14.05 T) by sweeping the bias voltage in the positive and negative directions at a sweep rate of 1 mV/s. Hysteresis is clearly observed in the I-V curves. Hysteretic transport properties are observed in the breakdown regime of odd-integer QHE where DNP occurs[4]. Therefore the

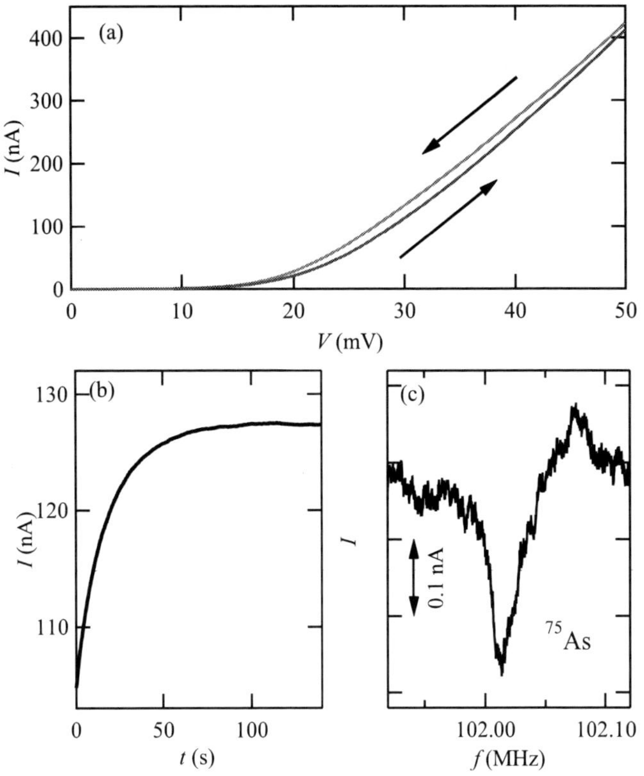

Figure 2. (a) I-V characteristic curves obtained at $B = 14.05$ T. The directions of the voltage sweep are indicated by arrows. (b) Evolution of I after a rapid increase of the bias voltage from 0 mV to 30 mV at $t = 0$. (c) Responses of I to the rf-magnetic field as a function of frequency. The data were obtained at $B = 14.05$ T and $V = 30$ mV.

observation of hysteresis in the present study suggests the occurrence of DNP in the Corbino disk in the breakdown regime of the $\nu = 2/3$ QHE. Figure 2(b) shows the time evolution of I at $\nu = 2/3$ after a rapid increase of the bias voltage from $V = 0$ mV to $V = 30$ mV. The value of I increased slowly over more than 60 s. The long relaxation time over 60 s is typical to the nuclear spin-related phenomena.

Involvement of nuclear spins to the hysteresis of the I-V curves and the slow evolution of I is directly shown by the NMR measurements, as shown in Fig. 2(c). The data shown in Fig. 2(c) were obtained by measuring I at bias voltage $V = 30$ mV under irradiation of continuous wave rf-magnetic field with input power of -15 dBm at the top of the cryostat. As the frequency f of the rf-magnetic field was scanned, the value of I dropped at 102.012 MHz which coincides with the NMR frequency of ^{75}As (gyro magnetic ratio $\gamma = 45.82$ rad·MHz/T) at $B = 14.05$ T. The detection of NMR signals by measuring the current flowing through the 2DES indicates that nuclear spins are dynamically polarized and detected in the Corbino disk in the breakdown regime of the $\nu = 2/3$ QHE. Since there is no edge channel in the Corbino-disk device, the occurrence of the DNP directly indicates that nuclear spin polarization occurs in the bulk part of the QH conductor. The occurrence of DNP in the bulk part of the QH conductor has been also reported in the case of DNP induced by the breakdown of odd-integer QHE[13]. This similarity may suggest the common underlying origin of DNP occurring in the breakdown regimes of both fractional and odd-integer QHE.

Acknowledgement
Transport measurements were carried out in collaboration with Prof. K. Hirakawa. This study is partially supported by a Grand-in-Aid Scientific Research and the Special Coordinate Funds for Promoting Science and Technology, and the CREST and PRESTO, Japan Science and Technology Agency.

References
[1] M. E. Cage, R. F. Dziuba, B. F. Field, E. R. Williams, S. M. Girvin, A. C. Gossard, D. C. Tsui, and R. J. Wagner, Phys. Rev. Lett. **51** 1374 (1983).
[2] G. Ebert, K. von Klitzing, K. Ploog, and G. Weimann, J. Phys. C **16**, 5441 (1983).
[3] A. M. Song and P. Omling, Phys. Rev. Lett. **84**, 3145 (2000).
[4] M. Kawamura, H. Takahashi, K. Sugihara, S. Masubuchi, K. Hamaya, and T. Machida, Appl. Phys. Lett. **90**, 022102 (2007).
[5] T. Machida, T. Yamazaki, K. Ikushima, and S. Komiyama, Appl. Phys. Lett. **82**, 409 (2003).
[6] G. Yusa, K. Muraki, K. Takashina, K. Hashimoto, and Y. Hirayama, Nature (London) **434**, 1001 (2005).
[7] H. Takahashi, M. Kawamura, S. Masubuchi, K. Hamaya, T. Machida, Y. Hashimoto, and S. Katsumoto, Appl. Phys. Lett. **91**, 092120 (2007).
[8] O. Stern, N. Freytag, A. Fay, W. Dietsche, J. H. Smet, K. von Klitzing, D. Schuh, and W. Wegscheider, Phys. Rev. B **70**, 075318 (2004).
[9] K. Hashimoto, K. Muraki, T. Saku, and Y. Hirayama, Phys. Rev. Lett. **88**, 176601 (2002).
[10] M. Kawamura, H. Takahashi, S. Masubuchi, Y. Hashimoto, S. Katsumoto, K. Hamaya, and T. Machida, Physica E **40**, 1389 (2008).
[11] M. Kawamura, M. Ono, Y. Hashimoto, S. Katsumoto, K. Hamaya, and T. Machida, Phys. Rev. B **79**, 193304 (2009).
[12] J. P. Eisenstein, H. L. Stormer, L. Pheiffer, and K. W. West, Phys. Rev. Lett. **62**, 1540 (1989).
[13] M. Kawamura. H. Takahashi, S. Masubuchi, Y. Hashimoto, S. Katsumoto, K. Hamaya, and T. Machida, J. Phys. Soc. Jpn. **77**, 023710 (2008).

Journal of Physics: Conference Series **334** (2011) 012029 doi:10.1088/1742-6596/334/1/012029

Resistively detected NMR with dispersive lineshape in single InSb quantum wells

K.F. Yang[1], H.W. Liu[1,2], T.D. Mishima[3], M.B. Santos[3], K. Nagase[1], Y. Hirayama[1,4]

[1]ERATO Nuclear Spin Electronics Project, Sendai, Miyagi 980-8578, Japan
[2]State Key Lab of Superhard Materials and Institute of Atomic and Molecular Physics, Jilin University, Changchun 130012, P. R. China
[3]Homer L. Dodge Department of Physics and Astronomy, University of Oklahoma, 440 West Brooks, Norman, OK 73019-2061, USA
[4]Department of Physics, Tohoku University, Sendai, Miyagi 980-8578, Japan

E-mail: yangkf@ncspin.jst.go.jp

Abstract. We present a DC measurement of resistively detected NMR (RDNMR) in a single InSb/AlInSb quantum well. Dynamic nuclear polarization has been demonstrated around the resistance spike, which is the signature of the pseudospin quantum Hall ferromagnet at the filling factor of two. The RDNMR signals with dispersive lineshape occur on each side of the spike, but show the shape inversion. The dispersionlike lineshape is found to be independent of measurement conditions (current intensity, temperature, and RF power), provided the polarized nuclei are in equilibrium.

1. Introduction

The contact hyperfine coupling between electrons and nuclei supports electron-nuclear flip-flop and thereby dynamic nuclear polarization (DNP). The polarized nuclei produce an effective magnetic field B_N, which modifies the Zeeman energy E_Z of electron spins in a two-dimensional electron gas (2DEG) via the Overhauser shift, and accordingly changes the longitudinal resistance R_{xx}. This simple picture is used to account for resistively detected nuclear magnetic resonance (RDNMR), an emerging high-sensitivity NMR approach for probing the dynamics of electron and nuclear spins in low-dimensional semiconductor structures [1-3]. The RDNMR signal usually shows a dip or a peak in R_{xx} response to the applied radio-frequency (RF) field (continuous wave mode). However, more complicated RDNMR signals with anomalous dispersive lineshape (a dip-peak structure) have been observed in the GaAs 2DEG recently [4-11]. This dispersionlike RDNMR lineshape was first demonstrated near the $\nu = 1$ quantum Hall (QH) state and was assigned to be a signature of Skyrmions [4], while similar RDNMR lineshape observed at other filling factors casted doubt on this interpretation [5, 6]. In particular, the RDNMR study of the $\nu = 1$ QH state in a tilted magnetic field strongly suggests that there is no correlation between the dispersionlike lineshape and the Skyrmion formation [7]. The origin of this anomalous RDNMR lineshape still remains unclear.

More recently, we have successfully demonstrated DNP and RDNMR in single InSb quantum wells (QWs) [12]. The large g factor of the InSb 2DEG enables construction of quantum Hall ferromagnet (QHF) with domain structures at $\nu = 2$ using tilted magnetic fields. The ferromagnetism formed at ν = 2 is characterized as the simplest Pseudospin QHF (PQHF) because only the pseudospin-down [(n, σ) = (0, ↓)] and pseudospin-up [(n, σ) = (1, ↑)] sublevels are involved, where n and σ are the orbit and spin indices, respectively. The PQHF is featured by a resistance spike (or peak) within persistent R_{xx} minima. A large AC current induces the resistance change of the spike with respect to DNP, allowing us to observe the RDNMR signals of both In and Sb with a dip or a peak feature. Domain structures are shown to play a dominant role in the DNP process. Here we focus on the DC measurement of RDNMR in the InSb 2DEG. In contrast to the results in the AC measurement, the DC-induced RDNMR shows a dip-peak structure. We describe properties of the dispersionlike RDNMR signal under different temperature, current, and RF conditions. A comparative study of this anomalous RDNMR lineshape in different 2DEG systems will enhance our understanding of its underlying mechanism.

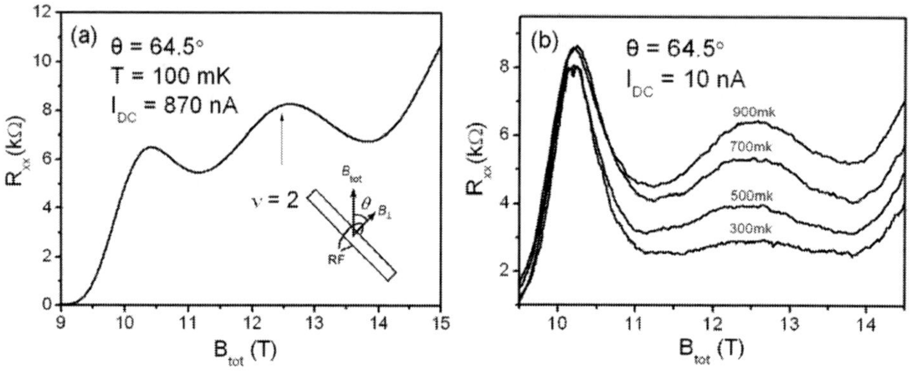

Figure 1. (a) Longitudinal resistance R_{xx} vs. total magnetic field B_{tot} at θ=64.5°, T = 100 mK, and I_{DC} = 870 nA. The resistance spike of the ν = 2 PQHF locates around 12.5 T. Inset: Setup of tilted-field experiments. Tilt angle θ was determined from the Hall resistance. (b) R_{xx} vs. B_{tot} at several temperatures.

2. Experimental details

Our 2DEG sample is a 20-nm-wide InSb QW with asymmetrically silicon delta-doped $Al_{0.2}In_{0.8}Sb$ barriers and has a low-temperature electron mobility of μ = 13.6 m^2/Vs and an electron density of n_s = 2.56 × 10^{15} m^{-2}. The RDNMR measurement in a tilted magnetic field is performed using a dilution refrigerator with an in-situ rotator. A DC bias is applied to measure R_{xx} and to polarize the nuclei. A small RF field ~ μT (continuous wave mode) matching the NMR frequency of individual nuclide is generated by a single turn coil wound around the Hall bar (inset of figure 1 (a)). To minimize heating effects, the sample is irradiated by the RF field with a small power of -6 dbm, unless noted otherwise.

3. Dispersive RDNMR lineshape

We perform the RDNMR measurement on the resistance spike of the ν = 2 PQHF. As a typical example, we present the results obtained from the resistance spike at θ = 64.5° (figure 1 (a)). A large DC current of I_{DC} = 870 nA is kept constant at fixed B_{tot} around the spike (11.6 T or 13.7 T) till R_{xx} becomes saturated. Then, the RF field is applied while sweeping the frequency through the Larmor resonance of ^{115}In at 6 kHz/min. The RDNMR signals are thus obtained, as shown in figures 2(a) and 2(b). It is clear that the RDNMR signal on each side of the spike exhibits the dispersive lineshape. More interestingly, these two RDNMR signals show inverted dip-peak structure. Similar results have already been obtained around the ν = 1 state of the GaAs 2DEG, where the shape inversion is attributed to the sign change of the derivative of temperature (T)-dependent R_{xx} (dR_{xx}/dT) [8]. In our case, however, the spike amplitude monotonously increases with increasing temperature (figure 1 (b)). Therefore, the shape inversion around the spike cannot be simply attributed to the thermal effect.

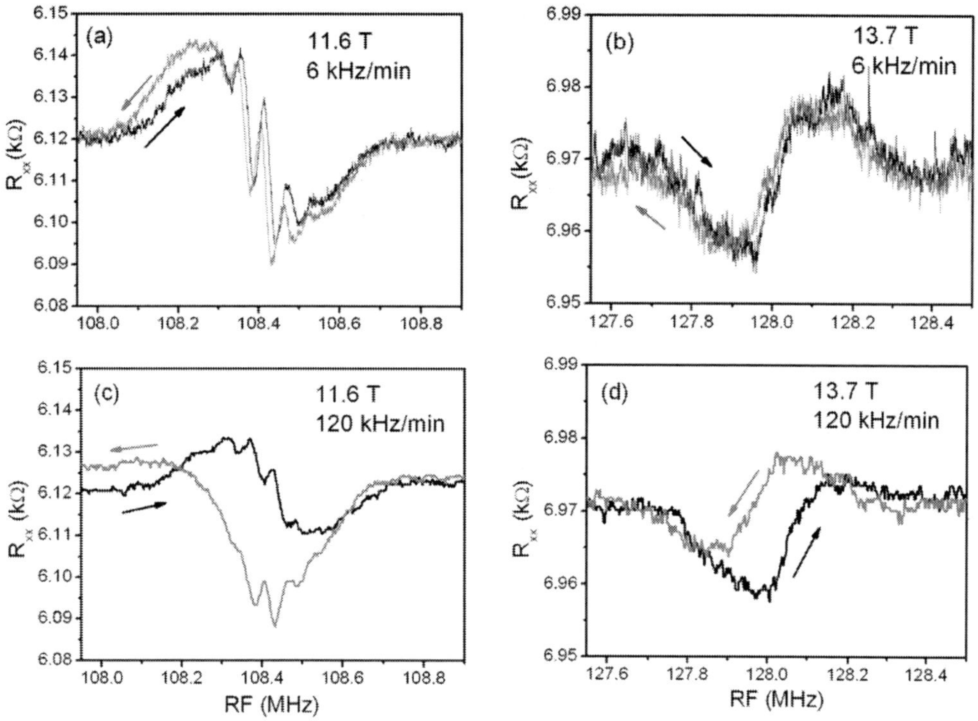

Figure 2. RDNMR signals of ^{115}In attained at 11.6 T and 13.7 T in figure 1 (a) with different RF sweep rates. The spectra of (a) and (b) were obtained at a slow rate of 6 kHz/min, while (c) and (d) at a high rate of 120 kHz/min. The quadrupole splitting of ^{115}In are partially resolved in (a) and (c). The arrow indicates the scan direction.

We systematically probe properties of the dispersive lineshape response to various measurement conditions (current intensity, temperature, RF power, and sweep rate). It is found that the RDNMR signal exhibits dispersionlike lineshape over a wide range of current intensity (from 300 nA to 3 μA), temperature (from 100 mK to 4 K), and RF power (from -30 dBm to 2 dBm), provided the signal is distinguishable. This result is different from the observation in the GaAs 2DEG, where the RDNMR lineshape strongly depends on the current intensity and the RF power [9, 10]. Nevertheless, the RDNMR spectra depend on the scan direction at a relatively high RF sweep rate of 120 kHz/min (figure 2 (c) and (d)). For instance, the RDNMR signal at 11.6 T shows the dispersionlike lineshape for the upward scan (black arrow, figure 2(c)), while a normal dip in R_{xx} appears for the downward scan (gray arrow, figure 2(c)). Similar findings are observed at 13.7 T (figure 2(d)). This result compared with that of the slow RF scan indicates that the RDNMR spectra obtained at relatively high sweep rate are out of equilibrium, i.e., R_{xx} is recorded at the level below its saturation value.

4. Conclusion

In summary, we have demonstrated the dispersive RDNMR lineshape in the simplest PQHF of the InSb 2DEG by a DC measurement. The RDNMR on each side of the resistance spike (signature of the

PQHF) exhibits inverted dispersionlike lineshape. The monotonous increase of the spike resistance with increasing temperature rules out the possibility that the thermal effect accounts for the shape inversion as described in [8]. At slow RF sweep rate, the dispersionlike RDNMR lineshape is robust irrespective of temperature, current, and RF power. This is different from the findings in the GaAs 2DEG. Although the origin of the anomalous RDNMR lineshape is still unknown, our results obtained from the 2DEG made of material other than GaAs will contribute to a better understanding of its underlying mechanism.

References

[1] Smet J H *et al.*, 2002 *Nature* 415 281.
[2] Hashimoto K *et al.*, 2002 *Phys. Rev. Lett.* 88 176601; Yusa G *et al.*, 2005 *Nature* 434 1001.
[3] Hirayama Y *et al.*, 2009 *Semicond. Sci. Technol.* 24 023001.
[4] Desrat W *et al.*, 2002 *Phys. Rev. Lett.* 88 256807.
[5] Gervais G *et al.*, 2005 *Phys. Rev. B* 72 041310(R).
[6] Stern O *et al.*, 2004 *Phys. Rev. B* 70 075318.
[7] Bowers C R *et al.*, 2010 *Phys. Rev. B* 81 073301.
[8] Tracy L A *et al.*, 2006 *Phys. Rev. B* 73 121306(R).
[9] Dean C R *et al.*, 2009 *Phys. Rev. B* 80 15331.
[10] Dean C R *et al.*, 2008 *Physica E* 40 990.
[11] Kodera K *et al.*, 2006 *phys. Stat. sol. (c)* 3 4380.
[12] Liu H W *et al.*, 2010 *Phys. Rev. B* 82 241304 (R).

Interlevel hybridization phenomena in the coincidence effect under quantum Hall regime in a HgTe quantum well

M.V. Yakunin[1], A.V. Suslov[2], S.M. Podgornykh[1], S.A. Dvoretsky[3] and N.N. Mikhailov[3]

[1]Institut of Metal Physics, Ekaterinburg 620041, Russia
[2]National High Magnetic Field Laboratory, Tallahassee, Florida 32310, USA
[3]Institute of Semiconductor Physics, Novosibirsk 630090, Russia

E-mail: yakunin@imp.uran.ru

Abstract. As found from quantum magnetotransport under tilted magnetic fields in a (013)-oriented n-$Cd_{0.7}Hg_{0.3}Te/HgTe/Cd_{0.7}Hg_{0.3}Te$ quantum well of ~20nm width, the positions of magnetic level coincidences agree quite well with a traditional description of the coincidence effect like in a simple Γ_6 band in spite of the Γ_8 character of the conduction band in HgTe. The coincidence angles yield the effective g-factor $g^*= 33$ while $g^*= 50\div60$ is obtained with different magnetotransport techniques under fields oriented perpendicular to the layers. The difference is settled when assuming the g-factor anisotropy of $g_\perp/g_\parallel \approx 2$. Level anticrossings are revealed at high values of perpendicular magnetic fields, corresponding to the quantum Hall regime. They manifest themselves in a splitting of the coincidence magnetoresistivity peaks and in a shift of the split-off peaks from the integer filling factors ν to the neighbouring half-integer filling factors. The anticrossing at $\nu = 3$ is found to be significantly stronger than the neighbouring ones at $\nu = 2$ and $\nu = 4$ that is tentatively explained in terms of formation of easy-axis quantum Hall ferromagnetic states in the vicinity of the expected crossings.

1. Introduction

In a HgTe quantum well (QW) within the $HgTe/Cd_xHg_{1-x}Te$ heterosystem, a unique inverted band diagram is realized when the well width d_w is larger than the critical value of 6.3nm so that the conduction band is formed of the p-like wave functions of the Γ_8 symmetry. Properties of the Γ_8 conduction band differ considerably from those of a traditional s-like Γ_6 conduction band, in particular, due to a strong spin-orbit interaction [1]. In spite of that the Γ_8 conduction band demonstrates coincidences of spin sublevels under tilted magnetic fields very much alike as in a Γ_6 band [2,3]. A large value of $g^*m^*/m_0 \geq 1$ (g^* is the effective g-factor, m^*/m_0 is a ratio of the effective to free electron masses) in HgTe layers makes it possible to observe the coincidences at perpendicular fields high enough to reach the conditions of the quantum Hall effect (QHE), while in most of the traditional heterostructures the coincidences were seen only on Shubnikov – de Haas oscillations. We report a study of specific manifestations of the coincidences in HgTe QW under QHE regime.

2. Experiment

The sample is a $HgTe/Cd_xHg_{1-x}Te$ (013)-oriented QW, $x = 0.6 \div 0.73$, $d_w = 20.3$nm, doped on both sides with 10nm spacers, with electron density $n_s = 1.5\times10^{15}m^{-2}$ and mobility of $20\div25 m^2/V\cdot s$ at low

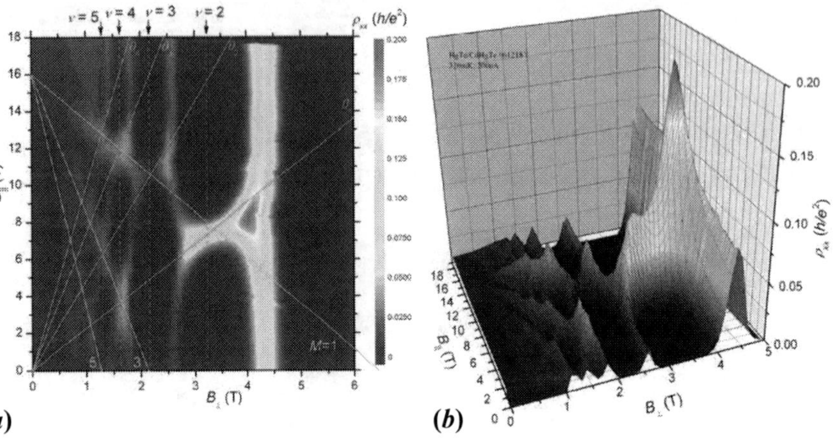

Figure 1. Magnetoresistivity $\rho_{xx}(B_{\perp},B_{\parallel})$ at $T = 0.32$K presented (*a*) as a map and (*b*) in 3D. The beams in (*a*) going from zero correspond to constant *r* values in eqn.(1) and those going down from $B_{\parallel}{}^{0} = 16$T represent varying *r* as described by eqn.(2).

temperatures. Electron density may be increased by about 20% via the IR illumination that is accompanied with significant improvement of the sample quality. Investigations under tilted magnetic fields were performed by measuring of the magnetoresistivity (MR) tensor components [4] as a function of the tilt angle in a series of fixed total magnetic fields. Then the data obtained were rebuilt into the continuous functions of two variables $\rho_{ij}(B_{\perp},B_{\parallel})$, where $B_{\perp}$ and $B_{\parallel}$ are the field components perpendicular and parallel to the layers.

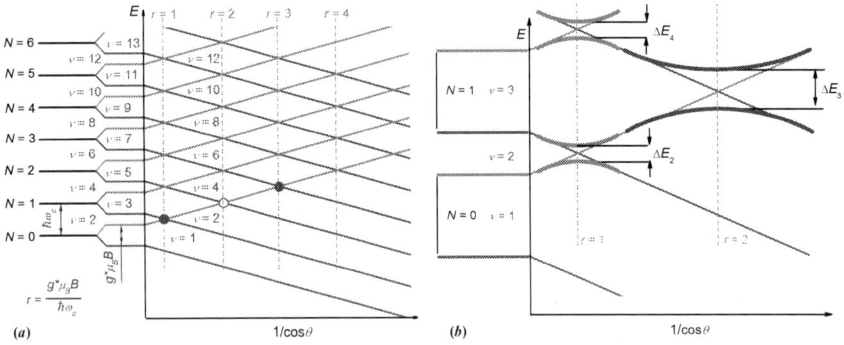

Figure 2. (*a*): Evolution of the magnetic levels with tilt θ in a simplest one-electron approach for a Γ_6-like conduction band. Coincidences occur at integer *r* values. Each interlevel gap is characterized by its filling factor ν. Traces for coincidences on the $(B_{\perp},B_{\parallel})$ plane at the fixed angles θ_r in fig.1a are built for motion along dashed verticals here while the down sloped traces described by (2) – for motion along upper spin sublevels with coincidences characterized by the fixed values of $\nu - r = 1,3,5\ldots$. (*b*): Anticrossings that substitute the crossings as revealed experimentally (unscaled).

3. Results and discussion

Large spin splittings in our sample are directly manifested in the prevailing of quantum Hall (QH) features with odd-numbered filling factors ν over the even-numbered ones [3]. In analogy with a simple Γ_6 band, it puts the limits to the g^*-factor values as $1 < g^*m^*/m_0 < 2$ and using $m^*/m_0 = 0.024$ as obtained from the cyclotron resonance in a similar sample [5] we get $42 < g^* < 84$. Another way for estimations is to compare the fields for onsets of the oscillations and of their splittings, which yields $g^*m^*/m_0 = 1.38$, in agreement with previous condition, and $g^* = 58$ for the same m^*. The gaps between magnetic levels were obtained from the activation temperature dependencies of the magnetoresistivity minima under perpendicular fields and yielded $g^* = 50.4$ from the line slope through the odd-numbered points and $m^*/m_0 = 0.022$ from that through the even-numbered points.

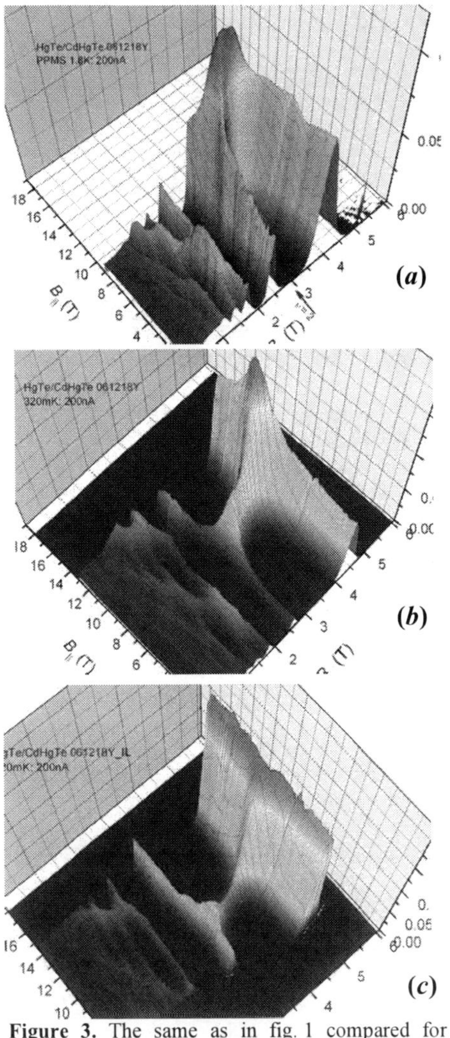

Figure 3. The same as in fig. 1 compared for different electron densities ($10^{15}m^{-2}$): (a) $n_s = 1.45$, (T= 1.8K), (b) 1.56, (0.32K), (c) 1.74, (0.32K).

A conventional technique to measure g^* by magnetotransport in a Γ_6 band is via the coincidence effect, when at tilts θ_r of magnetic field relative to the sample normal either all the even-numbered or all the odd-numbered gaps are successively closed according to the relation [6]:

$$g^* m^* / m_0 = 2r\cos\theta_r, r = 1,2,3..., \quad (1)$$

with r being a ratio of the spin splitting $g^*\mu_B B$ (μ_B – Bohr magneton) to the cyclotron splitting $\hbar\omega_c$, $\omega_c \equiv eB_\perp/m^*$. Our results for $\rho_{xx}(B_\perp,B_\|)$, together with estimations by eqn.(1) are presented in fig.1a. An excellent agreement is seen between the beams drawn for the coincidence angles and the MR peaks. Moreover, when there are many coincidences one can see an additional system that they line up: note a set of beams going down from some finite value of $B_\| = B_\|^0$. This additional system is especially pronounced in the 3D presentation of the results (fig.1b). In a simple one-electron scheme of the level motion with a tilt, like in a Γ_6 band (fig.2a), this second system of coincidences is described when going along the upper spin sublevels. Thus each line of the second system corresponds to the constant value of the difference between the filling factor ν and r: $\nu - r \equiv M = 1,3,5...$, and the equation for the trajectories on the ($B_\perp$,$B_\|$) plane for the tilts far enough from the perpendicular configuration reads:

$$B_\| = \frac{2}{g^*(m^*/m_0)}\left(\frac{h}{e}n_s - MB_\perp\right), M = 1,3,5... \quad (2)$$

According to eqn.(2), all the beams of the second system should go from a single point:

$$B_{\parallel}^{0} = \frac{2}{g^*(m^*/m_0)}\frac{h}{e}n_s \qquad (3).$$

The second system for the coincidences (i) yields one more evidence that the interpretation of the coincidences within a simple scheme is valid for the Γ_8 conduction band, (ii) might be used for additional estimations of the sample parameters (g^*, m^* and n_s) from (3) and (iii) indicates the trajectories useful for taking cross-sections of the $\rho_{xx}(B_{\perp},B_{\parallel})$ surface to compare the coincidences for different values of ν and r (see below).

Thus, the coincidence effect seems to work quite well here in spite of the specific band structure. The problem is that θ_1 should be below 60° to agree with the odd-numbered QH features to prevail over the even-numbered ones at perpendicular fields while we measured $\theta_1 = 67^{\text{o}}$ [and $\theta_2 = 79^{\text{o}}$, $\theta_3 = 82^{\text{o}}$, in agreement with eqn.(1)]. The data in perpendicular and tilted fields are reconciled when assuming anisotropy of the g-factor [7] with $g_{\perp}/g_{\parallel} \approx 2$. The origin of the anisotropy is most likely due to the Γ_8 nature of this conduction band, which allows some admixture to the electron states of the heavy-hole states, which are insensitive to $B_{\parallel}$.

In a simple one-electron scheme (fig.2a), a coincidence under QH regime should manifest itself in a transformation of the $\rho_{xx}(B)$ minimum at integer ν into a peak corresponding to a gap shut down. This effect was indeed observed in ref.3 on the same sample. In the data presented here, the electron density is either slightly (less than 10%) or significantly (about 20%) increased by illumination that caused a dramatic changes in the structure of coincidences as shown in fig.3: while at initial density $n_s = 1.5\times10^{15}\text{m}^{-2}$ an only weakly splitted coincidence peak exists at $\nu = 2$ (fig.3a), almost in accordance with expectations, the peak splits strongly after a small increase in n_s with its components shifted from the integer ν to the neighboring half-integer values, leaving a decreased bridge at integer ν, and the effect is still enhanced with further increase in the n_s (figures 3b, 3c). These changes in the structure of the coincidences might reflect transformations in the magnetic level picture with their crossings at coincidences evolving into anticrossings (fig.2b). The existence of an anticrossing implies a gradual interchange in the characters of the two approaching magnetic levels within some range of magnetic fields where these levels acquire hybridized wave functions.

It is worth noting that the effect is nonmonotonously ν-dependent: while for $\nu = 3$ at low temperatures the coincidence elevation in $\rho_{xx}(B)$ is almost vanishing, those for its neighbors at $\nu = 2$ and 4 on the same trajectory for $\nu - r = 1$ are still quite pronounced, with the difference reaching an order of magnitude (figs.1,3). It means that the anticrossing at $\nu = 3$ is much larger than for $\nu = 2$ and 4, $\Delta E_3 >> \Delta E_2$, ΔE_4 (fig.2b). The related anticrossings are marked in fig.2a by filled circles for $\nu = 2$ and 4 and a hollow circle for $\nu = 3$. With increasing temperature the coincidence elevation for $\nu = 3$ increases faster than for its neighbors and the difference gradually disappears. The larger value of g^*m^* in HgTe creates more rigorous conditions for offering the QH regime at the coincidences within a single spatial subband than in previous studies [8-10] where only a monotonous depression of the anticrossings with increased ν has been observed.

Spontaneous formation of the quantum Hall ferromagnetic (QHF) state is utilized for explanation of anticrossings since their discovery in the first experiments with the conditions for coincidences in tilted fields reaching the range of high enough perpendicular fields where the QH regime is achieved [8-10] that is reviewed and analyzed in detail in papers [11,12]. According to theory [11] the easy-axis QHF state might be formed on approaching of two single subband spin sublevels in tilted fields. It was confirmed in the experiment [10] where the observed persistent existence of the $\nu = 2$ MR minimum under tilted fields in the GaAs QW was explained quantitatively by formation of the easy-axis QHF state while the simultaneous observation of the reentrant behavior of the $\nu = 4$ minimum explained as being due to a decrease in the magnetic anisotropy energy because of a tendency to similar spatial distributions across the QW of the wave functions of the two approaching levels at smaller fields. Our observation of the opposite tendency, i.e. that the $\nu = 3$ anticrossing is stronger than the $\nu = 2$ one, might be explained by the action of another factor in Eqn. (21) of [11] for the magnetic anisotropy energy: exp(-*d/l*) (*d* is the effective QW width, $l \equiv \sqrt{\hbar / eB_{\perp}}$), which increases with decreasing field. Additionally, the Landau number difference $\Delta N = 2$ for the coincidence at $\nu = 3$, as opposed to $\Delta N = 1$ for $\nu = 2$ and $\nu = 4$, might lead to a larger difference in spatial distributions of the wave functions of the crossing levels [10] that would support the same tendency. Sharp transformations in the observed oscillation structures, especially around the expected coincidences, with small changes in the system (changes in n_s and probably in the spatial distribution of defects created by illumination – compare figures 1 and 3a-c) support the phase-transition-like nature of the anticrossings inherent to the easy-axis QHF.

Acknowledgements
The work is done within RAS program 01.2.006 13395 with partial support of RFBR, projects 11-02-00427 and 09-02-96518. National High Magnetic Field Laboratory is supported by NSF Cooperative Agreement No. DMR-0654118, the State of Florida, and the DOE.

References
[1] Konig M, Buhman H, Molenkamp W et al. 2008 *J. Phys. Soc. Jpn.* **77,** 031007
[2] Zhang X C, Ortner K, Pfeuffer-Jeschke A et al. 2004 *Phys. Rev. B* **69** 115340
[3] Yakunin M V, Podgornyh S M, Mikhailov N N, Dvoretsky S A 2010 *Physica E* **42**, 948-951
[4] Suslov A V 2010 *Rev. Sci. Instr.* **81**, 075111
[5] Kvon Z-D, Danilov S N, Mikhailov N N, Dvoretsky S A et al. 2008 *Physica E* **40**, 1885-1887
[6] Nicholas R J, Brummel M A, Portal J C, Cheng A Y et al. 1983 *Solid State Comm.* **45**, 911-914
[7] Dorozhkin S I 1989 *Solid State Commun.* **72**, 211-214
Martin R W, Nicholas R J, Rees G J, Haywood S K et al. 1990 *Phys. Rev. B* **42**, 9237- 9240
[8] Koch S, Haug R J, v.Klitzing K, Razeghi M 1993 *Phys. Rev B.* **47,** 4048-4051
[9] Daneshvar A J, Ford C J B, Simmons M Y et al. 1997 *Phys. Rev Lett.* **79,** 4449-4452
[10] Jungwirth T, Shukla S P, Smrcka L, Shayegan M, MacDonald A H 1998 *Phys. Rev. Lett.* **81** 2328-2331
[11] Jungwirth T, MacDonald A H 2000 *Phys. Rev. B* **63**, 035305
[12] Girvin S M 2000 *Physics Today* **53**, 53-45

HMF-19 IOP Publishing
Journal of Physics: Conference Series **334** (2011) 012031 doi:10.1088/1742-6596/334/1/012031

Non-equilibrium transport in a quantum wire in the quantum Hall regime

K. Chida[1], M. Hashisaka[1†], Y. Yamauchi[1], S. Nakamura[1], T. Arakawa[1], T. Machida[2], K. Kobayashi[1], T. Ono[1]

[1]Institute for Chemical Research, Kyoto University, Uji, Japan
[2]IIS and INQIE, University of Tokyo, Tokyo, Japan
[†]Present address: Department of Physics, Tokyo Institute of Technology, Tokyo, Japan

E-mail: chida@scl.kyoto-u.ac.jp

Abstract. We investigate the equilibrium and non-equilibrium electron transport in a quantum wire (QW) in the quantum Hall regime. While the equilibrium conductance of the QW is quantized reflecting the dissipation-less quantum Hall edge transport, the quantization collapses when the QW is voltage-biased with the voltages larger than the critical ones. This phenomenon is accompanied by the dynamic nuclear polarization which causes the temporal variation of the QW conductance of the order of minutes.

1. Introduction

The quantum Hall effect (QHE) is the phenomenon peculiar to the two-dimensional electron gas (2DEG) systems, where the longitudinal resistance vanishes with a quantized Hall resistance [1]. This behavior is well explained by the formation of the dissipation-less quantum Hall (QH) edge state [2]. However, the dissipation-less state collapses in non-equilibrium [3, 4], which is called QHE breakdown. The mechanism of the breakdown still remains to be clarified in spite of intensive and extensive studies [5, 6]. Many studies on this typical non-equilibrium phenomenon have been conducted in the bulk Hall-bar samples, for example, to determine where the breakdown takes place.

Recently, a similar collapse of the conductance quantization is known to occur in a mesoscopic scale in the non-equilibrium quantum Hall regime. Transport measurements in mesoscopic devices have been done by several groups [7-12], where the non-equilibrium state is addressed by probing the associated dynamic nuclear polarization (DNP). The mechanism of the DNP was explained as the flip-flop process between electron spin and nuclear spin due to the electron scattering between spin-resolved Landau levels [7].

As both the QHE breakdown and the DNP are non-equilibrium phenomena due to the collapse of the QH edge transport, it is natural to expect that they coexist in some situations. Indeed, the QHE breakdown is recently reported to serve as a method to create DNP in 2DEG [13]. Moreover, Córcoles *et al.* observed DNP in a quantum wire (QW) and discussed that it is due to the QHE breakdown [11].

Here we report the experimental observation of the collapse of the conductance quantization in the non-equilibrium QW. In the quantum Hall regime the differential conductance of QW at zero-bias is quantized, while the dissipation-less state collapses when the source-drain bias voltage applied to the QW (V_{sd}) exceeds a critical value. We show that the collapse is induced by the electron tunneling from

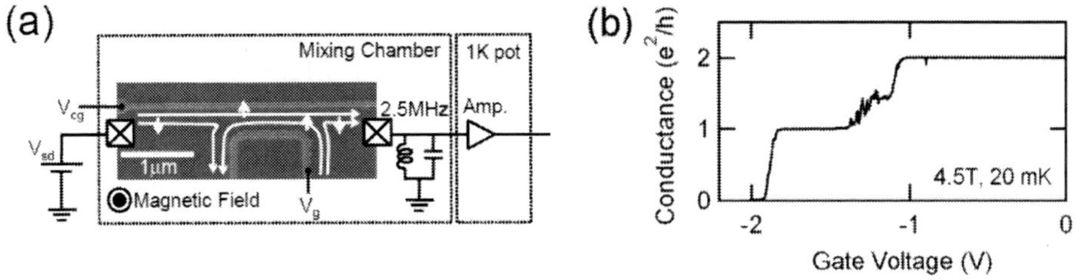

Figure 1. (a) Schematic diagram of the measurement setup with the scanning electron microscope image of the sample fabricated on the GaAs/AlGaAs 2DEG. In the present experimental situation, the two channels are fully spin-polarized as schematically shown by the arrows. (b) Gate voltage dependence of the quantum wire conductance. At the conductance plateau at e^2/h, only the outer edge channel transmits electrons while the inner edge channel is reflected (see Fig. 1 (a)).

the occupied lowest Landau level to the empty second Landau level. We also discuss the observation of the DNP accompanied by the electron tunneling between the Landau levels.

2. Experiments

Figure 1 shows the schematic measurement setup with the scanning electron micrograph image of the QW fabricated on the 2DEG (electron density: 2.3×10^{11} cm^{-2} and electron mobility: 1.1×10^6 cm^2/Vs) in the AlGaAs/GaAs heterostructure. The magnetic field perpendicular to the 2DEG (B) around 4 and 5 T was applied to make the QH state with the filling factor of the 2DEG (ν_{2DEG}) of two. We defined the QW by applying negative gate voltages (V_g and V_{cg}) to the gate electrodes on the 2DEG and control the local filling factor inside the QW (ν_{QW}). The differential conductance of the QW at a finite V_{sd} was measured by the standard lock-in technique at frequency of 37 Hz. The current noise spectral density centered around 2.5 MHz defined by the resonant (inductor-capacitor) circuit was measured through the homemade cryogenic amplifier [14, 15]. The measurements were performed in the dilution refrigerator with base temperature of 20 mK. The electron temperature of the QW was confirmed to be also 20 mK by measuring its thermal noise [16]. In this condition, the expected Zeeman splitting (~100 μeV at 4.5 T) is much larger than the thermal energy (~2 μeV at 20 mK). Thus, the edge states are fully spin-polarized. The equilibrium conductance of the QW as a function of V_g is shown in Fig. 1 (b). The conductance plateau at $2e^2$/h at small $|V_g|$ reflects $\nu_{2DEG} \sim 2$. By negatively applying the gate voltage, we tuned the conductance of the QW to be e^2/h, where $\nu_{QW} \sim 1$ with $\nu_{2DEG} \sim 2$ is realized. Below, we report the result of the measurement obtained in this situation.

3. Results and Discussion

Figure 2 (a) shows the image plot of the differential conductance of the QW as a function of V_{sd} and B at V_g = -1.4 V. The upper panel shows the conductance at B = 4.4 T. Clear conductance plateau exists around zero bias voltage reflecting the dissipation-less QH edge transport. In Fig. 2(a), the white region corresponds to the conductance plateau. The plateau, however, collapses by applying the bias voltage exceeding ~100 μV. This critical value (eV_{sd} ~100 μeV) falls in the same range as the Zeeman energy, while the Zeeman energy in the QW may be enhanced by the exchange interaction. As shown in the dotted line in the lower panel of Fig. 2 (a), the critical value of V_{sd} to induce the collapse depends on the magnetic field. This suggests that the collapse is induced by the electron tunneling between the spin-resolved Landau levels; As the second Landau level in QW is empty in the present situation, the origin of the collapse can be explained as due to the electron tunneling from the lowest Landau level to the second one, which is localized inside or in the vicinity of the QW.

Figure 2 (b) shows the image plot of the differential conductance of the QW as a function of V_{sd} and V_g. The conductance at V_g = -1.6 V is shown in the upper panel. As seen by the dotted curve in the image plot, the critical value is almost independent of V_g. By comparing between the results in Fig. 2(a) and 2(b), it is clear that the critical value is determined by the magnetic field, most probably by

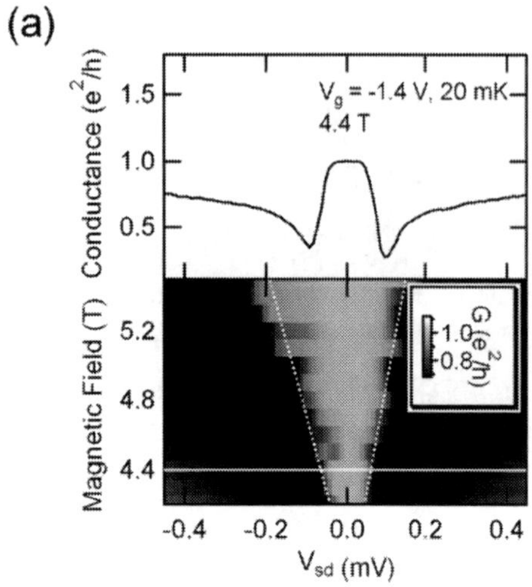

Figure 2. Differential conductance of the QW as a function of V_{sd} measured at various (a) magnetic fields and (b) gate voltages. Upper panels are the cross-sections of the conductance at a fixed field or a gate voltage shown by the solid lines in the lower panels. The critical value to induce the collapse is shown in the lower panels by the dotted line.

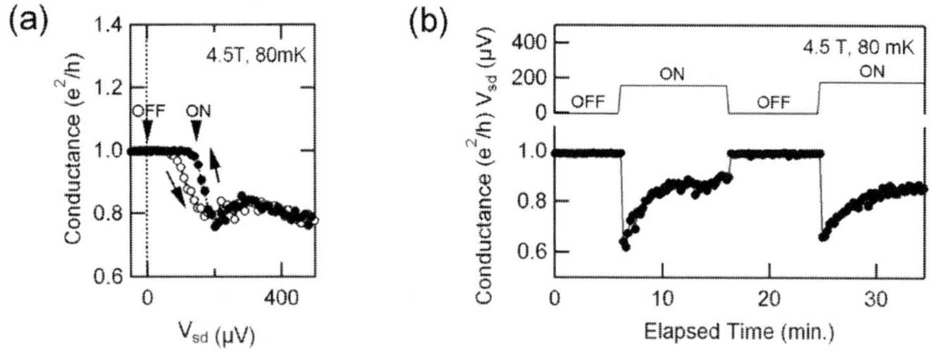

Figure 3. (a) The QW conductance as a function of V_{sd} showing the dependence on the sweep direction of V_{sd}. The arrows with "ON" and "OFF" indicate the source-drain bias voltage used in the measurement (b). (b) Observation of DNP development where the conductance varies in minutes. The relaxation time of DNP is also of the order of minutes.

the Zeeman splitting, being in agreement with the relevance of the electron tunneling between the spin-resolved Landau levels as discussed above.

The result shown in Figs. 2 (a) and 2 (b) is obtained when V_{sd} is swept from the negative to the positive values. Figure 3 (a) shows dependence of the conductance on the sweep direction of V_{sd}. The hysteresis is observed around the critical value. As is often the case in the previous studies [9], such a hysteresis behavior strongly suggests the DNP generation via the hyperfine interaction between the electron spins and the nuclear spins. Actually, Fig. 3 (b) shows that the conductance obtained at V_{sd}

around the critical value varies temporally with the scale of minutes, typical time scale of the phenomenon associated with DNP. The occurrence of the DNP supports that the origin of the collapse of the quantized conductance is the electron tunneling between the Landau levels, where the electron spins have to be flipped and some part of the spin degree of freedom is transferred to the nuclear spins. It was recently found that in bulk 2DEGs the skyrmions [17-19] and spin-textured edge states [20] affect the relaxation time of the DNP, although such a characteristic behavior of the relaxation time was not observed in the present study. The difference might be attributed to the difference between the bulk and the QW.To get further understandings, we have conducted the resistance-detected nuclear magnetic resonance and the shot noise measurement, which will be discussed elsewhere [21].

In conclusion, we have observed the collapse of the quantized conductance in the QW in the non-equilibrium QHE regime. The critical vale of the bias voltage to induce this phenomenon depends on the Zeeman energy and the accompanying DNP is observed, which suggests the presence of the electron tunneling between the Landau levels in the QW.

Acknowledgements

We appreciate fruitful discussions from Minoru Kawamura, Norio Kumada, Takashi Nakajima, Yasuhiro Tokura, and Koji Muraki. This work is partially supported by KAKENHI, Yamada Science Foundation and Matsuo Science Foundation.

References

[1] K. von Klitzing, G. Dorda, and M. Pepper, Phys. Rev. Lett. **45**, 449 (1980).
[2] M. Büttiker, Phys. Rev. **B38**, 9375 (1988).
[3] G. Ebert *et al.*, J. Phys. **C 16**, 5441 (1983).
[4] M. E. Cage *et al.*, Phys. Rev. Lett. **51**, 1374 (1983).
[5] G. Nachtwei, Physica **E 4**, 79 (1999).
[6] For example, S. Komiyama and Y. Kawaguchi Phys. Rev. **B 61**, 2014 (2000) and references therein.
[7] Keith R. Wald *et al.*, Phys. Rev. Lett. **73**, 1011 (1994).
[8] David C. Dixon *et al.*, Phys. Rev. **B 56**, 4743 (1997).
[9] T. Machida, T. Yamazaki and S. Komiyama, Appl. Phys. Lett. **80**, 4178 (2002).
[10] G. Yusa *et al.*, Nature (London) **434**, 1001 (2005).
[11] A. Córcoles, C. J. B. Ford, M. Pepper, G. A. C. Jones, H. E. Beere, and D. A. Ritchie, Phys. Rev. **B 80**, 115326 (2009).
[12] A. Kou *et al.*, Phys. Rev. Lett. **105**, 056804 (2010).
[13] M. Kawamura *et al.*, Appl. Phys. Lett. **90**, 022102 (2007).
[14] L. DiCarlo *et al.*, Phys. Rev. Lett. **97**, 036810 (2006).
[15] M. Hashisaka *et al.*, Phys. Rev. **B 78**, 241303(R) (2008); M. Hashisaka, *et al.*, Phys. Status Solidi C **5**, 182 (2008).
[16] M. Hashisaka *et al.*, Review of Scientific Instruments, **80** (2009) 096105.
[17] R. Tycko *et al.*, Science **268**, 1460 (1995).
[18] K. Hashimoto *et al.*, Phys. Rev.Lett. **88**, 176601 (2002).
[19] J. H. Smet *et al.*, Nature (London) **415**, 281 (2002).
[20] T. Nakajima, Y. Kobayashi, S. Komiyama, Phys. Rev. **B 82**, 201302(R) (2010).
[21] K. Chida *et al.*, (in preparation).

Journal of Physics: Conference Series **334** (2011) 012032 doi:10.1088/1742-6596/334/1/012032

Field and Density Dependence of Edge Magnetoplasmon Transport in a Quantum Hall System

N. Kumada[1], H. Kamata[1,2], K. Muraki[1], and T. Fujisawa[2]

[1] NTT Basic Research Laboratories, NTT Corporation, 3-1 Morinosato-Wakamiya, Atsugi, Kanagawa 243-0198, Japan

[2] Research Center for Low-Temperature Physics, Tokyo Institute of Technology, 2-12-1 Ookayama, Meguro, Tokyo 152-8551, Japan

Abstract. We investigate edge channel properties in integer quantum Hall regime through time-of-flight measurements of edge magnetoplasmons (EMPs). EMPs are injected by applying a voltage pulse to on Ohmic contact and detected by applying another voltage pulse to a quantum point contact. By controlling the time interval between the injection and detection pulses, the velocity of EMPs is determined. The width of edge channels calculated from the velocity of EMPs oscillates with filling factor: as the filling factor is decreased to an integer, the width increases almost by one order of magnitude from ~ 0.3 to $\sim 3\,\mu$m. Furthermore, we find that the width at a fixed filling factor increases with decreasing electron density in the bulk two-dimensional system.

1. Introduction

In a quantum Hall (QH) system, the interior of the two-dimensional electron system (2DES) is incompressible and the current flows in one-dimensional edge channels. Collective charge density excitations in edge channels are called edge magnetoplasmons (EMPs) [1]. EMPs have been investigated extensively both theoretically [1–3] and experimentally [4–7]. In a sample with a front gate, oscillations of the width l of edge channels as a function of the filling factor ν has been detected as the oscillations of the velocity of EMPs [5, 7]. Recently, the velocity control of EMPs has been demonstrated by changing the degree of screening by a front gate [8].

In this work, we investigate effects of electron density on the velocity of EMPs in a sample with a front gate. We carried out time-of-flight measurements of EMPs for three values of the density. For all the densities, the velocity takes minima at integer ν. At a fixed ν, the velocity decreases with decreasing density. These results show that l depends not only on ν but also on the density.

2. Experiments

The sample used contains a 20-nm-wide GaAs quantum well. The 2DES is 160 nm below the surface. The low-temperature mobility is $2.1 \times 10^6\,\mathrm{cm}^2/\mathrm{Vs}$ at an as grown density of $1.27 \times 10^{11}\,\mathrm{cm}^{-2}$. A schematic view of the sample structure and the experimental setup are shown in the inset of Fig. 1. A quantum point contact (QPC) was fabricated close to the drain contact D1. A front gate crosses the sample between the source contact S and the D1 contact.

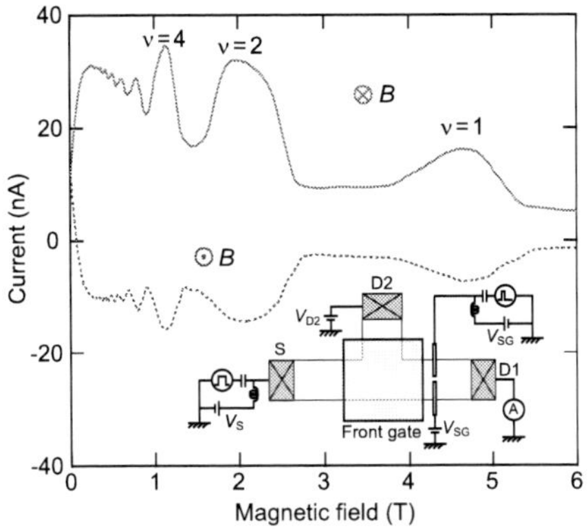

Figure 1. Current through the D1 contact as a function of the magnetic field for $V_{fg} = 0\,V$. Constant biases of 1 mV and −0.2 mV are applied to the S and D2 contacts, respectively. For the solid and dotted traces, the perpendicular field is applied in the direction pointing into and out of the page, respectively. Inset shows a schematic illustration, not to scale, of the sample structure and the experimental setup for the time-of-flight measurement.

For the time-of-flight measurement, the DC bias for the source and drain contacts is 0 V and the QPC is set close to the pinch-off condition. An EMP pulse is injected by applying a square voltage pulse to the S contact; the time width and the amplitude of the pulse is 5 ns and 5 mV, respectively. To detect the EMP pulse, another voltage pulse with 1 ns in time width and 100 mV in amplitude is applied to the QPC. If the QPC is open when the EMP pulse is arrived, EMPs are detected as a current through the D1 contact; otherwise they are reflected by the split gates and collected by the drain contact D2 [8]. By controlling the time interval t between the injection and detection pulses, a profile of the EMP pulse is obtained. The rising time of the pulses is 20 ps at the output of the pulse generator and increases to 50 ps at the sample; in the sample, the rising time is further increased by various mechanisms including high Ohmic resistance, capacitive coupling between EMPs and the front gate, and nonlinear dispersion of EMPs. The repetition time is 60 ns. All measurements are performed at 1.5 K.

3. Results and discussions

First, we show results of three-terminal DC measurements, demonstrating the chirality of the edge current. Figure 1 shows the current through the D1 contact as a function of the magnetic field B for the front gate bias $V_{fg} = 0\,V$, where the electron density in the gated region is $n = 1.08 \times 10^{11}\,cm^{-2}$. Constant biases of 1.0 mV and −0.2 mV are applied to the S and D2 contacts, respectively. When a perpendicular field is applied in the direction pointing into the page, the current shows peaks in the QH effect regime. In this case, the chirality of the edge channel is anticlockwise, where the current flows in the direction S→D1→D2→S. On the other hand, when the perpendicular field is in the opposite direction, the current flows clockwise and shows minima in the QH effect regime.

For the time-of-flight measurements, the chirality is fixed to be anticlockwise, so that EMPs injected from the S contact propagate along the lower boundary to the contact D1. The length of the edge channel between the S contact and the QPC in the regions with and without the front gate is 480 μm and 937 μm, respectively. Figure 2 presents results of time-of-flight measurements for $B = 0.2\,T$ and several magnetic fields around $\nu = 1$ at $V_{fg} = 0\,V$. For $B = 0.2\,T$, where injected electrons propagate as two-dimensional bulk plasmons in the 2DES rather than EMPs, the current start to rise at $t \sim 0$. For larger fields, where the edge channels are well developed, current rises with a time delay, which corresponds to the traveling time of EMPs from the S contact to the QPC along the edge channel. Around $\nu = 1$, the delay time increases as $\nu = 1$ is approached from lower field. At the same time, the EMP pulse becomes broad; for the field slightly larger than that for $\nu = 1$, the EMP pulse becomes broader than the repetition time 60 ns.

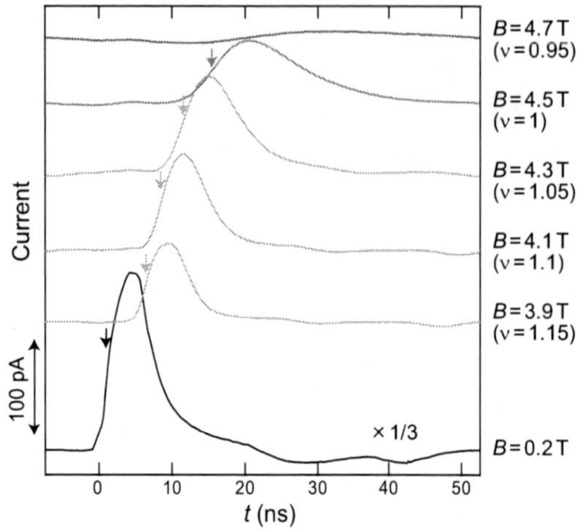

Figure 2. Current as a function of the time interval t between the injection and detection pulses for a low field $B = 0.2\,\mathrm{T}$ and around $\nu = 1$ at $n = 1.08 \times 10^{11}\,\mathrm{cm}^{-2}$. Data for $B = 0.2\,\mathrm{T}$ is reduced by a factor of three. Arrows represent the delay time t_d. Traces are vertically offset for clarity.

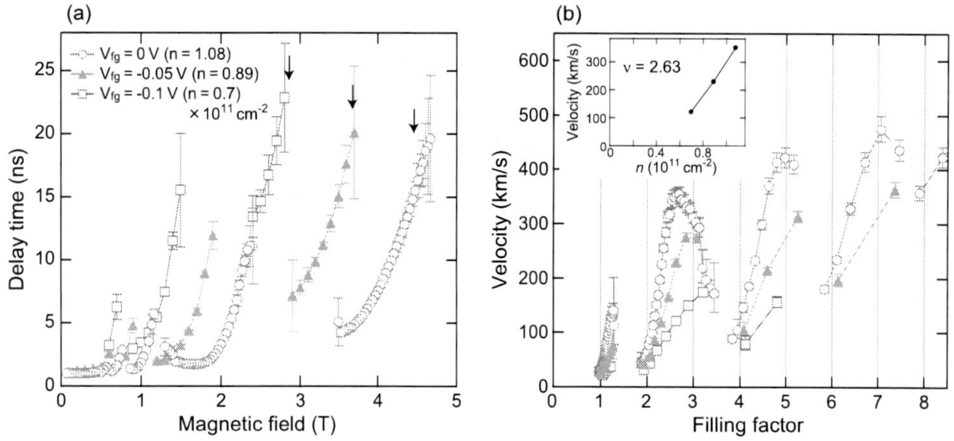

Figure 3. (a) Delay time t_d of EMPs as a function of the magnetic field for three values of the electron density in the gated region. Arrows represent the field for $\nu = 1$. (b) Velocity of EMPs as a function of filling factor. Inset shows the velocity at $\nu = 2.63$ as a function of n.

We carried out similar measurements over a wide range of the magnetic field and for three values of V_{fg}. Figure 3(a) shows the delay time t_d of the EMP pulse for three values of n as a function of the magnetic field; t_d is defined as the maximum of the derivative of the EMP pulse (arrows in Fig. 2). For all n, t_d increases as the field is increased to the value for even integer ν or $\nu = 1$. For slightly larger field, t_d can not be determined because the EMP pulse is too broad. Note that t_d contains contributions both from the gated and ungated regions; by a similar time-of-flight measurement using an ungated sample, we confirmed that the velocity of EMPs in the ungated region is 10 times faster and depends monotonically on the magnetic field (not shown here) as shown in Ref. [4]. Accordingly, t_d mainly reflects the contribution from the gated region.

We subtracted the contribution of the ungated region from t_d and calculated the velocity of EMPs in the gated region. Figure. 3(b) shows the velocity as a function of ν. The velocity

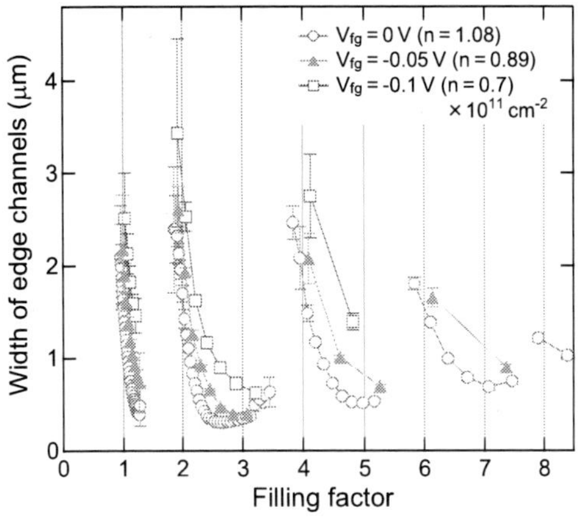

Figure 4. Width of the edge channel calculated by using Eq. (1) as a function of the filling factor.

shows, in addition to the asymmetric oscillations as a function of ν, a monotonic increase with increasing n [an example is shown in the inset of Fig. 3(b)].

Now, we calculate the width of the edge channels l using the equation [3, 6],

$$v = \frac{e^2 \nu_{in}}{h} \frac{d}{\epsilon_0 \epsilon l}, \tag{1}$$

where v is the velocity of EMPs, $d = 160$ nm is the distance between the 2DES and the front gate, $\epsilon_0 \epsilon$ is the dielectric constant of GaAs. ν_{in} is the filling factor of the innermost incompressible strip, e.g., $\nu_{in} = 1$ for $1 \leq \nu < 2$. Figure4 shows the calculated l as a function of ν. l varies over a wide range between 0.3 and 3.5 μm. The asymmetric oscillations of l as a function of ν has been observed [5] and explained as due to a periodic change in the position of the innermost incompressible stripe, which separates edge channels from the bulk 2DES. For instance, as ν is decreased from 2.5, the innermost incompressible stripe with $\nu_{in} = 2$ moves toward the interior of the 2DES, leading to the increase in l. When ν becomes smaller than 2, the $\nu_{in} = 2$ incompressible stripe vanishes and then the $\nu_{in} = 1$ incompressible stripe becomes the boundary between the edge channel and the 2DES, resulting in the abrupt decrease in l [9]. On the other hand, at a fixed ν, l increases with decreasing n. This indicates that the slope of the edge potential is slower for smaller n. This is consistent with an electrostatic calculation [10] of edge states.

Acknowledgments

The authors are grateful to T. Ota and M. Ueki for experimental support. This work was supported in part by Grant-in-Aid for Scientific Research (21000004) from MEXT of Japan.

References

[1] Volkov V A and Mikhilov S A 1988 *Zh. Eksp. Theor. Fiz.* **94** 217
[2] Aleiner I L and Glazman L I 1994 *Phys. Rev. Lett.* **72** 2935
[3] Johnson M D and Vignale G 2003 *Phys. Rev. B* **67** 205332
[4] Ashoori R C, Stormer H L, Pfeiffer L N, Baldwin K W and West K 1992 *Phys. Rev. B* **45** 3894
[5] Zhitenev N B, Haug R J, v Klitzing K and Eberl K 1994 *Phys. Rev. B* **49** 7809
[6] Zhitenev N B, Haug R J, v Klitzing K and Eberl K 1995 *Phys. Rev. B* **52** 11277
[7] Ernst G, Zhitenev N B, Haug R J and von Klitzing K 1997 *Phys. Rev. Lett.* **79** 3748
[8] Kamata H, Ota T, Muraki K and Fujisawa T 2010 *Phys. Rev. B* **81** 085329
[9] Ahlswede E, Weitz P, Weis J, von Klitzing K and Eberl K 2001 *Physica B* **298** 562
[10] Chkovskii D B, Shklovskii B I and Glazman L I 1992 *Phys. Rev. B* **46** 4026

HMF-19 IOP Publishing
Journal of Physics: Conference Series **334** (2011) 012033 doi:10.1088/1742-6596/334/1/012033

Quantum phase transition of dynamical resistance in a mesoscopic capacitor

Yuji Hamamoto[1], **Thibaut Jonckheere**[2], **Takeo Kato**[3] **and Thierry Martin**[2,4]

[1] Institute of Physics, University of Tsukuba, Tennodai, Tsukuba, Ibaraki 305-8571, Japan
[2] Centre de Physique Théorique, Case 907 Luminy, 13288 Marseille cedex 9, France
[3] Institute of Solid State Physics, University of Tokyo, Kashiwa, Chiba 277-8581, Japan
[4] Université de la Méditérannée, 13288 Marseille cedex 9, France

E-mail: `hamamoto.yuji.fp@u.tsukuba.ac.jp`

Abstract. We study theoretically dynamic response of a mesoscopic capacitor, which consists of a quantum dot connected to an electron reservoir via a point contact and capacitively coupled to a gate voltage. A quantum Hall edge state with a filling factor ν is realized in a strong magnetic field applied perpendicular to the two-dimensional electron gas. We discuss a noise-driven quantum phase transition of the transport property of the edge state by taking into account an ohmic bath connected to the gate voltage. Without the noise, the charge relaxation resitance for $\nu > 1/2$ is universally quantized at $R_q = h/(2e^2\nu)$, while for $\nu < 1/2$, the system undergoes the Kosterlitz-Thouless transition, which drastically changes the nature of the dynamical resistance. The phase transition is facilitated by the noisy gate voltage, and we see that it can occur even for an integer quantum Hall edge at $\nu = 1$. When the dissipation by the noise is sufficiently small, the quantized value of R_q is shifted by the bath impedance.

1. Introduction

In the long history of the mesoscopic physics, it is quite recent that it became possible to observe dynamic response of coherent systems in experiments [1, 2]. Theoretically, dynamical transport properties of the mesoscopic systems has been extensively investigated with the scattering theory [3, 4]. One of the most intriguing results of the theoretical efforts is the universal quantization of charge relaxation resistance in mesoscopic capacitors [4]. The mesoscopic

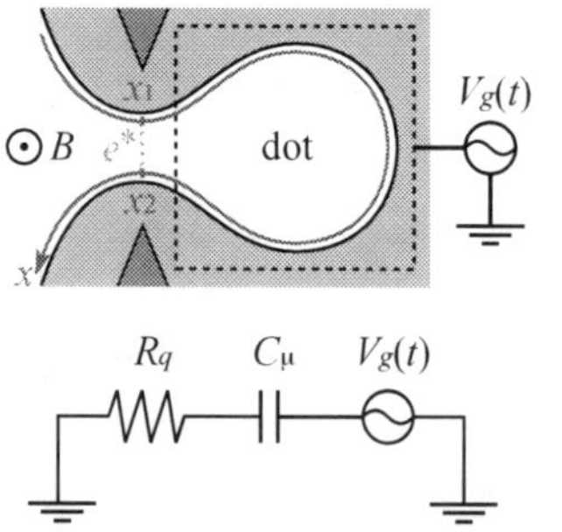

Figure 1. Schematic view of the mesoscopic capacitor and its equivalent RC circuit. The quantum dot is connected to an electron reservoir via a point contact and capacitively coupled to a gate (shown as a dotted square) with time-dependent voltage $V_g(t)$. A magnetic field applied perpendicular to the plane realizes a quantum Hall edge state moving along the x axis. The dotted line between x_1 and x_2 represents quasiparticle tunneling on the point contact.

capacitors are quantum analogs of classical RC circuits, among which the simplest one (Fig. 1) consists of a quantum dot connected to an electron reservoir via a quantum point contact and capacitively coupled to an ac gate voltage. An electron entering the quantum dot relaxes to equilibrium in the scale of charge relaxation time τ_{RC}. The dynamical resistance observed at this time scale is the so-called *charge relaxation resistance.* It has been predicted by Büttiker *et al.* that the charge relaxation resistance is quantized at half a resistance quantum $h/(2e^2)$ per channel irrespective of the property of the point contact [4]. Indeed, the universal behavior has been confirmed experimentally [1], which has further attracted the theorists' interest in the charge relaxation resistance [5, 6, 7]. Recently, we have investigated theoretically the charge relaxation resistance, treating electron interactions exactly [6]. We have considered a one-dimensional mesoscopic capacitor and predicted the followings; the charging effect of the quantum dot has no influence on the quantization, while short-range interactions described by the Luttinger parameter K drastically affects the nature of the dynamical resistance. In contrast to the universal quantization at $R_q = h/(2e^2K)$ for $K > 1/2$, the Kosterlitz-Thouless transition occurs for $K < 1/2$ and as a result one can not any more observe the universal charge relaxation resistance. In this paper, we extend our theory to discuss the quantum phase transition driven by a noisy gate voltage. How our treatment of the charging effect is related to the conventional picture of the scattering theory is also discussed by applying the mean field approximation.

2. Model

The system we have in mind is the mesoscopic capacitor depicted in Fig 1. Due to the magnetic field applied perpendicular to the 2DEG, a chiral mode called quantum Hall edge state is realized along the x axis. It has been shown by Wen [8] that the edge state at filling factor (1/ is an odd integer) is a chiral Luttinger liquid, and its Hamiltonian reads

$$H_0 = \frac{v}{4\pi} \int_{-\infty}^{\infty} \mathrm{d}x \left(\frac{\partial \phi}{\partial x} \right)^2 . \tag{1}$$

Here the bosonic field ϕ describes the gapless edge excitations, whose velocity is denoted by v. The narrow constriction on the point contact causes intra-edge tunneling of Laughlin quasiparticles between $x = x_1$ and $x = x_2$. Since a Laughlin quasiparticle at x is annihilated by the operator $\propto \exp[i\phi(x)]$, the tunneling process is written as

$$H_V = V \cos[\phi(x_1) - \phi(x_2)], \tag{2}$$

where V is the strength of quasiparticle tunneling [1]. Change in the dot charge Q in the range of $x_1 \leq x \leq x_2$ is strongly restricted by the long-range interaction between electrons, i.e., the charging effect. Upon using the fact that the electron density is bosonized as $= \partial_x \phi/(2\pi)$, one can see that the term describing the charging effect takes the form

$$H_C = \frac{Q^2}{2C} + QV_g(t) \qquad \left(Q = \frac{e}{2\pi}[\phi(x_2) - \phi(x_1)] \right), \tag{3}$$

where C is the geometrical capacitance essentially determined by the charging effect. When the gate voltage is oscillating at a low frequency ω $1/\tau_{RC}$, the admittance of the quantum RC circuit can be expanded as $G(\omega) = -i\omega C_\mu + \omega^2 {C_\mu}^2 R_q + \mathcal{O}(\omega^3)$, from which the charge relaxation resistance R_q is obtained. Note that the electrochemical capacitance defined by $C_\mu \equiv -\partial\langle Q \rangle / \partial V_g$ in general differs from C, since, e.g., the quasiparticle tunneling induces Coulomb blockade oscillation in C_μ. The above expansion of $G(\omega)$ is possible only when the τ_{RC} is finite: if the phase transition occurs, τ_{RC} diverges faster than $1/T$ as tempereture T is lowered, so that one can not even regard R_q as the charge relaxation resistance [6].

[1] In the argument of the cosine in Eq. (2), we have omitted the phase that an edge mode gains along the circumference of the dot by shifting zero of the field ϕ.

Journal of Physics: Conference Series **334** (2011) 012033 doi:10.1088/1742-6596/334/1/012033

3. Mean field theory

At a glance, our formalism described in the previous section looks slightly different from the scattering theory [3, 4], where the charging effect is incorporated through an effective gate voltage U_g in the quantum dot. Clearly, the latter treatment is justified when fluctuation of the dot charge Q is negligible enough that one can approximate the interaction of an electron with the others in the dot with an effective potential created by mean charge $\langle Q\rangle$. Thus one can expect that the two-capacitor picture obtained from the scattering theory [4] is reproduced by applying the mean field theory to our model. In this section, we first confirm this and compare the results of the scattering theory and our calculation where the interaction is exactly treated. Upon neglecting higher orders in $Q - \langle Q\rangle$ in the charging term (3), we obtain

$$H_C{}^{\mathrm{MF}} \simeq \frac{\langle Q\rangle}{C} + QV_g \equiv QU_g, \qquad \therefore \langle Q\rangle = C(U_g - V_g), \tag{4}$$

where we have dropped constant terms. Since an electron in the dot behaves as a free particle subject to the effective potential U_g, the mean charge can be also expressed using the density of states in the dot $\mathrm{d}N/\mathrm{d}\varepsilon$ as

$$\langle Q\rangle = e\frac{\mathrm{d}N}{\mathrm{d}\varepsilon}(-eU_g) \equiv -C_qU_g. \tag{5}$$

C_q is called quantum capacitance since it reflects the nature of standing wave of an electron in the dot. By eliminating U_g from Eqs. (4) and (5), we can calculate the electrochemical capacitance

$$C_\mu = -\frac{\partial\langle Q\rangle}{\partial V_g} = \frac{CC_q}{C + C_q}, \qquad \therefore \frac{1}{C_\mu} = \frac{1}{C} + \frac{1}{C_q}. \tag{6}$$

Thus in the mean field approximation C_μ is reduced to a series combination of the classical part C and the quantum part C_q, in agreement with the results of the scattering theory [4]. Note that $C_q \propto \mathrm{d}N/\mathrm{d}\varepsilon$ oscillates as V_g is varied. For an integer quantum Hall edge state at $\ = 1$, one can refermionized the Hamiltonian to solve the transport problem with the effective potential U_g with the scattering theory. Then the charge relaxation resistance quantized at $R_q = h/(2e^2)$ will be also reproduced.

One might conjecture that the mean field assumption is valid in the limit of weak quasiparticle tunneling $V \to 0$, where the potential that dot charge Q feels is almost harmonic, so that the charge fluctuation is suppressed upon lowering temperature. Indeed, we can exactly derive quantized charge relaxation resistance $R_q = h/(2e^2\ )$ up to second order in V [6], but a qualitative difference between the mean field theory and our exact result emerges in the capacitance. From the perturbative result [6], one can split C_μ into two parts mimicking Eq. (6):

$$\frac{1}{C_\mu} = \frac{1}{\gamma C} + \frac{1}{\gamma C_q'} \qquad \left(C_q' \equiv \frac{{}^2e^2L}{2\pi v}\right), \tag{7}$$

where γ is a numerical factor oscillating as a function of V_g, and $L \equiv x_2 - x_1$ is the circumference of the dot. For $\ = 1$, $\gamma C_q'$ can be identified as the quantum capacitance, since it coincides with the small V limit of C_q obtained from the scattering theory, while γC differs from the geometrical capacitance by the oscillating factor. The latter is because the mean field approximation underestimates the charge fluctuation (capacitance) due to electron tunneling between the dot and the reservoir, which should oscillate depending on V_g. Therefore, one can see from Eq. (7) that in the small V limit, C_μ is a series combination of γC and $\gamma C_q'$, which reflect the particle and wave natures of an electron, respectively.

It is clear that the mean field theory breaks down at charge degenerate points $\langle Q\rangle = (n+1/2)e$ (n is an integer) for large V, where the charge fluctuation can be finite down to low temperatures.

In Ref. [6], we have shown that the charge relaxation resistance for $\;> 1/2$ is always quantized at $R_q = h/(2e^2\;\;)$ owing to the so-called charge-Kondo effect [9], where two degenerate charge states ne and $(n+1)e$ play the role of up and down spins of a magnetic impurity [2]. More surprisingly, charge relaxation resistance becomes ill-defined due to the Kosterlitz-Thouless transition for $\;< 1/2$, and the observed dynamical resistance R_q diverges with lowering temperature.

4. Noise-driven quantum phase transition

The phase transition mentioned in the previous section occurs for a fractional quantum Hall edge state such as $\;= 1/3$. In the rest of this paper, we discuss another possibility for the phase transition by considering a noisy gate voltage connected to an ohmic bath. Such a situation can be realized, e.g., by connecting a long LC transmission line in series with the gate voltage [10]. The Hamiltonian describing the Caldeira-Leggett bath of an infinite number of harmonic oscillators reads [11]

$$H_B = \sum_j \left[\frac{P_j^2}{2M_j} + \frac{M_j \Omega_j^2}{2} \left(X_j - \frac{\lambda_j}{M_j \Omega_j^2} Q \right)^2 \right], \tag{8}$$

where λ_j is the strength of interaction between the jth oscillator and the dot charge Q. X_j, P_j, M_j, Ω_j denote the normal coordinate, momentum, mass, and frequency of the jth oscillator. For an ohmic bath with impedance R_B, the spectral density of the environmental coupling satisfies

$$J(\omega) \equiv \frac{\pi}{2} \sum_j \frac{\lambda_j^2}{M_j \Omega_j} \delta(\omega - \Omega_j) = R_B \omega. \tag{9}$$

Below, we see how dissipation influences the behavior of charge relaxation resistance. To this end, we derive the euclidian effective action for the system consisting of the mesoscopic capacitor and the ohmic bath. By integrating out the quadratic degrees of freedom except $\phi(x_1)$ and $\phi(x_2)$ from the kinetic term (1), we first obtain the effective action for the mesoscopic capacitor

$$S_{MC} = \frac{1}{\pi} \sum_{\omega_n} \frac{|\omega_n|}{1 - \mathrm{e}^{-|\omega_n| L/v}} |\tilde{\phi}(\omega_n)|^2 + V \int \mathrm{d}\tau \cos[2\phi(\tau)] + \int \mathrm{d}\tau \left[\frac{E_C}{\pi^2} \{\phi(\tau)\}^2 - \frac{eV_g}{\pi} \phi(\tau) \right]. \tag{10}$$

Here $\phi \equiv [\phi(x_1) - \phi(x_2)]/2$, $\tilde{\phi}(\omega_n)$ is the Fourier transform of $\phi(\tau)$, and $E_C \equiv e^2/(2C)$ is the charging energy. Similarly, we can derive the effective action for the bath from Eq. (8) as

$$S_B = \frac{\alpha}{\pi} \sum_{\omega_n} |\omega_n| |\tilde{\phi}(\omega_n)|^2 \qquad \left(\alpha = \frac{R_B}{h/e^2} \right), \tag{11}$$

where α is the strength of dissipation. One can expect from Eqs. (10) and (11) that the coupling to the ohmic bath renormalizes the filling factor , which influences the critical point. To discuss the quantum phase transition of the mesoscopic capacitor, let us focus on the degeneracy point in the large V region. In this case, electron tunneling occurs between the dot and the reservoir with small tunneling strength t. Following Ref. [6], we can straightforwardly identify the scaling equations at low frequencies $|\omega_n| \quad v/L$:

$$\frac{\mathrm{d}t}{\mathrm{d}l} = \left[1 - \left(\alpha + \frac{1}{2} \right) s^2 \right] t, \qquad \frac{\mathrm{d}s^2}{\mathrm{d}l} = -4s^2 t^2 \qquad (0 < s \lesssim 1), \tag{12}$$

[2] Recently, an equivalent Kondo physics at $\nu = 1$ has been also discussed in Ref. [7].

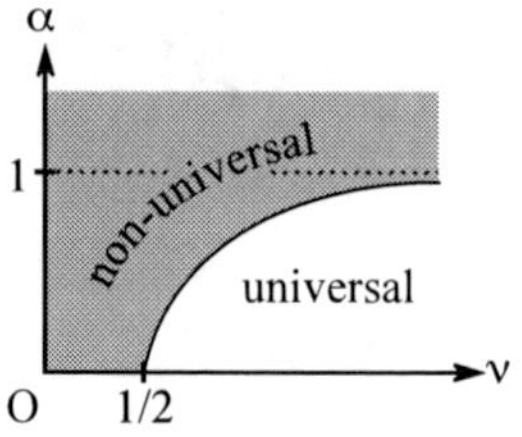

Figure 2. Phase diagram of the universal quantization of the charge relaxation resistance of the mesoscopic capacitor with a noisy gate voltage.   is the filling factor, and $\alpha \equiv R_B/(h/e^2)$ is the strength of dissipation, where R_B is the bath impedance. The charge relaxation resistance is universally quantized in the white region, while it can undergoes a phase transition for su ciently strong quasiparticle tunneling in the gray region.

which describes the Kosterlitz-Thouless transition. One can see from these equations that, for $\alpha < 1 - 1/(2\ )$, the tunneling strength always grows upon decreasing temperature. For $\alpha > 1 - 1/(2\ )$, on the other hand, it is possible that for su ciently large V (small t) that the system is renormalized to a weak coupling configuration with specified charge. It should be noted that, unlike in Ref. [6], the critical phase transition occurs even in the case of an integer quantum Hall edge ($= 1$) when the bath impedance exceeds half a resistance quantum, $R_B > h/(2e^2)$. Otherwise, the system flows towards the Kondo fixed point as $T \to 0$, and the charge relaxation resistance is given by the value for $V \to 0$ as $R_q = h/(2e^2) + R_B$, i.e., the noisy gate voltage results in the shift of the quantized value by the bath impedance R_B. The phase diagram of the universal quantization of the charge relaxation resistance is shown in Fig. 2.

5. Summary

We have theoretically studied the phase transition of dynamical resistance in a mesoscopic capacitor applied a perpendicular magnetic field. For weak quasiparticle tunneling, the electrochemical capacitance can be expressed as a series combination of two capacitors, one of which shows a behavior qualitatively different from the result of the scattering theory. We have also discussed the phase transition driven by noise with dissipation strength α. When $\alpha < 1 - 1/(2\ )$, the charge relaxation resistance is universally quantized at $h/(2e^2\ ) + R_B$, with the bath impedance R_B.

Acknowledgements

Y.H. and T.K. are grateful to T. Fujii for valuable discussions. This research was partially supported by JSPS and MAE under the Japan-France Integrated Action Program (SAKURA) and by a Grant-in-Aid for Young Scientists B No. 21740220 from the Ministry of Education, Science, Sports and Culture. It was also supported by ANR-PNANO Contract MolSpin- Tronics, No. ANR-06-NANO-27.

References

[1] Gabelli J *et al.* 2006 *Science* **313** 499
[2] Gabelli J *et al.* 2007 *Phys. Rev. Lett.* **98** 166806; Fève G *et al.* 2007 *Science* **316** 1169
[3] Buttiker M, Prêtre A and Thomas H 1993 *Phys. Rev. Lett.* **70** 4114
[4] Buttiker M, Thomas H and Prêtre A 1993 *Phys. Lett.* A **180** 364; Prêtre A, Thomas H and Buttiker M 1996 *Phys. Rev.* B **54** 8130
[5] Nigg S E, López R and Buttiker M 2006 *Phys. Rev. Lett.* **97** 206804; Nigg S E and Buttiker M 2008 **77** 085312
[6] Hamamoto Y, Jonckheere T, Kato T and Martin T 2010 *Phys. Rev.* B **81** 153305
[7] Mora C and Le Hur K 2010 *Nat. Phys.* **6** 697
[8] Wen X G 1990 *Phys. Rev. Lett.* **64** 2206; 1991 *Phys. Rev.* B **43** 11025; 1991 *ibid.* **44** 5708
[9] Glazman L I and Matveev K A 1990 *Zh. Eksp. Teor. Fiz.* **98** 1834; Matveev K A 1991 *Zh. Eksp. Teor. Fiz.* **99** 1598; Matveev K A 1994 *Phys. Rev.* B **51** 1743
[10] Le Hur K 2004 *Phys. Rev. Lett.* **92** 196804
[11] See, e.g., Weiss U 2008 *Quantum Dissipative Systems 3rd ed* (Singapore: World Scientific)

HMF-19 IOP Publishing
Journal of Physics: Conference Series **334** (2011) 012034 doi:10.1088/1742-6596/334/1/012034

Edge state distribution in an Aharonov-Bohm electron interferometer in the integer quantum Hall regime

T Ozturk , A E Kavruk , A Ozturk , U Atav and H Yuksel

[1]Selcuk University, Faculty of Science, Department of Physics, 42075 Kampus, Konya Turkey

E-mail: teozturk@gmail.com

Abstract. In this study we analyze the density distributions of the two dimensional electron system for an experimental geometry which is topologically equivalent of an Aharonov-Bohm interferometer in three dimensions in the quantum Hall regime and obtain the spatial distribution of the edge states. We employ the Thomas-Fermi approximation in our analysis and solve the Poisson equation in three dimensional using a multi grid technique. Also we obtain the distribution of incompressible strips for a wide range of magnetic fields strengths and comment on their relation with experimental results in literature.

1. Introduction

Aharonov-Bohm (AB) effect is a phenomenon which exhibits observable quantum mechanical effects with using a vector potential in a region where the magnetic field is zero [1]. The low dimensional experiments related to "AB effect" do not entirely display this effect because of the use of uniform magnetic field everywhere [2]. Similarities with the original AB effect are just geometry and a phase event in the presence of a magnetic field. One of these experimental studies is a quantum electron interferometer in the quantum Hall regime which is studied by Camino et al. [3]. They have used a ring shaped gate structure to experimentally observe resistivity oscillations. In this experimental setup, the incompressible strips (ISs) formed in two dimensional electron gas (2DEG) take place of the electron beams in the original AB effect. Therefore it is very important to have a knowledge of the distribution of the ISs in order to understand these effects.

In our study, we theoretically investigate the ISs formed in the structure used by Camino et al.[3] in their experiments. We obtain the distribution of the ISs for various magnetic field strengths using Thomas-Fermi-Poisson(TFP) approach. We consider a trenched structure which is illustrated in Fig. 1, where the gates are formed by applying a front gate potential to this trenched places. Under the gates there is a donor layer and 2DEG is placed under the donor layer. For such a geometry Poisson and Thomas-Fermi equations are solved self-consistently in three dimensions (3D) so the electron density and the screened potential are obtained then the position and structure of ISs are determined.

It was first proposed by Chklovskii et al.[4] that the electrostatics should be considered for determining the edge states. They also proposed the compressible and incompressible strips which identify the location of the current flowing in a 2DEG in a strong magnetic field. Self-consistent generalizations of model of Chklovskii et al.[4] have been done within the Thomas-Fermi [5-7] and Hartree [8] approximations during the recent two decades.

The electrostatics of a 2DEG within TFP approach consists of a simultaneous solution of Thomas-Fermi and Poisson equations. In a realistic approximation, Poisson equation must be solved in 3D. Solutions of Poisson equation in 3D has been obtained under open boundary conditions and applied to a quantum Hall bar [9] and an AB interferometer [10]. However, in these studies the Poisson solved only once without considering the distribution of the charges in the 2DEG and the gate potential obtained in this way is assumed to be constant throughout the calculations. This approach is called as frozen gate approximation. In the rest of the calculations the electrostatic potential is obtained by using a 2D Green's function with periodic boundary conditions to solve the Poisson's equation.

In order to obtain a completely self consistent solution in 3D, a solution of the Poisson equation should be obtained at every step throughout the calculations. Recently such an approach was developed for quantum Hall bar geometry[11] and in this study we have extended that work to AB geometry.

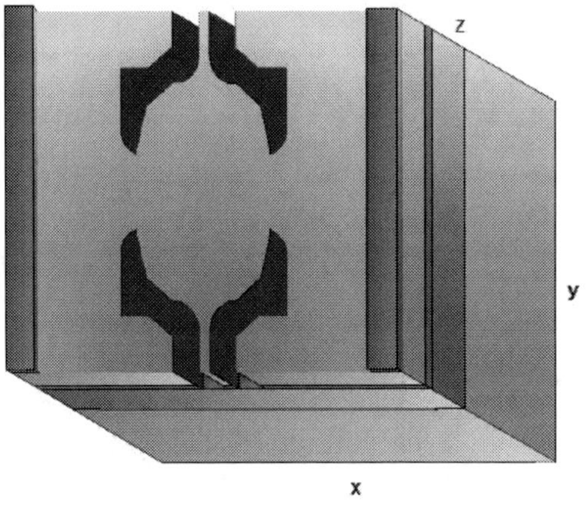

Figure 1. A schematic layout of an AB interferometer in 3D

2. Electrostatics of the system and numerical procedure

We can write 3D Poisson equation for the structure in Fig. 1 as

$$\nabla^2 V(x,y,z) = -\frac{\rho(x,y,z)}{\varepsilon} \tag{1}$$

here $V(x,y,z)$ is the screened potential, $\rho(x,y,z)$ is the charge density and ε is the dielectric constant of the material.

The Poisson equation is solved numerically in 3D using a multigrid technique. Because we solve the Poisson equation in 3D with all the gates, donors and 2DEG we do not introduce a frozen gate approximation. We use two bar shaped top gates on each side of the interferometer for confining the system. These bar shaped gates are chosen to be wide enough so that the potential heals under the gates, i.e. $\partial V / \partial x = 0$ at the edges.

For the electron density which is the other stand of the self-consistent calculation, we use Thomas-Fermi approximation. Within this approximation, in the presence of a strong magnetic field B we can write the electron density as

$$n_{el}(\vec{r}) = \frac{g_s}{2\pi l^2} \sum_{n=0}^{\infty} f\left[\frac{\left(E_n + V(r) - \mu\right)}{k_B T}\right] \tag{2}$$

where $l = \sqrt{\hbar / eB}$ is the magnetic length, μ is the chemical potential, E_n is the energy of the nth Landau level, and T is the temperature. The degeneracy of Landau levels are given by $\frac{g_s}{2\pi l^2}$, where $g_s = 2$ is the spin degeneracy.

3. Results and Discussion

The structure considered in this study is based on a recent experimental study of an AB interferometer fabricated on AlGaAs/GaAs heterojunction with etching techniques [3]. Here the gates are deposited in trenches. In our study for the similarity with experimental study we consider the trenched structure in Fig 1. The gates are obtained by filling these trenches down to 80 nm from the surface. The applied potential on all the gates is -0.7 V. We placed two gates to the edge of the structure in x direction. The widths of these edge gates are 50 nm, the length of the structure is 2000 nm in y direction. The 2DEG is at 210 nm below the surface. The average electron density n_e and the ionized donor density n_d are taken as equal $n_d = n_e = 0.97 \times 10^{11} \text{cm}^{-2}$ so that overall charge neutrality is maintained. In x, y and z directions the dimensions of the structure are taken 2600 nm, 2000 nm and 1000 nm, respectively. For numerical calculations a mesh is formed with 10 nm steps in all three directions.

Combining the Poisson and Thomas-Fermi equations one obtains a nonlinear differential equation for the electron density distribution. We have obtained a self consistent solution of this nonlinear equation using an iterative approach. The numerical procedure used to obtain the density distribution is as follows. First, we solve the equation to obtain electron density distribution for $T = 0$ and $B = 0$. Using this solution as the initial distribution, the electron density distribution is calculated iteratively for some sufficiently high temperature and zero magnetic field. Then we set the magnetic field and repeat iterative procedure for the self consistent solution. After this, the temperature is reduced slowly obtaining a self consistent density distribution at each temperature step until the system reaches to aimed temperature (T=1.4 K).

The ISs which are formed by completely filled Landau levels are located where the filling factor $v = n_{el} h / eB$ is equal to an even integer number. In order to obtain an AB like structure the ISs should form a circular path enclosing a magnetic flux. The results are illustrated Fig 2, where filling factors are shown as a function of the spatial coordinates for different magnetic field values. As shown in Fig. 2, for different values of magnetic field the ISs are easily seen in the electron distribution, the ISs are indicated with yellow colour and correspond to filling factor two. In Fig. 2(a) the magnetic field at 2.8 T, the ISs are observed but do not contact with each other. We do not expect observation of any interference pattern at this value of magnetic field. In Fig 2(b) the magnetic field strength is 2.9 T, here the ISs become closer, however there is still no overlap and again no interference is expected. Figs. 2(c) and 2(d) correspond to 3.0 T and 3.1 T respectively, at these values of magnetic field the ISs touch to each other and we expect observation of an interference pattern. If we increase the magnetic field still further then the ISs will split again (this time in the vertical direction) and again no interference is expected. The general behaviour of the ISs obtained in this study is consistent with the experimental findings of Camino et al. [3].

4. Conclusion

We have calculated the spatial electron distributions for a quantum Hall effect based electronic AB interferometer with using a self-consistent TFP approach. We determined the spatial distribution of ISs and discussed their relation with the interference phenomena.

Acknowledgments
This work is supported by Selcuk University BAP Grant No. 07101037.

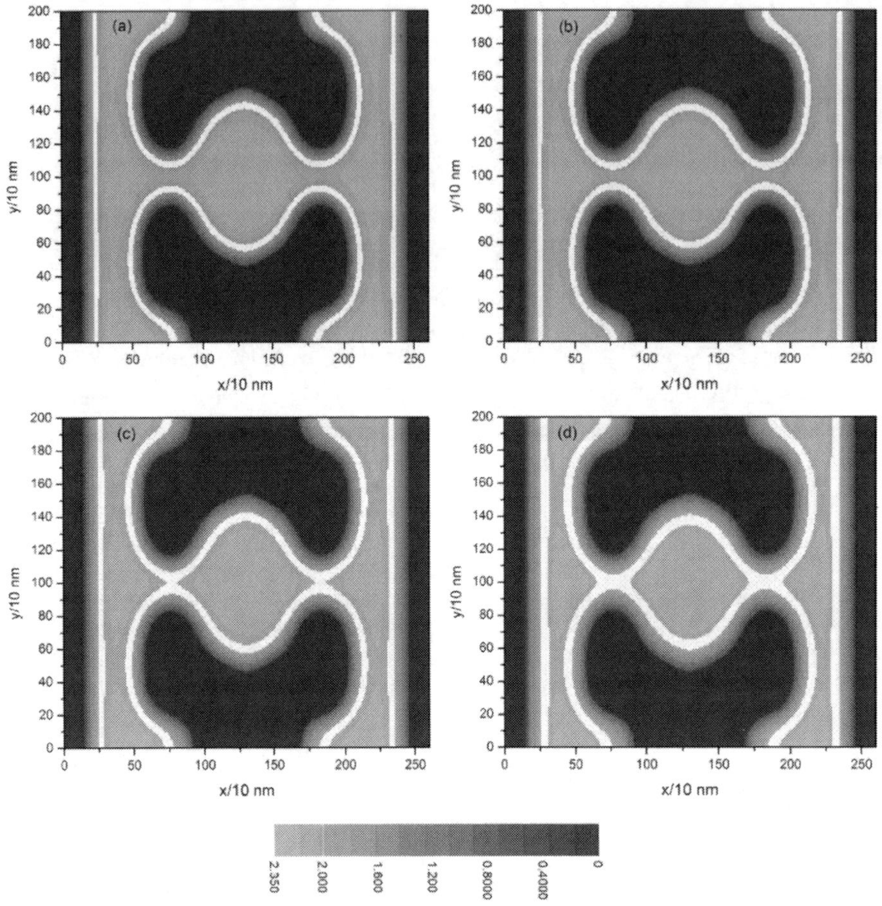

Figure 2. Contour graphs of the filling factors for different magnetic field values. The yellow colour indicates the ISs for filling factor two.

References

[1] Aharonov Y and Bohm D 1959 *Phys. Rev.* **115** 485-491
[2] Datta S 1995 *Electronic Transport in Mesoscopic Systems* (Cambridge: Cambridge University Press)
[3] Camino F E, Zhou W and Goldman V J 2005 *Phys. Rev.* B **72** 155313
[4] Chklovskii D B, Shklovskii B I and Glazman L I 1992 *Phys. Rev.* B **46** 4026
[5] Lier K and Gerhardts R R 1994 *Phys. Rev.* B **50** 7757
[6] Oh J H and Gerhardts R R 1997 *Phys. Rev.* B **56** 13519
[7] Güven K and Gerhardts R R 2003 *Phys. Rev.* B **67** 115327
[8] Siddiki A and Gerhardts R R 2004 *Phys. Rev.* B **70** 195335
[9] Arslan S, Cicek E, Eksi D, Aktas S, Weichselbaum A and Siddiki A 2008 *Phys. Rev.* B **78** 125423
[10] Cicek E, Mese A I, Ulas M and Siddiki A 2010 *Physica* E **42** 1095
[11] Kavruk A E 2010 *Investigation of quantum Hall effect in nanostructures* Ph.D. thesis (Turkey: Selcuk University)

Magneto-transport in the zero-energy Landau level of single-layer and bilayer graphene

U Zeitler[1], A J M Giesbers[2], H J van Elferen, E V Kurganova, A McCollam and J C Maan
Radboud University Nijmegen, Institute for Molecules and Materials, High Field Magnet Laboratory, Toernooiveld 7, 6525 ED Nijmegen, The Netherlands.

Abstract. We present recent low-temperature magnetotransport experiments on single-layer and bilayer graphene in high magnetic field up to 33 T. In single layer graphene the fourfold degeneracy of the zero-energy Landau level is lifted by a gap opening at filling factor $\nu = 0$. In bilayer graphene, we observe a partial lifting of the degeneracy of the eightfold degenerate zero-energy Landau level.

1. Introduction

Graphene is the purely two dimensional form of carbon arranged in a honeycomb-lattice [1]. Initially, this material was believed not to be stable thermodynamically [2]. However, in 2004 Novoselov *et al.* succeeded in producing an ambipolar field effect transistor made from one layer of graphene using mechanical exfoliation of graphite [3].

Maybe even more spectacular than the sole existence of a material which should not exist are its unique physical properties [1]. Charge carriers in graphene behave relativistically like massless chiral Dirac fermions rather than massive electrons in conventional materials. These relativistic properties of graphene are revealed most clearly in high magnetic fields where the energy spectrum splits up into non-equidistant Landau levels with energies given by

$$E_N^{SLG} = \mathrm{sgn}(N)\sqrt{2\hbar c^2 eB|N|} \quad \text{and} \quad E_N^{BLG} = \mathrm{sgn}(N)\frac{\hbar eB}{m^*}\sqrt{N(N-1)} \qquad (1)$$

for single-layer graphene (SLG) [4] and bilayer graphene (BLG) [5], respectively. The velocity $c = \sqrt{3}\gamma_0 a / 2\hbar \approx 10^6$ ms^{-1} is given by the intra-layer coupling $\gamma_0 = 3.2$ eV and the lattice constant a = 2.46 Å. The effective mass m^* in BLG is determined by c and the inter-layer coupling $\gamma_1 = 0.4$ eV as $m^* = \gamma_1 / c^2 \approx 0.035\, m_e$. In both cases, N is a negative integer for holes and positive for electrons. Due to these relativistic Landau-level structures, the well-known integer quantum Hall effect (IQHE) [6] observed in conventional two-dimensional electron systems transforms to a relativistic half-integer quantum Hall effect in SLG [7,8] and an new type of unconventional IQHE in BLG [9].

[1] To whom any correspondence should be addressed, E-mail: u.zeitler@science.ru.nl
[2] Present address: Max Planck Institute for Solid State Research, Heisenbergstrasse 1, 70569 Stuttgart, Germany

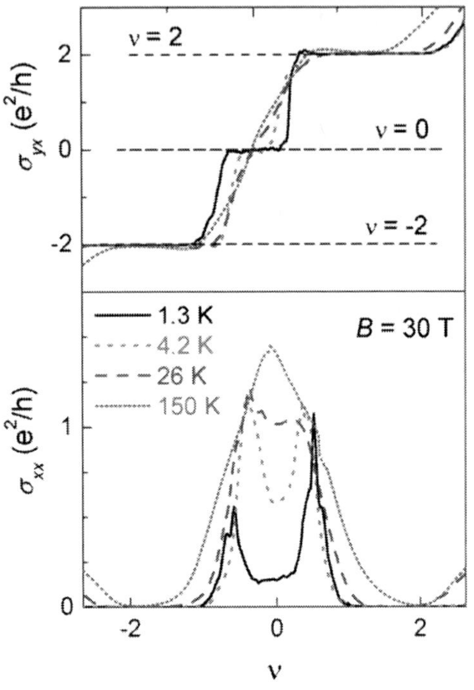

Figure 1. Hall conductivity σ_{yx} (top) and longitudinal conductivity σ_{xx} (bottom) in the zero-energy Landau level of a SLG sample at B = 30 T and different temperatures.

A particularity of this Landau level spectrum is the existence of a Landau level at zero energy (independent on the magnetic field) which is shared equally between electrons and holes. In SLG this level is fourfold degenerate, just like all other Landau levels with $|N| \geq 1$. In BLG, the zero-energy Landau level is eightfold degenerate whereas higher Landau levels with $|N| \geq 2$ are fourfold degenerate.

In this paper we present recent magnetotransport data on SLG and BLG performed at the High Field Magnet Laboratory in Nijmegen. For SLG we will show a narrowing of the this level in high fields [10] and the opening of a gap at $\nu = 0$ [11]. In BLG we start to lift the eightfold degeneracy in the highest-fields and observe minima in the conductance at $\nu = 3$ and 2.

2. Gap opening at ν = 0

The fourfold degeneracy of the zero-energy Landau level around the charge neutrality point (CNP) is visualized in magnetotransport experiments by a peak in ρ_{xx} at $\nu = 0$ neighboured by zero-minima at $\nu = \pm 2$ [7,8]. At the same time, the Hall resistivity ρ_{xy} passes through zero at $\nu = 0$ from ρ_{xy} ($\nu = -2$) = $-h/2e^2$ to ρ_{xy} (ν = 2) = $+h/2e^2$. Using the tensor inversions $\sigma_{xx} = \rho_{xx}/(\rho_{xx}^2 + \rho_{xy}^2)$ and $\sigma_{yx} = -\rho_{xy}/(\rho_{xx}^2 + \rho_{xy}^2)$ this implies a maximum in the conductivity σ_{xx} accompanied by a zero-crossing of the Hall conductivity σ_{yx}, a behaviour which is indeed found back in the high temperature data of Figure 1.

The occurrence of a zero ρ_{xy} at the CNP already proposes that the lowest Landau level of graphene has to be treated as a (partially) compensated semiconductor with a *finite* number of electrons and holes present at the CNP. In contrast, for a conventional system, with one type of carriers present simultaneously, the carrier concentration at the CNP is zero and, as a consequence, ρ_{xy} diverges to $\pm\infty$ when approaching the CNP from the electron side or from the hole side, respectively.

We will first concentrate on the high-field behaviour of ρ_{xx}, an issue still subject to intensive debate [12]. Experiments performed in the zero-energy Landau level of high-mobility SLG (made from Kish-graphite) have shown a lifting of its chiral degeneracy by the appearance of a $\nu = 1$ state [13, 14] and recently also the lifting the eightfold degeneracy in BLG has been observed [15]. Most other experiments, however, did not find a $\nu = 1$ state in SLG [11,16,17]. Possibly, the disorder in these experiments was still too large, preventing a spontaneous symmetry breaking in the lowest Landau level [18]. Instead, only a divergence of ρ_{xx} at the CNP is observed.

With the above tensor inversions a diverging ρ_{xx} at the CNP, together with a zero crossing of ρ_{xy}, inevitably leads to a zero minimum in σ_{xx} and a zero-value quantized plateau in σ_{yx}. Figure 1 visualizes the appearance of such a $\nu = 0$ quantum-Hall state where the Hall conductivity σ_{yx} and the longitudinal conductivity σ_{xx} of a SLG sample, deposed on a Si/SiO_2 substrate, are plotted as a function of filling factor. We performed these experiments for different temperatures at a magnetic field of 30 T [11]. The conductivities shown were calculated by tensor inversion from the measured resistivity ρ_{xx} and ρ_{xy} in a Hall-bar sample. The filling factor was varied by sweeping the carrier concentration at a constant magnetic field using the doped Si of the Si/SiO_2 substrate as a back-gate. The zero-energy Landau level is completely empty at $\nu = -2$ and is completely filled at $\nu = 2$.

As can been seen in the figure, σ_{xx} starts to develop a zero-minimum with decreasing temperature and a quantized zero-plateau appears in σ_{yx}. This new quantum Hall state, however is quite different compared to the conventional QHE [19]. It is caused by the zero crossing of ρ_{xy} at the CNP together with a diverging ρ_{xx}, whereas conventionally $\rho_{xy} \gg \rho_{xx}$ and quantum Hall states are characterized by a zero-minima of *both* the resistivity ρ_{xx} and the conductivity σ_{xx} and by *non-zero* quantized plateaus in ρ_{xy} and σ_{yx}.

The nature of the new $\nu = 0$ quantum Hall state can be elucidated further by investigating the temperature dependence of the σ_{xx} minimum at the CNP. It shows an activated behaviour, with a gap roughly scaling linearly with magnetic field [11]. This observation might suggest that the splitting is spin related. And indeed, comparison with a simple Zeeman splitting (using a free-electron g-factor $g = 2$) and an offset due to a final level width yield a reasonable agreement between this simple expectation and the experimentally measured gaps.

3. Lifting of the eightfold degeneracy in BLG

Having observed a splitting of the zero-energy Landau level in SLG it is certainly of interest to also investigate the transport properties of BLG. Experimental results on a bilayer sample are shown in Figure 2. In high magnetic fields ($B > 20$ T) and at low temperatures ($T < 1$ K) a partial lifting of the eightfold degenerate zero-energy Landau level is indeed observed and clearly pronounced minima in the conductivity σ_{xx} appear at filling factors $\nu = \pm3$ and ±2. They are accompanied by quantum Hall plateaus in $\sigma_{yx} = \nu e^2/h$ (not shown). However, the quality of the samples is not yet high enough to observe a full splitting of the Landau level, and, interestingly no $\nu = 0$ quantum-Hall state appears yet, or, more specifically, no strongly diverging ρ_{xx} can be observed at the CNP. Apparently, no narrowing of the zero-energy Landau level as in SLG [10,20] takes place and the finite level width still masks the effect of a possible Zeeman splitting.

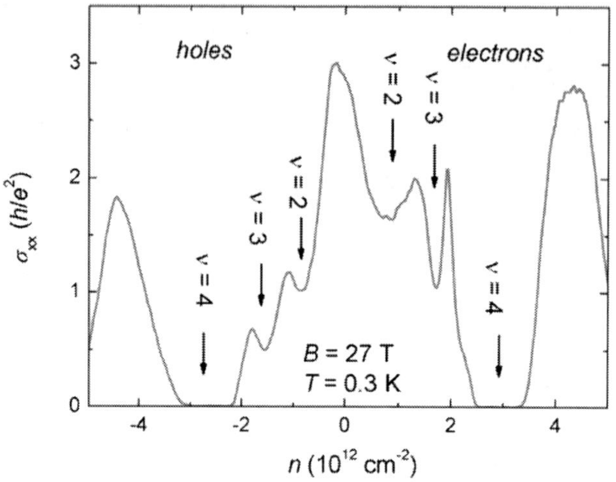

Figure 2. Conductivity σ_{xx} in the zero-energy Landau level of a BLG sample at B = 30 T as function of carrier concentrations. At ν = ±4 the level is entirely full or empty, respectively. The arrows at ν = ±3 and ±2 indicate how the eightfold degeneracy of the Landau level starts to be lifted.

4. Acknowledgements

We would like to thank K.S. Novoselov and A.K. Geim (Manchester) for providing samples and A. Veligura and B.J. van Wees (Groningen) for valuable assistance in fabricating samples at the University of Groningen.

This work is part of the research programme of the Foundation for Fundamental Research on Matter (FOM), which is part of the Netherlands Organisation for Scientific Research (NWO). Part of this work was also supported by EuroMagNET II under the EU contract number 228043.

5. References

[1] Geim AK and Novoselov KS 2007 Nature Materials **6** 183.
[2] Mermin ND 1968 Phys. Rev. **176** 250.
[3] Novoselov KS et al. 2004 Science **306** 666.
[4] McClure JW 1956 Phys. Rev. **104** 666.
[5] McCann E and Fal'ko VI 2006 Phys. Rev. Lett. **96** 086805.
[6] v. Klitzing K, Dorda H and Pepper M 1980 Phys. Rev. Lett. **45** 494.
[7] Novoselov KS et al. 2005 Nature **438** 197.
[8] Zhang Y, Tan Y, Stormer HL and Kim P 2005 Nature **438**, 201.
[9] Novoselov KS et al. 2006 Nature Physics **2** 177.
[10] Giesbers AJM et al. 2007 Phys. Rev. Lett. **99** 206803.

[11] Giesbers AJM et al. 2009 Phys. Rev. B **80** 201403(R).
[12] Castro Neto AH *et al.* 2009 *Rev. Mod. Phys.* **81** 109.
[13] Zhang Y *et al.* 2006 *Phys. Rev. Lett.* **96** 136806.
[14] Jiang Z, Zhang Y, Stormer HL, and Kim P 2007 *Phys. Rev. Lett.* **99**, 106802.
[15] Zhao Y, Cadden-Zimansky P, Jiang Z. and Kim P 2010 *Phys. Rev. Lett.* **104** 066801.
[16] Checkelsky JG, Li L, and Ong NP 2008 *Phys. Rev. Lett.* **100** 206801.
[17] Amado M *et al.* 2010 *New Journal of Physics* **12** 053004.
[18] Nomura K and MacDonald AH 2006 *Phys. Rev. Lett.* **96**, 256602.
[19] Das Sarma S and Yang K 2009, *Solid State Commun.* **149** 1502.
[20] Giesbers AJM et al. 2009 *Phys. Rev. B* **80** 241411(R).
[21] Bolotin KI *et al.* 2008 *Solid State Commun.* **146** 351.

High-field magneto-transport at charge neutrality point in monolayer graphene

T Yoshida and K Oto

Graduate School of Science, Chiba University, 1-33 Yayoi, Inage, Chiba 263-8522, Japan

E-mail: k-oto@faculty.chiba-u.jp

Abstract. We have investigated the magneto-transport around filling factor ν=0 in monolayer graphene sheet. Zero Hall plateau of Hall conductivity and minimum of longitudinal conductivity have been clearly observed despite the highly fluctuated behaviour of magneto-resistance above 10 T. These phenomena can be understood by the transport of counter-propagating electron and hole edge states at the charge neutrality point. We tried to explain the origin of remarkable fluctuations in magneto-resistance by the energy relaxations among the counter-propagating edge channels.

1. Introduction

Recently, the unique phenomena in a monolayer graphene around the charge neutrality point (CNP) have been reported [1, 2]. In the monolayer graphene, the lowest Landau level (LL) appears at the CNP corresponding to the zero energy in magnetic fields. In high magnetic fields, the lowest LL shows spin splitting and the confinement potential at the edge of graphene sheet splits the remaining valley degeneracy. Hence, the two band branches due to the valley splitting at the edge can be regarded as an electron-like and a hole-like edge states which are formed only in the lowest LL. In this case, the fluctuated magneto-resistance is observed, which is similar behavior reported in the electron-hole coexistence system, such as an InAs/GaSb type-III quantum well (QW) [3]. In the InAs/GaSb QW, the observed highly fluctuated magneto-resistance was explained by the energy relaxation of current carrying electron and hole edge channels. Motivated by the analogy of the properties of the electron transport, we have discussed the origin of the fluctuations in the magneto-resistance at the CNP in the monolayer graphene.

2. Sample

Monolayer graphene sheet was made from Kish graphite (Covalent material co.) by using a micro-mechanical cleavage method as shown in Ref. [4], and was mounted on an n-doped Si/SiO_2 (280 nm) substrate, which was used as a back-gate to control the carrier density by applying gate voltage (V_g). The annealing process for cleaning graphene sheet was omitted. A Hall bar shaped sample with Au/NiCr electrodes was fabricated by using an electron beam lithography, as shown in Fig. 1(a).

The magneto-transport measurement was performed in a top loading ^{3}He-^{4}He dilution refrigerator with a superconducting magnet up to 15 T, by using low frequency (12.5 Hz) ac excitation current of 30 nA. The carrier density n_s can be adjusted from -2.0×10^{16} (hole) to 8.0×10^{15} (electron) m^{-2} by tuning the back-gate voltage, as shown in Fig. 1(b). The dotted line in Fig. 1(b) is calculated carrier

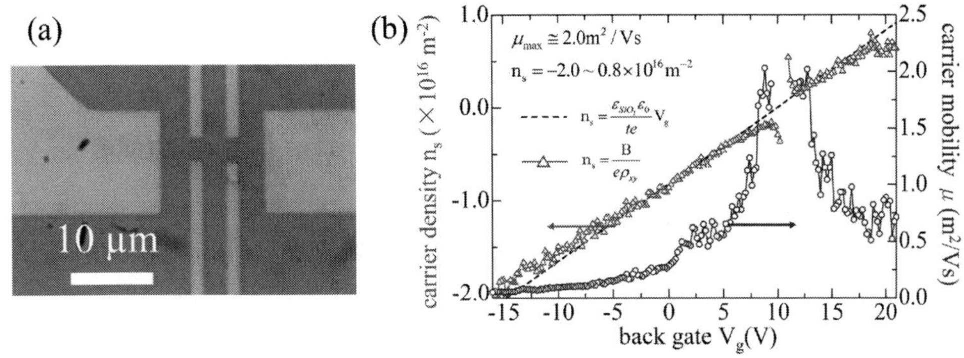

Fig. 1. (a) The optical microscope image of monolayer graphene sample. (b) Carrier density and mobility as a function of V_g evaluated from the low-field transport. The dotted line is a carrier density n_g calculated from the geometry of the sample.

density with the simple capacitor model of $n_g=\varepsilon_r\,\varepsilon_0\,V_g/te$, where the dielectric constant ε_r of SiO_2 is 4.2 and t is the thickness of SiO_2 layer. The calculated n_g agrees with the measured one, as shown in Fig. 1(b). The typical carrier mobility μ based on the Drude model was below ~2.0 m^2/Vs around the CNP.

3. Results and discussion

Figure 2 (a) shows Hall resistivity (ρ_{xy}) and longitudinal resistivity (ρ_{xx}) as a function of V_g at B= 4 T. Half-integer quantum Hall effect (QHE) and Shubnikov de-Haas oscillations corresponding to the filling factor ν=±2, 6, 10, 14, which are unique numbers in the monolayer graphene QHE plateaus, are clearly observed. In Fig. 2(b), the analysis of LL index plots at V_g=−8 and −12 V pass on the LL index n=1/2, which indicates the phase shift due to the Berry's phase π.

Next, we focus on the resistivity in higher magnetic field at B=15 T and at low temperature T=0.5 K, as shown in Fig. 3 (a). The observed ρ_{xy} and ρ_{xx} show remarkable fluctuations around the CNP (V_g=9.8 V) corresponding to the filling factor ν=0, and these fluctuations are reproducible. On the other hand, the Hall conductivity σ_{xy} and longitudinal conductivity σ_{xx} obtained from ρ_{xy} and ρ_{xx} data show the zero Hall plateau of σ_{xy} and the minimum of σ_{xx}, as shown in Fig. 3 (b). The value of σ_{xx} minimum and the width of the zero Hall plateau strongly depend on the value of magnetic field.

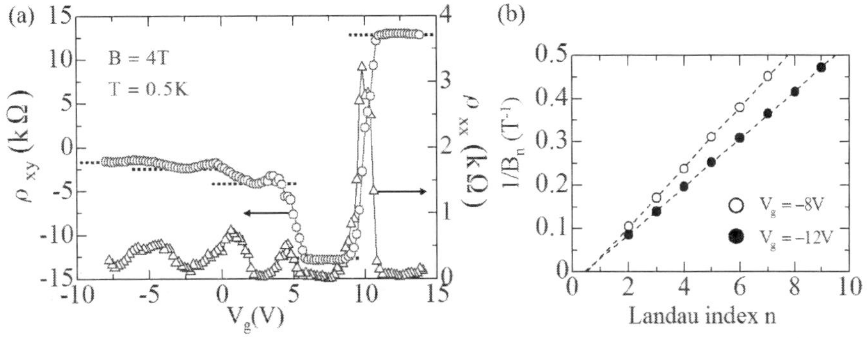

Fig. 2. (a) Resistivity at B=4 T as a function of V_g. (b) Landau level index plot of ρ_{xx}.

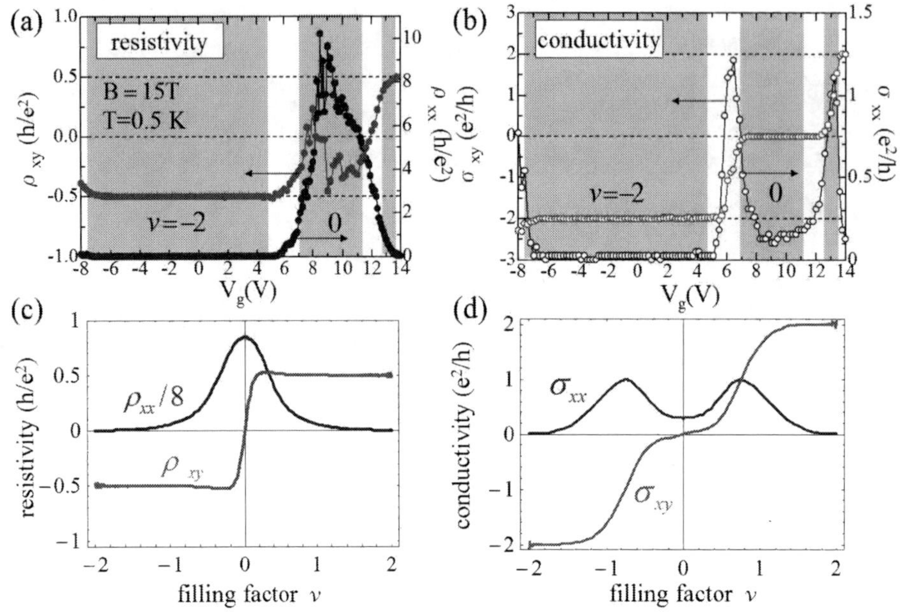

Fig. 3. Experimental (a,b) and simulated (c,d) results of the magneto-resistivity and magneto-conductivity in the monolayer graphene around the CNP at 15 T.

By using the counter-propagating edge (CPE) channel [1], we can explain the relation between Fig. 3(a) and 3(b) with the calculated curves as shown in 3(c) and 3(d). In the CPE channel model, the magneto-transport is influenced by the intra-edge scattering γ between two counter-propagating edge channels and the inter-edge scattering rate g through the edge-bulk-edge path. The analysis with the CPE channel model gives the scattering parameters $w\gamma$=15, gw=1.5 in our system (w: channel width), which roughly agree with ones in ref. [1]. Note that the fluctuation pattern of the magneto-resistance indicates the existence of edge channel branches at the CNP condition, which supports the spin splitting of the lowest LL in the bulk regime.

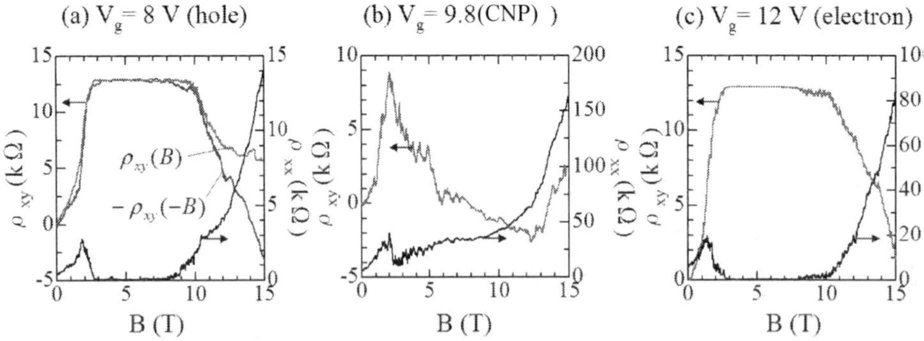

Fig. 4. The magneto-resistance at (a) V_g=8 V(hole), (b) 9.8 V(CNP), and (c) 12 V(electron), measured at 0.5 K. Two lines of ρ_{xy} in (a) are measured in the reversal magnetic fields.

Figure 4 (a)~(c) show the magneto-resistivity ρ_{xy} and ρ_{xx} at the low carrier density condition of V_g=8.0, 9.8, and 12.0 V, which correspond to $n_s= -2.4\times 10^{15}$ (hole), 0.0 (CNP), and $+2.1\times 10^{15}$ (electron) m^{-2}, respectively. Each ρ_{xy} shows small and reproducible fluctuations, and ρ_{xx} drastically increases in high magnetic fields. As shown in Fig. 4 (a), the ρ_{xy} indicates normal behavior with the reversal of magnetic fields; $\rho_{xy}(B) = -\rho_{xy}(-B)$ at $|B|<10$ T, on the other hand, at $|B|>10$ T, the remarkable asymmetry $\rho_{xy}(B) \neq -\rho_{xy}(-B)$ was observed. These fluctuation and asymmetry were observed clearly around the CNP condition and at low temperatures below 1 K.

We tried to understand the fluctuation behavior in the magneto-resistance near the CNP by using the fluctuated edge channel model based on the CPE channel [1] and the concept of electron transport in the electron-hole coexisting type-III QW [3]. Figure 5 shows the schematic cross-sectional view of the LL along the sample width direction. The edge channels in the graphene sheet are expected to be spatially fluctuated (winding) as schematically shown in Fig. 5. The electron scattering rate of the inter-edge channels is enhanced at the overlapping points of counter-propagating edge channels, where the chemical potential for each channel should be equilibrated. The number and the position of the overlapping point of the CPE channels may change with sweeping V_g and/or magnetic fields, since the spatial distribution of edge channels also changes as shown in Fig. 5. Thus the reproducible magneto-resistance fluctuation observed around the CNP condition can be understood by considering the distribution of the potential disorder in the graphene sample and the electron scattering among the overlapping CPE channels due to the potential fluctuation near the sample edge. In the high magnetic fields more than 10 T, the Hall resistance also shows fluctuation and it does not show symmetry (Fig. 4(a)) for magnetic field reversal $\rho_{xy}(B) \neq -\rho_{xy}(-B)$. Moreover, the observed ρ_{xy} deviates remarkably from the normal value estimated from the carrier density. These results are consistent with the electron transport model around the CNP in high magnetic fields, as mentioned above. Around the CNP, the distribution of the electrochemical potential along the sample edge is mainly governed by the energy equilibration among CPE channels due to the potential fluctuation in the graphene sample. So, the measured ρ_{xy} and ρ_{xx} do not show the 'bulk resistivity' but show the reproducible fine fluctuations as sweeping V_g and/or B.

Finally, we discuss about insulating behavior in ρ_{xx} in high magnetic fields and at low temperature. Figure 6 shows rapid increase from 6 to 200 kΩ in the resistivity ρ_{xx} with increasing magnetic fields measured at the gate voltage of V_g=9.8 V (CNP). The bulk conductivity σ_{xx} at the CNP decreases with

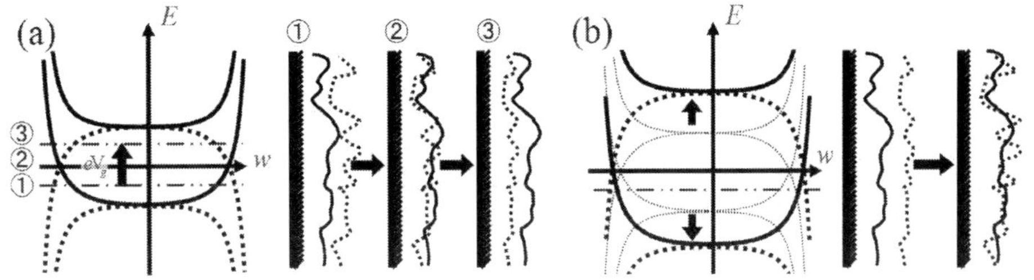

Fig. 5. The schematic explanation of the lowest LL and the edge channels at the CNP. (a) Electron and hole edge states (solid and broken curves) spatially overlap at the CNP(②) by tuning V_g. The overlapped CPE channels can be regarded as 'micro-loop array' and the adjacent micro-loops are weakly coupled each other at the crossing point of the CPE channels. (b) In the high magnetic fields, the overlapped CPE channels at the CNP are pushed to the sample boundary, where the energy equilibration length of the edge channels may be short.

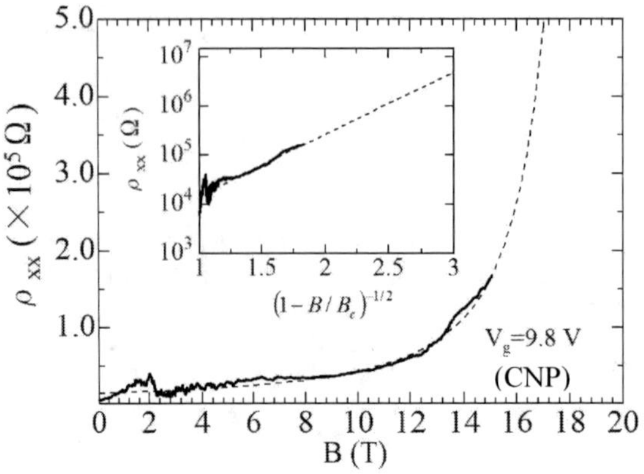

Fig. 6. Field induced insulating behavior of ρ_{xx} at the CNP at 0.5 K. Inset: ρ_{xx} as a function of $(1-B/B_c)^{-1/2}$, where the B_c is characteristic field.

lowering temperatures below 1 K, and the highly magnetic field dependent gap energy at the CNP was studied by the analysis of temperature dependence of the σ_{xx} [5]. Our experimental results also show the insulating behavior of ρ_{xx} at the CNP in high magnetic fields ($|B|$>10 T) which is understood by considering both the decreasing of the residual bulk σ_{xx} and the strong energy equilibration in the overlapped CPE channels at the CNP. Note that the energy equilibration in the CPE channels strongly reduces the electron transport through the CPE channels.

We also tried to plot our ρ_{xx} data taken at the CNP with the calculated curve by the theory of Kosterlitz-Thouless (KT) type function (inset of Fig. 6). The observed ρ_{xx} data can be fitted to the KT-type function with the parameter of B_c=21.4 T. Similar insulating behavior was reported, in which the possibility of the KT transition at the CNP was discussed [2]. And the origin of KT transition-like behavior [6,7] in the high mobility graphene was also discussed, however, more experimental and theoretical efforts are still needed to understand both the magneto-resistance fluctuation and the field-induced insulating behavior at the CNP in the low mobility samples.

Acknowledgements
We would like to thank Mr. Y. Shimane for experimental support. This work is partially supported by a Grant-in-Aid for Scientific Research on priority Areas "High Field Spin Science in 100T" (No.451), a Global COE in Physics at Chiba University (G03), and Scientific Research C (No. 21540315) from the MEXT, Japan.

References
[1] Abanin D A, Novoselov K S, Zeitler U, Lee P A, Geim A K, and Levitov L S, 2007 *Phys. Rev. Lett.* **98** 196806
[2] Checkelsky J G, Li L, and Ong N P, 2008 *Phys. Rev. Lett.* **100** 206801
[3] Takashina K, Nicholas R J, Mason N J, Maude D K, and Portal J C, 2003 *Physica E* **20** 160
[4] Novoselov K S, Geim A K, Morozov S V, Jiang D, Zhang Y, Dubonos S V, Grigorieva I V, and Firsov A A, 2004 *Science* **306** 666
[5] Giesbers A J M, Ponomarenko L A, Novoselov K S, Geim A K, Katsnelson M I, Maan J C, and Zeitler U, 2009 *Phys. Rev.B* **80** 201403
[6] Nomura K, and MacDonald A H, 2006 *Phys. Rev. Lett.* **96** 256602
[7] Nomura K, Ryu S, and Lee Dung-Hai, 2006 *Phys. Rev. Lett.* **103** 216801

Observation of quantum Hall e ect in mono- and bi-layer graphene using pulse magnet

S. Masubuchi[1,2], K. Suga[3], K. Kindo[3], S. Takeyama[3], and T. Machida[1,2,4]

[1]Institute of Industrial Science, University of Tokyo, Tokyo 153-8505, Japan
[2]Institute for Nano Quantum Information Electronics, University of Tokyo, Tokyo 153-8505, Japan
[3]Institute for Solid State Physics, University of Tokyo, Chiba 277-8581, Japan
[4]PRESTO, Japan Science and Technology Agency, Saitama 332-0012, Japan

E-mail: msatoru@iis.u-tokyo.ac.jp

Abstract. We report on the magnetotransport measurements of mono- and bi-layer graphene in pulsed high magnetic fields up to $B = 53$ T. In a mono-layer graphene, the Hall resistance R_{H} is quantized to $R_{\mathrm{H}} = (h/e^2)\nu^{-1}$ with $\nu = 2$, 6, and 10 for either hole or electron doping. In a bi-layer graphene, R_{H} is quantized to $R_{\mathrm{H}} = (h/e^2)\nu^{-1}$ with $\nu = 4, 8$, and 12. These results indicate that the precise magnetotransport measurements of the graphene can be conducted in the environment of pulse magnetic fields.

1. Introduction

The pulsed high magnetic field technique has been an effective experimental tool for the studies of the integer and fractional quantum Hall effect (QHE) in conventional semiconductor two-dimensional electron systems [1-3]. This is because strong magnetic fields B 730 T can be obtained by using the pulse magnet [4] whereas the static magnetic fields are limited to B 45 T. During the past few years, the QHE in mono-layer and bi-layer graphene have been attracting considerable interest [5-7]. In a mono-layer graphene, the charge carriers have a linear energy spectrum and show Berry's phase of π. The Landau quantization of those carriers in high magnetic fields results in the half-integer QHE with the shifted positions of the quantum Hall plateaus [6]. In a bi-layer graphene, its parabolic band structure and the chiral nature of charge carriers result in the integer QHE but the zero-level energy gap missing [8]. Recently, by using the static high magnetic field techniques, the lifting of valley and spin degeneracy of the Landau levels and the fractional QHE were reported [9-12]. For the further studies of the QHE in higher magnetic fields, it is of great importance to establish the measurement techniques for the QHE in graphene using pulse magnet.

In this work, we report on the magnetotransport measurements in mono- and bi-layer graphene in pulsed magnetic fields up to $B = 53$ T. In a mono-layer graphene, the Hall resistance is quantized into $R_{\mathrm{H}} = (h/e^2)\ ^{-1}$ with integer values $= \pm 2, \pm 6$, and ± 10 for either hole or electron doping, indicating the observation of half-integer QHE. In a bi-layer graphene, the R_{H} is quantized to $R_{\mathrm{H}} = (h/e^2)\ ^{-1}$ with $= \pm 4, \pm 8$, and ± 12. This is the observation of the integer QHE of massive chiral fermions in the bi-layer graphene. The measurement technique,

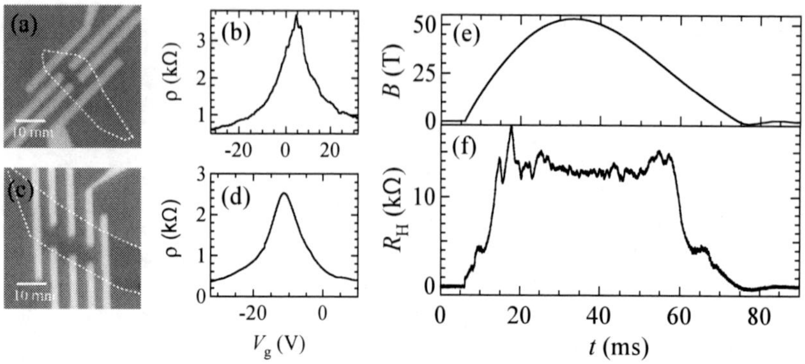

Figure 1. The optical images of (a) the mono-layer and (c) the bi-layer graphene. The outlines of the graphene flakes are highlighted by the dashed lines. The resistivity   of the (b) mono-layer and (d) bi-layer graphene as a function of the V_g at $T = 4.2$ K. (e) The magnetic field B generated by the pulse magnet. (f) Time evolution of the Hall resistance R_H of the mono-layer graphene with $V_g = 30$ V at $T = 4.2$ K.

established in this work, can be used effectively for the studies of QHE in mono- and bi-layer graphene in high magnetic fields.

2. Experiment

Mono-layer and bi-layer graphene flakes used in this study were extracted from Kish graphite and deposited onto a 300-nm-thick SiO_2 layer on top of heavily doped Si wafer using the conventional mechanical exfoliation technique of graphite. A large graphene flakes were selected using an optical microscope and the multiple electrodes were fabricated using electron-beam lithography (Elionix ELS 7500), followed by electron-beam evaporation of Au/Ti (40 nm /4 nm) [Fig. 1(a)]. The charge density n of the graphene flake was tuned by applying a gate bias voltage V_g to the heavily doped silicon wafer, which serves as global back gate [13, 14]. In order to remove the surface impurities, the devices were annealed in a measurement cryostat at T 400 K in vacuum (P 1×10^{-2} Pa) for several hours. Fig. 1 shows the resistivity   of the (b) mono-layer and (d) bi-layer graphene as a function of the V_g. The   shows a peak at (b) $V_g = 5$ V and (d) $V_g = -12$ V, indicating the charge neutrality points V_{Dirac}. The electron mobility of the mono-layer and bi-layer graphene are (b) $\mu = 4000$ cm^2/Vs and (d) $\mu = 3000$ cm^2/Vs at $V_g = 32$ V, respectively.

The magnetic field was generated by a pulse magnet immersed in liquid nitrogen. The magnet was energized by the 1 MJ capacitor bank. The value of the magnetic field was obtained by integrating the emf signal from a pick-up coil [Fig. 1(e)]. During the pulse, the voltage signals V_{xy} were recorded with DC currents of $I = \pm 200$ nA. The voltage signals were subtracted each other to eliminate the field-induced background voltage. The R_H was obtained by dividing the voltage signal by the current I [Fig. 1(f)]. The trace of R_H vs. B was studied in a downward sweep because the magnetic field changes more slowly. Here, we emphasize the technical di culties in measuring the QHE under the pulsed magnetic fields. When a pulsed magnetic field is applied, an additional transient current is induced in the sample and measurement circuits. This transient current results in the deformation of quantum Hall plateaus as observed in the conventional semiconductor two-dimensional electron gas systems. To obtain the well-developed Hall plateaus, Burgt *et al.* numerically compensated for the transient current flow in the device

Journal of Physics: Conference Series **334** (2011) 012037 doi:10.1088/1742-6596/334/1/012037

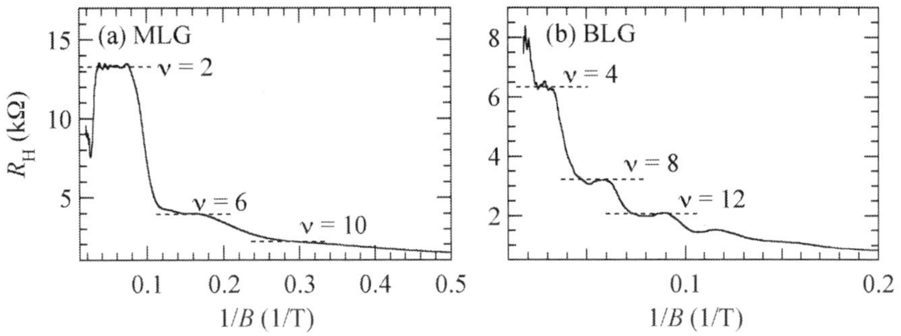

Figure 2. Hall resistance $R_{\rm H}$ as a function of the inverse of magnetic field B^{-1} measured for (a) mono- and (b) bi-layer graphene for gate-bias voltages (a) $V_{\rm g}$ = 17.5 and (b) -60 V at T = 4.2 K.

by assuming the appropriate capacitance [1]. Kido *et al.* used an active shielding technique; the voltages of the outer and inner conductors of the coaxial cables were equalized using the buffer amplifiers of unity gain thereby eliminating the capacitance of the circuits [15]. In both cases the reduction of the capacitance was a key for suppressing the deformation of quantum Hall plateaus. In our study, we employed copper wires inside the measurement cryostat thus reducing the capacitance of the circuit as less as possible, and all the other electric cables and measurement equipment including a computer were installed in a shield box set near the magnet to get rid of the electronic noise. This method has given su cient high signal-to-noise ratio and suppressed the deformation effects.

3. Results and discussion

Figure 2 shows the $R_{\rm H}$ as a function of the inverse of the magnetic field B^{-1} measured in (a) mono-layer and (b) bi-layer graphene for gate-bias voltages (a) $V_{\rm g}$ = 17.5 and (b) -60 V at T = 4.2 K. In these measurements, the magnetic field was swept from B = 53 T to 0 T. In the mono-layer graphene, the Hall plateaus are visible at $R_{\rm H}$ = 13.3, 4.1, and 2.2 kΩ [dashed lines in Fig. 2 (a)], and these values correspond to the quantum Hall resistances of $R_{\rm H} = h/2e^2$, $h/6e^2$, and $h/10e^2$ within 10% accuracy. In the bi-layer graphene, the Hall plateaus are observed at $R_{\rm H}$ = 6.7, 3.1, and 2.2 kΩ [dashed lines in Fig. 2 (b)], and these values correspond to the quantum Hall resistances of $R_{\rm H}$ = $-h/4e^2$, $-h/8e^2$, and $-h/12e^2$, respectively. This observation indicates the integer QHE of the massive chiral fermions in the bi-layer graphene in pulsed high magnetic fields. Here we emphasize that this is the first experimental observation of the integer QHE in the bi-layer graphene in pulsed high magnetic fields.

Next, we study the $R_{\rm H}$ vs. B up to 53 T at the various gate bias voltages. Fig. 3 shows the $R_{\rm H}$ vs. B measured for the mono-layer graphene at (a) $V_{\rm g}$ = 17.5 V, (b) 30.0 V, (c) 40.0 V, and (d) 48.0 V. The carrier densities in these conditions are calculated by using the standard formula for the graphene on SiO_2 layer: $n = \alpha(V_{\rm g} - V_{\rm Dirac})$ with the constant $\alpha = 7.2 \times 10^{10}$ cm^{-2}V^{-1}, as (a) $n = 9.0 \times 10^{11}$, (b) 1.8×10^{12}, (c) 2.5×10^{12}, and (d) 3.1×10^{12} cm^{-2}, respectively. The filling factor is proportional to the carrier density and the inverse of the magnetic field $=$ $n\phi_{0B^{-1}}$, where $\phi_0 = 4.14 \times 10^{-15}$ Tm^2 is the flux quantum. By using this relation we calculated the expected positions of the quantum Hall plateaus for the filling factors $=$ 2, 6, and 10 as indicated by the red, green, and black arrows in Fig. 3, respectively. The Hall plateaus are

Journal of Physics: Conference Series **334** (2011) 012037 doi:10.1088/1742-6596/334/1/012037

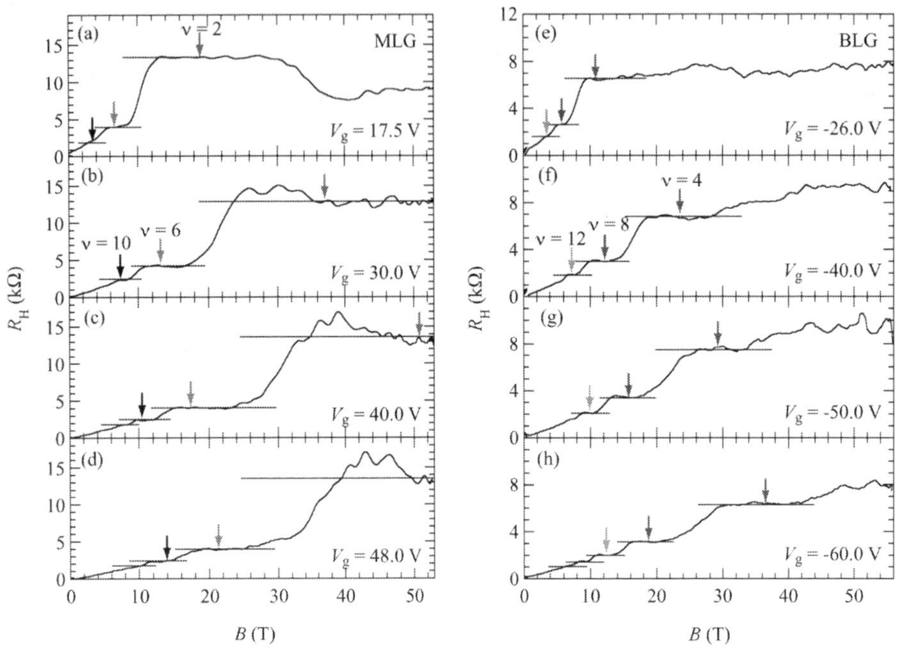

Figure 3. (Color online) Hall resistance $R_{\rm H}$ of a mono-layer graphene as a function of magnetic field B for gate bias voltages (a) $V_{\rm g}$ = 17.5, (b) 30.0, (c) 40.0, and (d) 48.0 V at T = 4.2 K. The $R_{\rm H}$ of a bi-layer graphene as a function of B for gate bias voltages (e) $V_{\rm g}$ = -26.0, (f) -40.0, (g) -50.0, and (h) -60.0 V at T = 4.2 K. The horizontal dotted lines indicate the $R_{\rm H}$ at each QH plateau. The arrows indicate the expected positions of the QH plateaus with (a)-(d) = 2, 6, and 10 (e)-(h) = 4, 8, and 12.

located exactly at the expected positions of the filling factors = 2, 6, and 10. In the bi-layer graphene, the $R_{\rm H}$ vs. B was measured at (e) $V_{\rm g}$ = -26.0 V, (f) -40.0 V, (g) -50.0 V, and (h) -60.0 V. The carrier densities are calculated to be (e) $n = 1.0 \times 10^{12}$, (f) 2.0×10^{12}, (g) 2.7×10^{12}, and (h) 3.4×10^{12} cm^{-2}, and the expected positions of the Hall plateaus at filling factors = 4, 8, and 12 are indicated by the blue, purple, and grey arrows in Fig. 3, respectively. The plateaus in the $R_{\rm H}$ are located at the expected positions of the filling factors = 4, 8, and 12. These results show that the QHE in mono-layer and bi-layer graphene are correctly measured for the various gate bias voltages in our experiments.

4. Summary

We have observed the integer QHEs in mono-layer and bi-layer graphene devices in magnetic fields up to B = 53 T by use of a pulse magnet. In a mono-layer graphene, Hall resistance $R_{\rm H}$ is quantized into $R_{\rm H} = (h/e^2)\ ^{-1}$ with filling factors $= \pm 2, \pm 6$, and ± 10. In a bi-layer graphene, the $R_{\rm H}$ is quantized to $R_{\rm H} = (h/e^2)\ ^{-1}$ with = 4, 8, and 12. We have shown that precise magnetotransport measurements of the graphene can be conducted in the environment of pulse magnetic fields. This work is a step forward to a new possibility for investigating electrical properties of graphene in ultrahigh magnetic fields $B > 100$ T.

5. Acknowledgements
The authors acknowledge Y. Hirayama, M. Kawamura, and T. Osada for helpful discussions and technical support. This work is supported by the Grant-in-Aid from MEXT, the Special Coordination Funds for Promoting Science and Technology, and the Grant-in-Aid for Scientific Research on priority areas, " High Field Spin Science in 100T " (No.451) from MEXT. One of the authors (S. M.) acknowledges JSPS Research Fellowship for Young Scientists.

[1] M. van der Burgt, P. Thoen, F. Herlach, F. M. Peeters, J. J. Harris, and C. T. Foxon, Physica B **177**, 409 (1992).
[2] T. Takamasu, H. Dodo, and N. Miura, Solid State Commun. **96**, 121 (1995).
[3] T. Takamasu, H. Dodo, M. Ohno, and N. Miura, Surface Science **361**, 95 (1996).
[4] S. Takeyama, H. Sawabe, and E. Kojima, J. Low Temp. Phys. **159**, 328 (2010).
[5] K. S. Novoselov, A. K. Geim, S. V. Morozov, D. Jiang, Y. Zhang, S. V. Dubonos, I. V. Grigorieva, and A. A. Firsov, Science **306**, 666 (2004).
[6] K. S. Novoselov, A. K. Geim, S. V. Morozov, D. Jiang, M. I. Katsnelson, I. V. Grigorieva, S. V. Dubonos, and A. A. Firsov, Nature **438**, 197 (2005).
[7] A. K. Geim, and K. S. Novoselov, Nature Materials **6**, 183 (2007).
[8] K. S. Novoselov, E. McCann, S. V. Morozov, V. I. Fal'ko, M. I. Katsnelson, U. Zeitler, D. Jiang, F. Schedin, and A. K. Geim, Nature Physics **2**, 177 (2006).
[9] Y. Zhang, Z. Jiang, J. P. Small, M. S. Purewal. Y.-W. Tan, M. Fazlollahi, J. D. Chudow, J. A. Jaszczak, H. L. Stormer, and P. Kim, Phys. Rev. Lett. **96**, 136806 (2006).
[10] K. I. Bolotin, F. Ghahari, M. D. Shulman, H. L. Stormer, and P. Kim, Nature **462**, 196 (2009).
[11] X. Du, I. Skachko, F. Duerr, A. Luican, and E. Y. Andrei, Nature **462**, 192 (2009).
[12] Z. Jiang, Y. Zhang, H. L. Stormer, and P. Kim, Phys. Rev. Lett. **99**, 106802 (2007).
[13] S. Masubuchi, M. Ono, K. Yoshida, K. Hirakawa, and T. Machida, Appl. Phys. Lett. **94**, 082107 (2009).
[14] S. Masubuchi, K. Suga, M. Ono, K. Kindo, S. Takeyama, and T. Machida, J. Phys. Soc. Jpn. **77**, 113707 (2008).
[15] G. Kido, N. Miura, H. Ohno, and H. Sakaki, J. Phys. Soc. Jpn. **51**, 2168 (1982).

Journal of Physics: Conference Series **334** (2011) 012038 doi:10.1088/1742-6596/334/1/012038

Quantum Hall Transport across Monolayer-Bilayer Boundary in Graphene

A Tsukuda, H Okunaga, D Nakahara, K Uchida, T Konoike, and T Osada
Institute for Solid State Physics, University of Tokyo,
5-1-5 Kashiwanoha, Kashiwa, Chiba 277-8581, Japan

E-mail: osada@issp.u-tokyo.ac.jp

Abstract. Magnetotransport across monolayer-bilayer boundary on a graphene piece has been studied in the quantum Hall state. The edge channel transport in monolayer-bilayer junctions has been discussed based on the Landauer-Buttiker picture following p-n bipolar junctions of monolayer graphene. We considered two extreme cases of transmission from input edge channels to output channels at the junction, the maximum transmission model and the full-mixing model. They give different two-terminal conductance as a function of carrier filling. We measured the two-terminal conductance in monolayer-bilayer junction devices, and the measured conductance agreed better with the full-mixing model. This result suggests that edge channels of monolayer and bilayer regions are almost fully mixed up at the monolayer-bilayer boundary.

1. Introduction: Quantum Hall transport in the p-n bipolar junction of graphene

In the graphene p-n bipolar junctions, the remarkable magnetotransport features have been observed in the quantum Hall state [1]. The two-probe conductance across the p-n junction shows fractional quantization. In addition to the single p-n junction, quantum Hall transport in p-n-p (or n-p-n) double junctions have been also studied [2, 3]. These samples are graphene FET devices with additional top gate electrode which partially covers the sample. The carrier filling of the top-gated region can be changed independently from the rest region. In graphene, the carrier filling can be controlled from electron-filling (n-type) to hole-filling (p-type), so that the p-n bipolar junction can be realized along the boundary between the top-gated part and the other part.

Under high magnetic fields, the n-type and p-type regions show electron-type and hole-type half-integer quantum Hall effect, respectively. Generally, in the quantum Hall state, the Fermi energy is located in the mobility gap between extended states of Landau levels, and the quantum Hall state is characterized by its plateau index, more precisely, the Chern number (the Thouless-Kohmoto-Nightingale-den Nijs number) [4]. This topological number is stable as long as the Fermi level exists in the mobility gap. When the system undergoes transition from one Hall plateau to another one with the change of external parameter such as magnetic field, in other words, when the topological number switches to another one, the Fermi level must pass through the extended states between mobility gaps. We apply this principle to the graphene p-n junction regarding the space position as the parameter. Since the topological number discontinuously changes at the boundary, there should exist one-dimensional extended states (metallic states) along the boundary. We call them the boundary states or the boundary channels.

Generally, quantum Hall transport can be explained as edge channel transport developed by Landauer and Buttiker [5]. In the quantum Hall state of graphene p-n junctions, the n-type and p-type regions have the edge channels with opposite chirality, namely, opposite propagation direction. So as to conserve the current at the edge points of the boundary, the boundary channel must carry the current, so that it is considered that the boundary channel has chirality in the same way as the edge channel. In fact, if edge channels surround both regions independently without any mixing at the boundary, they run in the same direction along the boundary. These chiral edge channels mix up forming the chiral boundary channels.

The fractional quantization of conductance in the graphene p-n junction [1] was well explained in terms of Landauer-Buttiker formalism by assuming the full mixing of n-type and p-type edge channels at the boundary [6]. The edge channels in n-type and p-type regions join together at the edge point of the boundary. Carriers enter to the boundary states and propagate along the boundary. At the other edge point of the boundary, the carrier flow diverges into edge channels in n-type and p-type regions. Here, it is assumed that carriers flow into all output edge channels with the same probability. Under this assumption, the two-terminal conductance across the boundary is given by

$$G^{(\mathrm{p-n})} = \frac{e^2}{h}\frac{|\nu_1||\nu_2|}{|\nu_1|+|\nu_2|}. \tag{1}$$

Here, ν_1 and ν_2 indicate the quantum Hall indices of n-type and p-type regions, respectively. This formula agreed with the experimental result [1] very well.

In the present work, we have studied the quantum Hall transport in another graphene junction strucure, the monolater-bilayer hetero-junction, that is, the monolayer step on graphene. This structure is often observed on graphene flakes on the substrate. The quantum Hall transport in this junction sysyem has also been studied by Zhao, Koshino, and Kim [7]. In below, we discuss possible scenarios on the edge channel mixing at the monolayer-bilayer boundary based on the Landauer-Buttiker formalism, and then, we compare them with our experimental results.

2. Model: Quantum Hall transport in the monolayer-bilayer graphene hetero-junction

We consider the monolayer-bilayer junction system of graphene illustrated schematically in Fig. 1. Two electrodes are located both in the monolayer side and in the bilayer side, and a single monolayer step exists between them. The carrier density in both sides is controlled by the common back-gate voltage, so that carrier numbers in both sides are equal. The filling factors in both sides are also equal

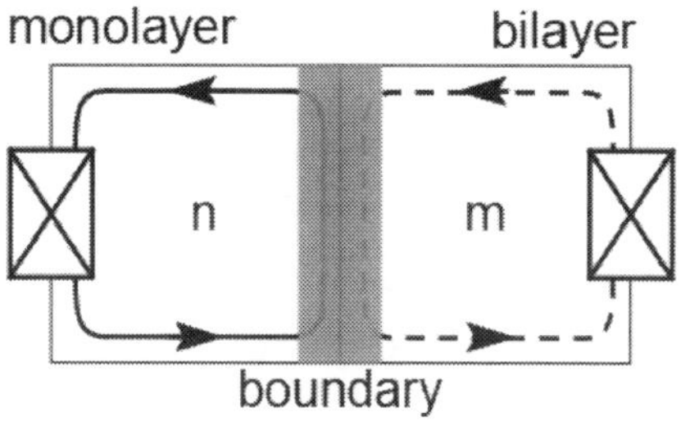

Figure 1. Schematic diagram of monolayer-bilayer junction of graphene. There are n edge channels in the monolayer region and m edge channels in the bilayer region. Along the boundary, one-dimensional metallic channels is formed.

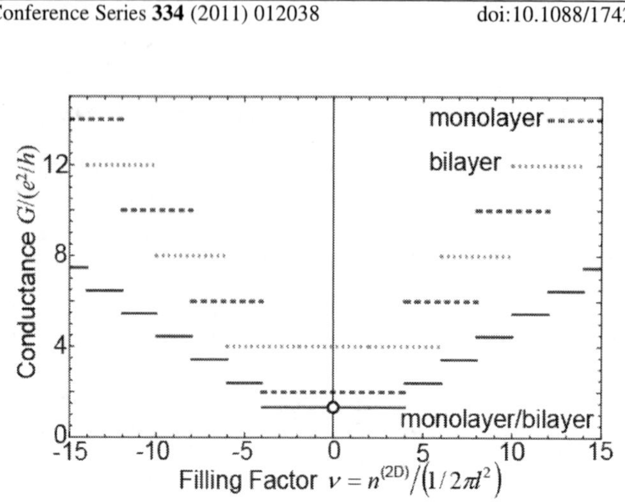

Figure 2. Two-terminal conductance of monolayer graphene (broken line), bilayer graphene (dotted line), and monolayer-bilayer junction (solid line) in the quantum Hall state as functions of the filling factor. Spin and valley degeneracy is included.

under magnetic fields. Although monolayer and bilayer graphene exhibit quantum Hall effect, it appears at different filling factors with different plateau values: The plateaus appear around filling factors of $n=\pm 2$, ± 6, ± 10, ... with plateaus values $\sigma_{xy}=(-e^2/h)n$ in monleyer, and around $m=\pm 4$, ± 8, ± 12, ... with $\sigma_{xy}=(-e^2/h)m$ in bilayer. This difference originates from different degeneracy of the n=0 Landau level. In below, we consider the situation where the monolayer side and the bilayer side show the quantum Hall effect simultaneously. This situation is expected when the transition regions between the quantum Hall states are narrow enough in both sides.

In order to discuss the quantum Hall transport across the monolayer-bilayer junction, we employ the similar logic as that used for the p-n junction mentioned above. The quantum Hall states of the monolayer and bilayer regions have different Chern topological numbers n and m, which never become equal. So, there must exist one-dimensional metallic states along the boundary between two quantum Hall regions. We call them the boundary states or the boundary channels following the p-n junction. However, contrary to the case of the p-n junction, the edge channels of the monolayer and bilayer regions have the same chirality, namely, the same propagation direction as shown in Fig. 1. Since both regions have different numbers of edge channels n and m, the boundary state must carry the current so as to conserve the current at two edge points of the boundary. In this case, however, it is considered that the boundary channels do not have the same chirality. If edge channels surround both regions independently without any mixing at the boundary, they run in the counter direction along the boundary. So, the boundary channels must carry the current in both directions.

We consider two-terminal conductance across the monolayer-bilayer boundary in the quantum Hall state in terms of the Landauer-Buttiker picture. At the lower (upper) edge point of the boundary in Fig. 1, n channels of monolayer (m channels of bilayer) enter into the boundary and m channels of bilayer (n channels of monolayer) leave the boundary. As the detail of carrier transmission from each input channel to each output channel is unclear, here, we consider two extreme cases. In the first case, we assume the maximum transmission from input channels of monolayer (bilayer) to output channels of bilayer (monolayer) at the lower (upper) edge point of the boundary. In this model, two-terminal conductance across the boundary takes the maximum value:

$$G^{(\text{monolayer-bilayer})} = \frac{e^2}{h}\min(n,m)\,. \qquad (2)$$

So, the even-integer plateau values are expected. In the second case, we assume full-mixing of edge channels at the boundary in the same way as the p-n bipolar junctions. In this model, it is assumed that electrons injected from all input channels flow out to all output channels with the equal probability. The two-terminal conductance is given by the same formula as p-n junctions:

$$G^{(\text{monolayer-bilayer})} = \frac{e^2}{h}\frac{|n||m|}{|n|+|m|}. \tag{3}$$

Figure 2 shows the filling factor dependence of the two-terminal conductance of the monolayer-bilayer junction in the quantum Hall state. The maximum transmission model gives lower plateaus of monolayer and bilayer indicated by dashed lines. The full-mixing model gives fractional plateaus indicated by solid lines. Note that the gate-voltage dependence has different slope in two cases.

3. Experimental confirmation

We have performed the magnetotransport measurement of monolayer-bilayer graphene junctions. Graphene pieces have been prepared from Kish graphite by mechanical exfoliation technique on SiO_2/n-Si substrate, which works as a back-gate electrode in FET devices. Graphene pieces with monolayer-bilayer boundary were picked up and electrodes were fabricated on them using electron-beam lithography technique. The graphene thickness was checked by the micro-Raman measurement. The microscope image of the typical device is shown in the inset of Fig. 3.

Figure 3 shows the two-terminal conductance measured by e^2/h across monolayer-bilayer boundary at B=10T and T=4.2K. A few fractional plateau values given by eq. (3) are indicated by dashed lines. As a function of the gate voltage, two-terminal conductance increases almost linearly with weak oscillatory (plateau-like) modulation, which corresponds to the transition between different quantum Hall states. Its slope agrees with that predicted by the full-mixing model much better than the maximum transmission model. Since the perfect quantum Hall states are not realized over the whole gate voltage range in the experiment, the data has not sharp staircase shape but rather smeared one.

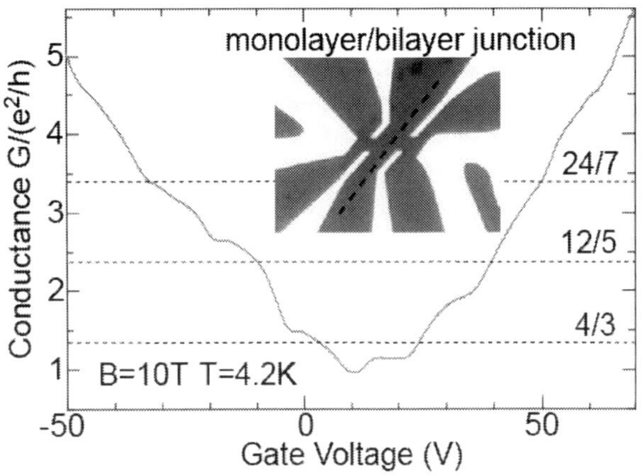

Figure 3. Back-gate-voltage dependence of two-terminal conductance across the monolayer-bilayer junction. Inset shows the microscope image of the grahene FET device with monolyer-bilayer junction indicated by broken line.

Nevertheless, the observed plateaus are in the same order of fractional values (4/3, 12/5, 24/7) expected from the full-mixing model. They are hardly explained as even integers deduced from the maximum transmission model. Therefore, the experimental result seems to support the full-mixing of edge states of monolayer and bilayer regions at the junction more strongly than the maximum transmission of edge states.

References

[1] Williams J R, DiCarlo L and Marcus C M 2007 *Science* **317** 638
[2] Ozyilmas B, Jarillo-Herrero P, Efetov D, Abanin D A, Levitov L S and Kim P 2007 *Phys. Rev. Lett.* **99** 166804
[3] Ki D-K and Lee H-J 2009 *Phys. Rev.* **B79** 195327
[4] Thouless D J, Kohmoto M, Nightingale M P and den Nijs M 1982 *Phys. Rev. Lett.* **49** 405; Kohmoto M 1985 *Ann. Phys. (N. Y.)* **160** 343
[5] Buttiker M 1988 *Phys. Rev.* **B38** 9375
[6] Avanin D A and Levitov L S 2007 *Science* **317** 641
[7] Zhao Y, Koshino M and Kim P 2009 B25.12 *Bull. Ame. Phys. Soc.* **54** *No.1 2009 APS March Meeting (Pittsburgh, USA, 16–20 Mar., 2009)*

HMF-19 IOP Publishing
Journal of Physics: Conference Series **334** (2011) 012039 doi:10.1088/1742-6596/334/1/012039

Magnetoresistance of Rippled Graphene in a Parallel Magnetic Field

Junichi Wakabayashi and Tomoya Sano
Department of Physics, Chuo University, 1-13-27 Kasuga, Bunkyo-ku, Tokyo 112-8551, Japan
E-mail: waka@phys.chuo-u.ac.jp

Abstract. The magnetoresistance of a monolayer graphene in a random magnetic eld(RMF) with zero mean has been investigated. The RMF was produced by applying a magnetic eld parallel to the graphene plane utilizing ripples. The magnetoresistance has shown the same magnetic eld dependence and, unexpectedly, the same carrier density dependence as the conventional two-dimensional electron systems in random magnetic elds. The relation between the characteristic length of ripples and the magnitude of the magnetoresistance is discussed.

1. Introduction

Since the discovery of graphene[1], numerous theoretical and experimental papers studying the novel properties of graphene have been published.[2, 3] Among these studies, it is well known that there are ripples in a graphene suspended in space or placed on a substrate. Ripples, microscopic corrugations of a graphene membrane, are considered inevitable to stabilize the graphene crystal in three-dimensional space, called intrinsic ripples,[4] and observed experimentally in a suspended graphene.[5] The graphene placed on a substrate, on the other hand, exhibits ripples basically following across the corrugations of the substrate underneath[6, 7, 8, 9, 10] except when it is placed on an ultra at substrate.[9]

The ripples are discussed to a ect the various electronic properties of graphene.[3, 11, 12] As for the transport properties, it has been shown experimentally rippling is the ultimate souce of scattering when extrinsic disorder is eliminated.[13] Owing to ripples, on the other hand, when a magnetic eld is applied parallel to the graphene plane(averaged two-dimensional plane), the magnetic eld component normal to the local surface of ripples is created according to the gradient of the local surface against the magnetic eld. Because ripples vary in size and in curvature from place to place, a random magnetic eld with zero mean is produced in the graphene sheet(corrugating membrane).

Dirac fermions in graphene exhibit exotic electrical properties such as a novel quantum Hall e ect, high carrier mobility, etc. Investigations of electrical transport properties of Dirac fermions in a random magnetic eld coupling with ripples are attractive. In the present paper we report the experimental results of the magnetoresistance in a random magnetic eld produced by applying a magnetic eld parallel to the graphene plane.

2. Experimental details

The sample is a graphene ake produced by mechanical exfoliation from a HOPG crystal and deposited on a Si/SiO_2(306nm) wafer. The sample was annealed in H_2/Ar ambience (1.5 ℓ/min,

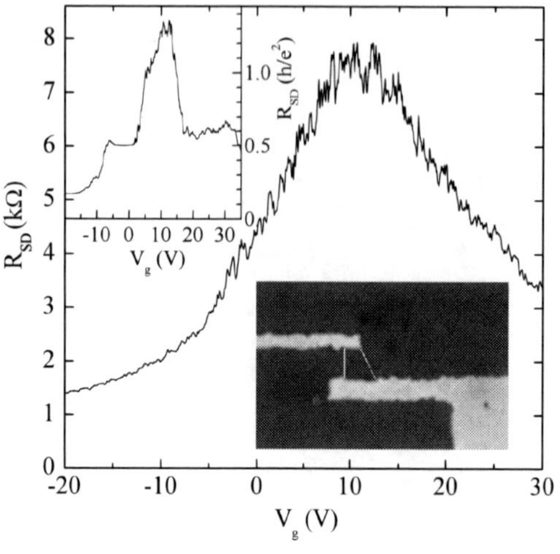

Figure 1. Main panel: Gate voltage dependence of two terminal resistance R_{SD} in zero magnetic eld. Upper left inset: Gate voltage dependence of R_{SD} in a magnetic eld of 15T applied normal to the graphene plane. Quantized plateaus show the sample is a monolayer graphene. Lower right inset: Optical microscope image of the sample. The length of the parallel sides of the trapezium are 3.1 m and 1.7 m, respectively, and the distance between them is 2.8 m. Contacts are made by Au(30nm) deposition except the large pad area by Au/Cr(30nm/3nm) deposition.

respectively) at a temperature of 200 C for two hours before and after the electron beam lithography for contacts. The shape of the sample is trapezoidal and the parallel sides are the source and the drain contact, respectively. (See lower right inset of Fig.1) We use the averaged value of 2.4 m as the sample width to calculate the resistivity from the resistance. The carrier density n was controlled applying a gate voltage V_g to n-type Si substrate(0.01 cm) and was calculated according to the equation $n = 7.1 \quad 10^{10}\ (V_g \quad V_{CNP})\mathrm{cm}^{\ 2}$, where $V_{CNP} \cong 11$ V is the gate voltage corresponding to the charge neutrality point (CNP). The coefficient was estimated from the thickness of the SiO_2 layer. All the measurements have been done using a standard lock-in technique with 87 Hz and the sample current of 100 nA at temperature $T \simeq 80$ mK.

Figure 1 shows the gate voltage dependence of the two-terminal resistance R_{SD}. The upper left inset shows the gate voltage dependence of R_{SD} measured in a magnetic eld of 15 T applied normal to the sample. Quantized plateaus of $N = 0$ for electrons and holes and that of $N = 1$ for holes con rm that the present sample is a monolayer graphene. The plateau value of $N = 0$ for electrons shows higher values than expected. Because the present sample is a two terminal device, this is considered as a result of contact resistance. We consider this contact resistance does not change the results in the present experiment.

3. Results and discussions

Figure 2 shows the parallel magnetic eld $B_{||}$ dependence of the magnetoresistance $\quad R_{SD} = R_{SD}(B) \quad R_{SD}(0)$ at typical gate voltage in the hole region, around CNP, and in the electron region, respectively. The magnetic eld $B_{||}$ was applied along the source-drain direction to minimize the additional e ect of the parallel magnetic eld to the magnetoresistance.[14] The miss orientation of the magnetic eld from the graphene plane was estimated to be less than 0.04 degree. Although the parabolic dependence is observed in the overall behavior, negative magnetoresistance appears in low magnetic elds. This negative magnetoresistance comes from the weak localization e ect due to normal component of the random magnetic eld. In order to subtract the contribution of the weak localization to the resistance, we have measured magnetoresistance in low magnetic elds normal to the graphene plane at each gate voltages. We assume the negative magnetoresistance in parallel magnetic elds is governed by the rms value of the normal component of the random magnetic led: $\overline{B_\perp}$. We have made try and error calculation so that the resultant resistance shows a parabolic dependence on $B_{||}$ after the

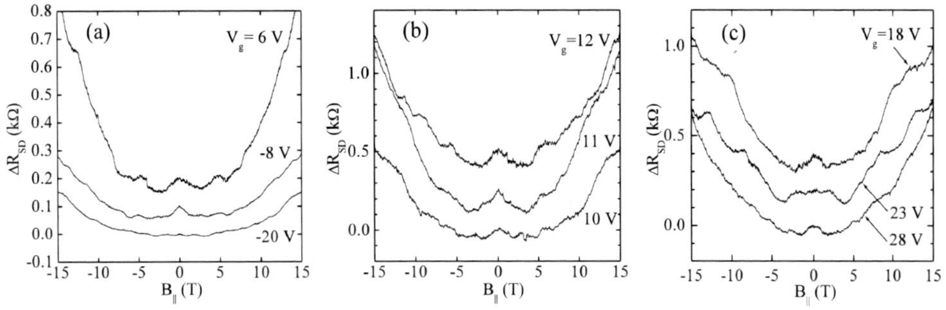

Figure 2. Magnetoresistance $R_{\mathrm{SD}} = R_{\mathrm{SD}}(B_{||})$ $R_{\mathrm{SD}}(0)$ against parallel magnetic eld $B_{||}$ at typical gate voltage V_{g} (a) in the hole region, (b) around CNP and (c) in the electron region, respectively. Upper two traces are vertically displaced for clarity. The negative magnetoresistance due to the weak localization is observed in the low magnetic elds.

subtraction of the negative magnetoresistance. We found $\overline{B_{\perp}}$ =40 mT for $B_{||} = 15T$ satis es with the data of all gate voltages. Figure 3 shows an example how the correction has been done.

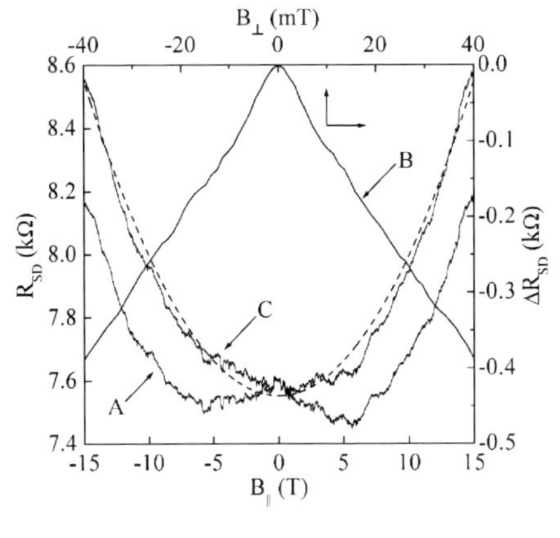

Figure 3. A: Magnetic eld dependence of resistance R_{SD} at gate voltage V_{g} = 9 V in parallel magnetic eld $B_{||}$. B: Negative magnetoresistance R_{SD} measured in weak magnetic eld applied normal to the graphene plane. C: The corrected resistance R_{SD} after subtraction of the data B from the data A. Parabolic dashed line is a guide to the eye.

After the correction of the e ect of weak localization, the magnetoresistance shows parabolic dependence against $B_{||}$. We have done the same correction on the data of all other gate voltages and calculated $\rho = \rho(B_{||})$ $\rho(0)$. Figure 4 shows ρ against $B_{||}^2$ for typical gate voltages for the hole region, around CNP, and the electron region, respectively. Although large uctuations are observed in some data, we have done linear t to all the data and obtained the coefficient C, where $\rho = CB_{||}^2$. Figure 5 shows the gate voltage dependence of the coefficient C. The value of C has a maximum at CNP and decreases rapidly as the carrier density increases.

For the conventional two-dimensional electron system(2DES), the magnetoresistance ρ in a weak random magnetic eld with zero mean is predicted theoretically[15] as $\rho \propto B_{\mathrm{amp}}^2 n_{\mathrm{s}}{}^{3/2}$ and con rmed experimentally,[14] where $B_{\mathrm{amp}}$is the amplitude of the random magnetic eld and n_{s} is the electron density. Recently, the magnetotransport measurements of graphene in parallel magnetic elds have been reported by Lundeberg and Folk.[16] They derived an equation for

Journal of Physics: Conference Series **334** (2011) 012039 doi:10.1088/1742-6596/334/1/012039

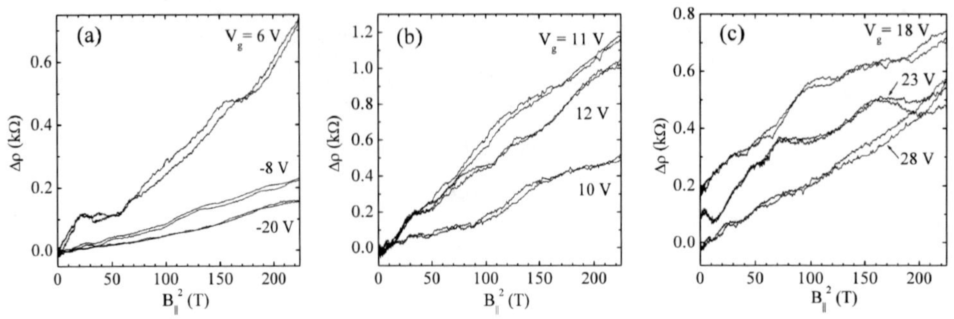

Figure 4. Magnetoresistivity $\rho = \rho(B_{||})$ $\rho(0)$ against $B_{||}^2$ obtained from the corrected magnetoresistance at gate voltage V_g corresponding to Fig.2. Upper two traces in gure (c) are vertically displaced for clarity.

magnetoresistance due to random magnetic elds as,

$$\rho(n, B_{||}) = \frac{Z^2}{R} \frac{1}{\hbar |n|^{\frac{3}{2}}} B_{||}^2, \tag{1}$$

where is a numerical factor which includes all the uncertainty in the approximation used in the derivation, Z and R are the rms height and the correlation length of the graphene ripples, respectively. The equation (1) is basically a modi ed version of the equation for conventional 2DES taking the valley degeneracy into account and approximating the B_{amp} $B_{||}Z/R$.

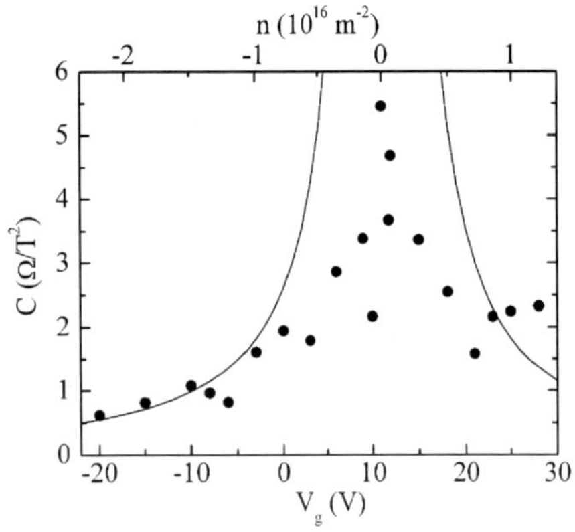

Figure 5. Gate voltage dependence of the coefficient C which is obtained by linear t to the data in Fig.4 as $\rho = CB_{||}^2$. The upper scale is the carrier density calculated assuming the charge neutrality point $V_{\mathrm{CNP}} = 11V$. The solid lines show $|n|$ $^{3/2}$ dependence ted in high carrier density region.

According to the equation (1), we get $C = Z^2/(R\hbar|n|^{3/2})$. The solid lines in Fig.5 show $|n|$ $^{3/2}$ dependence for $Z^2/R = 1.9$ 10 ^{10} m. In the high hole density region, the coefficient C shows $|n|$ $^{3/2}$ dependence. This is rather surprising result when we consider the di erence in energy dispersion relation between the graphene and the conventional 2DES. In Ref. 16, they also observed $\rho \propto |n|$ $^{3/2}$ in the hole region of 1 10^{16} m ^{2} $|n|$ 4 10^{16} m ^{2} at $B_{||} = 8$T.

Around CNP, on the other hand, the values of C is apparently smaller than the value expected. It is reported that electron-hole paddles are formed at carrier density near CNP and the e ective carrier density is higher than the value estimated using the capacitance.[17] The value of C, as a result, will be smaller than the value expected. The carrier density region where C deviates from $|n|^{\ 3/2}$ dependence shown in Fig.5 is approximately $|n| \quad 1 \quad 10^{16}$ m $^{\ 2}$. This value is an order of magnitude larger than the previous experiments.[17] We consider this result represents the quality of the present sample shown by the broad peak of the gate voltage dependence in Fig.1.

Finally, we discuss on the value of . We have done AFM measurements on two monolayer graphene exfoliated from the same HOPG ake using the same method as the sample in Fig.1. Using the $Z(x,y)$ data, we calculated Z, the rms value of the deviation of $Z(x,y)$ from the mean value, and the correlation length $R = (R_x + R_y)/2$ calculating the correlation $< Z(x,y)Z(x + \quad x, y) >$ for R_x and $< Z(x,y)Z(x, y + \quad y) >$ for R_y. We obtained $Z = 0.236$ nm/0.232 nm and $R = 14$ nm/25 nm for the two samples, respectively. The values of Z agree with each other and correspond to that of the bare SiO_2[10]. The variation of R suggests that the corrugation of SiO_2 is not homogeneous and it is re ected in the graphene ripples on top.[8, 9] When we use the values of Z and R shown above, we get $\quad = 48$ and 88 for the two samples, respectively. In Ref.16, they estimated Z and R combining measurements of the magneto-resistance and the daphasing time of the weak localization. They observed $\quad \rho$ against $B_{||}^2$ but without the correction of weak localization. In addition, the magnetic eld $B_{||}$ is not parallel to the direction of the sample current. The coefficient C in their calculation, hence, is presumably smaller than the correct one and they obtain $\quad \simeq 2.8$. In the present experiment, on the other hand, the value of Z and R are not for the sample used in the transport measurements. In order to discuss further the relation between the ripples and the magnitude of the magnetoresistance in a parallel magnetic eld, more experimental and theoretical studies are inevitable.

4. Summary

We have measured the magnetoresistance of a graphene in a random magnetic eld created by applying a magnetic eld parallel to the graphene sheet with ripples. The magnetoresistance has exhibited the same magnetic eld dependence, $\propto B_{||}^2$, and, contrary to our expectation, the same carrier density dependence, $\propto n^{\ 3/2}$, as the conventional 2DES in random magnetic elds.

References

[1] Novoselov K S, Geim A K, Morozov S V, Jiang D, Zhang Y, Dubonos S V,y Grigorieva S V and Firsov A A 2004 *Science* **306** 666
[2] Geim A K 2009 *Science* **324** 1530
[3] Castro Neto A H, Guinea F, Peres N M R, Novoselov K S and Geim A K 2009 *Rev. Mod. Phys.* **81** 109
[4] Fasolino A, Los J H and Katsnelson M I 2007 *Nature Mater.* **6** 858
[5] Meyer J C, Geim A K, Katsnelson M I, Booth T, Novoselov K S and Roth S 2007 *Nature* **446** 60
[6] Ishigami M, Chen J H, Cullen W G, Fuhrer M S and Williams E D 2007 *Nano Lett.* **7** 1643
[7] Stolyarova E, Rim K T, Ryu S, Maultzsch J, Kim P, Brus L E, Heinz T F, Hybertsen M S and Flynn G W 2007 *Proceedings of the National Academy of Sciences*(PNAS) **104** 9209
[8] Stöberl U, Wurstbauer U, W. Wegscheider A W , Weiss D and Eroms J 2008 *Appl. Phys. Lett.* **93** 051906
[9] Lui C H , Liu L , Mak K F , Flynn G W and Heinz T F 2009 *Nature* **462** 339
[10] Geringer V, Liebmann M, Echtermeyer T, Runte S, Schmidt M, Rükamp R, Lemme M C and Morgenstern M 2009 *Phys. Rev. Lett.* **102** 076102
[11] Katsnelson M I and Geim A K 2008 *Phil. Trans. R. Soc. A* **366** 195
[12] Deshpande A, Bao W, Miao F, Lau C N and LeRoy B J 2009 *Phys. Rev.* B **79** 205411
[13] Morozov S V, Novoselov K S, Katsnelson M I, Schedin F, Elias D C, Jaszczak J A and Geim A K 2008 *Phys. Rev. Lett.* **100** 016602
[14] Wada S, Okuda N, Wakabayashi J 2010 *Physica* E **42** 1138
[15] Evers F, Mirlin A D, Polyakov D G and Wölfle P 1999 *Phys. Rev.* B **60** 8951
[16] Lundeberg M B and Folk J A 2010 *Phys. Rev. Lett.* **105** 146804 (Revised in nal manuscript. Original manuscript citation; 2009 *Preprint* arXive:0910.4413v1)
[17] Martin J, Akerman N, Ulbricht G, Lohmann T, Smet J H, von Klitzing K and Yacoby A 2008 *Nat. Phys.* **4** 144

Journal of Physics: Conference Series **334** (2011) 012040 doi:10.1088/1742-6596/334/1/012040

ESR in a disordered network of nanographene sheets

M Hagiwara[1], H Tsugeno[1], H Yamaguchi[1], V L J Joly[2], K Takai[2] and T Enoki[2]

[1] KYOKUGEN, Osaka University, Toyonaka, Osaka 560-8531, Japan

[2] Department of Chemistry, Tokyo Institute of Technology, Meguro, Tokyo 152-8550, Japan

E-mail: hagiwara@cqst.osaka-u.ac.jp

Abstract. Randomly networked nanographene sheets have been studied by electron spin resonance (ESR) technique at 20 GHz (K-band) and 35 GHz (Q-band). Nanographene has spin-polarized non-bonding π-electron states (edge-state spins) localized in the zigzag edge region. We have investigated the temperature dependence of an ESR signal of activated carbon fibers at two different microwave powers for each frequency. The signal intensity smoothly increases with decreasing temperature at any microwave power. The line width of ESR signal with a Lorentzian line shape decreases linearly upon cooling, and then increases steeply after taking a minimum at about 20 K irrespective of microwave power. The former is interpreted as the Korringa relation in the localized edge spins and conduction π carriers. The latter may be caused by inhomogeneous line broadening of the ESR signal from randomly distributed nanographene sheets with different sizes due to the suppression of electron hopping between nanographene sheets, *i.e.* electron localization. The discontinuous line broadening and the signal intensity drop at around 20 K reported in the previous X-band ESR at 1 μW were not observed, probably because of either higher microwave power than 1 μW or some amount of oxygen adsorption in our sample in the present study.

1. Introduction

Nanosized materials have attracted considerable attention due to their unconventional electronic features depending on the sizes, shapes, surface conditions and so on. In particular, nanosized graphite (nanographene) with open edges is intriguing [1], because the spin-polarized non-bonding π-electronic states, namely edge-state spins, were predicted to appear at the zigzag edge region and not at the arm-chair edge region [2, 3], and were confirmed by scanning tunneling microscope (STM) and scanning tunneling spectroscopy (STS) experiments [1, 4, 5, 6, 7]. The presence of edge-state spins, therefore, results in a variety of magnetism in nanographene, depending on its shape.

The edge-state spins in the zigzag edge region are ferromagnetically coupled with the interaction of the order of 10^3 K (J_0), and the inter-zigzag chain interaction via the arm-chair edges mediated by conduction π electrons has a strength of $10^{-1}\sim10^{-2}$ of J_0 (the order of 10 $\sim$ 100 K) with either positive or negative sign depending on the mutual relation between the zigzag edges, resulting in forming ferrimagnetic spins in each nanographene sheet.

Activated carbon nanofiber (ACF), which consists of a three-dimensional disordered network of loosely stacked 3-4 nanographene sheets with a mean in-plane size of 2-3 nm, has internanographene domain interactions with the magnitude of a few Kelvin originating from hopping integral among domains. This electron transport in the network is governed by the

Coulomb-gap type variable range hopping [8], and thus strong electron localization was expected due to the suppression of electron hopping at low temperatures and was observed below about 20 K in the ACFs [9]. In Ref. 9, the electron-localization-induced magnetic state of the edge-state spins in the ACFs sample were observed below 20 K from X-band electron spin resonance (ESR), magnetic susceptibility and electrical conductivity experiments. In the present study, we further investigate whether this behavior is observed or not in high-frequency ESR.

2. Experimental

Phenol-based ACFs (FR-20, Kuraray Chemicals) were heat-treated at 473 K for 48 h in vacuum (10^{-8} torr) after evacuating at room temperature for five days to remove the adsorbed foreign gas species. The ACF sample was sealed into a high quality quartz (Sprasil) tube in vacuum. K- and Q-band ESR measurements were carried out at temperatures between 1.5 K and 80 K by utilizing a 16 T superconducting magnet (Oxford Instruments, UK), a vector network analyzer (ABmm, France) and home-made ESR cryostats with cylindrical resonant cavities at KYOKUGEN, Osaka University. We used the second (K-band) and the third (Q-band) harmonics of the fundamental microwave (8-18 GHz) generated with an internal YIG oscillator in the vector network analyzer equipped with Shottoky diode devices and attenuators. Since we do not have a power meter, we only know the relative power reduction when using the attenuators.

3. Results and discussion

Figures 1 (a) and (b) show the temperature dependences of Q-band (34.8 GHz) ESR signals without($\times$1) and with power attenuation ($\times$1/3), respectively. All the ESR signals can be fitted with a Lorentzian function, and hence the line widths of the signals are extracted from the fittings. For the sake of simplicity, the fittings on several signals are drawn with open circles in the figures. The resonance field does not shift much and the g-value is close to 2.0 which is associated with a small spin-orbit interaction of carbon. Similar ESR signals at K-band (19.7 GHz) without ($\times$1) and with power attenuation ($\times$1/10) are obtained as shown in Figs. 2 (a) and (b). A hole-burning effect, namely irregular feature in the ESR signal, which appeared in the previous X-band ESR, was not observed at 15 K.

The temperature dependences of the signal intensity (relative intensity normalized by the magnitude at 1.5 K) and the line width at Q-band are shown in Figs. 3 (a) and (b), respectively. The signal intensity increases monotonically with decreasing temperature like magnetic susceptibility without a sudden drop below 20 K observed in the previous X-band ESR at 1 μW. The line width decreases linearly upon cooling from 80 K and increase steeply after taking a minimum at around 20 K. Similar temperature dependences of the signal intensity and the line width are obtained at K-band as depicted in Figs. 4 (a) and (b).

Since the magnetic susceptibility of our ACFs obeys the Curie-Weiss law with a small negative Weiss temperature of -2$\sim$-3 K [9], a monotonous increase of the ESR signal intensity upon cooling is consistent with our expectation in the paramagnetic state. The line width above 20 K proportional to the temperature is probably caused by the Korringa relation $1/T_{1eff} \propto T$, where T_{1eff} represents the effective spin-lattice relaxation time, in the localized edge-state spins and conduction π carriers. Here, the line width ΔH has the relation of $\Delta H \propto 1/\mathrm{T}_{1eff}$ and the $1/T_{1eff}$ is given as $1/T_{1eff}=1/T_{n\pi}(T_{\pi n}/T_{\pi L})$, where $T_{n\pi}$, $T_{\pi n}$, and $T_{\pi L}$ are the relaxation times of from the edge-state spin to the π carrier, from the π carrier to the edge-state spin, and from the π carrier to the lattice, respectively [10]. The $T_{n\pi}$ is give by the Korringa relation and the other two relaxation times are nearly temperature independent as described in Ref. 9. A considerably long $T_{\pi L}$ causes the bottleneck effect and makes the ESR signal sharpen.

The conductivity due to the Coulomb-gap type variable range hopping decreases abruptly below about 20 K [9], and then the suppression of carrier hopping may cause weakening the internanographene domain interactions among the nanographene sheets. As a consequence,

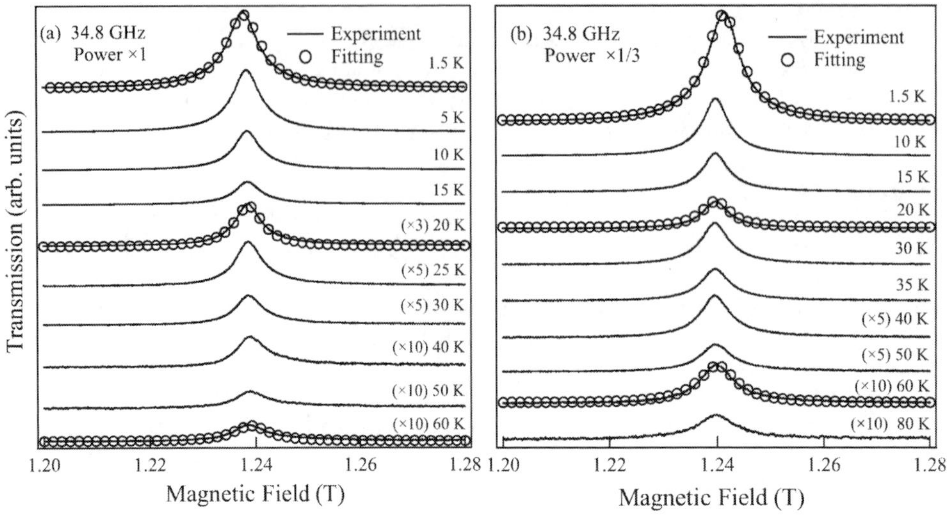

Figure 1. Temperature dependence of ESR signals at Q-band (a) without and (b) with power attenuation (×1/3). Open circles represent a fitting of the experimental data to a Lorentzian function.

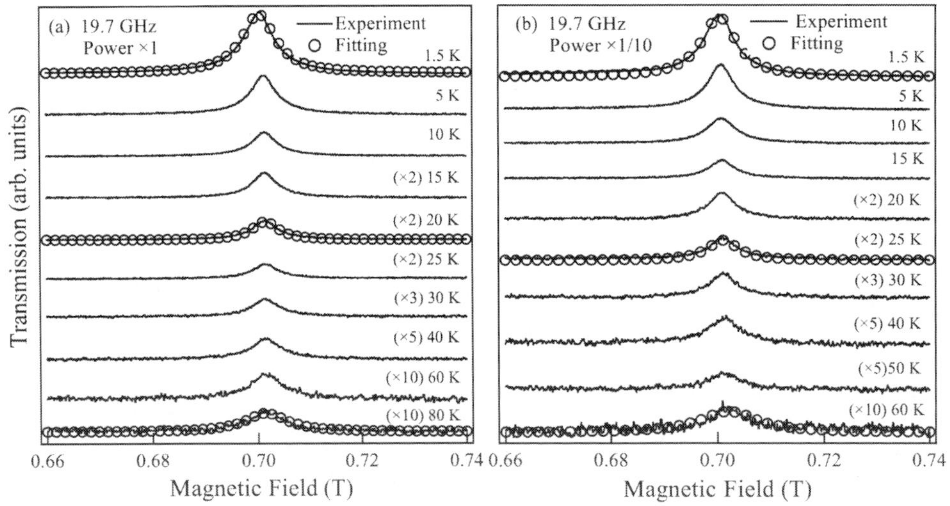

Figure 2. Temperature dependence of ESR signals at K-band (a) without and (b) with power attenuation (×1/10). Open circles represent a fitting of the experimental data to a Lorentzian function.

ferrimagnetic edge-state spins on each nanographene sheet behave independently. Kittle and Abrahams [11] calculated ESR resonance lines in crystals having lattice points populated at random by identical paramagnetic ions. Then, it was found that for fractional magnetic population $f<0.01$ the line shape is approximately Lorentzian with a width proportional to f. In our sample, the fractional magnetic population of edge-state spins on each graphene sheet

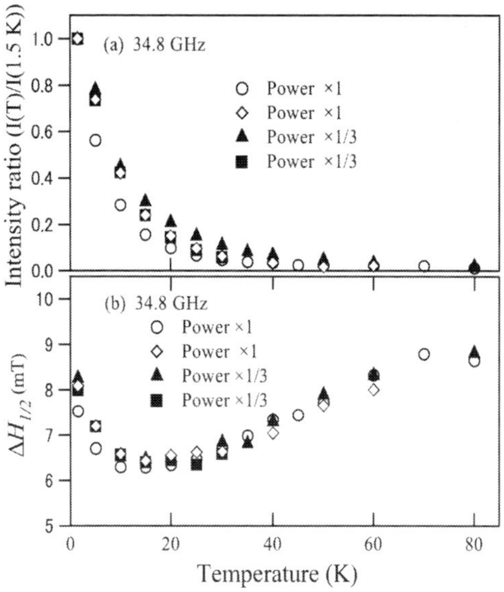

Figure 3. Temperature dependences of (a) ESR signal intensity ratio against that at 1.5 K and (b) ESR line width measured at K-band.

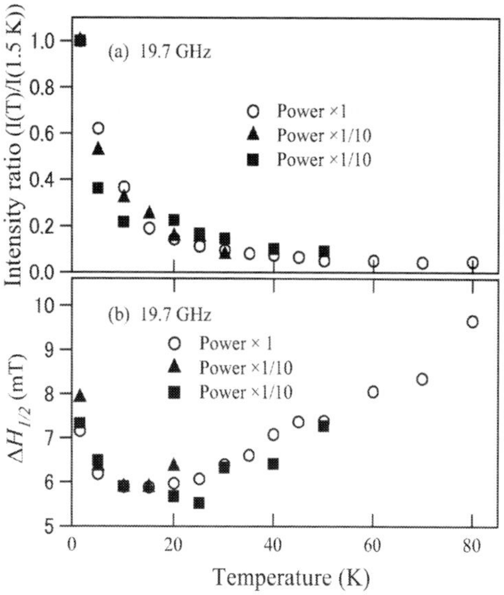

Figure 4. Temperature dependences of (a) ESR signal intensity ratio against that at 1.5 K and (b) ESR line width measured at Q-band.

is evaluated to be ~0.001 (0.25 μ_B magnetic moment, namely a quarter of spin 1/2 with g=2, for 300 carbon sites on each nanographene sheet) [12]. This probably causes a line broadening owing to the superposition of the ESR signals from the edge-state spins on randomly distributed nanographene sheets with different sizes.

In the present high-frequency ESR studies, we did not observe any sudden changes of the line width and the signal intensity at about 20 K which were observed in the previous X-band ESR at 1 μW [9]. This discrepancy is probably caused by two plausible reasons. At sufficiently low microwave power, the relaxation time $T_{\pi L}$ may become short and temperature dependent. Then, the bottleneck effect is probably broken and thus the signal becomes broad at low temperatures. From this consideration, microwave power in our high-frequency ESR experiments may not be small enough to break the bottleneck effect. On the other hand, the feature of the ESR signal is extremely sensitive to the presence of oxygen molecules [9, 13] and the transition at about 20 K shifts to high temperature side and smears out at sufficiently low oxygen concentration. Therefore, our ACF sample in the present study may contain some absorbed oxygen molecules even though the careful treatment of evacuation and heating.

4. Conclusions

We have studied randomly networked nanographene sheets realized in activated carbon fibers by K- and Q-band ESR. The temperature dependence of ESR signal intensity shows a smooth increase as the temperature is decreased. The line width of ESR signal with a Lorentzian line shape decreases linearly upon cooling and then increases steeply after taking a minimum at about 20 K. The former behavior above 20 K is interpreted as the Korringa relation in the localized edge spins and conduction π carriers, and the latter one below 20 K may be caused by the superposition of the ESR signals from ferrimagnetic edge-state spins on randomly distributed nanographene sheets due to the suppression of electron hopping among the nanographene sheets. The temperature dependence of the ESR line width in this study is similar to that of the X-band data above 20 K at 1 μW, but the discontinuous line broadening was not observed at about 20 K.

Acknowledgments
We thank Dr. T. Kashiwagi for carrying out ESR measurements on ACFs at the early stage of this work and fruitful discussion. This work was supported in part by Grant-in-Aid for Scientific Research on Priority Area (No. 19051010) and also by Global COE program "Core Research and Engineering of Advanced Materials-Interdisciplinary Education Center for Materials Science" from MEXT, Japan.

References
[1] Enoki T, Kobayashi Y and Fukui K 2007 *Int. Rev. Phys. Chem.* **26** 609
[2] Fujita M, Wakabayashi K, Nakada K and Kusakabe K 1996 *J. Phys. Soc. Jpn.*. **65** 1920
[3] Nakada K, Fujita M, Dresselhaus G and Dresselhaus M S 1996 *Phys. Rev.* B **54** 17954
[4] Kobayashi Y, Fukui K I, Enoki T, Kusakabe K and Kaburagi Y 2005 *Phys. Rev.* B **71** 193406
[5] Kobayashi Y, Fukui K I, Enoki T and Kusakabe K 2006 *Phys. Rev.* B **73** 125415
[6] Niimi Y, Matsui T, Kambara H, Tagami K, Tsukada M and Fukuyama H 2006 *Phys. Rev.* B **73** 085421
[7] Enoki T and Takai K 2009 *Solid State Commun.* **149** 1144
[8] Fung A W P, Wang Z H, Dresselhaus M S, Dresselhaus G, Pekala R W and Endo M 1994 *Phys. Rev.* B **49** 17325
[9] Joly V LK, Takahara K, Takai K, Sugihara K, Enoki T, Koshino M and Tanaka H 2010 *Phys. Rev.* B **81** 115408
[10] Hasegawa H 1959 *Prog. Theor. Phys.* **21** 483
[11] Kittel C and Abrahams E 1953 *Phys. Rev.* **90** 238
[12] Shibayama Y, Sato H, Enoki T, Bi X -X, Dresselhaus M S and Endo M 2000 *J. Phys. Soc. Jpn.* **69** 754
[13] Sumanasekera G U, Chen G, Takai K, Joly J, Kobayashi N, Enoki T and Eklund P C 2010 *J. Phys.: Condens. Matter.* **22** 334208

Journal of Physics: Conference Series **334** (2011) 012041 doi:10.1088/1742-6596/334/1/012041

Single-electron switching e ect in graphene parallel-coupled double quantum dots

M. Arai[1], **S. Masubuchi**[1,2], **and T. Machida**[1,2,3]
[1]Institute of Industrial Science, University of Tokyo, Tokyo 153-8505, Japan
[2]Institute for Nano Quantum Information Electronics, University of Tokyo, Tokyo 153-8505, Japan
[3]PRESTO, Japan Science and Technology Agency, Saitama 332-0012, Japan
E-mail: msatoru@iis.u-tokyo.ac.jp

Abstract. We have fabricated parallel-coupled quantum dots on single-layer graphene. The tunnel coupling between the quantum dots can be tuned by a graphene in-plane gate. Owing to the tunnel coupling, the Coulomb blockade oscillation peaks exhibit periodic shifts as the number of electron in the non-conducting side-coupled QD is changed. The result suggests the observation of the single electron switching effect, which is a prerequisite for a single photon detection scheme using parallel-coupled quantum dots.

1. Introduction

In a parallel-coupled double quantum dot (QD) system, where a non-conducting QD is coupled to a main QD on a current path through a tunable tunnel barrier, the electrostatic energy of the main conducting QD is modulated by the charge state of the other QD[1, 2]. When the two QDs are weakly coupled, the addition of an extra electron to the side-coupled QD leads to an abrupt change in the conductance of the main QD. This phenomenon is called as the single electron switching effect. Since the main conducting QD acts as an ultra-sensitive electrometer of the non-conducting QD, the single electron switching effect has been utilized for the detection of a far-infrared single photon absorption in GaAs/AlGaAs parallel QDs[3, 4].

In order to extend the range of detection wavelength and operation temperature in parallel QD single photon detectors, graphene can be a promising material[5]. This is because (i) the large Fermi velocity of the Dirac fermion induces large energy-level spacing in a quantum dot[6], which makes high-temperature single electron transistor (SET) operation of the quantum dot easier as compared to the conventional materials with massive carriers; (ii) the dipole-allowed transition energy in a graphene QD is expected to be about ten times larger than that in a GaAs quantum dot with the same size, which implies the detection wavelength of mid-infrared range for the parallel graphene QDs[7]. To date, however, the parallel QDs on a single-layer graphene have not been realized yet, although a single QD and series double QDs have been realized on a single-layer graphene[6, 8, 9, 10, 11]. Therefore, the fabrication of parallel QD system will be a fundamental step necessary for the development of graphene parallel QD single photon detectors.

In this work, we fabricated parallel-coupled quantum dots in a single-layer graphene. The tunnel coupling between the quantum dots can be tuned by a graphene in-plane gate. In a

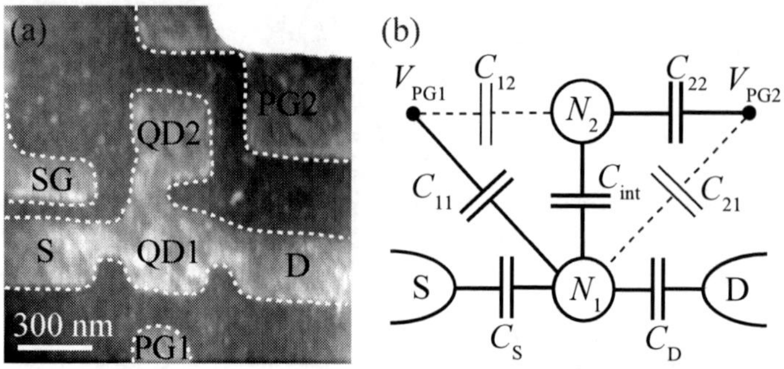

Figure 1. Atomic force microscopy image of the device studied in this work. The outlines of graphene regions are highlighted by white dashed lines. The size of the two quantum dots labeled QD1 and QD2 are 200×200 nm^2. The QD1 is connected to the source (S) and drain (D) regions through the two 100-nm-wide graphene constrictions. The QD2 is connected to QD1 through another constriction, which serves as a tunable tunnel barrier. The graphene in-plane gates labeled SG, PG1, and PG2 are defined 100 nm away from the graphene conducting channel. (b) The equivalent circuit of the parallel QD system using classical capacitance model. C_{int} is the interdot capacitance between QD1 and QD2. C_{12}, C_{11}, C_{22}, and C_{21} are the gate capacitances. C_S (C_D) is the capacitance between QD1 and source (drain) region.

weak tunnel coupling regime, the observed Coulomb blockade oscillation peaks exhibit periodic shifts as the number of electron in the non-conducting side-coupled QD is changed. The result suggests the observation of the single electron switching effect, which is a prerequisite for a single photon detection scheme using parallel-coupled quantum dots.

2. Experimental

Single-layer graphene flakes were deposited on a heavily-doped silicon substrate covered with a 290-nm-thick SiO_2 layer by the mechanical exfoliation of Kish graphite. The number of graphene layers was identified by the optical microscopy and the Raman spectroscopy. The graphene flakes were patterned by electron-beam lithography (Elionix ELS-7500) and reactive ion etching using argon and oxygen. The metal electrodes Au/Ti (40 nm/4 nm) were fabricated by the electron-beam lithography and lift-off process.

Figure 1(a) shows an atomic force microscopy image of the device studied in this work. The device consists of two quantum dots labeled QD1 and QD2. The lithographic size of each QD is 200×200 nm^2. The main QD1 is connected to the source and drain electrodes through the 100-nm-wide constrictions. The side-coupled QD2 is connected to QD1 through another constriction. In the constrictions, lateral confinement of charge carriers induces transport gap [12], hence the constrictions function as tunable tunnel barriers. The graphene in-plane gates labeled PG1 and PG2 are used to change the electrostatic potential of each QD by applying gate bias voltages V_{PG1} and V_{PG2}, respectively. The in-plane gate labeled SG is used to tune the tunnel coupling between QD1 and QD2. The measurements were performed in a dilution refrigerator at T = 100 mK. The back-gate bias voltage of V_{BG} = +3.5 V was applied to adjust the Fermi level of the graphene flake to the charge neutrality point. The current through QD1 is recorded by using the conventional low-frequency (18 Hz) lock-in technique with a small excitation voltage of V_{ac} = 10 μV.

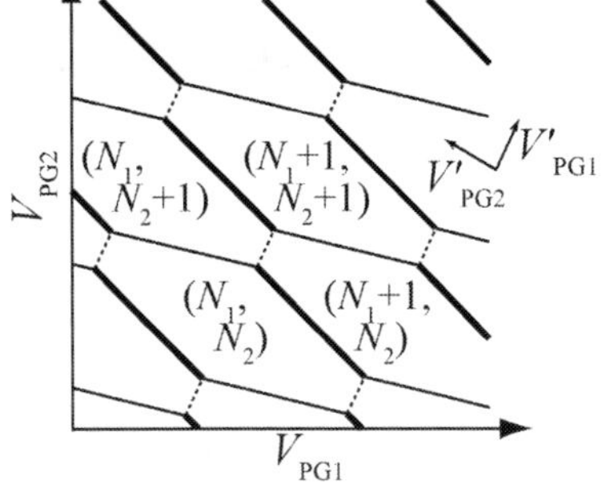

Figure 2. Charging diagram for the side-coupled quantum dot system. N_1 and N_2 indicate the electron numbers in QD1 and QD2, respectively. Thick solid (thin solid) lines correspond to the boundaries associated with the changes of the electron number in QD1 (QD2). Dotted boundaries indicate internal redistribution of the electrons between QD1 and QD2. Direction of the (V'_{PG1}, V'_{PG2}) axis is shown by the black arrows.

The parallel-coupled QD system can be modeled using classical capacitance picture as illustrated in Fig. 1(b)[1, 2]. By minimizing the total electrostatic energy of the classical capacitance model, we can construct the phase diagram that yields ground states of the parallel QD system as shown in Fig. 2. Each honeycomb cell corresponds to a different charging state of the parallel-coupled QD system with independently quantized electron numbers N_1 and N_2 in QD1 and QD2, respectively. The shapes of the honeycomb cells are determined by the capacitance values C_S, C_D, C_{int}, C_{11}, C_{12}, C_{21}, and C_{22} defined in Fig. 1(b). When V_{PG1} and V_{PG2} are in a region within a honeycomb cell, the QD1 and QD2 are in a Coulomb blockade regime, hence the conductance of the parallel QD system is completely suppressed. When V_{PG1} and V_{PG2} are changed across a boundary between honeycomb cells shown by thick solid lines, the number of electrons in QD1 is changed by one and the Coulomb blockade effect of the QD1 is lifted. Due to the tunnel coupling between QD1 and QD2, the position of the resonance line changes periodically as the number of electrons in QD2 is changed. Thus, the conductance of the main QD1 can be switched by the addition of a single electron to the non-conducting QD2, *i.e.* the single-electron switching effect, which is a characteristic transport phenomenon observed in GaAs/AlGaAs parallel-coupled quantum dot systems[1, 2].

3. Results and discussion

Figure 3(a) shows the differential conductance of the main conducting QD1 as a function of V'_{PG1} at incremented values of V'_{PG2}. The side-gate bias voltage applied to SG is V_{SG} = -2.65 V, which corresponds to the case in which the two quantum dots are very weakly coupled. V'_{PG1} and V'_{PG2} are defined by $V'_{PG1} = 0.86V_{PG1} + 0.51V_{PG2}$, $V'_{PG2} = +0.51V_{PG1}$ - $0.86V_{PG2}$, which correspond to the directions indicated by the black arrows in Fig. 2. In this plot, V'_{PG2} is changed from -55 mV to +40 mV from the bottom curve to the top curve. In the scanned range of V'_{PG1} and V'_{PG2}, the change in the tunnel coupling between QD1 and QD2 due to the capacitive crosstalk between PG1/PG2 and 100-nm-wide graphene constriction is negligibly small.

For V'_{PG2} = -55 mV (the bottom curve), we observe periodic modulation of the differential conductance as a function of V'_{PG1}. The periodic oscillation of the differential conductance is the Coulomb blockade oscillation in QD1. The observed peak corresponds to the thick solid phase boundaries between honeycomb cells in Fig. 2, across which the electron number in the main

Journal of Physics: Conference Series **334** (2011) 012041 doi:10.1088/1742-6596/334/1/012041

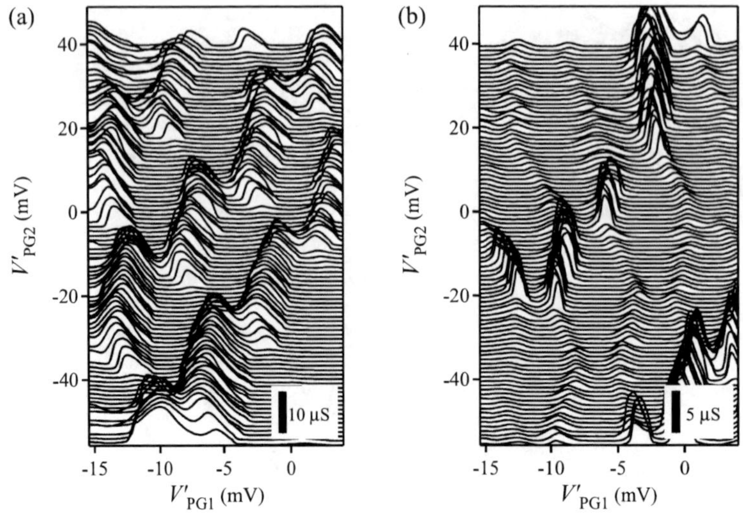

Figure 3. (a) Differential conductance dI/dV of the main conducting quantum dot as a function of V'_{PG1} and V'_{PG2} for (a) V_{SG} = -2.65 and (b) -2.35 V at T = 100 mK.

QD1 is changed by one.

When V'_{PG2} is increased, the positions of the Coulomb blockade oscillation peaks shift discontinuously toward the larger V'_{PG1} direction as the number of electron in QD2 is changed. The peak shifts occur periodically for further increasing of V'_{PG2}. The successive shifts of the Coulomb blockade oscillation peaks in the main conducting QD1 correspond to the pronounced periodic jumps of the resonance line in Fig. 2. The result suggests that the conductance of the main QD1 can be switched by the addition of a single electron to the non-conducting QD2, *i.e.* the single-electron switching effect, which is a characteristic transport phenomenon expected in parallel-coupled quantum dot systems. Here we emphasize that this is the first experimental demonstration of the single-electron switching effect in graphene-based parallel-coupled quantum dots.

Figure 3(b) shows the differential conductance of the main conducting QD1 as a function of V'_{PG1} at incremented values of V'_{PG2} from -55 mV (bottom) to +40 mV (top). The side-gate bias voltage is V_{SG} = -2.35 V, which corresponds to the case where the tunnel coupling between QD1 and QD2 is strong. We observe the Coulomb-blockade oscillation in the differential conductance as a function of V'_{PG1}. As we change V'_{PG2}, the position of conductance peaks changes continuously. The result indicates that the interdot coupling between the QDs is strong and that the observed conductance peak occurs when the total number of electron in QD1 and QD2 is changed by one. Thus, the change of the charge stability diagram suggests that the coupling between two quantum dots is modulated by the in-plane graphene side gate.

4. Summary

We have fabricated parallel-coupled quantum dots in a single-layer graphene. The tunnel coupling between the quantum dots can be tuned by a graphene in-plane gate. In a weak tunnel coupling regime, the observed Coulomb blockade oscillation peaks exhibit pronounced periodic shifts as the number of electron in the non-conducting side-coupled QD is changed, which is the first experimental observation of single-electron switching effect in graphene-based parallel-coupled quantum dots. Since a single-electron switching effect is a prerequisite for a single photon detection scheme using parallel-coupled quantum dots, this is a fundamental step for developing graphene-based parallel QD single photon detectors.

5. Acknowledgements
The authors acknowledge S. Katsumoto and T. Otsuka for helpful discussions. This work is supported by the Grant-in-Aid from MEXT, the Special Coordination Funds for Promoting Science and Technology, and Japan Science and Technology Agency.

[1] F. Hofmann, T. Heinzel, D. A. Wharam, J. P. Kotthaus, G. Bohm, W. Klein, G. Trankle, and G. Weimann, Physical Review B **51**, 13872 (1995).
[2] A. S. Adourian, C. Livermore, R. M. Westervelt, K. L. Campman, and A. C. Gossard, Superlattices and Microstructures **20**, 411 (1996).
[3] O. Astafiev, S. Komiyama, and T. Kutsuwa, Applied Physics Letters **79**, 1199 (2001).
[4] O. Astafiev, S. Komiyama, T. Kutsuwa, V. Antonov, Y. Kawaguchi, and K. Hirakawa, Applied Physics Letters **80**, 4250 (2002).
[5] A. H. Castro Neto, F. Guinea, N. M. R. Peres, K. S. Novoselov, and A. K. Geim, Reviews of Modern Physics **81**, 109 (2009).
[6] L. A. Ponomarenko, F. Schedin, M. I. Katsnelson, R. Yang, E. W. Hill, K. S. Novoselov, and A. K. Geim, Science **320**, 356 (2008).
[7] H-Y. Chen, V. Apalkov, and T. Chakraborty, Physical Review Letters **98**, 186803 (2007).
[8] C. Stampfer, J. G"uttinger, F. Molitor, D. Graf, T. Ihn, and K. Ensslin, Applied Physics Letters **92**, 012102 (2008).
[9] J. G"uttinger, C. Stampfer, S. Hellmuller, F. Molitor, T. Ihn, and K. Ensslin, Applied Physics Letters **93**, 212102 (2008).
[10] F. Molitor, S. Droscher, J. G"uttinger, A. Jacobsen, C. Stampfer, T. Ihn, and K. Ensslin, Applied Physics Letters **94**, 222107 (2009).
[11] J. G"uttinger, C. Stampfer, F. Libisch, T. Frey, J. Burgdorfer, T. Ihn, and K. Ensslin, Physical Review Letters **103**, 046810 (2009).
[12] M. Y. Han, B. Ozyilmaz, Y. Zhang, and P. Kim, Physical Review Letters **98**, 206805 (2007).

Journal of Physics: Conference Series **334** (2011) 012042 doi:10.1088/1742-6596/334/1/012042

Numerical study of electronic structure under uniform magnetic field and quantized Hall conductance for multi-band tight-binding models

Masao Arai[1] and Yasuhiro Hatsugai[2]

[1] Computational Materials Science Center, National Institute for Materials Science, Tsukuba, Japan

[2] Institute of Physics, University of Tsukuba, Tsukuba, Japan

E-mail: [1] arai.masao@nims.go.jp, [2] hatsugai@sakura.cc.tsukuba.ac.jp

Abstract. The electronic structure of periodic lattice under uniform magnetic field was studied numerically for multi-band tight-binding models with non-orthogonal basis sets. When magnetic translational symmetry is fully taken into account, computational time can be greatly reduced. Quantized Hall conductance was evaluated by robust multi-band formulation of Chern number. We found that calculated quantized Hall conductance coincides with the semi-classical results. Discontinuous jumps of Hall conductance occur at van-Hove singularities and correspond to mod q ambiguity of the Diophantine equation of Chern number.

1. Introduction

The electronic structure of periodic lattice under uniform magnetic field has been attracted continued interest for several decades. Early numerical calculations have revealed exotic characters such as self-similar energy spectrum, which is called Hofstadter's butterfly [1]. The quantized Hall conductance of periodic system has been explained from Chern number defined for occupied energy bands [2]. In addition, topological correspondence between bulk and edge states has been identified in this context. While these studies open new paradigm in electronic states under magnetic field, most of numerical calculations have been performed for simplified models under strong magnetic field which far exceeds experimental range. Therefore, these calculations have not been directly compared with experimental observation. This limitation partly arises from the symmetry restriction of magnetic translational group [3] as we must treat a large magnetic unit cell instead of the original cell. When magnetic flux penetrating into two dimensional unit cell is $\varphi = p/q$, where p and q are relatively prime integer, the magnetic unit cell becomes q times larger than the original cell. Thus, computational efforts increase significantly for small magnetic field with a large denominator q.

In this paper, we present detailed formulation for the electronic structure under uniform magnetic field. General tight-binding approximations with overlap integrals are employed to model the realistic energy bands. Uniform magnetic field is included as phase factors of both hopping and overlap matrix elements. We show that computational time can be greatly reduced by magnetic translational symmetry. As an example of such calculation, we present the calculated results for graphene with magnetic field of $\varphi = 1/200$.

2. Calculation method

Let us consider two-dimensional periodic system with primitive translational vectors $\boldsymbol{a}_1$ and $\boldsymbol{a}_2$. We adopt general tight-binding models because fully self-consistent calculations under uniform magnetic field [5] are not feasible for realistic compounds with current computer facilities. The electronic structure is calculated non self-consistently under uniform magnetic field perpendicular to the two-dimensional plane. This treatment may be justified if the magnetic field does not alter charge density significantly.

We assume that energy bands in the absence of magnetic field can be constructed from localized orbitals $\phi_{\nu\boldsymbol{R}}(\boldsymbol{r})$, which may not be orthogonal. Here, $\boldsymbol{R}$ and ν denote atomic positions and other quantum numbers such as angular momentum, respectively. Multi-band models are needed in general since energy bands degenerate typically at several symmetric points in Brillouin zone. The tight-binding models are parameterized by following matrix elements:

$$H_0(\nu, \boldsymbol{R}; \nu', \boldsymbol{R}') = \langle \phi_{\nu\boldsymbol{R}} | \hat{H} | \phi_{\nu'\boldsymbol{R}'} \rangle \tag{1}$$

$$S_0(\nu\boldsymbol{R}; \nu', \boldsymbol{R}') = \langle \phi_{\nu\boldsymbol{R}} | \phi_{\nu'\boldsymbol{R}'} \rangle . \tag{2}$$

They may be obtained from first-principles calculations by fitting of energy bands or several other methods which directly generate tight-binding parameters.

Under uniform magnetic field represented by the vector potential $\boldsymbol{A}(\boldsymbol{r}) = \frac{1}{2}\boldsymbol{B}\times\boldsymbol{r}$, the localized basis sets are modified as [6, 7]

$$\tilde{\phi}_{\nu\boldsymbol{R}}(\boldsymbol{r}) = \exp\{iI_{\boldsymbol{r},\boldsymbol{R}}\}\phi_{\nu\boldsymbol{R}}, \tag{3}$$

where

$$I_{\boldsymbol{r}',\boldsymbol{r}} \equiv \frac{ie}{\hbar c}\int_{\boldsymbol{r}}^{\boldsymbol{r}'} \boldsymbol{A}(\boldsymbol{s})d\boldsymbol{s}. \tag{4}$$

Then, the matrix elements under uniform magnetic field may be evaluated as

$$H(\nu', \boldsymbol{R}'; \nu, \boldsymbol{R}) \approx \exp\{iI_{\boldsymbol{R}',\boldsymbol{R}}\}H_0(\nu', \boldsymbol{R}'; \nu, \boldsymbol{R}), \tag{5}$$

$$S(\nu', \boldsymbol{R}'; \nu, \boldsymbol{R}) \approx \exp\{iI_{\boldsymbol{R}',\boldsymbol{R}}\}S_0(\nu', \boldsymbol{R}'; \nu, \boldsymbol{R}). \tag{6}$$

Here, we should note that above approximation is considered valid only if basis sets $\phi_{\nu\boldsymbol{R}}(\boldsymbol{r})$ are well localized around $\boldsymbol{R}$. Therefore, non-orthogonal orbitals are suitable for present calculations because they can be more localized than orthogonalized orbitals.

Since $\boldsymbol{A}(\boldsymbol{r})$ is not invariant with translation $\hat{T}(\boldsymbol{a}_i)$, we must consider magnetic translation [3] defined as

$$\hat{T}_M(\boldsymbol{a}_i) \equiv \exp\left\{\frac{ie}{2c}\boldsymbol{B}\cdot(\boldsymbol{a}_i \times \boldsymbol{r})\right\}\hat{T}(\boldsymbol{a}_i). \tag{7}$$

Because of additional phase factors, the magnetic translation operators do not commute each other and satisfy

$$\hat{T}_M(\boldsymbol{a}_1)\hat{T}_M(\boldsymbol{a}_2) = c^{i\varphi}\hat{T}_M(\boldsymbol{a}_2)\hat{T}_M(\boldsymbol{a}_2), \tag{8}$$

where φ is flux quanta penetrating into the unit cell spanned by $\boldsymbol{a}_1$ and $\boldsymbol{a}_2$. When $\varphi = p/q$ is a rational number, we can define enlarged magnetic unit cell with $\tilde{\boldsymbol{a}}_1 = q_1\boldsymbol{a}_1$ and $\tilde{\boldsymbol{a}}_2 = q_2\boldsymbol{a}_2$, where q_1 and q_2 are integers which satisfy $q_1q_2 = q$. Since the magnetic translation operators for the enlarged cell commute each other, eigen functions can be chosen to the form of magnetic Bloch function:

$$\hat{T}_M(\boldsymbol{T})\Psi_{\boldsymbol{k}} = (-1)^{pn_1n_2}e^{-i\boldsymbol{k}\cdot\boldsymbol{T}}\Psi_{\boldsymbol{k}}, \tag{9}$$

where $\boldsymbol{T} = n_1\tilde{\boldsymbol{a}}_1 + n_2\tilde{\boldsymbol{a}}_2$ is a translational vector and $\boldsymbol{k}$ is in magnetic Brillouin zone defined by reciprocal vectors $\tilde{\boldsymbol{g}}_1 = \boldsymbol{g}_1/q_1$ and $\tilde{\boldsymbol{g}}_2 = \boldsymbol{g}_2/q_2$. The additional prefactor $(-1)^{pn_1n_2}$ arises from the relation $\hat{T}_M(\boldsymbol{T}) = (-1)^{pn_1n_2}\hat{T}_M(n_1\tilde{\boldsymbol{a}}_1)\hat{T}_M(n_2\tilde{\boldsymbol{a}}_2)$. Furthermore, the eigen function Ψ satisfies

$$\hat{T}_M(\boldsymbol{a}_1)\Psi_{\boldsymbol{k}} = \Psi_{\boldsymbol{k}+\frac{p}{q_1}\tilde{\boldsymbol{g}}_2}, \tag{10}$$

$$\hat{T}_M(\boldsymbol{a}_2)\Psi_{\boldsymbol{k}} = \Psi_{\boldsymbol{k}+\frac{p}{q_2}\tilde{\boldsymbol{g}}_1}. \tag{11}$$

Namely, magnetic translations for $\boldsymbol{a}_1$ and $\boldsymbol{a}_2$ generate eigen functions at different k-points with the same energy eigenvalue. Using this relation, we need to calculate eigen functions only in $1/q$ area of magnetic Brillouin zone and can generate eigen functions in other area by above formula. This relation reduces the computational requirement to $1/q$ and enables us to treat larger magnetic unit cell, *i. e.*, small flux quanta φ.

Under uniform magnetic field $\varphi = p/q$, each band splits into q Landau subbands. When the chemical potential is located within an energy gap between Landau subbands, quantized Hall conductance σ_{xy} is given as [2, 4, 8, 9]

$$\sigma_{xy} = -\frac{e^2}{h}c_F(\mu), \tag{12}$$

where c_F is a topological integer called Chern number defined for the filled bands. The c_F can be calculated numerically by discretizing the magnetic Brillouin zone into mesh $\{\boldsymbol{k}_\ell\}$ as [9]

$$c_F(\mu) = \frac{1}{2\pi}\sum_\ell F(\boldsymbol{k}_\ell), \tag{13}$$

$$F(\boldsymbol{k}) = \mathrm{Arg}\left[u_1(\boldsymbol{k})u_2(\boldsymbol{k}+\Delta\boldsymbol{k}_1)u_2(\boldsymbol{k}+\Delta\boldsymbol{k}_2)^* u_2(\boldsymbol{k})^*\right], \tag{14}$$

$$u_\mu(\boldsymbol{k}) = \det_{n,m\in\text{occupied}}\left[\langle\Psi_{n,\boldsymbol{k}}|\Psi_{m,\boldsymbol{k}+\Delta\boldsymbol{k}_\mu}\rangle\right], \tag{15}$$

where n and m are band indices and $\Delta\boldsymbol{k}_\mu$ is a discretized momentum along $\tilde{\boldsymbol{g}}_\mu$ direction. As the c_F is integer, this expression rapidly converges with the Brillouin zone mesh and is thus suitable for numerical calculations.

3. Results and discussion

We have calculated energy bands under uniform magnetic field for several models [10, 11]. For graphene, we have treated a realistic multi-band model which consists of 4 non-orthogonal orbitals per carbon atom [10]. In Fig. 1, we show calculated Chern numbers for $\varphi = 1/200$. The energy bands split into $8 \times 200 = 1600$ Landau subbands. Since the energy gaps are small, the Chern numbers look almost continuous and are indistinguishable from semi-classical estimation of Hall conductance [10]. The Chern number changes discontinuously at the van-Hove singularities of energy bands without magnetic field.

We found that the pronounced Landau quantization occurs in whole energy region. The band widths of individual Landau subbands are negligibly small compared with the energy gaps between them except near van-Hove singularities. In some cases, several consecutive Landau subbands appear at almost same energies. Such pseudo degeneracy corresponds to the existence of multiple equivalent segments in Fermi surface without magnetic field. For example, two Landau subbands degenerate at Dirac particle region near $\varepsilon = 0$. The degeneracy of Landau subbands also occurs in other energy regions where electron or hole pockets of Fermi surface appear around K or M points.

Each Landau subband contributes to Chern number with $c_F = 1$ except at the van-Hove singularities. This is consistent with the requirement of Diophantine equation that each Landau

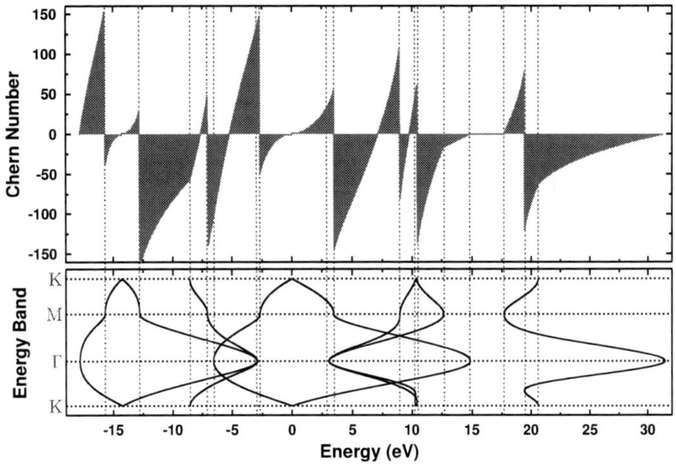

Figure 1. Calculated Chern numbers for graphene with $\varphi = 1/200$ are plotted as a function of chemical potential in upper panel. The Chern numbers for adjacent energy gaps are connected with strait lines. For comparison, lower panel shows energy bands without magnetic field with vertical dashed lines at van-Hove singularities.

subband gives $c_F = 1 + Nq$ for $\varphi = 1/q$ where N is an integer [12]. We found that the large jumps of Chern number at the van-Hove singularities can be explained by the additional factor $-q$ which corresponds to $N = -1$ solution of the Diophantine equation.

4. Summary

In this paper, computational methods for energy bands under uniform magnetic field were explained. Realistic energy bands can be treated easily with multi-band tight-binding model formulated by non-orthogonal basis sets. The calculated results for graphene with $\varphi = 1/200$ was shown as an example.

References

[1] Hofstadter D 1976 *Phys. Rev. B* **14** 2239
[2] Thouless D J, Kohmoto M, Nightingale M P and den Nijs M 1982 *Phys. Rev. Lett.* **49** 405
[3] Zak J 1964 *Phys. Rev.* **134** A1602
[4] Hatsugai Y, Fukui T and Aoki H 2006 *Phys. Rev. B* **74** 205414
[5] Cai W and Galli G 2004 *Phys. Rev. Lett.* **92** 186402
[6] Luttinger J M 1951 *Phys. Rev.* **84** 814
[7] Kita T and Arai M 2004 *J. Phys. Soc. Jpn.* **74** 2813
[8] Kohmoto M 1985 *Ann. Phys.* **160** 343–354
[9] Fukui T, Hatsugai Y and Suzuki H 2005 *J. Phys. Soc. Jpn.* **74** 1674
[10] Arai M and Hatsugai Y 2009 *Phys. Rev. B* **79** 075429
[11] Arai M and Hatsugai Y 2010 *Physica E* **42** 740
[12] Dana I Y, Avron I Y and Zak J 1985 *J. Phys. C* **18** L679

Edge states in graphene quantum Hall system with bond vs potential disorder

Mitsuhiro Arikawa[1,2], Hideo Aoki[3], Yasuhiro Hatsugai[4]

[1]Center for Computational Sciences, University of Tsukuba, Tsukuba 305-8571, Japan
[2]CREST, Japan Science and Technology Agency, Chiyoda-ku, Tokyo, 102-0075, Japan
[3]Department of Physics, University of Tokyo, Hongo, Tokyo 113-0033, Japan
[4]Institute of Physics, University of Tsukuba, Tsukuba 305-8571, Japan

E-mail: `arikawa@comas.frsc.tsukuba.ac.jp`

Abstract. While usual edge states in the quantum Hall effect (QHE) reside between adjacent Landau levels, QHE in graphene has a peculiar edge mode at $E = 0$ that resides right within the $n = 0$ Landau level as protected by the chiral symmetry. In real graphene, disorder always exists, and here we study how the edge states appear in disordered graphene QHE. We have to discriminate bond disorder and potential disorder, since the former respects the chiral symmetry in the bipartite lattice while the latter does not. We have found that, for a bond disorder, the charge accumulation along zigzag edges persists to occur, while a potential disorder destroys them. The charge accumulation along zigzag edges should be measured with an STM imaging in magnetic fields.

Quantum Hall effect (QHE) in graphene is interesting[1, 2], particularly because the chiral symmetry for the honeycomb lattice dictates the way in which the "massless Dirac" dispersion and the associated $n = 0$ Landau level in magnetic fields appear. Due to the intimate relation between the edge and bulk states in QHE[3], which is a topological effect, edge states are of crucial interest. Usual edge states in the QHE reside between adjacent Landau levels, which is related to an important general question is how the bulk and edge QHE conductions are related for finite samples. Many authors have addressed this question [4, 5], where one of the present authors has shown that the bulk QHE conductivity, a topological quantity, coincides with the edge QHE conductivity, itself another topological quantity. This exemplifies phenomena that, when a bulk system has a topological order[6, 7, 8, 9] that reflects the geometrical phase of the system[10], this should be reflected and become visible in the edge states in a bounded system [3, 11]. We have previously shown for a clean graphene that an edge mode exists at $E = 0$, and that, although the mode is embedded in the bulk Landau level, the charge is accumulated along a zigzag edge with an appreciable charge re-distribution involving the bulk states which we have called a "topological compensation"[12].

In real graphene, disorder should always exist. Especially, even when a single layer is atomically clean, there are ripples, i.e., long-ranged corrugation of the graphene plane[13]. Since the ripples consist of random bending of the honeycomb lattice, its main effect should be a modification of the hopping integral between neighboring sites in the tight-binding model[14]. The quantum Hall transition at $n = 0$ Landau level in the graphene has been shown to be anomalously sensitive to the spatial correlation of the random bonds, where a bond disorder correlated over a few lattice constatns recovers a step-function like QHE transition[15]. Here

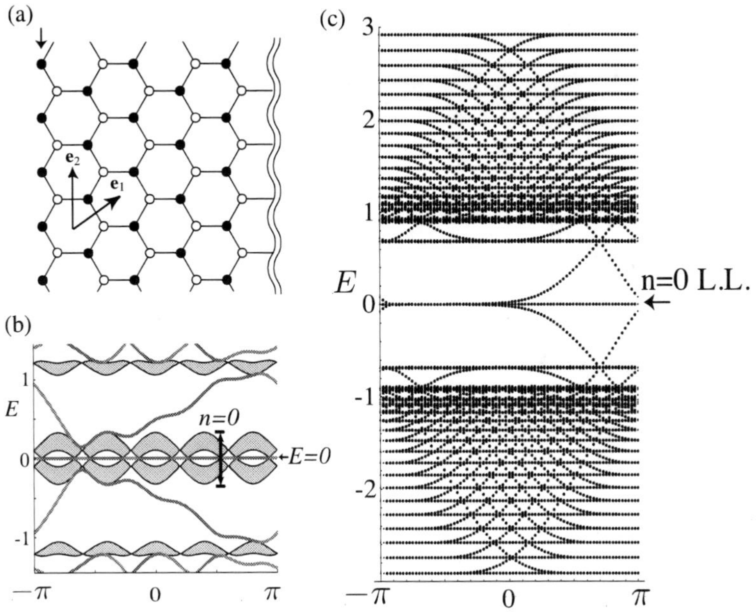

Figure 1. (a) Honeycomb lattice with zigzag edge with $\mathbf{e}_1$, $\mathbf{e}_2$ being unit translation vectors. Energy dispersion for a zigzag edge against momentum along the edge in magnetic fields $\phi = 1/5$ (b) and $\phi = 1/21$ (c) for a *clean* sample. For $\phi = 1/5$, here for clarity, with shaded regions represent the bulk energy spectra, while red (blue) curves are the modes localized on the zigzag (bearded) edge.

we combine these to ask a question: how the edge states appear in disordered graphene QHE. We have to discriminate bond disorder and potential disorder, since the former respects the chiral symmetry in the bipartite lattice, while the latter does not. So we have examined the dependence on the nature of disorder for the charge density along a zigzag edge. We consider the tight-binding model with nearest-neighbor hopping on the honeycomb lattice in a magnetic field with bond or potential disorder:

$$\mathcal{H} = -\sum_{<\mathbf{i},\mathbf{j}>}\left(t_{\mathbf{ij}}e^{\mathrm{i}\phi_{\mathbf{ij}}}c^{\dagger}_{\bullet,\mathbf{i}}c_{\circ,\mathbf{j}} + \mathrm{H.c.}\right) + \sum_{\alpha=\bullet,\circ}\sum_{\mathbf{i}}\mu_{\alpha,\mathbf{i}}c^{\dagger}_{\alpha,\mathbf{i}}c_{\alpha,\mathbf{i}}.$$

The summation $<\mathbf{i},\mathbf{j}>$ is taken over the nearest neighbor sites. Since honeycomb is a non-Bravais, bipartite lattice with two sublattice sites $\bullet$ and $\circ$ per unit cell, we can define two fermion operators $c_{\bullet,\mathbf{j}}$ and $c_{\circ,\mathbf{j}}$ with $\mathbf{j} = j_1\mathbf{e}_1 + j_2\mathbf{e}_2$ defined in Fig. 1(a) specifies the position of unit cell. The j_1 is taken as $j_1 = 0, \cdots, L_1$ for $\bullet$ lattice, and $j_1 = 0, \cdots, L_1 - 1$ for $\circ$ lattice, where the opposite side of the zigzag edge is bearded one. We apply the periodic boundary condition along the direction ($\mathbf{e}_2$) parallel to the zigzag edge. The magnetic field is introduced in terms of the Peierls phase $\phi_{\mathbf{ij}}$, so that the flux in units of the magnetic flux quantum is $\phi \equiv BS_6/(2\pi)$ in each hexagon with an area $S_6 = (3\sqrt{3}/2)a^2$. When $t_{\mathbf{ij}} = t$ and $\mu_{\alpha,\mathbf{i}} = 0$, the system is reduced to the *clean* system. Hereafter we choose t as the unit of energy. Figure 1 shows the energy spectrum for a zigzag edge in magnetic fields $\phi = 1/5$ (b) or $\phi = 1/21$ (c). For a magnetic field ($\phi = 1/21$) the $n = 0$ Landau level around $E = 0$, with a narrow width, almost looks like a line spectrum on

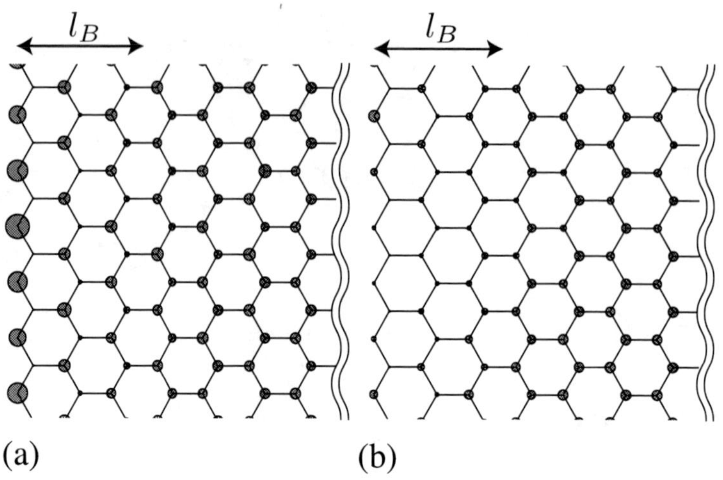

Figure 2. Charge density profile ($\propto$ area of each circle) for the energy window $|E| < 0.05$ in a magnetic field $\phi = 1/21$ with bond disorder $(W_1, \sigma) = (0.5, 0.2)$ (a) and potential disorder $(W_2, \sigma) = (0.5, 0.2)$ (b). For both cases, the system size is $(L_1, L_2) = (21, 21)$.

this plot, but we can see in (b) that an edge mode exists at $E = 0$, whose spectrum is embedded in the bulk Landau level[16, 17]. Without disorder, the $n = 0$ Landau level, including the edge mode, has a wave function amplitude accumulated along zigzag edges whose width scales with the magnetic length, l_B[12].

The bond disorder is realized when $\mu_{\alpha,\mathbf{i}} = 0$ and $t_{\mathbf{ij}} = t + \delta t_{\mathbf{ij}}$, where $\delta t_{\mathbf{ij}} = W_1 f_{\mathbf{ij}}(\sigma)$ is the strength of bond randomness. The $f_{\mathbf{ij}}(\sigma)$ is the normal Gaussian random variable with standard deviation σ. The potential order is realized when $t_{\mathbf{ij}} = t$ and $\mu_{\alpha,\mathbf{i}} = W_2 f_{\mathbf{ii}}(\sigma)$ is the strength of potential randomness. We define the operator $\gamma = \exp[\mathrm{i}\pi \sum_{\mathbf{j}} c^{\dagger}_{\bullet,\mathbf{j}} c_{\bullet,\mathbf{j}}]$, which satisfies $\gamma^2 = 1$. For a bond disorder we retain the chiral symmetry, $\{\gamma, \mathcal{H}\} = 0$, where $\{,\}$ is an anticommutation, whereas a potential disorder destroys the symmetry. Here we consider spatially uncorrelated disorders.

The charge density profiles are displayed in Fig. 2. The energy window $|E| < 0.05$ is set to just cover the $n = 0$ bulk Landau level with bond and potential disorders. For a bond disorder, the charge is accumulated along the zigzag edge (see Fig.2(a)), whose behavior is similar to those in the clean system. By contrast, for a potential disorder the charge density is depleted in the edge region, whose width scales with the magnetic length l_B (see Fig.2(b)) as in the usual QHE. Figure 3 quantifies this in terms of the local density of state at a zigzag site against energy. For a bond disorder, we can see a sharp peak at $E = 0$ in the local density of state. The local density of state also has a particle-hole symmetry, as expected for the preserved chiral symmetry for the bond disorder. These features are absent for a potential disorder. When we include the second-neighbor hopping, which also degrades the chiral symmetry in the bipartite lattice, nevertheless, the charge in $n = 0$ Landau level is accumulated along the zigzag edge[18].

In summary, we have found that, for a bond disorder, the edge mode, being protected by the chiral symmetry, still exists around $E = 0$ and embedded in the $n = 0$ bulk Landau level, where the charge is accumulated along the zigzag edge. This is no longer the case with a potential disorder. The dependence of the disorder in the charge density accumulated along the zigzag edge may be examined with the STM imaging in magnetic fields[19].

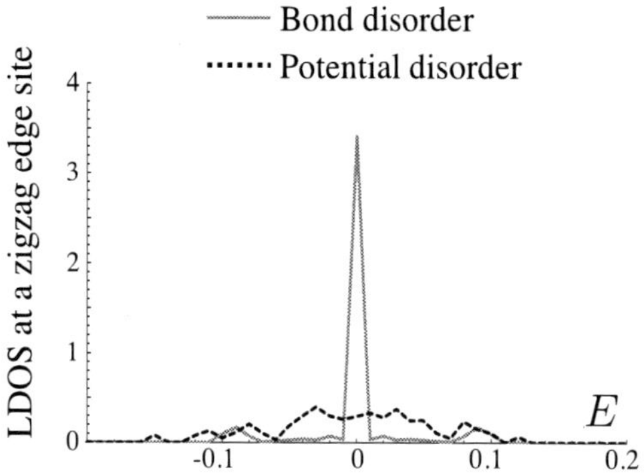

Figure 3. Local density of states at a zigzag edge site in a magnetic field $\phi = 1/21$ with bond/potential disorder. The parameters and the system size are same as in Fig.2.

Acknowledgments

This work has been supported in part by Grants-in-Aid for Scientific Research, No.20340098 (MA, HA, YH) from JSPS and 22014002 (Novel States of Matter Induced by Frustration) (MA, YH) on Priority Areas from MEXT.

References

[1] Novoselov K S, McCann E, Morozov S V, Fal'ko V I, Katsnelson M I, Zeitler V, Jiang D, Schedin F and Geim A K 2005 *Nature* (London) **438** 197
[2] Zhang Y, Tan Y W, Stormer H L, and Kim P 2005 *Nature* (London) **438** 201
[3] Hatsugai Y 1993 *Phys. Rev. Lett.* **71** 3697
[4] Laughlin R B 1981 *Phys. Rev.* B **23** 5632
[5] Halperin B I 1982 *Phys. Rev.* B **25** 2185
[6] Wen X G 1989 *Phys. Rev.* B **40** 7387
[7] Shapere A and Wilczek F ed. 1989 *Geometric Phases in Physics* (Singapore: World Scientific)
[8] Hatsugai Y 2004 *J. Phys. Soc. Jpn.* **73** 2604
[9] Hatsugai Y 2004 *J. Phys. Soc. Jpn.* **74** 1374
[10] Berry M V 1984 *Proc. R. Soc. London,* Ser. A **392** 45
[11] Hatsugai Y 1993 *Phys. Rev.* B **48** 11851
[12] Arikawa M, Hatsugai Y, and Aoki H 2008 *Phys. Rev.* B **78** 205401
[13] Meyer J C, Geim A K, Katsnelson M I, Novoselov K S, Booth T J, and Roth S 2007 *Nature* **446** 60
[14] Castro Neto A H, Guinea F, Peres N M R, Novoselov K S, and Geim A K 2009 *Rev. Mod. Phys.* **81**, 109
[15] Kawarabayashi T, Hatsugai Y, and Aoki H 2009 Phys. Rev. Lett. **103**, 156804
[16] Hatsugai Y, Fukui T and Aoki H 2006 *Phys. Rev.* B **74** 205414
[17] Hatsugai Y, Fukui T and Aoki H 2007 *Eur. Phys. J. Special Topics* **148** 133
[18] Arikawa M, Aoki H and Hatsugai Y, unpublished.
[19] Miller D L, Kubista K D, Rutter G M, Ruan M, de Heer W A, First P N and Stroscio J A 2009 *Science* **324** 924

Journal of Physics: Conference Series **334** (2011) 012044 doi:10.1088/1742-6596/334/1/012044

Manipulation of the Dirac cones and the anomaly in the graphene related quantum Hall effect

H Watanabe[1], Y Hatsugai[2] and H Aoki[1]

[1] Department of Physics, University of Tokyo, Hongo, Tokyo 113-0033, Japan
[2] Institute of Physics, University of Tsukuba, Tsukuba, 305-8571, Japan

E-mail: watanabe@cms.s.t-tokyo.ac.jp

Abstract. The quantum Hall effect in graphene is regarded to be involving half-integer topological numbers associated with the massless Dirac particle, this is usually not apparent due to the doubling of the Dirac cones. Here we theoretically consider two classes of lattice models in which we manipulate the Dirac cones with either (a) two Dirac points that have mutually different energies, or (b) multiple Dirac cones having different Fermi velocities. We have shown, with an explicit calculation of the topological (Chern) number for case (a) and with an adiabatic argument for case (b) that the results are consistent with the picture that a single Dirac fermion contributes the half-odd integer series ($\cdots$ -3/2, -1/2, 1/2, 3/2, $\cdots$) to the Hall conductivity when the Fermi energy traverses the Landau levels.

1. Introduction

In the graphene quantum Hall effect (QHE)[1, 2, 3], a most striking point is that the Hall conductivity $\sigma_{xy} = 2n + 1 (n = 0, \pm 1, ...)$ in units of $-e^2/h$ with the spin degrees of freedom dropped here, becomes $\sigma_{xy} = n + 1/2$ when we divide by two to have the contribution from each Dirac cone, since there are two of them at K and K' in graphene. The appearance of doubled Dirac cones is consistent with the Nielsen-Ninomiya theorem[4], which dictates that Dirac cones should appear in pairs as far as the chiral symmetry is present. However, a natural question is: can we do better than just dividing by two to convince ourselves on the half-integer topological numbers.

The quantization into half odd integers is of fundamental interest, but as far as lattice models are concerned, we always have a periodicity in the Brillouin zone, and the TKNN formula[5] dictates that integer topological numbers are imperative. For instance, we can conceive lattice models that is outside the applicability of Nielsen-Ninomiya, but we still end up with integer Hall conductivity. Here we take another approach to explore this problem: We theoretically consider two classes of lattice models in which we manipulate the Dirac cones with either (a) the two Dirac points that have mutually different energies[6], or (b) multiple Dirac cones with different Fermi velocities. With explicit calculations of the topological (Chern) number, we shall show that the results for systematically manipulated Dirac cones are consistent with the picture that a single Dirac fermion contributes the half odd integer series ($\cdots$ -3/2, -1/2, 1/2, 3/2, $\cdots$) to the Hall conductivity when the Fermi energy traverses the Landau levels.

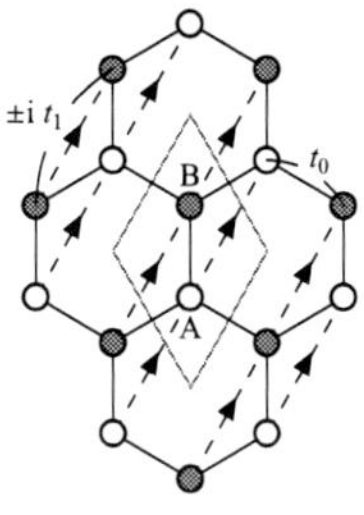

(a)

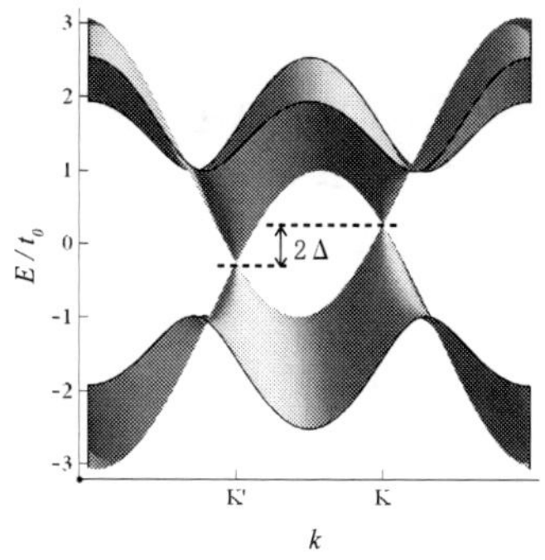

(b)

Figure 1. (a)The honeycomb lattice model considered here with nearest-neighbor hopping t_0 (solid lines) and second-neighbor hopping it_1 along each dashed arrow, $-it_1$ in the opposite direction. A unit cell and A, B sublattices are indicated. (b) The dispersion relation in the present model, seen horizontally, with Dirac cones shifted in energy by $\Delta = \sqrt{3}t_1$.

2. Shifted Dirac cones

A simplest way to realize a model in which two Dirac cones are preserved but have shifted energies is to add a term that is proportional to σ_0 (unit matrix) with a k-dependent coefficient in the space spanned by Pauli matrices[6]. So we introduce a lattice model with a Hamiltonian,

$$\mathcal{H} = \sum_{\boldsymbol{k}} \hat{c}^\dagger_{\boldsymbol{k}\alpha}[h^{\rm gr}_{\boldsymbol{k}} + 2t_1(\sin k_1)\sigma_0]_{\alpha,\beta}\hat{c}_{\boldsymbol{k}\beta}, \tag{1}$$

$$h^{\rm gr}_{\boldsymbol{k}} = t_0\left[(1+\cos k_1 + \cos k_2)\sigma_1 + (\sin k_1 + \sin k_2)\sigma_2\right],$$

where σ_i's are Pauli matrices and α, β denote their components. Namely, we have added, on top of the nearest-neighbor hopping t_0, an extra hopping involving σ_0 times a $\boldsymbol{k}$-dependent function. We then lift the degeneracy of the energies of K and K' if the $\boldsymbol{k}$-dependent term has different values between them. A simplest choice is $\propto \sin k_1$. If we go back to the real space, the tight-binding model is as depicted in Fig.1(a), which has extra second-neighbor hoppings on top of the nearest-neighbor ones. The added hoppings has to be only between A-A and B-B for the Dirac cone to be preserved, and they have to be complex for the degeneracy between K and K', mutually time-reversal pairs, to be broken. The model with a complex hopping is fictitious, but we do accomplish shifted Dirac cones as depicted in Fig.1(b). In this model the chiral symmetry is broken, since σ_0-term in $\mathcal{H}$ invalidates $\{\mathcal{H}, \sigma_z\} = 0$. Nevertheless, the addition of σ_0 preserves the shape of Dirac cones, along with the species doubling. If we expand the Hamiltonian (1) around $\boldsymbol{k}_0$ (K or K'), we have $h_{\boldsymbol{k}} \simeq \chi\Delta\,\sigma_0 - \hbar v_F\left[(-1)^{\frac{1-\chi}{2}}\delta k_x\sigma_1 + \delta k_y\sigma_2\right]$, where $\chi = +1(-1)$ correspond to K (K'), $\Delta = \sqrt{3}t_1$ is (half) the shift, $v_F = \sqrt{3}a\,t_0/2\hbar$ the Fermi velocity, and $\delta\boldsymbol{k} = \boldsymbol{k} - \boldsymbol{k}_0$.

3. Chern numbers

The calculation of the Hall conductivity, which is a topological Chern number, has to be calculated carefully, since we have to sum up over the contribution from the "Dirac sea". We can overcome the difficulty with a method employing non-commutative Berry's connection[7] and its integration over the Brillouin zone to estimate the Chern number with a technique developed in the lattice gauge theory.[8] The result for the Chern number in the present model is shown in Fig.2(b), while Fig.2(a) is for the the ordinary graphene for comparison.

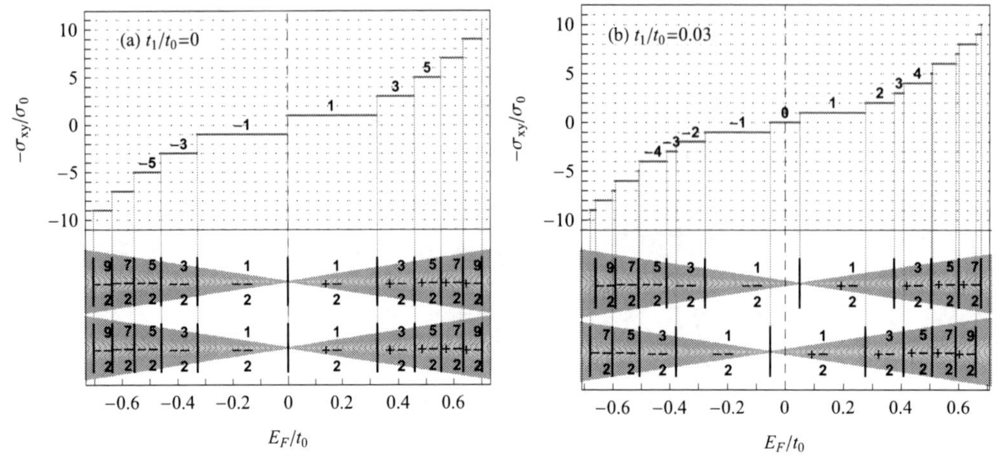

Figure 2. Numerically calculated Chern numbers against E_F (upper parts), along with the Landau levels for each of the two Dirac cones with the Chern numbers in each gap displayed (lower parts) for the ordinary graphene ($t_1 = 0$) (a) and for the present model with shifted Dirac cones ($t_1 \neq 0$) (b). The result is for the magnetic flux $\phi = 1/100$.

Figure 2(b) shows that the result is exactly what we expect when we sum the two half-odd-integer series (..., -3/2, -1/2, 1/2, 3/2, $\cdots$) with a shift in energy as E_F traverses shifted sets of Landau levels (as shown in the lower part in the figure). In other words, in a striking contrast to the ordinary graphene where each QHE step has a jump of 2 in the Chern number, the present model exhibits a jump of 1 at each step[6]. The agreement is rather surprising (since there is no apriori reason why the superposition of effective field theory for the vicinities of K and K' and the lattice model should have the same Chern numbers). Thus, although we have still no half integers for the total Hall conductivity (since a sum of two half-odd integers is an integer), we have indirectly confirmed the half integers. Since no half integers should appear according to TKNN, this is indeed as best as we can confirm the half-integer property of each Dirac cone.

4. Multiple Dirac cones as higher-pseudospin representations of SU(2)

The effective theory for the Dirac cones in graphene is expressed in terms of Pauli matrices, which is a representation of SU(2) with pseudospin 1/2. Now we pose a question: can we extend this to examine whether the generalized model has half-integer contributions to the Chern number as well? For the two-band case (which may be regarded as having a pseudospin 1/2) the each matrix is 2×2, which corresponds to two (A, B) sublattices. Here we show that a realization of higher pseudospin S Dirac cones is possible, where each matrix is $(2S+1) \times (2S+1)$ corresponding to $(2S+1)$ sublattices. By a pseudospin S Dirac cone we mean an effective Hamiltonian around the Dirac point,

$$\mathcal{H} = k_1\Sigma_1 + k_2\Sigma_2 + \mathcal{O}(k^2), \quad (2)$$

where Σ_i's are $(2S+1)$-dimensional representations of SU(2).

One such model actually appears in the "flat-band model", where one band is flat in a mult-band model.[9] Namely, in a Lieb's model for the flat band, $S = 1$ SU(2) is realized for a three-band case, where two bands form a Dirac cone, while the other is flat. We can generalize this to realize for general S with $(2S+1)$ sites in a unit cell, although the models can again be unrealistic, but does serve as examining half-integer Chern numbers.

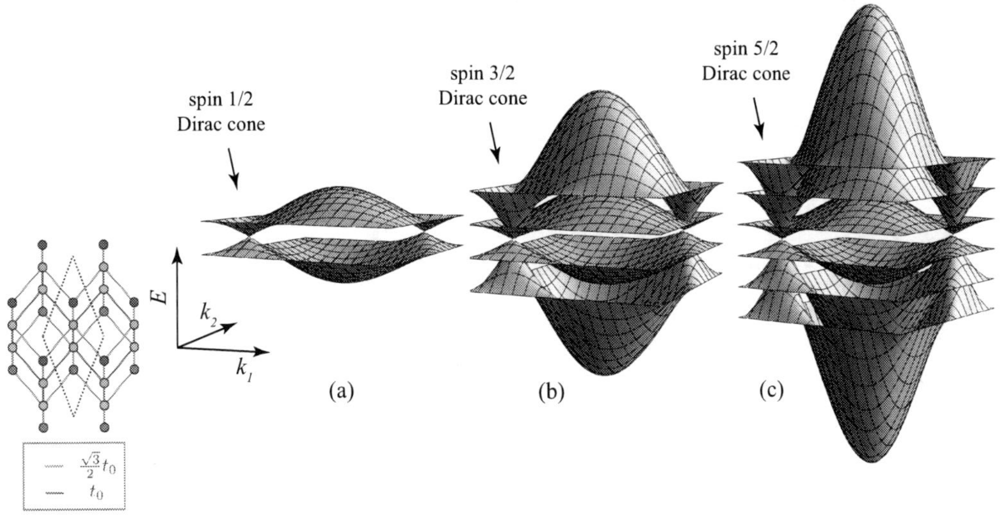

Figure 3. Dispersions for the pseudospin S SU(2) models that contain $(S+1/2)$ Dirac cones for $S=1/2$ (usual graphene) (a), $S=3/2$ (b), and $S=5/2$ (c). Inset depicts the lattice model for (b).

Although general values of S are permissable, here we focus on half-integer $S = 1/2, 3/2, 5/2, \cdots$ with $2S+1$ even (while the case of an integer S contains a flat band in the band structure). We can construct the lattice models that realize these pseudspins by inserting extra atoms on edges of each hexagon with real hoppings between them. In this case $(2S+1)/2$ cones appear around each of K and K' points as displayed in Fig. 3. Let us first look at the Landau level structure (E_n against n) in Fig. 4. This can be obtained by quantizing eqn.(2) with k replaced with $k+eA$. The Landau levels have $(2S+1)$ sequences that correspond to electron and hole Landau levels originating from the $(S+1/2)$ Dirac cones with different Fermi velocities, and they asymptotically approach $\hbar\omega_c\sqrt{n}S_z$, $(S_z = -S, -S+1, \cdots, S)$ for large n. The question is the Hall conductivity contributed by the multiple Dirac cones around each of K and K'. Figure 5 displays the result for $S=5/2$, which is obtained from an effective field theory as follows. We first add a mass term $m\Sigma_3$ to calculate the Chern number in zero magnetic field, following the idea of Haldane[10], and then apply a magnetic field to have Landau levels and let $m \to 0$. Topological numbers should be invariant in such an adiabatic process. In Fig.5 for the contribution to the Chern numbers of each single $S=5/2$ valley from the three sets of Landau levels associated with the three Dirac cones with different Fermi velocities, we do have half-odd integer series, $\cdots, -5/2, -3/2, 3/2, 5/2, \cdots$ as another generalization of the half-integer contribution from each Dirac cone. For the original lattice model we have two Diract points, so we can apply the shift between the two Dirac points to resolve them as discussed in the first half of this paper, which is a future problem.

Recently it is recognized[11] that massless Dirac fermions can be realized as surface states of the three-dimensional topological (quantum spin Hall) systems as one manifestation of the bulk-edge correspondence[13, 12]. There the doubling partner exists at the other side of the system, so that the decomposition of the topological numbers into contributions from each Dirac cone may become a realistic as well as intriguing question. The work was supported in part by grants-in-aid for scientific research No. 20340098 (YH and HA) from JSPS and No. 22014002(YH) on priority areas from MEXT.

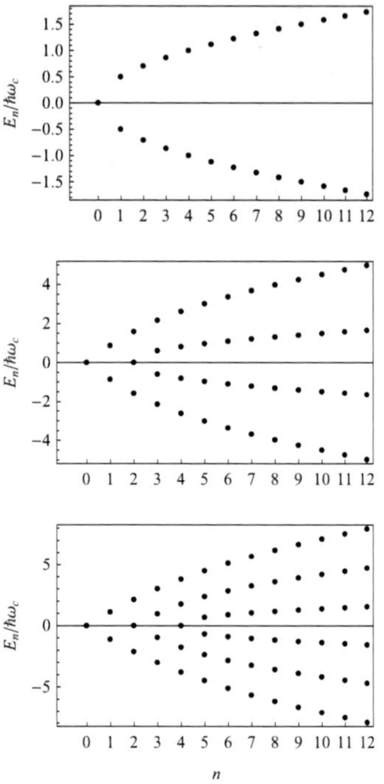

Figure 4. Landau levels E_n against n for the pseudspin $S = 1/2$ (ordinary graphene) (a), $S = 3/2$ (b), and $S = 5/2$ (c).

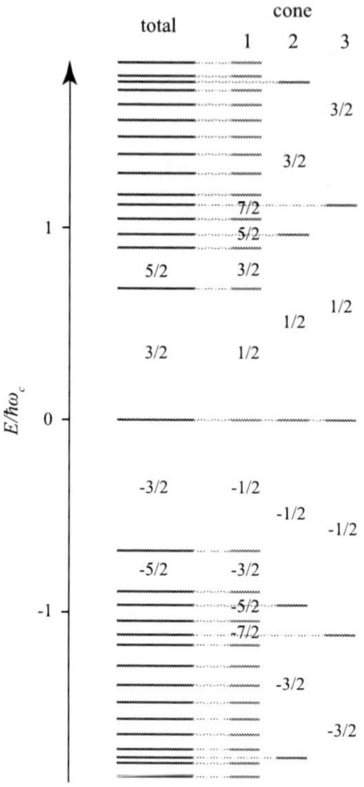

Figure 5. The Hall conductivity from each single valley for $S = 5/2$ (left column) and its decomposition into the contributions from three Dirac cones (right).

References

[1] Novoselov K S et al, *Nature* **438**, 197 (2005); Geim A K and Novoselov K S, *Nature Materials* **6**, 183 (2007).
[2] Zhang Y, Tan Y W, Stormer H L, and Kim P, *Nature* **438**, 201 (2005).
[3] Castro Neto A H et al, *Rev. Mod. Phys.* **81**, 109 (2009).
[4] Nielsen H B and Ninomiya M, *Nucl. Phys.* B **185** 20 (1981), **193** 173 (1981).
[5] Thouless D J, Kohmoto M, Nightingale M P, and den Nijs M, *Phys. Rev. Lett.* **49**, 405 (1982).
[6] Watanabe H, Hatsugai Y and Aoki H, submitted.
[7] Hatsugai Y, *J. Phys. Soc. Jpn.* **73**, 2604 (2004).
[8] Fukui T, Hatsugai Y, and Suzuki H, *J. Phys. Soc. Jpn.* **74**, 1674 (2005).
[9] Aoki H, Ando M, and Matsumura H, *Phys. Rev. B* **54**, R17296 (1996).
[10] Haldane F D M, *Phys. Rev. Lett.* **61**, 2015 (1988).
[11] Hasan M Z and Kane C L, arXiv:1002.3895.
[12] Hatsugai Y, Fukui T, and Aoki H, *Phys. Rev.* B **74**, 205414 (2006); *Eur. Phys. J.* Special topics **148**, 133 (2007).
[13] Hatsugai Y, *Phys. Rev.* B **48**, 11851 (1993).

HMF-19 IOP Publishing
Journal of Physics: Conference Series **334** (2011) 012045 doi:10.1088/1742-6596/334/1/012045

Dynamical scaling analysis of the optical Hall conductivity in the graphene quantum Hall system with various types of disorder

Takahiro Morimoto[1], Yshai Avishai[2] and Hideo Aoki[1]

[1] Department of Physics, University of Tokyo, Hongo, Tokyo 113-0033, Japan
[2] Department of Physics, Ben Gurion University, Beer Sheva 84105, Israel

E-mail: `morimoto@cms.phys.s.u-tokyo.ac.jp`

Abstract. Dynamical scaling of the optical Hall conductivity $\sigma_{xy}(\varepsilon_F, \omega)$ at the $n = 0$ Landau level in graphene is analyzed for the 2D e ective Dirac fermion and honeycomb lattice models with various types of disorder. In the Dirac fermion model with potential disorder, $\sigma_{xy}(\varepsilon_F, \omega)$ obeys a well-de ned dynamical scaling, characterized by the localization exponent   and the dynamical critical exponent z. In sharp distinction, scaling behavior of $\sigma_{xy}(\varepsilon_F, \omega)$ in the honeycomb lattice model with bond disorder (preserving chiral symmetry), becomes anomalous.

1. Motivation

Study of the quantum Hall e ect (QHE) has been primarily focused on static properties. Yet, dynamical aspects are expected to display a wealth of interesting phenomena. For the QHE in graphene, investigation of dynamical response is rather scarce. Two of the present authors (in collaboration with Hatsugai), studied the ac Hall e ect in the THz frequency regime. The Hall conductance (although not quantized in ac), is shown to display plateau structure resulting from the localization of states. [1] While the localization dominates the QHE already in the static case, here we are talking about the ac Hall conductivity, so that the most reliable way to examine the dynamics of localization is to apply the dynamical scaling argument. [2]

The QHE in graphene is peculiar due to the occurrence of $n = 0$ Landau level (LL) around the Dirac point where electron and hole branches intersect. It has no counterpart in the ordinary QHE[3, 4], and hence, static and dynamic scaling are of special interest in the $n = 0$ LL. Analysis of localization delocalization transition in QHE has been mostly focused on the scaling behavior of the dc *longitudinal* conductivity $_{xx}(\varepsilon_F)$, where ε_F is the Fermi energy. [5] On the other hand, *dynamical* scaling properties of the longitudinal ac conductivity $_{xx}(\varepsilon_F, \omega)$ and ac Hall conductivity $_{xy}(\varepsilon_F, \omega)$ have not yet been thoroughly addressed, less so in the graphene QHE.

Experimentally, scaling properties of $_{xx}(\varepsilon_F, \omega)$ was investigated in frequency range $\omega = 0$ [6] up to the GHz regime [7], while the relevant (i.e., $\sim$ cyclotron energy) frequency range is THz. Recent advances in THz measurements (e.g., Faraday rotation in magnetic elds) enable us to study the dynamical response functions [8, 9]. Study of optical properties of graphene develops rather intensely, including experimental transmission spectra[10], or theoretical examination of the cyclotron emission[11]. Thus, the physics of dynamical scaling in graphene QHE is experimentally accessible in the THz regime.

Theoretically, an intriguing question is how the static Hall conductivity, which may be regarded as a topological quantity[12, 13], evolves into the optical Hall conductivity, especially in the THz regime where the relevant energy scale is the cyclotron energy. [14, 1, 15] It has recently been shown that the plateau structure of $\sigma_{xy}(\omega)$ is retained in the ac ($\sim$ THz) regime in both the ordinary two-dimensional electron gas and the massless Dirac model, although the plateau height in the ac conductivity deviates from the quantized values.[1] The plateaus remain remarkably robust against disorder, which can be attributed to localization. To gain a deeper understanding of the robust step structures in the ac regime, we need to elucidate the static and dynamical scaling behavior of the response functions near the plateau to plateau transition. For finite frequencies we have a new length scale, $L_\omega \sim \omega^{-\frac{1}{z}}$, where z is the dynamical critical exponent. A proper quantity for carrying scaling analysis is the width W of the transition region, which generically depends on ω through L_ω and on the sample size L.

An interest in graphene QHE also comes from the relationship between the chiral symmetry or sublattice symmetry in graphene and the anomalous $n = 0$ LL. If the disorder respects the chiral symmetry which is the case with bond disorder or random magnetic fields, $n = 0$ LL remains anomalously sharp [16]. Hence, for the first time in this paper, we also discuss what happens to the scaling behavior in the honeycomb lattice (tight-binding) model, with bond disorder that preserves chiral symmetry.

2. Formalism

To study graphene QHE we first employ the two-dimensional effective 2D Dirac fermion model in a magnetic field with random potential,

$$H = v_F \boldsymbol{\sigma}\cdot\boldsymbol{\pi} + V(\mathbf{r}), \tag{1}$$

where $\boldsymbol{\sigma} = (\sigma_x, \sigma_y)$ is a two-component vector of Pauli matrices, and $\boldsymbol{\pi} = \mathbf{p} + e\mathbf{A}$ is the covariant momentum. Disorder is introduced by a random potential $V(\mathbf{r})$ composed of Gaussian scattering centers of range d and amplitude $\pm u$ with a density $n_{\rm imp}$ placed at randomly chosen positions $\mathbf{R}_j$. A measure of disorder (the Landau level broadening) is $\Gamma = 2u[n_{\rm imp}/2\pi(\ell^2 + d^2)]^{1/2}$, where ℓ is the magnetic length and we chose $d = 0.7\ell$.

Diagonalization of the Hamiltonian is carried out within subspace spanned by the five LLs around $n = 0$ LL. The system's geometry is a square of size $L \times L$ with $L/\ell = 25, 30, 35, 40$. In terms of the eigenfunctions and the energy eigenvalues ε_a, the optical Hall conductivity is evaluated in terms of the Kubo formula,

$$\sigma_{xy}(\varepsilon_F, \omega) = \frac{i\hbar e^2}{L^2} \sum_{\epsilon_a < \varepsilon_F} \sum_{\epsilon_b \geq \varepsilon_F} \frac{1}{\varepsilon_b - \varepsilon_a} \left(\frac{j_x^{ab} j_y^{ba}}{\varepsilon_b - \varepsilon_a - \hbar\omega} - \frac{j_y^{ab} j_x^{ba}}{\varepsilon_b - \varepsilon_a + \hbar\omega} \right), \tag{2}$$

where j_x^{ab} is the current matrix element[1]. The conductance is then averaged over a few thousands samples with different disorder potential realizations.

The optical Hall conductivity $\sigma_{xy}(\varepsilon_F, \omega)$ as a function of the Fermi energy ε_F and frequency ω is displayed for the graphene (Fig.1(a)) QHE systems, which clearly exhibits plateau structure discussed in [1].

3. Dynamical Scaling Analysis

For a finite system size L and a finite frequency ω, dynamical scaling analysis of the optical Hall conductivity $\sigma_{xy}(\varepsilon_F, \omega, L)$ and the width $W(\omega, L)$ of the plateau to plateau transition are carried out by inspecting the *ac step structure* as follows:[1, 2] The steepness of the transition

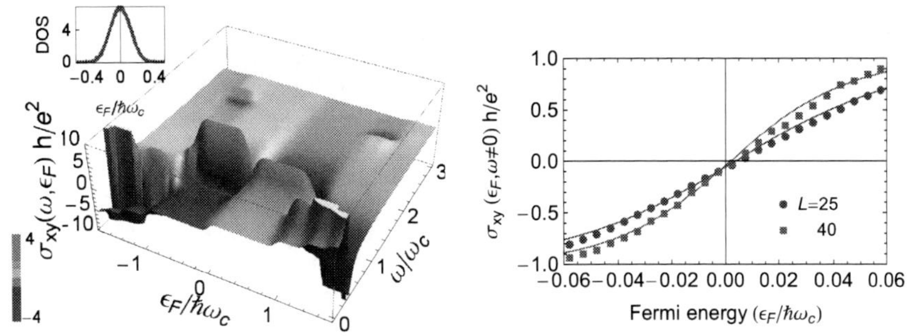

(a) (b)

Figure 1. (a) Optical Hall conductivity $\sigma_{xy}(\varepsilon_F, \omega)$ plotted against Fermi energy ε_F and frequency ω for graphene quantum Hall system with disorder parameter $\Gamma = 0.4\hbar\omega_c$. Inset is the density of states for $n = 0$ LL. (b) $\sigma_{xy}(\varepsilon_F, \omega)$ plotted against ε_F for the graphene QHE system of sizes $L = 25, 40$ at a fixed $\omega = 6\omega_c/L^2$). The solid lines represent a fit with eqn.(3).

can be quantified by fitting $\sigma_{xy}(\varepsilon_F, \omega, L)$ around the transition region for a given LL to a smooth function of ε_F. For the transition $\sigma_{xy}/(e^2/h) = -1 \to 1$ in graphene QHE we chose

$$\sigma_{xy}(\varepsilon_F, \omega, L)_{\text{graphene}} = \tanh\left(\frac{\varepsilon_F - \varepsilon_0}{W(\omega, L)}\right), \tag{3}$$

where ε_0 is a very small shift of the center of the tanh function to achieve a better fit. The quality of fitting of the plateau to plateau transition by the tanh function is quite satisfactory as can be deduced from Fig.1(b).

Dynamical scaling analysis for $\sigma_{xy}(\varepsilon_F, \omega, L)$ is carried out in a similar manner as that for the longitudinal conductivity [17]. In this ansatz the optical Hall conductivity depends on the Fermi energy and frequency only through the ratios L/ξ and L_ω/ξ. The relevant exponents are the localization exponent ν that describes the localization length $\xi \sim 1/|\varepsilon_F - \varepsilon_c|^{\nu}$ where ε_c is the critical energy which coincides with the center of the LL, while a frequency dependent length scale L_ω behaves like $L_\omega \sim 1/\omega^{1/z}$ with the dynamical critical exponent z. Then the dynamical scaling ansatz for the optical Hall conductivity reads

$$\sigma_{xy}(\varepsilon_F, \omega, L) = \frac{e^2}{h} F((\varepsilon_F - \varepsilon_c)L^{1/\nu}, \omega L^z), \tag{4}$$

where F is a universal scaling function. This implies that the width of the plateau to plateau transition scales as,

$$W(\omega, L) = L^{-1/\nu} f(\omega L^z), \tag{5}$$

where f is another universal function deduced from F. The first factor, which is responsible for the plateau structure in the static QHE, makes the step sharper for larger systems, while the second factor f describes the dynamical behavior.

Fitting the conductivity with eqn.(3) yields W for each system size L and frequency ω. In Fig.2 we show the scaling of the inverse transition width for the graphene QHE system. If we first inspect the inverse width $1/W(\omega = 0)$ for the static case as a function of L, the localization critical exponent is obtained from the expression $\log 1/W(\omega = 0) = 1/\nu \log L + f(0)$, with the result $\nu = 2.1 \pm 0.1$. This agrees with (albeit slightly smaller than) the accepted value of the static critical exponent in the integer 2DEG QHE.

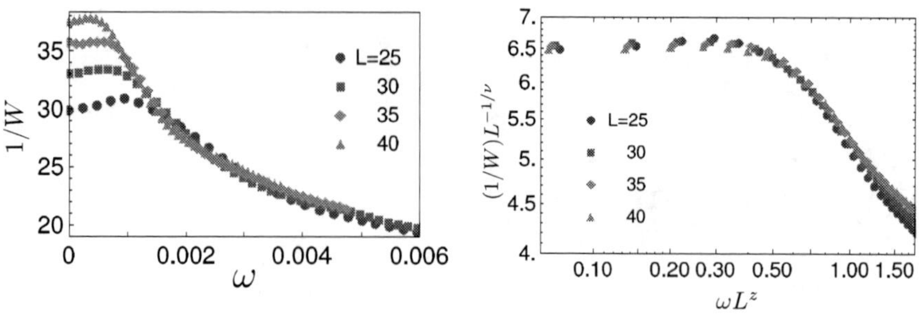

(a) (b)

Figure 2. Dynamical scaling analysis for the graphene QHE system with $\Gamma = 0.4\hbar\omega_c$ and $L = 25 \quad 40$: (a) The inverse width $1/W$ plotted against the frequency ω, and (c) the rescaled inverse width $\frac{1}{W}L^{\ 1/}$ plotted against the rescaled frequency ωL^z with a tted dynamical critical exponent $z = 1.8 \quad 0.2$.

The frequency dependence of the inverse width for a xed system size L is displayed in (Fig.2(a)). We can immediately notice that there are two regions: In the rst one, $1/W$ stays nearly constant up to some critical frequency that depends on the system size L, while in the higher-frequency region it begins to decrease monotonically with ω. In the latter range, $1/W$ assumes similar values for all the sample sizes considered here as shown in (Fig.2(a)). Roughly speaking then, in the rst region $L < L_\omega$ (L_ω is a typical distance over which an electron travels during one cycle of the ac eld), W is almost independent on ω, while in the second region $L > L_\omega$ W is virtually independent of L. Inspecting eqn.5, and assuming a power-law for the scaling function f, the L-independent inverse width in this region takes a form $1/W(\omega) \propto \omega^{\ 1/z}$. Evaluation of the dynamical exponent should be carried out near the critical point at the crossing region where $1/W$ begins to decrease. This happens, typically, for $\omega < 0.002\omega_c$. With a least-square tting, $\log 1/W(\omega) = \text{const} \quad \frac{1}{z}\log\omega$, the dynamical critical exponent for the graphene QHE system is obtained as $z = 1.8 \quad 0.2$.

Having derived the static and dynamic exponents, we are in position to look at the scaling plot against the rescaled inverse width against rescaled frequency in Fig.2(b) for graphene QHE. It can be judged that the scaling is obeyed within the numerical accuracy, which indicates that the form of the universal function assumed in eqn(5) is adequate. In the rst region (with small ωL^z) we basically have a constant value of $\frac{1}{W}L^{\ 1/}$, while in the second region we have monotonic decrease of $1/W$ for increasing ωL^z

4. Bond disorder, chiral symmetry and anomalous $n = 0$ LL

In order to consider the e ect of another type of disorder, i.e. bond disorder, we now study the honeycomb tight-binding model with a Hamiltonian,

$$H = \sum_{\langle i,j\rangle} (\ t + \ t_{ij}) e^{\ 2 \ i \ ij} c_i^\dagger c_j + \text{h.c.}.$$

Here t_{ij} is the random component in the transfer energy, for which we assume a spatially correlated disorder to represent the ripples in graphene. The random-bond disorder respects the chiral symmetry[16].

Figure 3 displays $_{xy}(\varepsilon_F, \omega)$ for a bond disorder $\sqrt{\langle \ t^2 \rangle} = 0.1t$. While the overall structure of the $_{xy}(\varepsilon_F, \omega)$ is similar to the case of potential disorder, a closer look at the $n = 0$ step (Fig.3(b)) immediately reveals that $_{xy}(\varepsilon_F, \omega)$ has an anomalously sharp step structure at the

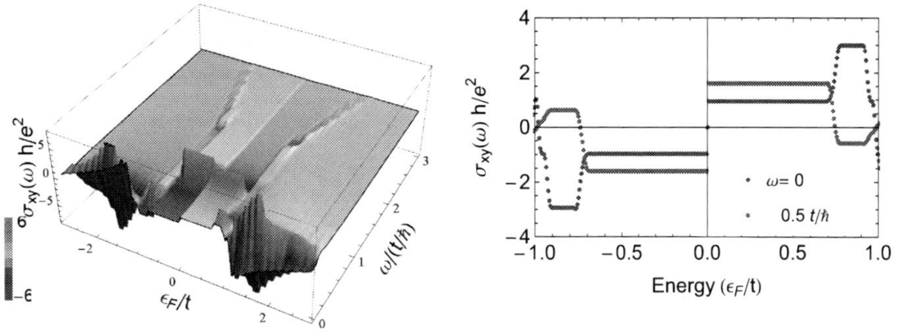

(a) (b)

Figure 3. (a) $_{xy}(\varepsilon_F, \omega)$ plotted against Fermi energy ε_F and frequency ω for the honeycomb tight-binding model with random hopping $\sqrt{\langle \ t^2 \rangle} = 0.1t$ which respects the chiral symmetry. (b) $_{xy}(\varepsilon_F, \omega)$ with $\omega = 0, 0.5t/\hbar$ is displayed to show the anomalous step structure at $n = 0$ LL in the ac Hall conductivity. The parameters are $\eta/a = 2.0$, $/ _0 = 1/18$, $(L_x/(\sqrt{3}a/2), L_y/(3a/2)) = (27, 24)$

$n = 0$ LL. So the step-function like anomaly persists for the optical Hall conductivity as in the dc Hall conductivity for the chiral-symmetric disordered honeycomb system. Indeed, Fig.3(a) shows that $_{xy}(\varepsilon_F = 0, \omega)$ continues to be a sharp step for all the value of ω, which may be thought of as coming from the sharp (delta-function like) $n = 0$ LL due to the chiral symmetry. This means that $W(\omega, L) = 0$ for all the value of ω for $n = 0$ LL, which implies that the dynamical scaling becomes ill-de ned in this case. So the applicability of the dynamical scaling argument depends on the type of disorder, and the presence or otherwise of the chiral symmetry in the disorder in particular.

To summarize, a dynamical scaling analysis of the optical Hall conductivity in graphene quantum Hall systems has been carried out in the presence of potential disorder, and the e ect of bond disorder which is chiral symmetric is also argued in terms of anomalous $n = 0$ LL.

References

[1] Morimoto T, Hatsugai Y and Aoki H 2009 *Phys. Rev. Lett.* **103** 116803
[2] Morimoto T, Avishai Y and Aoki H 2010 *Phys. Rev. B* **82** 081404
[3] Novoselov K, Geim A, Morozov S, Jiang D, Katsnelson M, Grigorieva I, Dubonos S and Firsov A 2005 *Nature* **438** 197–200
[4] Zhang Y, Tan Y, Stormer H and Kim P 2005 *Nature* **438** 201
[5] Huckestein B 1995 *Rev. Mod. Phys.* **67** 357–396
[6] Koch S, Haug R, Klitzing K and Ploog K 1991 *Phys. Rev. Lett.* **67** 883–886
[7] Hohls F, Zeitler U, Haug R, Meisels R, Dybko K and Kuchar F 2002 *Phys. Rev. Lett.* **89** 276801
[8] Ikebe Y and Shimano R 2008 *Appl. Phys. Lett.* **92** 012111
[9] Ikebe Y, Morimoto T, Masutomi R, Okamoto T, Aoki H and Shimano R 2010 *Phys. Rev. Lett.* **104** 256802
[10] Sadowski M L, Martinez G, Potemski M, Berger C and de Heer W A 2006 *Phys. Rev. Lett.* **97** 266405
[11] Morimoto T, Hatsugai Y and Aoki H 2008 *Phys. Rev. B* **78** 073406
[12] Thouless D J, Kohmoto M, Nightingale M P and den Nijs M 1982 *Phys. Rev. Lett.* **49** 405–408
[13] Hatsugai Y 1993 *Phys. Rev. Lett.* **71** 3697–3700
[14] Volkov V and Mikhailov S 1985 *JETP Letters* **41** 476
[15] Fialkovsky I V and Vassilevich D V 2009 *J. Phys. A: Math. Theor.* **42** 442001
[16] Kawarabayashi T, Hatsugai Y and Aoki H 2009 *Phys. Rev. Lett.* **103** 156804
[17] Avishai Y and Luck J 1996 *arXiv:cond-mat/9609265*

Journal of Physics: Conference Series **334** (2011) 012046 doi:10.1088/1742-6596/334/1/012046

Many-body corrections to cyclotron resonance in graphene

K. Shizuya

Yukawa Institute for Theoretical Physics, Kyoto University, Kyoto 606-8502, Japan

E-mail: shizuya@yukawa.kyoto-u.ac.jp

Abstract. We report our recent study of cyclotron resonance in graphene, with focus on the many-body e ect on the resonance energies. The genuine many-body corrections turn out to derive from vacuum polarization, speci c to graphene, which diverges at short wavelengths and which requires renormalization of velocity and, for bilayer graphene, interlayer coupling as well. As a result, the renormalized velocity and interlayer coupling strength run with the magnetic eld, and many-body corrections are uniquely determined from one resonance to another. Theory is compared with the experimental data for both monolayer and bilayer graphene.

1. Introduction

Graphene and its multilayer give rise to rich spectra of cyclotron resonance, with resonance energies varying from one transition to another within the electron band or the hole band, and, notably, even between the two bands [1]. This is in sharp contrast with conventional quantum Hall (QH) systems with a parabolic dispersion, where cyclotron resonance takes place between adjacent Landau levels, hence at a single frequency $\omega_c = eB/m$ which, according to Kohn's theorem [2], is una ected by Coulomb interactions [3]. The nonparabolic electronic spectra in graphene evade Kohn's theorem and o er the possibility to detect the many-body corrections to cyclotron resonance [4, 5, 6].

Experiment has so far veri ed, via infrared spectroscopy, some characteristic features of cyclotron resonance in monolayer [7, 8] and bilayer [9] graphene. Data generally show no clear sign of the many-body e ect, except for one [7] on monolayer graphene.

In this paper we would like to report our recent study of many-body corrections to cyclotron resonance in graphene and its bilayers [10]. We construct an e ective theory of cyclotron resonance within the single-mode approximation and point out the following:

(1) The genuine nonzero many-body corrections arise from vacuum polarization, speci c to graphene, which diverges logarithmically at short wavelengths.

(2) Both intralayer and interlayer coupling strengths require renormalization, which then allows one to determine many-body corrections uniquely from one resonance to another. The renormalized velocity $v_0^{\rm ren}$ and inter-layer coupling $_1^{\rm ren}$ thereby run with the magnetic eld B, decreasing gradually with increasing B.

(3) Theory is favorably compared with the experimental data on graphene.

2. Monolayer graphene

Monolayer graphene supports as charge carriers massless Dirac fermions at two inequivalent valleys (K and K'), and they, in a magnetic eld, lead to a tower of unequally-spaced Landau levels $|n, y_0\rangle$ of energy

$$ _n = s_n\, \omega_c \sqrt{|n|} \text{ with } \omega_c \quad \sqrt{2}\, v_0/\ell, \tag{1}$$

labeled by integers $n = \quad 0, \quad 1, \quad 2, \ldots$, and the center coordinate $y_0 \quad \ell^2 p_x$ with the magnetic length $\ell \quad 1/\sqrt{eB}$; $s_n \quad \mathrm{sgn}[n] \to \quad 1$; the "light velocity" $v_0 \sim 10^6$ m/s leads to the basic cyclotron energy $\omega_c \quad \sqrt{2}\, v_0/\ell \quad 36.3 \quad v_0[10^6 \mathrm{m/s}]\sqrt{B[\mathrm{T}]}\,\mathrm{meV}$. Each Landau level is fourfold degenerate (in spin and valley) when Zeeman coupling is ignored. Supplying a tiny asymmetry in sublattices reveals that the $n = 0$ level is doubly degenerate and electron-like with $_0 = 0_+$ at one valley while it is hole-like with $_0 = \quad 0_+$ at another valley. See Fig. 1(a).

Let us expand the electron eld $(\mathbf{x}, t)$ in terms of the Landau-level eigenmodes $_n(y_0, t)$ and write the charge density $\rho \; _{\mathbf{p}}(t) = \int d^2\mathbf{x}\, e^{i\mathbf{p}\ \mathbf{x}} \quad ^{\dagger} \quad$ as [11]

$$\rho \; _{\mathbf{p}} \; = \sum_{k,n= \quad \infty}^{\infty} g_{\mathbf{p}}^{kn}\, R_{\mathbf{p}}^{kn}, \qquad R_{\mathbf{p}}^{kn} = e^{\ \frac{1}{4}\ell^2\mathbf{p}^2} \int dy_0 \quad ^{\dagger}_{k}(y_0, t)\, e^{i\mathbf{p}\ \mathbf{r}} \quad _n(y_0, t), \tag{2}$$

where $\mathbf{r} = (i\ell^2\partial/\partial y_0, y_0)$ stands for the center coordinate. The coefficient matrix $g_{\mathbf{p}}^{kn}$ is given by

$$g_{\mathbf{p}}^{kn} = \frac{1}{2}\, b_k b_n (f_{\mathbf{p}}^{|k| \ 1,|n| \ 1} + s_k s_n\, f_{\mathbf{p}}^{|k|,|n|}) \text{ with } b_0 = \sqrt{2} \text{ and } b_n = 1 \text{ for } n \neq 0, \tag{3}$$

where $s_n = \mathrm{sgn}[n] \to \quad 1$; $f_{\mathbf{p}}^{kn} = \sqrt{n!/k!}\,(\quad \ell p/\sqrt{2})^{k \quad n} L_n^{(k \quad n)}(\frac{1}{2}\ell^2\mathbf{p}^2)$ for $k \quad n \quad 0$, and $f_{\mathbf{p}}^{nk} = (f^{kn}_{\quad \mathbf{p}})^{\dagger}$; $p = p_x \quad ip_y$. The Coulomb interaction is written as

$$H^{\mathrm{Coul}} = \frac{1}{2}\sum_{\mathbf{p}} v_{\mathbf{p}} \; : \rho \; _{\mathbf{p}}\, \rho_{\mathbf{p}} \; :, \tag{4}$$

where $v_{\mathbf{p}} = 2 \quad /(\; _b|\mathbf{p}|)$ with $\quad = e^2/(4 \quad _0) \quad 1/137$ and the substrate dielectric constant $_b$.

3. Cyclotron resonance and renormalization

Let us note that the charge operator $\rho^{kn}_{\quad \mathbf{p}} = g_{\mathbf{p}}^{kn}\, R_{\mathbf{p}}^{kn}$ in Eq. (2) annihilates an electron at the nth level L_n and creates one at the kth level L_k. One may thus associate it with the inter-Landau-level transition $L_n \to L_k$, and calculate the Coulombic corrections ($\propto H^{\mathrm{Coul}}$) to the excitation energy using the single-mode approximation in the standard way. Actually, the calculation is done in a purely algebraic manner if one notes that the charge operators $R_{\mathbf{p}}^{kn}$ obey the W_∞ algebra.

The result is the following [10]: The cyclotron-resonance energy for a general $L_a \to L_b$ channel with the Landau levels lled up to $n = n_{\mathrm{f}}$ is written as $^{\mathrm{exc}}_{\mathbf{k}=0} = \ _b \quad _a + \triangle \ ^{b\leftarrow a}_{\mathbf{k}=0}$ with

$$\triangle \ ^{b\leftarrow a}_{\mathbf{k}=0} = \sum_{\mathbf{p}} v_{\mathbf{p}}\, e^{\ \frac{1}{2}\ell^2\mathbf{p}^2}\Big[\sum_{n\leq n_{\mathrm{f}}} (|g^{an}_{\quad \mathbf{p}}|^2 \quad |g_{\mathbf{p}}^{bn}|^2) \quad g_{\mathbf{p}}^{bb}\, g^{aa}_{\quad \mathbf{p}}\Big]. \tag{5}$$

Note that $n_{\mathrm{f}} = 1, 0_+, \quad 1, \quad 2, \quad 3$ correspond to the lling factors $\nu = 4\,n_{\mathrm{f}}+2 = 6, 2, \quad 2, \quad 6, \quad 10$, respectively. The $\sum_{n\leq n_{\mathrm{f}}}(|g_{\mathbf{p}}^{an}|^2 \quad |g^{bn}_{\quad \mathbf{p}}|^2)$ term refers to the change in quantum uctuations, via the $a \to b$ transition, of the lled states.

For standard QH systems this correction $\triangle \ ^{b\leftarrow a}_{\mathbf{k}=0}$ vanishes for each transition to the adjacent level, $L_n \to L_{n+1}$, according to Kohn's theorem. Let us consider it for the $0 \to 1$ transition in graphene. We use $g_{\mathbf{p}}^{00} = 1$, $g_{\mathbf{p}}^{10} = \quad \ell p/2$ and $g_{\mathbf{p}}^{11} = 1 \quad \ell^2\mathbf{p}^2/4$, and note the following:

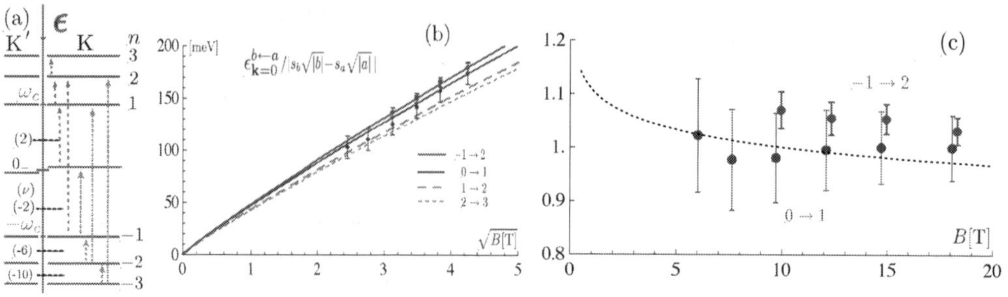

Figure 1. (a) Cyclotron resonance; circularly-polarized light can distinguish between transitions indicated by di erent types of arrows. (b) Rescaled cyclotron-resonance energies as a function of $\sqrt{B}$, with $v_0^{\mathrm{ren}}|_{B=10\mathrm{T}} \quad 1.13 \quad 10^6$ m/s and, as a typical value, $V_c \quad 12\sqrt{B[\mathrm{T}]}$ meV ($_b$ 5). The experimental data are quoted from Ref. [7] with error bars inferred from the original data. (c) The same data plotted in units of $\omega_c = \sqrt{2}\,v_0^{\mathrm{ren}}|_{B=10\mathrm{T}}/\ell$ as a function of B. The dotted curve represents a possible pro le of the running of $v_0^{\mathrm{ren}}|_B$, normalized to 1 at $B = 10$T.

Any nonzero shift $\triangle\ _{\mathbf{k}=0}^{1\leftarrow 0}$ comes from the quantum uctuations of the Dirac sea, and actually diverges logarithmically with the number N_{L} of lled Landau levels in the sea,

$$\triangle\ _{\mathbf{k}=0}^{1\leftarrow 0} = \sum_{\mathbf{p}} v_{\mathbf{p}}\, e^{\ \frac{1}{2}\ell^2\mathbf{p}^2} \sum_{N_L\le n\le\ 1} (|g^{0n}_{\ \mathbf{p}}|^2 \quad |g^{1n}_{\mathbf{p}}|^2) = V_c\,\mathcal{C}_N, \tag{6}$$

$$\mathcal{C}_N \quad (\sqrt{2}/8)\,(\log N_L \quad 1.017),$$
$$V_c \quad /(\ _b\ell) \quad (56.1/\ _b)\sqrt{B[\mathrm{T}]}\,\mathrm{meV}. \tag{7}$$

This divergence in $\mathcal{C}_N$ derives from short-wavelength vacuum polarization and is present even for $B = 0$, i.e., in free space [12]. It thus does not make sense to discuss the magnitude of $\mathcal{C}_N$. The legitimate procedure is to renormalize v_0 by rescaling

$$v_0 = Z_v\, v_0^{\mathrm{ren}} \tag{8}$$

and to put reference to the cuto into Z_v.

The renormalized velocity v_0^{ren} is de ned by referring to a speci c resonance. Let us take the $0 \to 1$ resonance and choose to absorb the entire $O(V_c)$ correction at some reference scale (e.g., at magnetic eld B_0) into Z_v, i.e., we write

$$\ _{\mathbf{k}=0}^{1\leftarrow 0}|\ _{=2} = \ _1 + \triangle\ _{\mathbf{k}=0}^{1\leftarrow 0} = \sqrt{2}\,v_0^{\mathrm{ren}}|_B/\ell \quad \omega_c^{\mathrm{ren}}|_B. \tag{9}$$

The renormalized velocity then depends on B, or runs with B,

$$v_0^{\mathrm{ren}}|_B = v_0^{\mathrm{ren}}|_{B_0} \quad \frac{}{8\ _b}\log(B/B_0), \tag{10}$$

decreasing gradually for $B > B_0$. We denote $v_0^{\mathrm{ren}}|_B$ as v_0^{ren} and $\omega_c^{\mathrm{ren}}|_B$ as ω_c^{ren} for short.

The resonance energies $\ _{\mathbf{k}=0}^{b\leftarrow a}$ for all other channels, as illustrated in Fig. 1 (a), are made nite by this renormalization. The resulting corrections $\propto V_c$ then make sense as genuine observable corrections. In particular, for several intraband channels one nds

$$\ _{\mathbf{k}=0}^{2\leftarrow 1}|\ _{=6} = \ _{\mathbf{k}=0}^{1\leftarrow\ \ 2}|\ _{=\ 6} = (\sqrt{2} \quad 1)\,\{\omega_c^{\mathrm{ren}} \quad 0.264\,V_c\},$$
$$\ _{\mathbf{k}=0}^{3\leftarrow 2}|\ _{=10} = \ _{\mathbf{k}=0}^{2\leftarrow\ \ 3}|\ _{=\ 10} = (\sqrt{3} \quad \sqrt{2})\,\{\omega_c^{\mathrm{ren}} \quad 0.358\,V_c\}. \tag{11}$$

Cyclotron resonance in graphene is governed by the selection rule $\triangle|n| =\ \ 1$, and, as a result, graphene supports interband cyclotron resonances. Some typical channels yield

$$
\begin{aligned}
{}^{2\leftarrow\ 1}_{\mathbf{k}=0}| = 2 &= (\sqrt{2}+1)\,\{\omega_c^{\rm ren} + 0.122\,V_c\},\\
{}^{1\leftarrow\ 2}_{\mathbf{k}=0}| = 2 &= (\sqrt{2}+1)\,\{\omega_c^{\rm ren} + 0.155\,V_c\}.\\
{}^{1\leftarrow\ 2}_{\mathbf{k}=0}| = 6 &= (\sqrt{2}+1)\,\{\omega_c^{\rm ren} + 0.084\,V_c\},\\
{}^{2\leftarrow\ 3}_{\mathbf{k}=0}| = 10 &= (\sqrt{2}+\sqrt{3})\,\{\omega_c^{\rm ren} + 0.044\,V_c\}.
\end{aligned}
\tag{12}
$$

It is now clear that cyclotron resonance is best analyzed by plotting the rescaled energies ${}^{b\leftarrow a}_{\mathbf{k}=0}/|s_b\sqrt{|b|}\ \ s_a\sqrt{|a|}|$ as a function of $\sqrt{B}$ or B. The Coulombic many-body e ect will be seen as a variation of the characteristic velocity $v_0^{\rm ren}[1+O(V_c)]$ from one resonance to another, and a deviation of $\omega_c^{\rm ren}$ from the $\sqrt{B}$ behavior would indicate the running of $v_0^{\rm ren}$ with B.

Actually, experiment by Jiang *et al.* [7] did observe a small deviation of the $1:(1+\sqrt{2})$ ratio of ${}^{1\leftarrow 0}_{\mathbf{k}=0}$ to ${}^{2\leftarrow\ 1}_{\mathbf{k}=0}$ well outside of the experimental errors under magnetic elds $B = (6 \sim 18)$ T. Such data are included in Fig. 1 (b), which plots ${}^{1\leftarrow 0}_{\mathbf{k}=0}$ and ${}^{2\leftarrow\ 1}_{\mathbf{k}=0}/(\sqrt{2}+1)$ as a function of $\sqrt{B}$. A small increase of $v_0^{\rm ren}$ in ${}^{2\leftarrow\ 1}_{\mathbf{k}=0}/(\sqrt{2}+1)$, relative to ${}^{1\leftarrow 0}_{\mathbf{k}=0}$, is roughly consistent with Eq. (12) which suggests a 0.122 $V_c/\omega_c^{\rm ren} \sim 4\%$ increase in $v_0^{\rm ren}$ (since $V_c/\omega_c^{\rm ren} \sim 0.3$).

This feature is more pronounced in Fig. 1 (c), which plots the same data in units of $\omega_c = \sqrt{2}\,v_0^{\rm ren}|_{B=10{\rm T}}/\ell \propto \sqrt{B}$. There a dotted curve represents a possible pro le of the running of $v_0^{\rm ren}$ with B, and, especially, the ($1\rightarrow 2$) data (with smaller error bars) suggests such running.

The data are thus consistent (in sign and magnitude) with the present estimate of the many-body e ect. In this connection, let us note that an earlier experiment on thin epitaxial graphite [13] also observed the same pair of resonances, with apparently no deviation from the $1:(1+\sqrt{2})$ ratio. This measurement was done under relatively low magnetic elds $B = (0.4 \sim 4)$ T, and it could be that a small deviation, under larger error bars, simply escaped detection, apart from the potential di erence between thin graphite and graphene.

More precise measurements of cyclotron resonance, especially in the high B domain, would be required to pin down the many-body e ect in graphene. In this respect, the comparison between interband and intraband resonances from the *same* initial state, such as the ($2\rightarrow\ 1$) resonances and ($3\rightarrow\ 2$) resonances, would provide a clearer signal for the many-body e ect. One would expect sizable variations in $v_0^{\rm ren}$ for such pairs, $\triangle R(\ 2\rightarrow\ 1)\ ^{=\ 6}\ 0.34\,V_c/\omega_c^{\rm ren} \sim$ 10%, and $\triangle R(\ 3\rightarrow\ 2)\ ^{=\ 10}\ 0.40\,V_c/\omega_c^{\rm ren} \sim 12\%$.

4. Cyclotron resonances in bilayer graphene

In bilayer graphene interlayer coupling $_1\ \ \ _{A'B} \sim (0.3\ \ 0.4)$ eV modi es the intralayer linear spectra to yield, in the low-energy branches $|\ | < \ _1$, quasiparticles with a parabolic dispersion. They, in a magnetic eld, support a variety of cyclotron resonances, as listed in Fig. 2 (a).

One can again use the SMA formula (5) to calculate the interlevel excitation energies. It turns out that not only v_0 but also $_1$ undergo in nite renormalization and, rather unexpectedly, the divergent terms are the same for both of them to $O(V_c)$ at least. We thus make the common rescaling $v_0 = Z\,v_0^{\rm ren}$ and $_1 = Z\ \ _1^{\rm ren}$, and eventually learn that the excitation spectrum is cast in the form

$$
{}^{b\leftarrow a}_{\mathbf{k}=0} = (s_b\eta_b\ \ \ s_a\eta_a)\,(\omega_c^{\rm ren} + \triangle c^{ba}\,V_c),
\tag{13}
$$

where η_k are some given functions of $v_0^{\rm ren}/\ \ _1^{\rm ren}$ and $\triangle c^{ba}$ are uniquely xed as genuine quantum corrections.

The many-body e ect is expected to be sizable in bilayer graphene. The existing data by Henriksen *et al.* [9] generally appear to defy good t by theory but certainly suggest nontrivial features of many-body corrections, such as running with B; see Fig. 2.

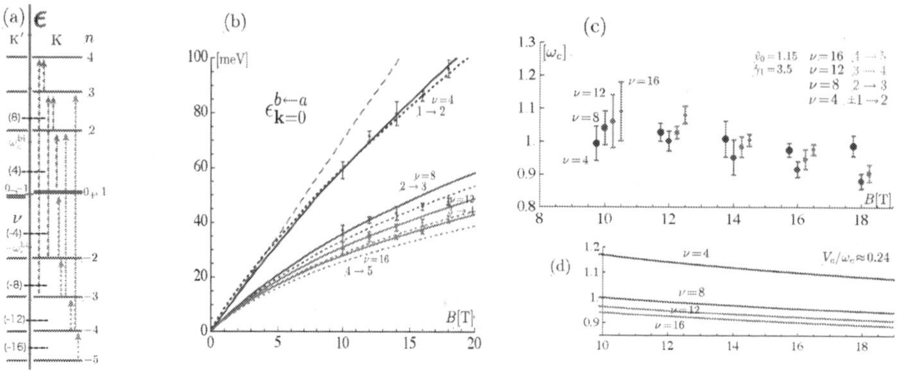

Figure 2. (a) Cyclotron resonance in bilayer graphene. (b) Resonance energies; real curves, with $v_0=1.15$ 10^6m/s, $_1=0.35$ eV and $V_c=0$; dotted curves, with $_1=0.38$ eV and V_c $5.6\sqrt{B[\mathrm{T}]}$ meV (or $_b$ 10). The experimental data are reproduced from Ref. [9]. (c) Electron data, reorganized in the form $^{b\leftarrow a}_{\mathbf{k}=0}/|s_b\eta_b \quad s_a\eta_a|$ and plotted in units of $\omega_c = \sqrt{2}\,v_0/\ell^2$; for clarity the data points are slightly shifted in B. (d) Theoretical expectation according to Eq. (13).

5. Concluding remarks

In this paper we have studied many-body corrections to cyclotron resonance in graphene, with emphasis on the need for renormalization. More precise measurements of cyclotron resonance are highly desired. Of particular interest are experiments that compare interband and intraband resonances from the same initial states, which would clarify the many-body e ect with minimal uncertainties.

Acknowledgments

This work was supported in part by a Grant-in-Aid for Scienti c Research from the Ministry of Education, Science, Sports and Culture of Japan (Grant No. 21540265).

References

[1] Abergel D S L and Fal'ko V I 2007 *Phys. Rev.* B **75** 155430
[2] Kohn W 1961*Phys. Rev.* **123** 1242
[3] For many-body corrections to cyclotron resonance in quantum Hall systems see, Asano K and Ando T 1998 *Phys. Rev.* B **58** 1485
[4] Iyenga A, Wang J, Fertig H A and Brey L 2007 *Phys. Rev.* B **75** 125430
[5] Bychkov Yu A and Martinez G 2008 *Phys. Rev.* B **77** 125417
[6] Viola Kusminskiy S, Campbell D K and Castro Neto A H 2009 *Euro. Phys. Lett.* **85** 58005
[7] Jiang Z, Henriksen E A, Tung L C, Wang Y-J, Schwartz M E, Han M Y, Kim P and Stormer H L 2007 *Phys. Rev. Lett.* **98** 197403
[8] Deacon R S, Chuang K-C, Nicholas R J, Novoselov K S and Geim A K 2007 *Phys. Rev.* B **76** 081406(R)
[9] Henriksen E A, Jiang Z, Tung L-C, Schwartz M E, Takita M, Wang Y-J, Kim P and H. L. Stormer 2008*Phys. Rev. Lett.* **100** 087403
[10] Shizuya K 2010 *Phys. Rev.* B **81** 075407
[11] Shizuya K 2007 *Phys. Rev.* B **75** 245417
[12] Gonzalez J, Guinea F and Vozmediano M A H 1994 *Nucl. Phys.* B **424** [FS] 595
[13] Sadowski M L, Martinez G, Potemski M, Berger C and de Heer H A 2006 *Phys. Rev. Lett.* **97** 266405

Dirac cones on the generalized honeycomb lattice

Keita Kishigi[1], Kai Ueno[1], Eriko Miyamoto[1], and Yasumasa Hasegawa[2]

[1] Faculty of Education, Kumamoto University, Kurokami 2-40-1, Kumamoto, 860-8555, Japan
[2] Department of Material Science, Graduate School of Material Science, University of Hyogo, Hyogo, 678-1297, Japan

E-mail: kishigi@educ.kumamoto-u.ac.jp

Abstract.
We study the tight-binding model on the generalized honeycomb lattice, where nearest-neighbor and next-nearest-neighbor transfer integrals and the on-site potentials are taken as parameters. We obtain the condition for the untilted Dirac cone. We discuss the effects of the uniaxial strain on the opening of the gap and the tilting of the Dirac cone.

1. Introduction

It is well-known that "massless Dirac fermions" are realized in graphene [1, 2, 3], the organic conductor α-(BEDT-TTF)$_2$I$_3$ [4, 5, 6, 7, 8] and iron-based superconductors, BaFe$_2$As$_2$[9]. The massless Dirac fermions are the origin of anomalous phenomena such as temperature-independent conductivity and unusual quantum Hall effect[2].

The electron band of graphene is obtained by fully first-principles band calculations[10, 11, 12]. It can be reproduced by a tight-binding model on a honeycomb lattice with nearest-neighbor transfer integrals[13, 14, 15, 16]. As the electrons are half-filled, a lower band is fully occupied. The energy gap between two bands is zero at the points, $\mathbf{k}^* = (k_x^*, k_y^*)$[13, 14]. The linearlized energy dispersion is called the Dirac cone, as seen in Fig. 1 (a).

The opening of the gap has been observed on an epitaxial graphene on a SiC substrate[17]. The finite gap is necessary for the application in electronic devices. When only the nearest-neighbor transfer integrals are present, the condition for the zero gap is obtained as [14]

$$\left|\frac{|t_b|}{|t_a|} - 1\right| \le \frac{|t_c|}{|t_a|} \le \left|\frac{|t_b|}{|t_a|} + 1\right|. \tag{1}$$

The large asymmetry (for example, $t_c > 2t_a$ when $t_b = t_a$) is necessary to open the gap. The band calculations[16] have shown that the deformation of the order of 20% is indispensable to open the gap.

We consider the next-nearest-neighbor transfer integrals ($t'_a, t'_b, t'_c, t''_a, t''_b$ and t''_c), where t'_a, t'_b and t'_c are transfer integrals within sublattice A and t''_a, t''_b and t''_c are those within sublattice B and the on-site potentials ϵ_A and ϵ_B on sublattices A and B as seen in Fig. 1 (b). Then, the condition of the zero gap is given by Eq. (1) and Eq. (12) in this paper[18, 19]. If $t_a = t_b = t_c$, the gap is opened due to at least one of the inequalities, $t'_a \neq t''_a$, $t'_b \neq t''_b$, $t'_c \neq t''_c$ or $\epsilon_A \neq \epsilon_B$. Thus, even if the asymmetries of next-nearest-neighbor transfer integrals or on-site potentials

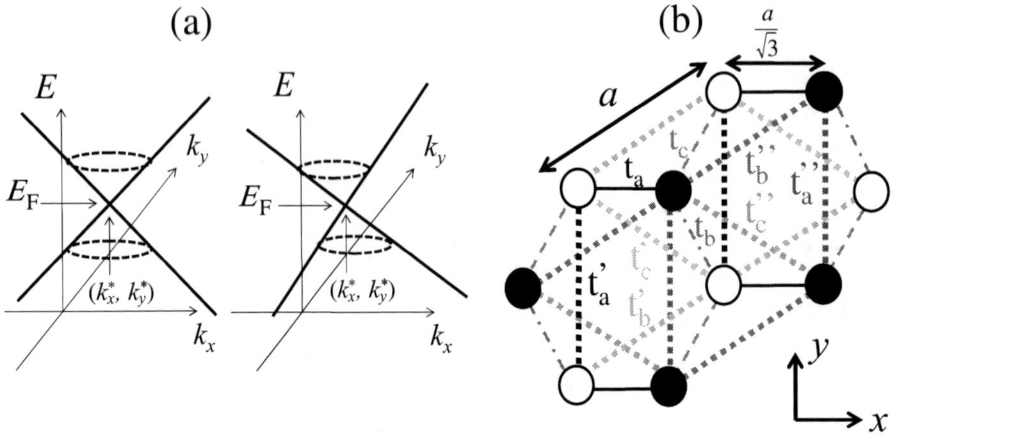

Figure 1. (a) Schematic figures of the untilted (the left figure) and the tilted (the right figure) Dirac cones. (b) Honeycomb lattice and transfer integrals. •: sublattice A. ○: sublattice B.

are very small, a finite gap is opened. It is contrary to the results in the systems with only nearest-neighbor transfer integrals. Furthermore, when the uniaxial strain is applied to the tight-binding model such that the condition of $t'_a = t'_b = t'_c = t''_a = t''_b = t''_c$ is no longer satisfied, the Dirac cone of graphene is shown to be tilted[6] as in the right figure in Fig. 1 (a). The tilted Dirac cone means the anisotropy of the Fermi velocity. In α-(BEDT-TTF)$_2$I$_3$ with the triclinic lattice symmetry, the tilted Dirac cone is realized[4, 5, 6, 7]. The anomalous Kosterlitz-Thouless transition due to the tilting of the Dirac cone in α-(BEDT-TTF)$_2$I$_3$ is also predicted[8]. In this paper, we show the condition for the untilted Dirac cone in the tight-binding model with the generalized honeycomb lattice symmetry.

2. Model

The Hamiltonian is given by

$$\mathcal{H} = \sum_{\mathbf{k}} A^\dagger_{\mathbf{k}} \sum_{\mu=0,1,2,3} \epsilon_\mu(\mathbf{k})\sigma_\mu A_{\mathbf{k}}, \tag{2}$$

where $A^\dagger_{\mathbf{k}} = (a^\dagger_{\mathbf{k}}\ b^\dagger_{\mathbf{k}})$, σ_0 is the 2×2 unit matrix, σ_j $(j = 1, 2, 3)$ are Pauli matrixes, $a^\dagger_{\mathbf{k}}(b^\dagger_{\mathbf{k}})$ are creation operators for A (B) sites, respectively, and

$$\begin{aligned}
\epsilon_0(\mathbf{k}) &= \frac{\epsilon_A + \epsilon_B}{2} - (t'_c + t''_c)\cos(\mathbf{a}_1\cdot\mathbf{k}) - (t'_b + t''_b)\cos(\mathbf{a}_2\cdot\mathbf{k}) \\
&- (t'_a + t''_a)\cos((\mathbf{a}_2 - \mathbf{a}_1)\cdot\mathbf{k}), && (3)\\
\epsilon_1(\mathbf{k}) &= -t_a\cos(\frac{1}{3}(\mathbf{a}_1 + \mathbf{a}_2)\cdot\mathbf{k}) - t_b\cos(\frac{1}{3}(2\mathbf{a}_1 - \mathbf{a}_2)\cdot\mathbf{k}) - t_c\cos(\frac{1}{3}(2\mathbf{a}_2 - \mathbf{a}_1)\cdot\mathbf{k}), && (4)\\
\epsilon_2(\mathbf{k}) &= -t_a\sin(\frac{1}{3}(\mathbf{a}_1 + \mathbf{a}_2)\cdot\mathbf{k}) + t_b\sin(\frac{1}{3}(2\mathbf{a}_1 - \mathbf{a}_2)\cdot\mathbf{k}) + t_c\sin(\frac{1}{3}(2\mathbf{a}_2 - \mathbf{a}_1)\cdot\mathbf{k}), && (5)\\
\epsilon_3(\mathbf{k}) &= \frac{\epsilon_A - \epsilon_B}{2} - (t'_c - t''_c)\cos(\mathbf{a}_1\cdot\mathbf{k}) - (t'_b - t''_b)\cos(\mathbf{a}_2\cdot\mathbf{k}) \\
&- (t'_a - t''_a)\cos((\mathbf{a}_2 - \mathbf{a}_1)\cdot\mathbf{k}), && (6)
\end{aligned}$$

where the unit vectors are given by

$$\mathbf{a}_1 = (\frac{\sqrt{3}}{2}a, -\frac{1}{2}a), \quad \mathbf{a}_2 = (\frac{\sqrt{3}}{2}a, \frac{1}{2}a),$$

and a is the lattice constant. The off-diagonal elements of $\epsilon_1(\mathbf{k})$ and $\epsilon_2(\mathbf{k})$ contain the nearest-neighbor transfer integrals, and the diagonal elements of $\epsilon_0(\mathbf{k})$ and $\epsilon_3(\mathbf{k})$ contain the next-nearest-neighbor transfer integrals, which connect sublattice A (B) and A (B). We show in this paper that the tilting of the Dirac cone is caused by $\epsilon_0(\mathbf{k})$ and that the deformation of the Dirac cone from a circle to an ellipsoid is caused by $\epsilon_1(\mathbf{k})$, $\epsilon_2(\mathbf{k})$ and $\epsilon_3(\mathbf{k})$. While the nearest-neighbor transfer integrals, t_a, t_b and t_c for graphene are estimated to be $2.7 \sim 3.1$ eV[20, 21], the values of the next-nearest-neighbor transfer integrals are not known. In some studies [15, 22], the values of t'_a, t'_b, t'_c, t''_a, t''_b and t''_c have been taken to be the order of 0.1 t_a.

3. Eigenvalues

Eigenvalues of Eq. (2) are obtained as

$$E^{\pm}(k_x, k_y) = \epsilon_0(\mathbf{k}) \pm \Delta_{\mathbf{k}}, \tag{7}$$

where

$$\Delta_{\mathbf{k}} = \sqrt{\sum_{j=1}^{3} (\epsilon_j(\mathbf{k}))^2}. \tag{8}$$

One of the authors[14] has obtained that $(\epsilon_1(\mathbf{k}^*))^2 + (\epsilon_2(\mathbf{k}^*))^2$ becomes zero when the following three equations are satisfied,

$$\cos(\mathbf{a}_1 \cdot \mathbf{k}^*) = \frac{t_c^2 - t_a^2 - t_b^2}{2t_a t_b}, \tag{9}$$

$$\cos(\mathbf{a}_2 \cdot \mathbf{k}^*) = \frac{t_b^2 - t_a^2 - t_c^2}{2t_a t_c}, \tag{10}$$

$$\cos((\mathbf{a}_1 - \mathbf{a}_2) \cdot \mathbf{k}^*) = \frac{t_a^2 - t_b^2 - t_c^2}{2t_b t_c}. \tag{11}$$

In general, there exist $\mathbf{k}^*$ and $-\mathbf{k}^*$, which are located at the positions of inversion symmetry and time reversal symmetry. These points are merged when one of the equalities in Eq. (1) is satisfied[14].

When $\epsilon_3(\mathbf{k}^*)$ becomes zero, the energy gap, $\Delta_{\mathbf{k}^*}$, is zero. Thus, by putting Eqs. (9), (10) and (11) into Eq (6), we obtain the condition for the zero gap at $\mathbf{k}^*$ as following[18, 19];

$$\epsilon_A - \epsilon_B - (t'_a - t''_a)\left(\frac{t_a^2 - t_b^2 - t_c^2}{t_b t_c}\right) - (t'_b - t''_b)\left(\frac{t_b^2 - t_a^2 - t_c^2}{t_a t_c}\right) - (t'_c - t''_c)\left(\frac{t_c^2 - t_a^2 - t_b^2}{t_a t_b}\right) = 0. \tag{12}$$

We call the condition given in Eq. (12) the averaged inversion symmetry. When only the nearest-neighbor integrals are considered, the system is invariant with respect to inversion. On the other hand, when the next-nearest-neighbor integrals are finite, the gap is not opened as long as the averaged inversion symmetry is conserved, even if the inversion symmetry is broken[18, 19].

4. Linearization for eigenvalues near $\mathbf{k}^*$

In this section we study the energy around $\mathbf{k}^*$ when Eqs. (1) and (12) are satisfied, i.e., $\Delta_{\mathbf{k}^*} = 0$.

$$\epsilon_0(\mathbf{k}) \simeq \epsilon_0(\mathbf{k}^*) + v_{0x}(\mathbf{k}^*)q_x + v_{0y}(\mathbf{k}^*)q_y + O(q_x^2, q_y^2, q_x q_y), \tag{13}$$

$$\epsilon_j(\mathbf{k}) \simeq v_{jx}(\mathbf{k}^*)q_x + v_{jy}(\mathbf{k}^*)q_y + O(q_x^2, q_y^2, q_x q_y), \quad j = 1, 2, 3 \tag{14}$$

where $q_x = k_x - k_x^*$, $q_y = k_y - k_y^*$ and $v_{\mu i}(\mathbf{k}^*) = \frac{\partial \epsilon_\mu(\mathbf{k})}{\partial k_i}\Big|_{\mathbf{k}=\mathbf{k}^*}$ $(\mu = 0,1,2,3, i = x,y)$. For example, we obtain

$$v_{0x}(\mathbf{k}^*) = \frac{\sqrt{3}a}{2}\Big\{(t_c' + t_c'')\sin(\mathbf{a}_1\cdot\mathbf{k}^*) + (t_b' + t_b'')\sin(\mathbf{a}_2\cdot\mathbf{k}^*)\Big\}, \tag{15}$$

$$\begin{aligned} v_{0y}(\mathbf{k}^*) &= -\frac{a}{2}\Big\{(t_c' + t_c'')\sin(\mathbf{a}_1\cdot\mathbf{k}^*) - (t_b' + t_b'')\sin(\mathbf{a}_2\cdot\mathbf{k}^*) \\ &- 2(t_a' + t_a'')\sin((\mathbf{a}_2 - \mathbf{a}_1)\cdot\mathbf{k}^*)\Big\}. \end{aligned} \tag{16}$$

We get from Eqs. (8) and (14)

$$\Delta_{\mathbf{k}}^2 \approx (q_x\ q_y)\begin{pmatrix} |\vec{v}_x(\mathbf{k}^*)|^2 & (\vec{v}_x(\mathbf{k}^*)\cdot\vec{v}_y(\mathbf{k}^*)) \\ (\vec{v}_x(\mathbf{k}^*)\cdot\vec{v}_y(\mathbf{k}^*)) & |\vec{v}_y(\mathbf{k}^*)|^2 \end{pmatrix}\begin{pmatrix} q_x \\ q_y \end{pmatrix}, \tag{17}$$

where $\vec{v}_i(\mathbf{k}^*) = (v_{1i}(\mathbf{k}^*), v_{2i}(\mathbf{k}^*), v_{3i}(\mathbf{k}^*))$. The 2×2 matrix in Eq. (17) can be diagonalized by the rotation[6, 23] in the q_x-q_y plane, i.e.

$$\begin{pmatrix} q_x \\ q_y \end{pmatrix} = U\begin{pmatrix} \tilde{q}_x \\ \tilde{q}_y \end{pmatrix}, \tag{18}$$

with the unitary matrix

$$U = \begin{pmatrix} \cos\frac{\theta}{2} & -\sin\frac{\theta}{2} \\ \sin\frac{\theta}{2} & \cos\frac{\theta}{2} \end{pmatrix}, \tag{19}$$

where θ is given by

$$\tan\theta = \frac{2(\vec{v}_x(\mathbf{k}^*)\cdot\vec{v}_y(\mathbf{k}^*))}{|\vec{v}_x(\mathbf{k}^*)|^2 - |\vec{v}_y(\mathbf{k}^*)|^2}. \tag{20}$$

Then we obtain

$$E^{\pm}(\mathbf{k}) \approx \epsilon_0(\mathbf{k}^*) + w_{0x}\tilde{q}_x + w_{0y}\tilde{q}_y \pm \sqrt{w_x^2\tilde{q}_x^2 + w_y^2\tilde{q}_y^2}, \tag{21}$$

where

$$w_x^2 = \frac{|\vec{v}_x(\mathbf{k}^*)|^2 + |\vec{v}_y(\mathbf{k}^*)|^2}{2} + \sqrt{\left(\frac{|\vec{v}_x(\mathbf{k}^*)|^2 - |\vec{v}_y(\mathbf{k}^*)|^2}{2}\right)^2 + (\vec{v}_x(\mathbf{k}^*)\cdot\vec{v}_y(\mathbf{k}^*))^2}, \tag{22}$$

$$w_y^2 = \frac{|\vec{v}_x(\mathbf{k}^*)|^2 + |\vec{v}_y(\mathbf{k}^*)|^2}{2} - \sqrt{\left(\frac{|\vec{v}_x(\mathbf{k}^*)|^2 - |\vec{v}_y(\mathbf{k}^*)|^2}{2}\right)^2 + (\vec{v}_x(\mathbf{k}^*)\cdot\vec{v}_y(\mathbf{k}^*))^2}, \tag{23}$$

and

$$(w_{0x}\ w_{0y}) = (v_{0x}(\mathbf{k}^*)\ v_{0y}(\mathbf{k}^*))U. \tag{24}$$

Note that $w_x^2 \geq w_y^2 \geq 0$ and we obtain $w_y = 0$ only when $\vec{v}_x(\mathbf{k}^*) \parallel \vec{v}_y(\mathbf{k}^*)$. When $w_y = 0$, the Dirac cone is not formed. The conditions for the isotropic ($|w_x| = |w_y|$) and untilted ($w_{0x} = w_{0y} = 0$) Dirac cone are obtained as follows. The Dirac cone is isotropic, if and only if the vectors $\vec{v}_x(\mathbf{k}^*)$ and $\vec{v}_y(\mathbf{k}^*)$ are perpendicular each other and have the same length. If these conditions are not satisfied, the cross-section of the Dirac cone at the constant energy is elliptic. The Dirac cone is untilted, if and only if

$$v_{0x}(\mathbf{k}^*) = v_{0y}(\mathbf{k}^*) = 0. \tag{25}$$

Since the nearest-neighbor transfer integrals (t_a, t_b and t_c) do not appear in $\epsilon_0(\mathbf{k})$ of Eq. (3), the next-nearest-neighbor transfer integrals are required to tilt the Dirac cone[6]. The equations for untilted Dirac cone (Eq. (25)) can be written as

$$(t'_a + t''_a)\sin((\mathbf{a}_2 - \mathbf{a}_1)\cdot\mathbf{k}^*) = -(t'_b + t''_b)\sin(\mathbf{a}_2\cdot\mathbf{k}^*) = (t'_c + t''_c)\sin(\mathbf{a}_1\cdot\mathbf{k}^*). \tag{26}$$

By taking square of each terms and inserting Eqs. (9), (10) and (11), we obtain the condition for the untilted Dirac cone as

$$t_a(t'_a + t''_a) = t_b(t'_b + t''_b) = t_c(t'_c + t''_c). \tag{27}$$

5. Discussion

Here, we discuss the effects of strain on graphene. When there is no strain, $t_a = t_b = t_c$, $t'_a = t''_a = t'_b = t''_b = t'_c = t''_c$ and $\epsilon_A = \epsilon_B$. In this case the energy gap is zero and the Dirac cone is untilted. If the uniaxial strain is applied, transfer integrals are changed, but the parallel pairs of the next-nearest-neighbor transfer integrals are the same, i.e., $t'_a - t''_a = t'_b - t''_b = t'_c - t''_c = 0$. In that case, the energy gap is zero, since Eq. (12) is satisfied. On the other hand, Eq. (27) is not satisfied, i.e., the Dirac cone is tilted.

6. Conclusion

We obtain the band dispersion near $\mathbf{k}^*$ in the tight-binding model on generalized honeycomb lattice, Eq. (21). The condition for the untilted Dirac cone is given by Eq. (27). The tilted Dirac cone of graphene is realized in the weak uniaxial strain although the gap is not opened. In that case, the peculiar property of α-(BEDT-TTF)$_2$I$_3$ such as Kosterlitz-Thouless transition at lower temperature[8] may also be observed in graphene.

[1] K. S. Novoselov, A. K. Geim, S. M. Morozov, D. Jiang, Y. Zhang, S. V. Dubonos, I. V. Grigorieva, and A. A. Firsov, Science **306**, 666 (2004).
[2] K. S. Novoselov, A. K. Geim, S. V. Morozev, D. Jiang, M. I. Katsnelson, I. V. Grigorieva, S. V. Dubonos, and A. A. Firsov, Nature (London) **438**, 197 (2005).
[3] Y. Zhang, Y. W. Tan, H. L. Stormer, and P. Kim, Nature **438**, 201 (2005).
[4] S. Katayama, A. Kobayashi, and Y. Suzumura, J. Phys. Soc. Jpn. **75**, 023708 (2006).
[5] A. Kobayashi, S. Katayama, Y. Suzumura, and H. Fukuyama, J. Phys. Soc. Jpn. **76**, 034711 (2007).
[6] M. O. Goerbig, J.-N. Fuchs, G. Montambaux, and F. Piechon, Phys. Rev. B **78**, 045415 (2008).
[7] N. Tajima, S. Sugawara, R. Kato, Y. Nishino and K. Kajita, Phys. Rev. Lett. **102**, 176403 (2009).
[8] A. Kobayashi, Y. Suzumura, H. Fukuyama and M. O. Goerbig, J. Phys. Soc. Jpn. **78**, 114711 (2009).
[9] P. Richard, K. Nakayama, T. Sato, M. Neupane, Y.-M. Xu, J. H. Bowen, G. F. Chen, J. L. Luo, N. L. Wang, X. Dai, Z. Fang, H. Ding and T. Takahashi, Phys. Rev. Lett. **104**, 137001 (2010).
[10] M. Farjam and H. Rafii-Tabar, Phys. Rev. B **80**, 167401 (2009).
[11] Gui Gui, Jin Li, and Jianxin Zhong, Phys. Rev. B **80**, 167402 (2009).
[12] Seon-Myeong Choi, Seung-Hoon Jhi and Young-Woo Son, Phys. Rev. B **81**, 081407 (2010).
[13] P. R. Wallace, Phys. Rev. **71**, 622 (1947).
[14] Y. Hasegawa, R. Konno, H. Nakano, and M. Kohmoto, Phys. Rev. B **74**, 033413 (2006).
[15] V. M. Pereira, J. M. B. Lopes dos Santos, and A. H. Castro Neto, Phys. Rev. B **77**, 115109 (2008).
[16] Victor M. Pereira, A. H. Castro Neto and N. M. R. Peres, Phys. Rev. B **80**, 045401 (2009).
[17] S. Y. Zou, G.-H. Gweon, A. V. Fedorov, P. N. First, W. A. de Heer, D.-H. Lee, F. Guinea, A. H. Castro Neto, and A. Lanzara, Nat. Phys. **6**, 770 (2007).
[18] K. Kishigi, H. Hanada and Y. Hasegawa, J. Phys. Soc. Jpn. **77**, 074707 (2008).
[19] K. Kishigi, R. Takeda and Y. Hasegawa, J. Phys., Conf. Ser. vol. 132, 012005 (2008).
[20] M. S. Dresselhaus, G. Dresselhaus, and J. E. Fischer, Phys. Rev. B **15**, 3180 (1977).
[21] W. W. Toy, M. S. Dresselhaus, and G. Dresselhaus, Phys. Rev. B **15**, 4077 (1977).
[22] N. M. R. Peres, F. Guinea, and A. H. Castro Neto, Phys. Rev. B **73**, 125411 (2006).
[23] T. Morinari, T. Himura, T. Tohyama, J. Phys. Soc. Jpn. **78**, 023704 (2009).

Search for non-Abelian statistics in half-filled Landau levels of graphene

A Wójs[1,2], G Möller[1], and N R Cooper[1]

[1]TCM Group, Cavendish Laboratory, University of Cambridge, Cambridge, UK
[2]Institute of Physics, Wrocław University of Technology, Wrocław, Poland

E-mail: arkadiusz.wojs@pwr.wroc.pl

Abstract. We have employed large scale exact numerical diagonalization in Haldane spherical geometry in a comparative analysis of the correlated many-electron states in the half-filled low Landau levels of graphene and such conventional semiconductors as GaAs, including both spin and valley (i.e., pseudospin) degrees of freedom. We present evidence that the polarized Fermi sea of essentially non-interacting composite fermions remains stable against a pairing transition in both lowest Landau levels of graphene. However, it undergoes spontaneous depolarization, which in (ideal) graphene is unprotected for the lack of a single-particle pseudospin splitting. These results point to the absence of the non-Abelian Pfaffian phase in graphene.

1. Introduction

The fundamental properties of the enigmatic incompressible quantum liquid formed by the quasi-two-dimensional electrons in a half-filled second Landau level (LL) of such conventional semiconductors as GaAs, and thus responsible for the fractional quantum Hall (FQH) effect observed at the LL filling factor $\nu=5/2$ [1] in such materials, continue to attract attention. A somewhat trivial reason is that $\nu=5/2$ is unique as the only known even-denominator FQH state observed so far. More importantly, although the nature of the $\nu=5/2$ FQH state is not yet decisively proven, in light of the accumulating evidence [2] it has become commonly believed that this state is adiabatically connected to the so-called Pfaffian wave function introduced by Moore and Read [3]. The Pfaffian wave function represents a condensed paired state of composite fermions (CFs) in the absence of an effective magnetic field B^* (the CFs are weakly interacting quasiparticles formed by the electrons through capturing vortices of the many-body wave function and thus experiencing a reduced effective field B^* [4]). It has been almost immediately realized that the quasihole (QH) excitations of the Pfaffian ground state obey fractional and nonabelian braiding statistics [5]. More recently, it has been suggested that employing the statistical degeneracy of the multi-QH states for encoding quantum information might pave the way to the essentially error-free topological quantum computation [6]. One of the problems with the practical realization of this idea is a small excitation gap of this state measured in the actual GaAs structures (of the order of hundreds of mK) [7] and not much larger values predicted theoretically for the idealized systems (defined by the absence of disorder, vanishing layer thickness, etc.). It is therefore natural to extend the search for the nonabelian Pfaffian phase to other materials, where the excitation gap might be enhanced either due to a higher overall Coulomb energy scale or due to a more favorable form of the pseudopotential (defined as the dependence of the interaction energy V of a pair of electrons inside a given LL on their relative pair angular momentum m [8]), to which the half-filled ground state is known to be sensitive.

In this work, we investigate the possibility of a hypothetical non-abelian Pfaffian phase appearing in half-filled LLs of graphene [9], characterized by a significantly (up to an order of magnitude) larger Coulomb energy scale than GaAs due to a lower dielectric constant (down to unity, the precise value depending on the coupling to the background) and literally atomic layer width [10]. This question has been earlier considered theoretically [11-13] and now became more realistic in view of experimental demonstration of the $\nu=1/3$ FQH effect in graphene [14]. Our results come from exact diagonalization or the largest systems studied so far and using a more accurate form of the Coulomb pseudopotential.

Emergence of the Pfaffian state requires a particular form of the interaction among the electrons in a given LL. In addition to supporting the formation of CFs, it must also induce an effective (residual) interaction among them – such as to lead to the Cooper pairing instability of the CF Fermi sphere at $B^*=0$. Hence, a broader question we address in this paper is that of CF dynamics in different LLs of graphene. The lack of immediate analogy between graphene and the (better explored) two-dimensional GaAs is the consequence of the differing single-particle orbitals defining all but the lowest LL (caused by the presence of two atomic sublattices spanning the crystallographic structure of graphene), and of an additional "pseudospin" degree of freedom associated with the valley degeneracy in graphene. Our main conclusions are that the CF-CF interaction in graphene is insufficient to induce pairing at $B^*=0$, and that the polarized compressible CF Fermi sea is unstable toward the depolarization of pseudospin.

2. Interaction matrix elements and pseudopotentials on a sphere

Interaction in an isolated LL is completely determined by its pseudopotentials V_m, which (except for the lowest LL) are different in graphene and GaAs. Their analytic expressions for an arbitrary (nth) LL in the planar geometry have been derived earlier [11]. These "planar" pseudopotentials $V_{\text{plane},n,m}$ were used in (to our best knowledge) all previous many-body calculations (e.g., [13]) even though these calculations were most often carried out in a more convenient spherical geometry (with N electrons confined to the surface of a sphere and exposed to a magnetic flux $2Q\ hc/e$). The series $V_{\text{plane},n,m}$ were simply truncated at $m=N_\phi$, where $N_\phi=2(Q+n)$ represents the magnetic flux of a corresponding system in the lowest LL, with the same LL degeneracy of $N_\phi+1$). This approach has an obvious disadvantage in the disconnection of V_m from the Coulomb potential $V(r)\sim1/r$ from which it ought to be derived. This is especially problematic at a longer range r, comparable to the sphere radius $R=Q^{1/2}$ (in units of the magnetic length $\lambda=(hc/eB)^{1/2}$), i.e., for m not much smaller than Q. Therefore, in the following we instead adopted a direct analytical solution of the Dirac problem on a sphere provided by Jellal [15].

In the excited LLs of graphene, the single-particle states $\|n,m\rangle\rangle$ are spinors, whose two components represent standard LL wave functions, with equal angular momenta m but different LL indices: $|n,m\rangle$ and $|n\text{-}1,m\rangle$. It is essential that in the spherical geometry both components of $\|n,m\rangle\rangle$ must have the same LL degeneracy (i.e., equal N_ϕ) and thus different flux $2Q$ (i.e., different magnetic length scales λ). Using this convention for the meaning of "$|n,m\rangle$", the Coulomb matrix elements of graphene $\langle\langle\ldots\rangle\rangle$ take the form of simple averages over the corresponding two-body matrix elements of GaAs $\langle\ldots\rangle$ [16].

The "spherical" pseudopotentials V_m (describing graphene in the spherical geometry and correctly linked to the Coulomb potential $\sim1/r$) are then obtained from the above matrix elements by standard diagonalization of the two-electron problem; cf. Fig. 1. Comparison of pseudopotentials of different LLs of graphene with the "reference" ones of the first and second LL of GaAs (for which the ground state at a half-filling is known/believed to be the nearly free CF Fermi sea and the paired CF superfluid described by the Pfaffian, respectively) is best illustrated in Fig. 2, in which we present the symmetric first-order differences defined as $\Delta^{(s)}V_m=V_m-(V_{m-s}+V_{m+s})/2$. For $s=2$ (top frames), $\Delta^{(2)}V_m$ considered for the odd values of m (as appropriate for the spin-polarized electrons) reveals the deviation of V_m from a straight line, which is its most important feature in producing the particular form of the pair electron correlation in a spin-polarized ground state [17]. Clearly, the pseudopotentials in the lowest two LLs of graphene appear similar in this respect. It is also remarkable that none of the LLs of graphene has a pseudopotential similar to that of the second LL of GaAs. For $s=1$ (bottom frames), $\Delta^{(1)}V_m$ considered for the even values of m tells about the stability of the ground state against spin depolarization. Here,

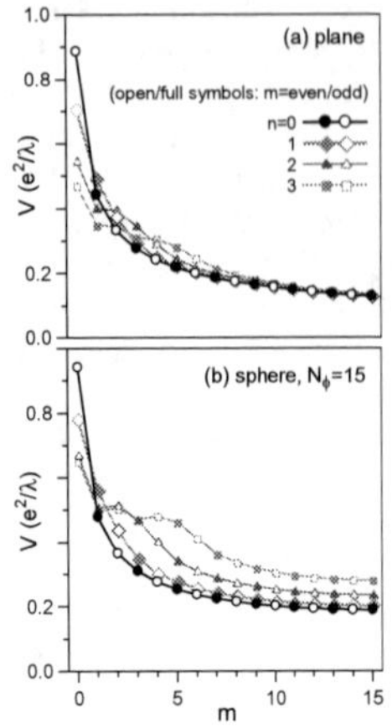

Figure 1. Coulomb interaction pseudopotentials V_m in lowest four LLs of graphene (plane vs finite flux N_ϕ=15 on a sphere).

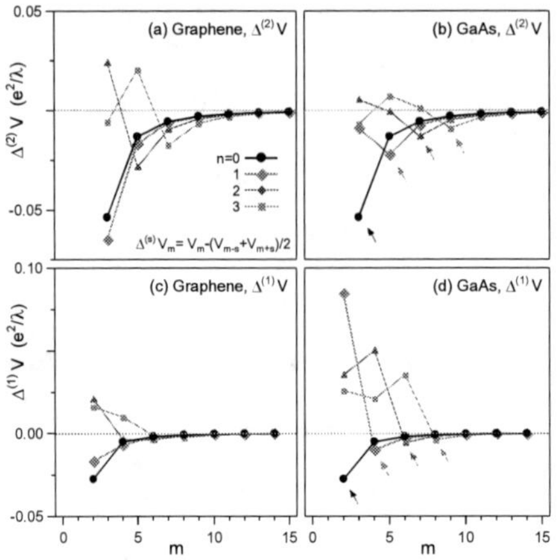

Figure 2. Comparison of the symmetric first-order differences of Coulomb pseudopotentials V_m in the lowest four LLs of graphene and GaAs, calculated on a plane (i.e., for a large flux on a sphere).

we find that the lowest two LLs of graphene behave similarly (again), while the pseudopotential for n=2 in graphene resembles that of n=1 in GaAs (for s=2 these two differed in the sign of the leading term, $\Delta^{(2)}V_3$). This suggests that none of the graphene LLs is analogous to the second LL of GaAs.

3. Many-electron ground states with full spin and pseudospin polarization

We begin by studying the polarized N-electron ground states at different $N_\phi=2N-\sigma$ (corresponding to a half-filled LL). An extended incompressible liquid is represented on a sphere by a series of uniform (i.e., having zero total angular momentum, L=0) finite-size states (N, N_ϕ) with a constant shift σ (e.g., σ=3 for the Pfaffian). The correlation energy, pair correlation function, etc. should depend smoothly on N along the series (and extrapolate properly to the plane). Using an efficient Lanczos method we were able to diagonalize hamiltonians of up to N=20 electrons, with dimensions of the order of 10^9.

As anticipated from the analysis of pseudopotentials, the Pfaffian wave function was not found to have high squared overlaps $|\chi|^2$ with the corresponding (σ=3) exact N-electron Coulomb ground states in any LL of graphene. Actually, the oscillations in $|\chi|^2$ as a function of N reveals the emergence of a shell structure of the (essentially) non-interacting CFs at B^*=0 (the shells being the effective CF LLs) rather than the formation of a uniform phase, adiabatically connected to the Pfaffian. Specifically, the ground states of N=6, 12, and 20, corresponding to the number ν^*=2, 3, and 4 of completely filled CF LLs all have relatively small $|\chi|^2$ while the other ground states (corresponding to partial filling of the topmost CF LL) all have relatively high $|\chi|^2$. This is illustrated in Fig. 3, presenting also a comparison with the Coulomb eigenstates in the second LL in GaAs (for two different widths w=0 and 3λ, as the finite width is known to improve the overlap in such system [18]). Evidently, the decrease of $|\chi|^2(N)$ in graphene is much faster than in GaAs (especially for w=3λ), where also the N-dependence is different. In the case of GaAs, a significant enhancement of $|\chi|^2|$ was also recently demonstrated through the LL mixing [19]. However, this effect is not relevant in graphene due to a much smaller (constant) ratio of the inter-LL splitting to the Coulomb energy scale. Fig. 3 also compares the Coulomb energies of the exact Coulomb eigenstates and the Pfaffian states, also showing lack of a Pfaffian phase in graphene.

Emergence of a shell (CF-LL) structure in the lowest two LLs of graphene is most clear in Fig. 4(a) showing (on the example of n=1) N-dependence of the ground state correlation energy E (counted per

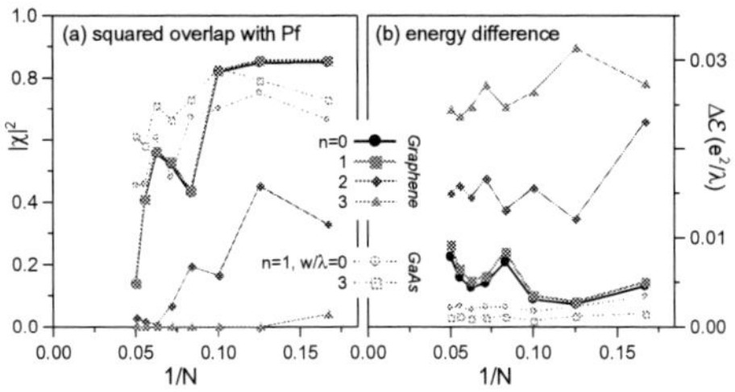

Figure 3. (a) Squared overlaps $|\chi|^2$ of the Pfaffian with the lowest L=0 Coulomb eigenstates at half-filling of different LLs in graphene and of the 2nd LL in GaAs, plotted as a function of an inverse electron number N^{-1}. (b) Difference $\Delta\mathcal{E}$ between the average Coulomb energy of the Pfaffian and the exact energy of the lowest L=0 Coulomb eigenstate, also at ν=1/2 and σ=3, and for different LLs of graphene and 2nd LL of GaAs, versus N^{-1}.

particle, including the charge compensating background, and calculated separately for different shifts). The dominant tendency is the CF shell filling, with low values of E coincident with exact filling of an indicated number of the CF-LLs. Let us illustrate this tendency on a few examples. At the half-filling, the (N,N_ϕ) states of strongly interacting electrons map onto the states of nearly non-interacting CFs at N_ϕ*= N_ϕ-2(N-1)=2-σ (negative sign meaning the effective magnetic flux acting on the CFs oriented opposite to the original flux 2Q [20]), with the n*th CF-LL degeneracy of |2-σ|+1+2n*. For σ=3 and N=20, this yields four lowest CF-LLs completely filled (i.e., this finite-size half-filled state is "aliased" with an incompressible ν*=-4 CF state, corresponding to the electron filling factor ν=4/7), leading to a relatively low correlation energy. For σ=4 and N=16, there are three filled CF-LLs and an additional single CF in the n*=3 level with the single-particle angular momentum equal to |N_ϕ*|/2+n*=4, yielding an L=4 state (aliased with a single CF-quasiparticle in an incompressible ν*=3 state) with a somewhat higher correlation energy than in the case of complete CF-LL filling. Finally, for σ=3 and N=16, there are three full CF LLs and additional four CFs in an 8-fold degenerate CF-LL with n*=3, leading to a relatively high correlation energy. The above behavior for the second LL in graphene repeats that of the lowest LL (not shown; studied earlier [13]), both hosting a CF Fermi sea rather than a CF liquid.

4. Spontaneous depolarization of pseudospin

Full spin polarization of any FQH state can in principle be enforced by the Zeeman effect. However, in graphene one must also consider an additional "pseudospin" degree of freedom associated with the valley degeneracy. Since valley depolarization is not protected by a pseudospin analog of the single-electron Zeeman splitting (at least, in the absence of external potentials or lattice distortions [10]), we have included in diagonalization the single-electron pseudospin (p_z=±1/2) and classified the N-electron eigenstates by its total length P and projection P_z. Dependence of the correlation energy E on size N for the pseudospin-unpolarized states (P=0), is shown in Fig. 4(b,c). Especially for n=0, the half-filled state evidently undergoes a spontaneous depolarization of pseudospin (e.g., the unpolarized series with σ=1, 3, and 4 clearly extrapolate to lower energies than any polarized series). This is not unexpected, given that spontaneous FQH ferromagnetism is a rare phenomenon (ν=1 and 1/3 in the lowest LL).

5. Conclusion

We used exact diagonalization to study many-electron ground states at a half-filling of different LLs in graphene, including pseudospin (i.e., valley) degeneracy. We found that even in the subspace of fully polarized spin and pseudospin the Pfaffian state is not realized in graphene. Instead, the Fermi sea of free CFs forms in the lowest two LLs (with non-liquid phases most likely prevailing in higher LLs). We also demonstrated that the CF Fermi sea in graphene undergoes the spontaneous (and unprotected)

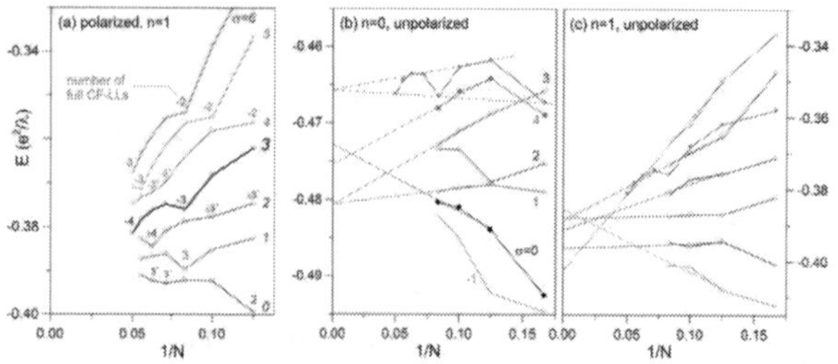

Figure 4. Correlation energies of the N-electron ground states in a half-filled lowest (b) and second (a,c) LLs in graphene. Panel (a) assumes full polarization and (b,c) – complete depolarization of pseudospin. Grey symbols show the polarized σ=3 series in (b,c).

depolarization of pseudospin. These results point to the absence of a Pfaffian phase in graphene.

The authors thank M. O. Goerbig and N. Regnault for the insightful comments. AW acknowledges support by Marie Curie PIEF-GA-2008-221701 and Polish MNiSW Grant No. N202179538.

References

[1] Willett R L, Eisenstein J P, Stormer H L, Tsui D C, Gossard A C, and English J H 1987 *Phys. Rev. Lett.* **59** 1776
[2] Morf R H 1998 *Phys. Rev. Lett.* **80** 1505
Rezayi E H and Haldane F D M 2000 *Phys. Rev. Lett.* **84** 4685
Möller G and Simon S H 2008 *Phys. Rev.* B **77** 075319
Wójs A 2009 *Phys. Rev.* B **80** 041104(R)
Storni M, Morf R H, and Das Sarma S 2010 *Phys. Rev. Lett.* **104** 076803
Wójs A, Möller G, Simon S H, and Cooper N R 2010 *Phys. Rev. Lett.* **104** 086801
[3] Moore G and Read N 1991 *Nucl. Phys.* B **360** 362
[4] Jain J K 1989 *Phys. Rev. Lett.* **63** 199
[5] Nayak C and Wilczek F 1996 *Nucl. Phys.* B **479** 529
[6] Nayak C, Simon S H, Stern A, Freedman M, and Das Sarma S 2008 *Rev. Mod. Phys.* **80** 1083
[7] Dean C R, Piot B A, Hayden P, Das Sarma S, Gervais G, Pfeiffer L N, and West K W 2008 *Phys. Rev. Lett.* **100** 146803
[8] Haldane F D M 1983 *Phys. Rev. Lett.* **51** 605
Quinn J J, Wójs A, Yi K S, and Simion G 2009 *Phys. Rep.* **481** 29
Wójs A and Quinn J J 2001 *Phys. Rev. B* **63** 045304
[9] Novoselov K S, Geim A K, Morozov S V, Jiang D, Katsnelson M I, Grigorieva I V, Dubonos S V, and Firsov A A 2005 *Nature* **438** 197
Zhang Y, Tan Y-W, Störmer H L, and Kim P 2005 *Nature* **438** 201
[10] Papić Z, Goerbig M O, and Regnault N 2009 *Solid. State Commun.* **149** 1056
[11] Goerbig M O, Moessner R, and Douçot B, 2006 *Phys. Rev.* B **74** 161407(R)
Nomura K and MacDonald A H 2006 *Phys. Rev. Lett.* **96** 256602
[12] Khveshchenko D V 2007 *Phys. Rev.* B **75** 153405
[13] Tőke C and Jain J K 2007 *Phys. Rev.* B **76** 081403(R)
[14] Bolotin K I, Ghahari F, Shulman M D, Störmer H L, and Kim P 2009 *Nature* **462**, 196
[15] Jellal A 2008 *Nucl. Phys.* B **804** 361
[16] Tőke C and Jain J K 2006 *Phys. Rev.* B **74** 235417
[17] Wójs A and Quinn J J 2000 *Philos. Mag.* B **80** 1405
[18] Peterson M R, Jolicoeur T, and Das Sarma S 2008 *Phys. Rev. Lett.* **101** 016807
[19] Wójs A, Tőke C, and Jain J K 2010 *Phys. Rev. Lett.* **105** 096802
[20] Möller G and Simon S H 2005 *Phys. Rev.* B **72** 045344

Journal of Physics: Conference Series **334** (2011) 012049 doi:10.1088/1742-6596/334/1/012049

Magnetotransport of Massless Dirac Fermions in Multilayer Organic Conductors

T Osada, K Uchida, and T Konoike
Institute for Solid State Physics, University of Tokyo,
5-1-5 Kashiwanoha, Kashiwa, Chiba 277-8581, Japan

E-mail: osada@issp.u-tokyo.ac.jp

Abstract. Magnetotransport has been considered in multilayer massless Dirac fermion systems, in which two-dimensional (2D) layers with a pair of Dirac-cone dispersion stack with weak interlayer coupling. The Fermi level is assumed to be fixed at the Dirac point resulting in the zero-gap conductor. At the high-field quantum limit, where only the n=0 Landau level at the Dirac point is partially occupied, all elements of conductivity tensor have been evaluated by the lowest order contribution of interlayer coupling. We have deduced saturation of in-plane resistance, local maxima of in-plane Hall resistance, negative interlayer magnetoresistance, and cotθ-type unusual angle-dependence of interlayer Hall resistance. These results propose good explanations for mysterious transport features observed in organic zero-gap conductors, α-(BEDT-TTF)$_2$I$_3$ and its related compounds.

1. Introduction

The massless Dirac fermion system in solids has attracted a great deal of attention since monolayer graphene devices were realized and the half-integer quantum Hall effect was demonstrated using them [1, 2]. The enormous amount of research has been done on magnetotransport of two-dimensional (2D) massless Dirac fermions in graphene [3]. On the other hand, just around the same time, another massless Dirac fermion system was discovered in layered organic conductors α-(BEDT-TTF)$_2$I$_3$ and its related compounds, where BEDT-TTF denotes bis(ethylenedithio)-tetrathiafulvalene. In α-(BEDT-TTF)$_2$I$_3$ crystals, BEDT-TTF molecules and I_3^- anions form conducting layers and insulating layers, respectively, and they stack alternatively. Since the interlayer coupling is very weak (the interlayer transfer energy t_c is much smaller than 1 meV), the electron system is regarded as a quasi-2D system with no coherent interlayer coupling down to low-temperature. Based on the band calculation using pressure-dependent band parameters, it was theoretically predicted that these compounds become anisotropic massless Dirac fermion systems under high pressures [4]. The Fermi level is always located at the band contact point (Dirac point) of 2D Dirac cone dispersion. In other words, these systems are zero-gap conductors [5]. In fact, the charge ordering transition at T=135 K observed at ambient pressure is suppressed at high pressures above 15 kbar, suggesting the change to the zero gap conductor with very small carrier density.

Before the discovery of Dirac fermion, it has been known that α-(BEDT-TTF)$_2$I$_3$ and its related compounds exhibit anomalous magnetotransport features at low temperatures. We set the z-axis along the stacking direction normal to the conducting plane (xy-plane). The in-plane diagonal resistivity ρ_{xx} exhibits saturation behavior in high magnetic fields [5], whereas the in-plane Hall resistivity ρ_{yx} deviates from B-linear behavior and shows a local maxima at high fields [6]. The interlayer diagonal resistivity ρ_{zz} shows remarkable decrease with increasing magnetic fields [7]. Namely, interlayer transport shows negative magnetoresistance at high magnetic fields. In addition, recently, Kajita and Sato *et al.* have reported another remarkable behavior of interlayer Hall resistance (ρ_{zx}) [8]. They have measured the off-diagonal resistivity ($\rho_{zx}=E_z/j_x$) of α-(BEDT-TTF)$_2$I$_3$ as a function of strength and direction of the magnetic field. Under sufficiently strong magnetic fields, ρ_{zx} seems to depend not on magnetic field strength but only on magnetic field direction. It shows unusual angle-dependence proportional to $\cot\theta=B_y/B_z$ ($B_z>0$), where θ denotes the elevation angle from the y-axis (Hall configuration) to the z-axis. The origins of above mysterious magnetotransport features have been unsolved questions for a long time.

In this paper, we clarify magnetotransport behaviors of multilayer massless Dirac fermion systems in the high-field quantum limit where only the n=0 Landau level exists at the Fermi level. Particularly, we concentrate on the interlayer transport which is additional characteristic of multilayer systems to those of monolayer systems like graphene. We employ the tunneling picture which treats interlayer coupling as a perturbation, and we discuss its lowest order contribution to conductivity. We show that almost all of magnetotransport features observed in α-(BEDT-TTF)$_2$I$_3$ can be explained as those of multilayer Dirac fermion systems at the quantum limit.

2. Tunneling picture for magnetotransport in multilayer massless Dirac fermion system

We consider the multilayer massless Dirac fermion system of which dispersion around the Fermi level $E_F=0$ is given by $H(\mathbf{k})=\hbar v(k_x\sigma_1+k_y\sigma_2)-2t_c\cos ck_z\sigma_0$ [9, 10]. Here, $\mathbf{k}=(k_x, k_y)$ is the 2D wave number measured from the band-contact point (Dirac point) $\mathbf{k}=\mathbf{k}_0$ (or $\mathbf{k}=-\mathbf{k}_0$). v is the in-plane group velocity of isotropic Dirac cones, and t_c and c denote interlayer transfer energy and interlayer distance, respectively. σ_0 and (σ_1, σ_2, σ_3) are the unit matrix and Pauli spin matrices. Under the general magnetic field $\mathbf{B}=(B_x, B_y, B_z)$, the effective mass Hamiltonian $H_{\text{eff}}=H_0+H'$ is given by

$$H_0=\hbar v\begin{pmatrix} 0 & \left(-i\partial_x+eB_yz/\hbar\right)-i\left\{-i\partial_y+e(B_zx-B_xz)/\hbar\right\} \\ \left(-i\partial_x+eB_yz/\hbar\right)+i\left\{-i\partial_y+e(B_zx-B_xz)/\hbar\right\} & 0 \end{pmatrix},$$

$$H'=-2t_c\cos\left(-ic\partial_z\right)\begin{pmatrix} 1 & 0 \\ 0 & 1 \end{pmatrix}. \tag{1}$$

Here, we have chosen a gauge as $\mathbf{A}=(B_yz, B_zx-B_xz, 0)$. We assume $B_z>0$ without any loss of generality. In the present model, spin splitting due to Zeeman effect is not taken into account. The unperturbed Hamiltonian H_0 gives the free Landau levels of 2D Dirac fermion system on each layer located at $z=z_i$. The quantum numbers of these unperturbed states are the Landau index n, the center coordinate x_0, and the layer position z_i. Their eigen energies are given by

$$E_{n,x_0,z_i}=\text{sgn}(n)\left(\sqrt{2}\hbar v/l\right)\sqrt{|n|}. \tag{2}$$

Here, $l=\sqrt{\hbar/eB_z}$ is the magnetic length. We also assume the existence of two Dirac cones at $\mathbf{k}_0$ and $-\mathbf{k}_0$ so as to satisfy time-reversal symmetry. So, each of Landau states has two-fold spin degeneracy and two-fold valley degeneracy. To reproduce the finite Hall effect, we consider very small but finite carrier imbalance $n^{(2D)}$ due to extrinsic origin such as charged impurities. The sheet carrier density per single layer $n^{(2D)}$ is defined positive for electrons ($E_F>0$) and negative for holes ($E_F<0$). When $|n^{(2D)}|$ becomes smaller than one half of the Landau level degeneracy $4/2\pi l^2$ (including spin and valley degeneracy), the system reaches the quantum limit where the Fermi level exists in the n=0 Landau level and only the n=0 Landau level is partially occupied. Note that in the present Dirac fermion

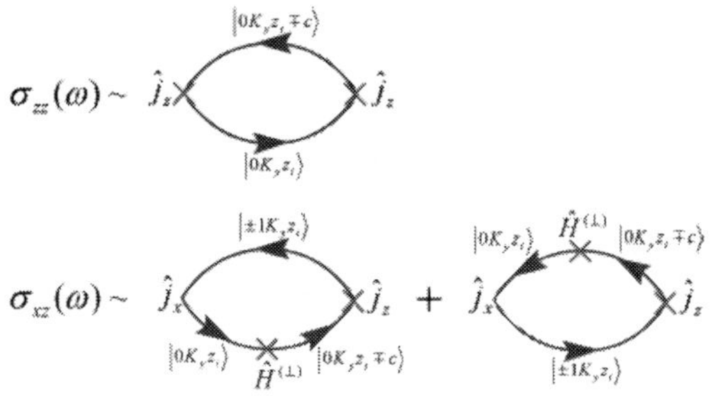

Figure 1. Lowest order contributions of interlayer coupling to diagonal and off-diagonal interlayer conductivity elements. Arrows and crosses indicate in-plane propagators with scattering and interlayer tunneling, respectively.

system, the quantum limit is easily realized even in low magnetic fields because of large Landau level spacing between n=0 and n=1 (or −1) levels and small carrier density $|n^{(2D)}|$.

We treat the interlayer coupling H' in (1) as a perturbation. We call this method the tunneling picture. This picture is effective when interlayer coupling is weaker than in-plane scattering resulting in incoherent interlayer coupling. The interlayer coupling is responsible for interlayer transport and it causes broadening of unperturbed Landau levels together with scattering. At the quantum limit, the perturbation matrix elements relating to the n=0 Landau level are important. They can be easily obtained using eigen states of H_0. They give the following selection rule of interlayer tunneling: Tunneling is only allowed between neighboring layers ($z_i'=z_i \pm c$), and the center coordinate of the final state on the layer at $z=z_i \pm c$ satisfies $x_0'=x_0 \pm (B_x/B_z)c$, where x_0 is the center coordinate of initial state on the layer at $z=z_i$. So, the direction connecting the center coordinates of the initial and final states is parallel to the projected magnetic field on the xz-plane. The tunneling amplitude is proportional to the overlap integral of the initial and final states. The matrix elements of interlayer current $j_z =(-e)(1/i\hbar)[z, H']$ are proportional to those of interlayer tunneling H'.

Each element of the conductivity tensor can be calculated using Kubo formula. In the tunneling picture, each conductivity element is expanded into a power series of interlayer coupling. In the case of weak interlayer coupling ($t_c \ll \hbar/\tau$: the dirty limit), the lowest order contribution gives a good approximation of the conductivity. So, we have evaluated all conductivity elements by the lowest order contribution, *i.e.*, the leading term of expansion at the quantum limit where only the n=0 Landau level is located at the Fermi level. Figure 1 shows the leading terms of interlayer diagonal conductivity σ_{zz} and off-diagonal conductivity σ_{xz}. Note that interlayer current j_z changes the layer z_i but in-plane current j_z and j_z do not change the layer. They give the following analytic formulae for d.c. interlayer conductivity elements (B_z>0) [9, 10].

$$\sigma_{zz}(0) = \frac{8t_c^{\,2} ce^3\tau^2}{\pi^2\hbar^4} B_z e^{-\frac{1}{2}\frac{ec^2}{\hbar}\frac{B_x^{\,2}+B_y^{\,2}}{B_z}} \tag{3}$$

$$\sigma_{xz}(0) = -\frac{t_c^{\,2} cen^{(2D)}}{\hbar^2 v^2}\frac{B_y}{B_z^{\,2}} e^{-\frac{1}{2}\frac{ec^2}{\hbar}\frac{B_x^{\,2}+B_y^{\,2}}{B_z}} \tag{4}$$

Here, we assumed low temperature ($k_B T \ll \hbar/\tau$) and the dirty limit ($t_c \ll \hbar/\tau$). The exponential factor results from the overlap of n=0 wave functions. Except this factor, σ_{zz} is proportional to the normal magnetic field B_z, reflecting the degeneracy of the n=0 Landau level. The factor B_y/B_z^2 in σ_{xz} also results from the overlap between n=0 and $n=\pm 1$ wave functions.

As for the in-plane conductivity elements σ_{xx} and σ_{xy}, their leading terms are the 0-th order of interlayer coupling. They have already been discussed for graphene using self-consistent Born approximation by Shon and Ando [11]. According to their results, 2D diagonal conductivity $\sigma_{xx}^{(2D)}$ takes a constant value $2e^2/\pi^2$ independent on temperature and magnetic field. This is called the universal conductance of graphene. So, using their results, we can set $\sigma_{xx}=2\alpha e^2/\pi^2\hbar c$ and $\sigma_{xy}=-en^{(2D)}/cB_z$, where α (~1) is an empirical correction factor for multilayers.

3. Comparison with experiments on organic Dirac fermion systems

Since we have obtained all elements of conductivity tensor, we can discuss magnetotransport of the multilayer Dirac fermion system in any configuration. Figure 2 shows the calculated field-dependence of in-plane resistivity. The in-plane diagonal resistivity ρ_{xx} shows saturation to the constant value of $1/\sigma_{xx}$. The amplitude of the in-plane Hall resistivity ρ_{yx} exhibits the local maximum corresponding to the crossover of σ_{xx} and σ_{xy}. On the other hand, as shown in Fig. 3, the interlayer diagonal resistivity ρ_{zz} shows remarkable negative magnetoresistance. This feature was already reported in our previous paper [9], and calculation details are described there.

One of the most remarkable magnetotransport features in the multilayer Dirac fermion system is anomalous dependence of interlayer off-diagonal resistivity ρ_{zx} on the magnetic field strength and orientation. ρ_{zx} at the quantum limit is analytically written as

$$\rho_{zx} = -\alpha \frac{\hbar c}{4e^2 v^2 n^{(2D)}\tau^2}\left(\frac{B_y}{B_z}\right)\frac{1}{1+(2\alpha e B_z/\pi^2\hbar n^{(2D)})^2}. \quad (5)$$

Here, we assumed $B_z>0$ and that the magnetic field is rotated in the yz-plane normal to the current. When the normal field component B_z is less than $\pi^2\hbar n^{(2D)}/2\alpha e$, which is the lower bound of quantum limit region, the last factor in (5) can be regarded as 1, so that ρ_{zx} is almost proportional to $B_y/B_z=\cot\theta$, where θ is the elevation angle of magnetic field from the conducting plane. This means that ρ_{zx} depends only on the field direction θ and it has no dependence on the field strength. In the field configuration parallel to the y-axis (θ=0), ρ_{zx} gives the interlayer Hall effect. So the absence of field-strength-dependence is very anomalous behavior. The cotθ-type angle-dependence is also anomalous

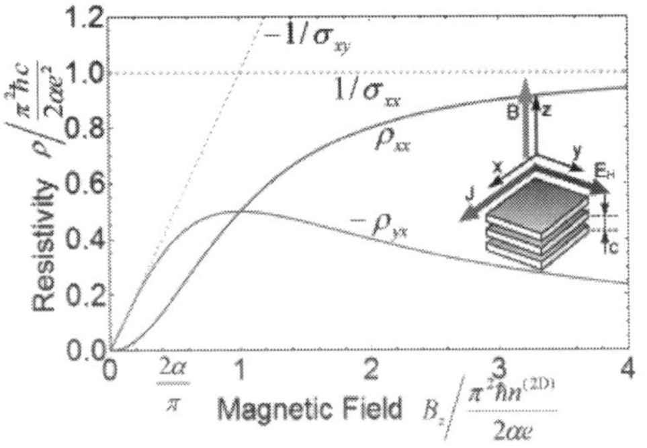

Figure 2. Calculated field-dependence of in-plane diagonal resistivity ρ_{xx} and Hall resistivity ρ_{yx} at the quantum limit. Dotted lines indicates the inverse of σ_{xx} and σ_{xy}.

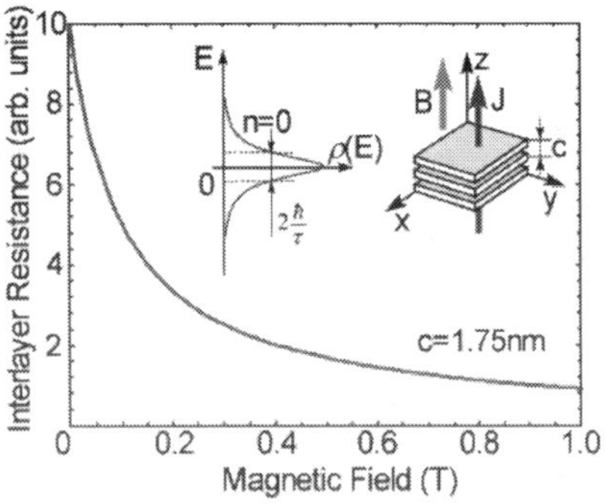

Figure 3. Calculated field-dependence of interlayer diagonal resistivity ρ_{zz} at the quantum limit. The interlayer spacing c=1.57nm is assumed. (after [9])

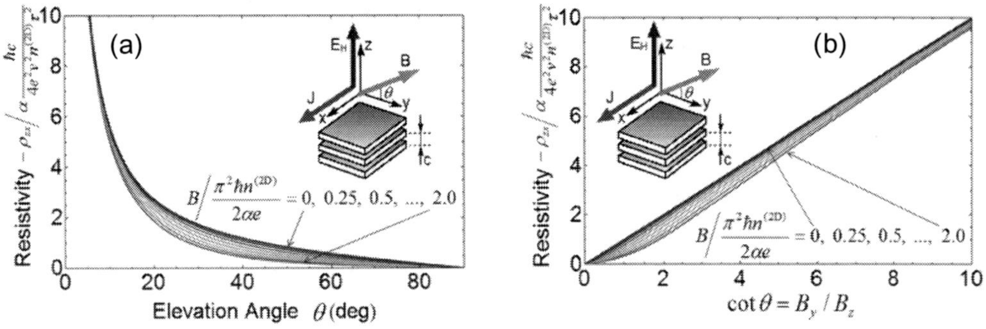

Figure 4. Calculated angle-dependence of interlayer off-diagonal resistivity ρ_{zx} at the quantum limit. (a) Angle-dependence of ρ_{zx} at several field values. Inset shows the configuration of the system. (b) cotθ-dependence of ρ_{zx} at several field values.

since semiclassically cosθ-type dependence is expected. In higher fields in the quantum limit region, ρ_{zx} shows field-strength dependence. Figure 4(a) illustrates the angle-dependence of ρ_{zx} for several field values. We can see that ρ_{zx} deviates from the exact cotθ-dependence (limit of B_z=0) in higher fields. However, the deviation is not large, so that the angle-dependence is approximately cotθ-like. This feature is more clearly seen in Fig. 4(b), in which ρ_{zx} is plotted as a function of cotθ.

The above results obtained in this work explain all of mysterious magnetotransport behaviors observed in α-(BEDT-TTF)$_2$I$_3$, which were mentioned in the first part, qualitatively very well. They are the characteristic phenomena of multilayer Dirac fermion systems, since the n=0 Landau level plays an important role for their appearance. Therefore, conversely, the observation of these magnetotransport phenomena gives strong evidences of realization of massless Dirac fermions in organic conductors, α-(BEDT-TTF)$_2$I$_3$ and its related compounds.

References

[1] Novoselov K S, Geim A K, Morozov S V, Jiang D, Katsnelson M I, Grigorieva I V, Dubonos S V and Firsov A A 2005 *Nature* **438** 197
[2] Zhang Y, Tan Y -W, Stormer H L and Kim P 2005 *Nature* **438** 201
[3] Castro Neto A H, Guinea F, Peres M M R, Novoselov K S and Geim A K 2009 *Rev. Mod. Phys.* **81** 109; its references
[4] Katayama S, Kobayashi A and Suzumura Y 2006 *J. Phys. Soc. Jpn.* **75** 054705; Kobayashi A, Katayama S, Suzumura Y and Fukuyama H 2007 *J. Phys. Soc. Jpn.* **76** 034711
[5] Tajima N, Sugawara S, Tamura M, Nishio Y and Kajita K 2006 *J. Phys. Soc. Jpn.* **75** 051010
[6] Miura K, Kajita K *et al.* 2009 25aYC2 *Meeting Abstract of Phys. Soc. Jpn.* **64** *Issue2 2009 Autumn Meeting (Kumamoto Univ., Japan, 25-28 Sept., 2009)* pp 730
[7] Tajima N, Sugawara S, Kato R, Nishio Y and Kajita K 2009 *Phys. Rev. Lett.* **102** 176403
[8] Sato M, Miura K, Endo S, Sugawara S, Tajima N, Murata K, Nishio Y, and Kajita K 2011 J. Phys. Soc. Jpn. **80** 023706
[9] Osada T 2008 *J. Phys. Soc. Jpn.* **77** 084711
[10] Osada T 2011 *J. Phys. Soc. Jpn.* **80** 033708
[11] Shon N H and Ando T 1998 *J. Phys. Soc. Jpn.* **67** 2421

Journal of Physics: Conference Series **334** (2011) 012050 doi:10.1088/1742-6596/334/1/012050

Strong temperature destabilization of free exciton recombination in a two-dimensional structures with hole gas

J. Jadczak[1], L. Bryja[1], A. Wójs[1], G. Bartsch[2], D. R. Yakovlev[2], M. Bayer[2], P. Plochocka[3], M. Potemski[3], D. Reuter [4] and A. Wieck[4]

[1]Institute of Physics, Wroclaw University of Technology, Wyb. Wyspianskiego 27, 50-370 Wroclaw, Poland
[2]Experimentelle Physik 2, Technische Universität Dortmund, D-44227 Dortmund, Germany
[3]Grenoble High Magnetic Field Laboratory, CNRS, 38-042 Grenoble, France
[4] Lehrstuhl fur Festkörperphysik, Ruhr Universität, Universitätstrasse, Bochum, Germany

E-mail: leszek.bryja@pwr.wroc.pl

Abstract. We have investigated polarisation resolved magneto-photoluminescence from 22 nm wide asymmetric GaAs quantum well. In contrast to previous studies we observed abrupt decrease of photoluminescence intensity of the neutral exciton line with increasing temperatures whereas photoluminescence intensity of the positively charged exciton remains unchanged. The effect is a clear evidence of different mechanisms of neutral and charged exciton recombination. We have also detected a coupling of two different radiative states: an acceptor-bound and an essentially free trions.

1. Introduction

The energy spectrum of quasi-two-dimensional excitons in quantum well semiconductor heterostructures has been extensively studied in recent decades [1]. Most experiments were conducted with 2D electrons with highest mobility. In our studies we have explored 2D holes. We have investigated low-temperature (T=2.0-18K), high-field (B<17T), polarization-resolved magneto- photoluminescence (PL) on a range of high-quality asymmetric GaAs quantum wells with different widths w, hole concentrations p and (high) mobility μ. In contrast to previous studies [2] we observed abrupt decrease of photoluminescence intensity of neutral exciton line (X) with increasing temperatures whereas PL intensities of positively charged excitons (X+) remain unchanged. We also detected a coupling of two different radiative states observed as an anti-crossing in the magneto- photoluminescence spectrum, of essentially-free and acceptor-bound trion states immersed in a magnetically quantized hole gas, aided by an additional cyclotron excitation.

2. Experimental results

Te effect reported here was most prominent in the asymmetric 22nm wide GaAs/Ga0.65Al0.35As quantum well. The sample was grown by molecular beam epitaxy on a (001) semi-insulating GaAs substrate and δ-Carbon doped in the barrier on one side. The well was separated from the doping layer by a 40nm wide spacer. The hole mobility measured at T=4.2K was equal to μ=1.71x105cm-2/Vs. The hole concentration measured at low temperatures (T=17 mK-4.2K) and estimated from the quantum Hall effect measured in van der Pauw configuration, was almost temperature independent. In the dark, it was equal to $p=1.43x10^{11}cm^{-2}$. Under laser illumination it decreased linearly with excitation power density, by up to several percent. The PL was excited by the 632.8nm line of a Helium-Neon laser (above the band gap of the barrier). The optical experiments were performed in temperatures from T=2.3K up to 18K, in the Faraday configuration, in magnetic fields up to B=17T varied with a small step ΔB=0.05T. To switch between the σ- and σ+ polarisations the direction of the field was reversed. The spectra were analyzed in a high-resolution monochromator, with a Nitrogen-cooled 2048-pixel CCD camera.

The magnetic field evolution of PL spectrum in σ^- polarization is presented in Fig. 1. In the absence of a field a single line is observed. When magnetic field is applied this line splits into two distinct components. The energy positions of both lines can be resolved in the fields above B=6T. After the detailed analysis and comparison with our previous experiments we attribute the observed lines to the neutral exciton X (higher energy) and the spin-singlet state of a positive trion X^+s (lower energy) [3].

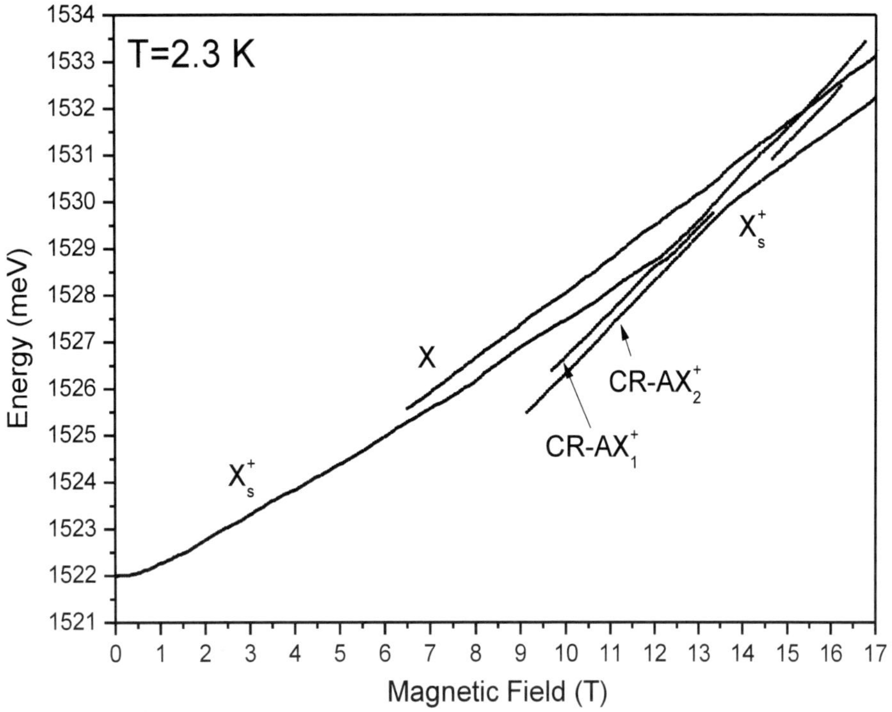

Fig.1 Energy position of all observed lines in photoluminescence in σ^- polarization.

In the lower energy part of the spectrum we detect the whole group of lines of very weak intensity. Two of them shift from the remaining ones toward higher energy, linearly with the increase of the magnetic field (with the slope of ~1.0meV/T). This effect is similar to the "shake-up" process [4] observed previously also by us [3] in a symmetric 15nm wide GaAs quantum well, but here the cyclotron shift is in the opposite energy direction. The magnetic field slope for the rest of weak lines (not marked in Fig. 1) is about 0.65meV/T. We interpreted the weak lines as follows: those lines with the smaller energy vs. field slope are due to the recombination of the trion bound on acceptor AX^+ (inside the well), whereas the two other lines with the higher slope are their hole cyclotron replicas. We named them CR-AX^+_1 and CR-AX^+_2 in analogy to a previous observation of an electron cyclotron replica of the free exciton line [5]. The presence of a pair of cyclotron replicas is a consequence of the fact that the allowed (by the conservation of angular momentum and spin) recombination of the AX+ ground state (labelled by the 3h spin of 1/2 and the total 3h+e angular momentum projection of -1) occurs to a pair of spin-singlet and triplet states of the A+, separated by an exchange gap [3]. Comparison of the slopes of CR-AX^+ and AX^+ gives a difference of 0.35meV/T, which agrees very well with the hole cyclotron energy determined earlier from the cyclotron resonance of 22 nm GaAs/AlGaAs quantum wells by Cole and co-workers [6] equal to 0.28meV/T). The effective mass determined from measured slope is m_{hh}=0.33. In σ^+ polarisation the cyclotron replicas are not observed in PL spectra. In this polarisation we detect only neutral and charged excitons lines.

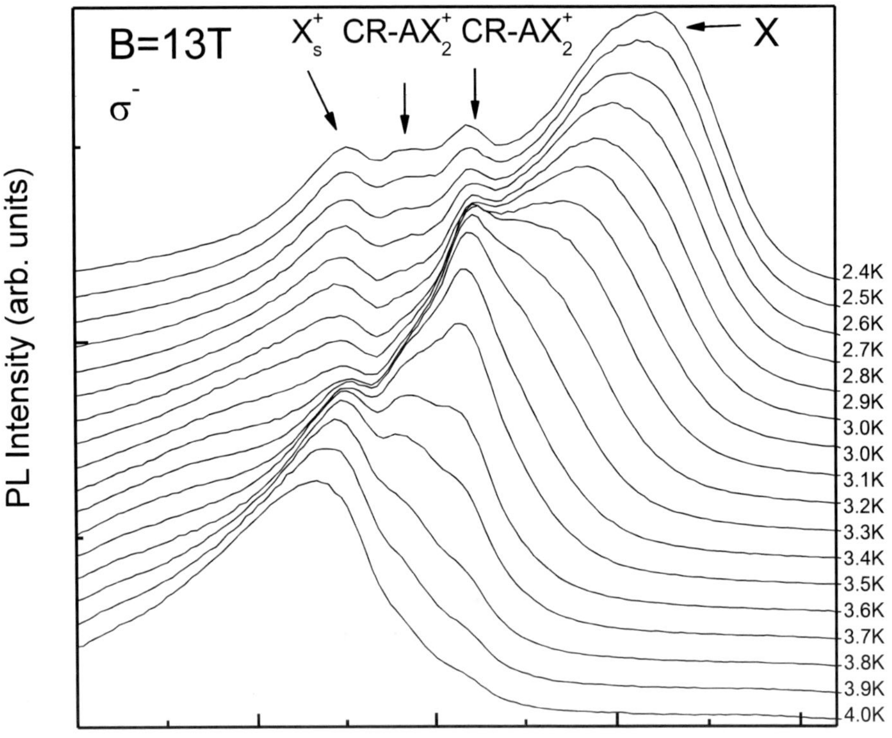

Fig.2. Temperature evolution of photoluminescence spectra.

The most intriguing effect is observed when the pair of CR-AX^+_1 and CR-AX^+_2 lines approach the positive trion X^+_s which results in a clear anticrossing. Furthermore, the intensity of both lines is greatly enhanced in the narrow anticrossing region. Beyond this region all lines return to a similar behaviour as in the lower fields (both in terms of energy and intensity).

Interpretation of this anticrossing was aided by the numerical calculation of the energy and recombination spectra of the involved radiative complexes [3, 7]. The conclusion is that the X^+ and AX^+ states are brought into resonance by an additional cyclotron excitation of a nearby free hole. The coupling mechanism for this pair of quasi-degenerate states is the resonant exciton transfer between the localized and (essentially) free trion states. The coupling strength (dependent on, e.g., the relative acceptor and hole concentrations) extracted from the energy splitting at the point of anticrossing is about 0.1meV.
The coincidence of the degeneracy of the initial radiative states X^+ and AX^+ (required for their efficient mixing) and of the transition energies (whose anticrossing has been directly observed in the PL spectrum) is explained by the near degeneracy of the optically active A^+ channel (corresponding to the angular momentum -1 of the AX^+ ground state) with the free hole state, the two serving as the final states in the radiative recombination of either X^+ or AX^+ [3]. The above interpretation is well supported by a simple calculation involving the set of constant X^+ and AX^+ oscillator strengths and the linear field dependences of the uncoupled initial and final states (extracted from the spectrum outside of the anticrossing region) and the constant coupling strength (determined at the point of anticrossing).
In the Figure 2 the temperature evolution of photoluminescence spectra at magnetic field B=13T are presented. In contrast to previous studies we observed abrupt decrease of photoluminescence intensity of neutral exciton line (X) with increasing temperature whereas PL intensities of positively charged excitons (X^+) remain unchanged [2]. We interpret this effect as a different mechanism of neutral and charged exciton recombination. During the recombination of charged exciton the total momentum can be transferred to left over hole whereas neutral exciton recombine with the total momentum close to zero. This is realized by the neutral exciton localization on a shallow lateral potential. With increasing temperature the recombination of excitons bound on lower potentials consecutively disappear from the PL spectra. This is observed as a low energy shift and narrowing of neutral exciton line and finally its complete disappearance from the PL spectra.

3. Conclusion

We have observed an anomalous effect of abrupt decrease of photoluminescence of neutral exciton recombination with increasing temperature, which is interpreted in terms of different dynamics of neutral and charged excitons. We have also detected the anticrossing of the nearly free trion lines with a pair of cyclotron replicas of an acceptor-bound trion line. The effect has been observed here in a 2D hole gas; it has not been earlier detected in an electron system. The coupling is interpreted as a resonant exciton exchange between free and bound trions assisted by the cyclotron hole excitation.

4. References

[1] Heiman D et al. 1998, Phys. Rev. Lett. **61**, 605; Goldberg B B et al. 1990, ibid. **65**, 641; Kheng K et al. 1993, ibid. **71**, 1752; Yusa G et al. 2001, ibid. **87**, 216402; Buhmann H et al. 1995, Phys. Rev. B **51**, 7969; Byszewski M et al. 2006, Nat. Phys. (London) **2**, 239.
[2] Esser A et al. 2000, Phys. Rev. B **59**, 8231.
[3] Bryja L et al. 2007, Phys. Rev. B **75**, 035308.
[4] Finkelstein G et al. 1996, Phys. Rev. B **53**, 12593; 1997 ibid. **56**, 10326; Dzyubenko A B, 2004 ibid. **69**, 115332.
[5] Yakovlev D et al. 1997, Phys. Rev. Lett. **79**, 3974.
[6] Cole B E et al. 1997, Phys. Rev. B **55**, 2503.
[7] Wójs A and Quinn J J 2007, Phys. Rev. B **75**, 085318; Wójs A, 2007 ibid. **76**, 085344.
Glasberg S et al. 1999, Phys. Rev. B **59**, R10425.

Acknowledgments
This work has been supported by the research Grant No. N202179538 of the Polish MNiSW, Marie Curie Grant No. PIEF-GA-2008-221701 of EU (AW) and EC-EuroMagNetII No. 228043-CNRS-PICS-4340.

Aharonov-Bohm Exciton Splittings in Carbon Nanotubes in Ultra-High Magnetic Fields

S Takeyama[1], H Suzuki[2], Y Otsubo[2], H Yokoi[3], Y Murakami[4], and S Maruyama[2]

[1] Institute for Solid State Physics, University of Tokyo, Kashiwa, Chiba, 277-8581, Japan

[2] Applied Phys. School of Engineering, University of Tokyo, Hongo, 113-8656, Japan

[3]Grad. School of Science and Technology, Kumamoto University, Japan

[4]Global Edge Institute, Tokyo Institute of Technology, Ookayama, 152-8550, Japan

E-mail: takeyama@issp.u-tokyo.ac.jp

Abstract. The Ajiki-Ando (A-A) splitting of single-walled carbon nanotubes was observed by the magneto-absorption measurements conducted up to a very high magnetic field, 78 T for PFO-samples. The well-resolved absorption spectra from the E_{11} transitions in the PFO-samples showed a clear A-A splitting. The electro-magnetic flux compression method was used for generation of the field up to 360 T, where the absorption spectra of the E_{22} transition of the HiPco samples were measured by the streak spectroscopy. Parameters for the A-A splitting were determined for different chiralty of each case.

1. Introduction

A single-walled carbon nanotube (SWNT) is well known for showing up splitting of the energy bands upon applying an external magnetic field parallel to the tube axis, which is known as the Aharanov-Bohm effect, and the splitting is also called Ajiki-Ando (A-A) splitting [1]. Observation of the A-A splitting has been reported in various SWNTs by many authors either by means of the absorption [2] or by the photoluminescence (PL) spectra [3].

Recently, excitonic effects are recognized as very crucial in understanding the fundamental optical transition. Owing to interplay between the inter- and intra-K-K valley short range scattering, the exciton states become very complicated with 16 split states of the bright and dark excitons [4]. Application of a magnetic field causes a mixing of both states, and the complicated exciton states are expected to be clarified experimentally. So far, the dark exciton states are identified by magneto-PL in ensemble samples [5,6] or by micro-magneto-PL [7] in a single SWNT.

Aiming at observing the A-A splitting and the behaviors of these exciton states, we attempted the magneto-optical absorption measurements up to an ultra-high magnetic field. An oscillator strength of the optical transition is directly deduced from the absorption spectra, but not necessarily from those of PL.

[1] To whom any correspondence should be addressed.

2. Experimental

The samples used are micelle-suspended HiPco or PFO-SWNTs dispersed in a liquid. We have performed measurements in ultra-high magnetic fields using the electro-magnetic flux compression (EMFC) techniques [8] up to 360 T, and also by a newly developed giant single-turn coil (GSTC) method up to 80 T [9].

2.1. PFO-SWNT up to 80 T

We attempted near-infrared magneto-optical absorption measurements to the PFO (poly(9,9-dioctylfluorenyl-2,7-diyl)) SWNTs [10] at a room temperature. The PFO-SWNTs show a very sharp and well-defined absorption spectra with respect to the chirality of SWNTs. We first focused on the first subband E_{11} absorption spectra showing sharp peaks, observed in the near-infrared region. We have performed this work to study the exciton absorption spectral change in the region of a magnetic field where the exciton exchange interaction is manifested below 100 T.

2.2. HiPco-SWNT up to 360 T

We have conducted magneto-absorption measurements of HiPco SWNT samples in extremely high magnetic fields above 300 T by the EMFC system. In principle 700 T is now possible to generate in a bore of 6 mm. Due to restriction of the optical alignments around the sample cell, the large diameter of the sample holder limits the highest achievable magnetic field for the measurements. Fast rise time of the magnetic field (2 μs from 100 T to 360 T) also restricts the optical exposure time. The streak spectroscopy (Hamamatsu Photonics Co. Ltd.) is only the available methods of detecting the optical transmission signal during such a short time. Only the visible light region is detectable, which enable us to measure the interband E_{22} transitions.

3. Results and Discussion

Well resolved absorption spectra are obtained for the PFO-SWNTs of the chirality (8,6), (7,6), and (7,5). The absorption peak is well resolved each other in the absence of a magnetic field. Upon applying magnetic field up to 53 T by a non-destructive pulse magnet, a new peak appeared at lower energy side of the main peak as seen in Fig.1. However, peak separation is not enough for the chirality (7,6) and (7,5) to undergo a process of reliable spectral deconvolution. The peak separation becomes more unambiguous by a further increase of the fields up to 78 T generated by the GSCT. Further increase of the field broadens the spectra due to a substantial contribution from those of the randomly oriented tubes with respect to the direction of magnetic field.

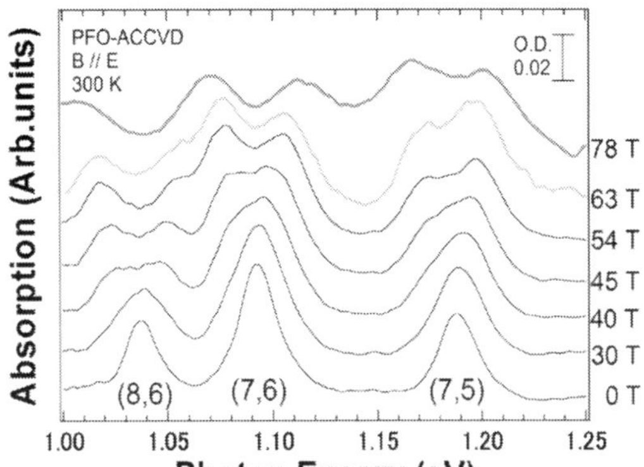

Figure 1. Magneto-absorption spectra showing well-defined AB splitting, measured up to 78 T for the chirality (8,6), (7,6), and (7,5), respectively, in PFO-SWNTs.

In order to estimate the peak splitting against an effective magnetic field $B_{//}$, we need to consider the randomness of the tube orientation against a magnetic field as well as the field induced dynamical orientation of the ensemble nano-tubes in an aqueous solution subjected to a pulse magnetic field [11]. Details of evaluation for $B_{//}$ from the observed peak separation will be discussed elsewhere. Figure 2 is a plot of the results of the peak splitting against $B_{//}$ after the corrections. Due to short time nature of the GSCT in comparison to that of the long pulse magnet, $B_{//}$ becomes much less. The peak splitting against $B_{//}$ was fitted by the relation employed by other authors [6,7], and the results are shown by dotted lines in Fig. 2. The values of the dark and bright exciton splitting Δ_{bd} and of the coefficient representing the A-A splitting μ_{ex} (defined by $\Delta_{A\text{-}A}(B)= \mu_{ex} B_{//}$) are listed in Table 1 together with the positions of the dark and bright exciton in energy known from the relative intensity of the split peaks.

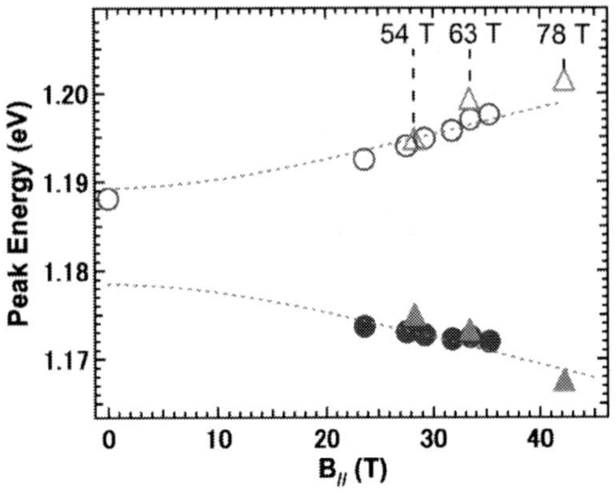

Figure 2. The A-A splitting observed for the chirality (7,5) of PFO-SWNTs. Circles are from the long pulse magnet, and triangles are from the GSTC. The dotted lines are from calculation. Values of magnetic field generated by the GSTC is denoted each by numbers for data of triangles.

Figure 3 is a plot of the time evolution of a magnetic field produced by the EMFC and an obtained streak image of the absorption spectra of the HiPco SWNTs for the chirality (10,2) (9,5) and (7,5). Owing to the broadness of the absorption spectra of the E_{22} transition in comparison to that of E_{11} and also to the peak overlaps of each chirality, spectral resolution of the peak splitting in magnetic fields is less qualified than that in Fig.1. However, the substantial peak splitting is observed, which amounts to about 160 meV at $B_{//}$=260 T. The results of the splitting against $B_{//}$ is summarized in Fig. 4, where the characteristic parameters are evaluated from the fitting, and results are tabulated in Table 1.

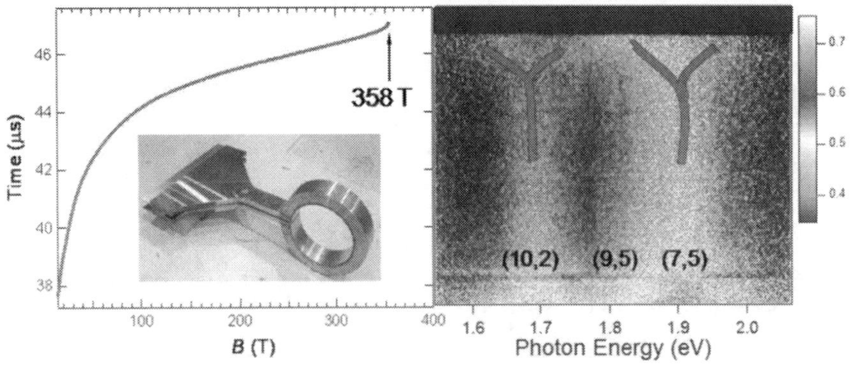

Figure 3. The signal of a magnetic field produced by the EMFC, and the magnet coil (inset). The streak image of the absorption spectra of the HiPco SWNTs with the chirality (10,2),(9,5), and (7,5).

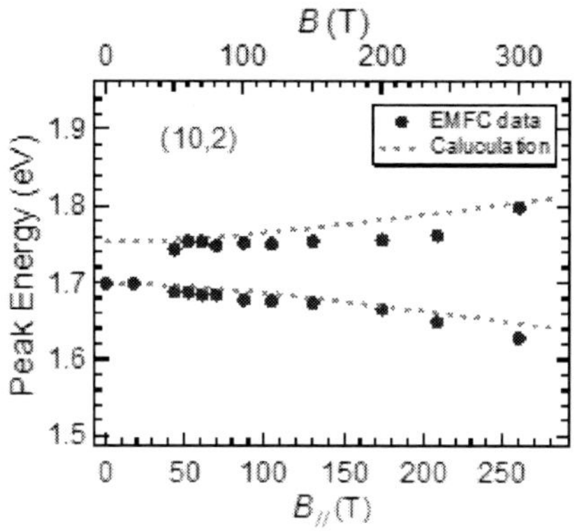

Figure 4. The A-A splitting for the HiPco SWNTs with the chirality (10,2). The upper abscissa shows the external magnetic field, and lower one is the effective magnetic field $B_{//}$. The dotted lines are from calculation.

4. Summary

We have shown the results of the magneto-absorption spectra of PFO- and HiPco-SWNTs. The absorption spectra of the E_{11} transition for PFO samples up to 78 T and those of the E_{22} for HiPco samples up to 360 T were measured. The important parameters associated with the A-A splitting were determined for several different chirality samples. The values of μ_{ex} determined from the experiments are always almost 70 % of those of the theory, μ_{th}. The values of the bright-dark exciton splitting as well as the position of each state were determined from the spectral fitting owing to the A-A splitting.

Table 1. Results of the parameters from the A-A splitting. *d* denotes the diameter of the nanotube.

Sample	Transition	Chirality	*d* [nm]	μ_{ex}	μ_{th}	ratio μ_{ex}/μ_{th}	Δ_{bd} [meV]	Lower State
PFO	E_{11}	(7,5)	0.83	0.66	0.93	0.71	10.4	Dark
		(7,6)	0.89	0.71	0.98	0.72	9.8	Dark
HiPco	E_{22}	(7,5)	0.83	0.56	0.75	0.75	32	Bright
		(10,2)	0.88	0.57	0.77	0.74	55	Bright
		(9,5)	0.97	0.68	0.98	0.69	51	Dark

References

[1] Ajiki H and Ando T 1993 *J. Phys. Soc. Jpn.* **62** 1225
[2] Zaric S, Ostojic G N, Kono J, Shaver J, Moore V C, Strano M S, Hauge R H, Smalley R E and Wei X 2004 *Science* **34** 1129
[3] Mortimer I B, Li L –J, Taylor R A, Rikken G L J A, Portugall O and Nicholas R J 2007 *Phys. Rev.* B **76** 085404-1
[4] Zhao H and Mazumdar S 2004 *Phys. Rev. Lett.* **93** 157402-1
[5] Mortimer I B and Nicholas R J 2007 *Phys. Rev. Lett.* **98** 027404
[6] Shaver J, Kono J, Portugall O, Krstic V, Rikken G L J A, Miyauchi Y, Maruyama S and Perebeinos V 2007 Nano Lett. **7** 1851
[7] Matsunaga R, Matsuda K and Kanemitsu Y 2008 Y *Phys. Rev. Lett.* **101** 147404
[8] Takeyama S, Sawabe H and Kojima E 2010 *J. of Low Temp. Phys.* 159 328
[9] Kindo K, Takeyama S, Tokunaga M, Matsuda Y H, Kojima E, Matsuo A, Kawaguchi K and Sawabe H 2010 *J. of Low Temp. Phys.* **159** 381
[10] Nish A, Hwang J-Y, Doig J and Nicholas J 2007 *Nature Nanotechnology* **2** 640
[11] Takeyama S, Nakamura S and Uchida K 2006 *J. of Phys. Conf. Series* **51** 446

Near-infrared magneto-optical study of excitonic states in single-walled carbon nanotubes under ultra-high magnetic fields

H Yokoi[1], Mukhtar Effendi[1], N Minami[2] and S Takeyama[3]

[1]Department of Materials Science and Technology, Kumamoto University, Kurokami, Kumamoto 860-8555, Japan
[2]Nanosystem Research Institute, National Institute of Advanced Industrial Science and Technology, Central 5 Higashi, Tsukuba, Ibaraki 305-8565, Japan
[3]Institute for Solid State Physics, University of Tokyo, Kashiwanoha, Kashiwa, Chiba 277-8581, Japan

yokoihr@kumamoto-u.ac.jp

Abstract. Singlet excitonic states at the first subband-edge in single-walled carbon nanotubes (SWCNTs) have been studied through near-infrared magneto-absorption spectroscopy under magnetic fields to 105.9 T. Well-resolved absorption spectra of stretch-aligned SWCNT(CoMoCAT)-gelatin films were obtained above 100 T. By the application of magnetic fields in parallel to the alignment of SWCNTs, peak shift toward the lower energy was observed for (8, 4) and (7, 6) tubes and the opposite behavior was observed for (7, 5) and (6, 5) tubes. Above 28.8 T, new peaks emerged at the higher energy side of the peak for the (8, 4) and (7, 6) tubes, and at the lower energy side of the peaks for the (7, 5) and (6, 5) tubes. The magnetic splitting between the existing peak and the new peak was symmetric for every tube, which is in line with the energy splitting due to the Aharonov-Bohm effect. Judging from the energetic positions where the new peaks emerged, the singlet dark excitonic state locates at the lower energy than the singlet bright one in the (7, 5) and (6, 5) tubes while it is suggested strongly that the bright one locates at the lower energy in the (8, 4) and (7, 6) tubes.

1. Introduction

Single-walled carbon nanotube (SWCNT) is one of the ideal materials to study one-dimensional excitonic states. The edges of the conduction bands and the valence bands in this material locate at K and K' points which are degenerate. This feature brings exctionic states at the points 16-fold degeneracy. It is predicted theoretically that the degenerate states is split into two singlet states and twelve triplet states and only the bonding state of the singlet exciton is optically active (bright exciton) and the others are all inactive (dark exciton) due to intra- and inter-valley mixing at K and K' points (K-K' coupling) [1]. As the optical properties of one-dimensional materials are governed strongly by the excitons, their energetic states, especially energetic configuration of the singlet bright and dark excitons, have been studied intensively.

Experimental investigations for this issue have been conducted through photoluminescence (PL) measurements including micro PL spectroscopy under magnetic fields at low temperatures [2, 3]. In these studies, lifting of the degeneracy at the K and K' points due to the Aharonov-Bohm effect by the application of magnetic fields along the tube axis [4] is utilized to cause brightening of the singlet dark exciton. The authors all concluded that the lower singlet state was assigned to a dark one for every investigated SWCNT, taking into consideration emergence of a new PL peak at the lower energy side of a PL peak dominant at zero tesla. However, PL is not a crucial approach to identify the excitonic states as the PL intensity does not reflect the density of states directly. While the magneto-absorption spectroscopy is more preferable, application of higher magnetic fields is required as the absorption peaks become broad to such an extent in SWCNTs that it is difficult to distinguish spectral changes at conventional magnetic fields.

We have recently developed a near-infrared (NIR) spectroscopy system for absorption measurements under very fast megagauss fields [5]. The system enables us to obtain absorption spectra associated with the first subband-edge excitons in SWCNTs above 100 T. In this paper, we report well-resolved infrared absorption spectra of SWCNTs in the megagauss region and discuss the energetic configuration of the excitonic states.

2. Experimental details

In the absorption measurement system, a xenon short-arc flash lamp (Eagle corp., 600 V, 30 J) was employed as a light source and operated at the discharge voltage of 450 V. The light was delivered to an optical probe through a visible-NIR optical fiber. Transmission measurements were conducted in the Voigt configuration by sandwiching a sample with two right-angled prisms made of BK7 at room temperature. Transmission light was conveyed to a 0.3 m spectrometer (Acton SpectraPro-2300i) with gold coated mirrors through another visible-NIR optical fiber. These employed fibers were of the graded-indexed type with a core of 800 μm in diameter. Spectral light intensities were detected with an InGaAs photodiode array (Xenics, XLIN 1.7, 0.9-1.7 μm, 512 pixels of 25 μm in pitch and 500 μm in height). The detector was cooled down to 260 K thermo-electrically during the measurements.

Magnetic fields were generated using a single-turn coil system with a fast capacitor bank (40 kV, 100 kJ) [6]. Though the single-turn coils made of copper were destroyed after discharge due to the strong Maxwell stress, the probe settled in the coil bore was not destroyed as the coil was destroyed outward. A pick-up coil for the measurement of magnetic fields was positioned away from the field center by 2.3 mm along the radial direction in the optical probe. The deviation of the measured field from the center field was estimated at 1.5% at most when a field coil with both the inner diameter and the width of 12 mm was used [7]. Magnetic fields with the peak field below 30 T were generated non-destructively with a single-turn coil reinforced with a collar made of iron.

A typical wave form of the magnetic fields is shown in Fig. 1. In this experiment, a pulsed field with the peak field of 108.3 T was generated using a coil with the diameter of 12 mm and a discharge

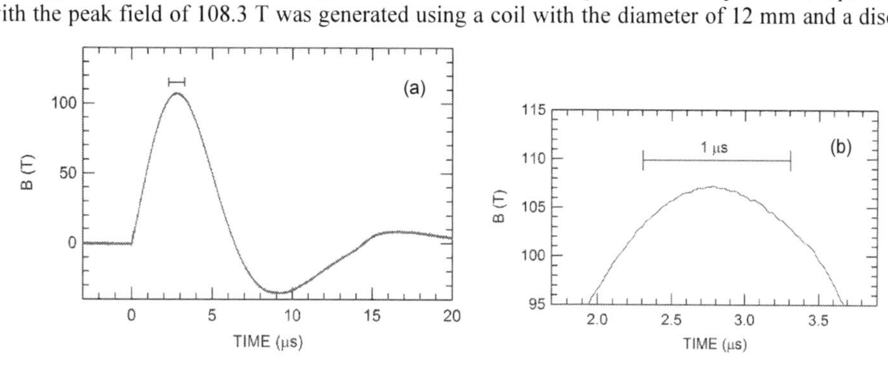

Figure 1. Wave form of a pulsed magnetic field generated using a copper coil with the diameter of 12 mm and a discharge voltage of 35 kV in the single-turn coil system. (a) The whole wave form of the pulsed field. (b) Enlarged view of the field pulse around the peak. The exposure period of the detector array is indicated with the horizontal lines.

voltage of 35 kV. In the absorption measurement, the InGaAs diode array was exposed to the spectral light for 1 μs while the pulsed field was almost maximal as exhibited in Fig. 1(b). Field variation during the exposure time was around 1.2%.

Absorption measurements were conducted for aligned SWCNT-gelatin composite films prepared in the same procedure reported previously. SWCNTs (CoMoCAT) were dispersed with sodium dodecylbenzene sulfonate (SDBS) in gelatin matrix and four times stretch-aligned along the plane of the films. Magnetic fields were applied either in parallel (hereafter, parallel configuration) or in perpendicular (perpendicular configuration) to the alignment of SWCNTs in the Voigt geometry. The light incident on the sample was also polarized in parallel to the alignment of SWCNTs with an NIR linear polarizer.

3. Results and discussion

Figure 2(a) displays magnetic field expansion of absorption spectra of a stretch-aligned SWCNT film in the parallel configuration. Three absorption peaks at the photon energies of 1.097, 1.203 and 1.256 eV were recognized at zero tesla. The peaks at 1.203 and 1.256 eV are assigned to the first subband-edge singlet bright excitons in SWCNTs with the chirality vectors of (7, 5) and (6, 5), respectively. As for the peak at 1.097 eV, those excitons in (8, 4) and (7, 6) tubes are thought to contribute to the absorption comparably. The small difference of about 9 meV between those exciton energies [8] made it difficult to distinguish the two corresponding absorption peaks.

With increasing the magnetic field, peak shift toward the lower energy was observed for the (8, 4) and (7, 6) tubes and that toward the higher energy was observed for the (7, 5) and (6, 5) tubes. Above 28.8 T, new peaks emerged at the higher energy side of the peak for the (8, 4) and (7, 6) tubes, and at the lower energy side of the peaks for the (7, 5) and (6, 5) tubes. In the perpendicular configuration, neither peak shift nor emergence of new peaks was recognized up to 105.2 T (Fig. 2(b)). Judging from these results, we have determined that the magnetically induced peak shift is attributed to the Aharonov-Bohm effect and that the emergence of new peaks is associated with brightening of the singlet dark exciton following the lifting of the degeneracy of the K and K' valleys due to the Aharonov-Bohm effect.

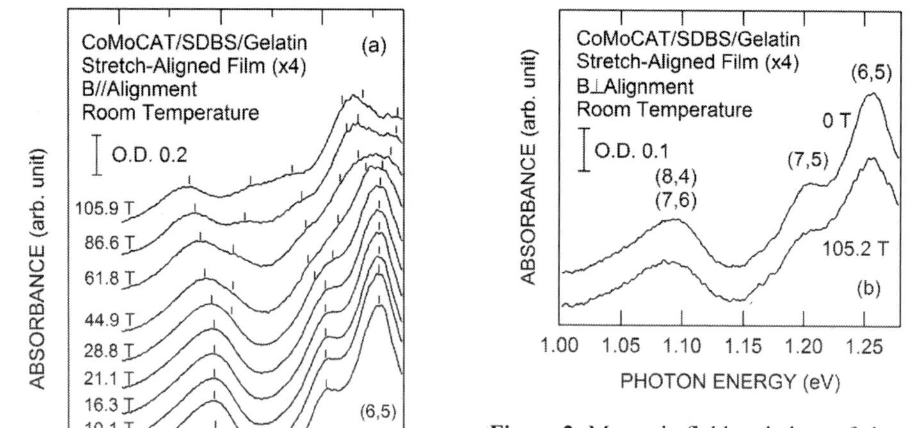

Figure 2. Magnetic field variations of absorption spectra of aligned SWCNT films in the configurations that the magnetic fields are applied in parallel to the alignment (a), and in perpendicular to the alignment (b). The vertical lines indicate peak positions.

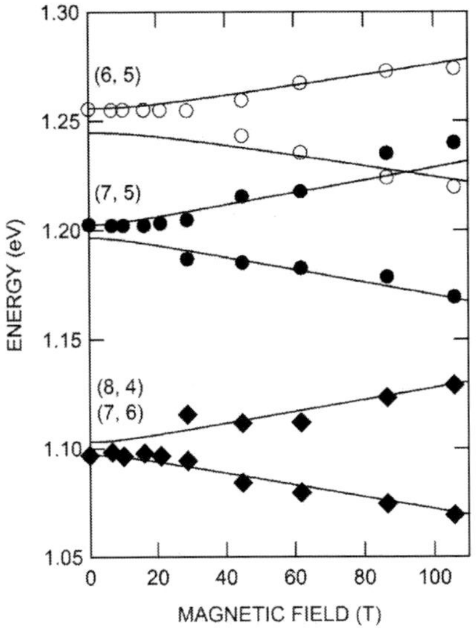

Figure 3. Magnetic field dependence of energies of absorption peaks in the configurations that the magnetic fields are applied in parallel to the alignment of SWCNTs. Lines indicate the results of the perturbation calculation fitted to the experimental data.

Table 1. The diameter calculated assuming the C-C bond length to be 0.144 nm, the energy gap between the first subbands without the K-K' coupling, the splitting energy between the singlet bright and dark excitonic states Δ_{bd}, lower singlet excitonic state in SWCNTs with chiral vectors investigated in this study.

Chirality	Diameter (nm)	Energy gap (eV)	Δ_{bd} (meV)	Lower state
(6, 5)	0.756	1.250	11	Dark
(7, 5)	0.827	1.200	6	Dark
(8, 4), (7, 6)	0.839, 0.893	1.100	6	Bright (tentative)

The absorption spectra for the parallel configuration were deconvolved with Gaussian functions to determine the peak energies precisely. Magnetic field dependence of the peak energies is exhibited in Fig. 3. It is clearly recognized that the splitting between the absorption peak dominant at zero tesla and the new peak emerging in higher fields is symmetric for the (8, 4), (7, 6), (7, 5) and (6, 5) tubes, which is theoretically predicted [1]. This feature was missing in the micro-PL experiments reported previously [2, 3]. Therefore, it would be more conclusive to determine the excitonic states using magneto-absorption techniques. Judging from the relative energy position of the emergence of new peaks, the energetic configuration of the singlet bright and dark excitonic states can be determined for each SWCNT as listed in Table 1. We have analyzed the magnetic field dependence of the peak energies using the perturbation theory where the splitting energy between the singlet bright and dark excitonic states Δ_{bd} (bright-dark splitting energy) is introduced to the off-diagonal terms of a 2×2 matrix and the splitting of energy levels due to the Aharonov-Bohm effect is incorporated. The experimental results were reproduced quantitatively as shown in Fig. 3, and the first subband gap energy without the K-K' coupling and the bright-dark splitting energy were derived for each SWCNT as listed in Table 1. The tendency that the bright-dark splitting energy becomes smaller with increasing the tube diameter is admitted for the (6, 5) and (7, 5) tubes, which is in line with the previous PL works [2, 3].

As for the (8, 4) and (7, 6) tubes, the magnetic behavior of their absorption peaks could not be distinguished in this study. The possibility that the two energy branches of the absorption is corresponding to a magnetic splitting of the bright excitonic states in these tubes would be denied. The magnetic enhancement of the upper energy branch of the absorption suggests strongly that the branch is corresponding to a dark excitonic state. We assign the lower singlet excitonic state as the bright state for both tubes tentatively. However, further study using, for example, samples in which the (8, 4) or (7, 6) tubes are isolated is required to confirm the assignment.

4. Conclusion

The successful observation of the symmetric splitting of the exciton energies has proved that the NIR magneto-absorption spectroscopy of aligned SWCNT films to the megagauss region is a conclusive approach to determine the energy states of the first subband-edge singlet excitons in SWCNT. As reported in the previous micro-PL studies, the tendency that the splitting energy between the singlet bright and dark excitonic states becomes smaller with increasing the tube diameter is admitted for the (6, 5) and (7, 5) tubes. The lower singlet excitonic state is also confirmed to be dark for the (6, 5) and (7, 5) tubes. As for the (8, 4) and (7, 6) tubes, the possibility that the lower state is the bright one is strongly suggested.

Acknowledgments

This work was supported by Grant-in-Aid for Scientific Research on priority Areas "High Field Spin Science in 100T" (No. 451) from the Ministry of Education, Culture, Sports, Science and Technology (MEXT).

References

[1] Ando T 2006 *J. Phys. Soc. Japan* **75** 024707
[2] Srivastava A, Htoon H, Klimov V I and Kono J 2008 *Phys. Rev. Lett.* **101** 087402
[3] Matsunaga R, Matsuda K and Kanemitsu Y *Phys. Rev. Lett.* **101** 147404
[4] Ajiki H and Ando T 1993 *J. Phys. Soc. Jpn.* **62** 1225
[5] Yokoi H, Effendi M, Kojima E and Takeyama S 2010 *J. Low Temp. Phys.* **159** 311
[6] Nakao K, Herlach F, Goto T, Takeyama S, Sakakibara T, Haruyama T and Kikuchi T 1985 *J. Phys. E, Sci. Instrum.* **18** 1018
[7] Takeyama S, Amaya K, Nakagawa T, Ishizuka M, Nakao K, Sakakibara T, Goto T, Miura N, Ajiro Y and Kikuchi H 1988 *J. Phys. E, Sci. Instrum.* **21** 1025
[8] Bachilo S M, Balzano L, Herrera J E, Pompeo F, Resasco D E and Weisman R B 2003 *J. Am. Chem. Soc.* **125** 11186

Photoluminescence anomalies of indirect excitons localized at interfaces in CdS/ZnSe MQWs in very high magnetic fields

H Saito [1,2], R Akimoto[3], K Kindo[1] and S Takeyama[1]

[1] ISSP, University of Tokyo, Kashiwa, Chiba, 277-8581, Japan

[2] Department of Applied Physics, University of Tokyo, Hongo, Tokyo, 113-8656, Japan

[3] AIST, Network Photonics Research Center, Tsukuba, Ibaraki, 305-8568, Japan

E-mail: saito@issp.u-tokyo.ac.jp

Abstract. CdS/ZnSe multiple quantum wells (MQWs) shows very strong photoluminescence at low temperatures. The strong photoluminescence (PL) arises from the interface localization of the type-II excitons. A systematic study was attempted on the exciton density dependence of magneto-PL behavior up to 53 T by using a pulse magnet. We obtained a strong excitation density dependence of the type-II exciton Bohr radius trapped in interface roughness potentials.

1. Introduction

The type-II quantum structures fabricated from group II-VI semiconductor alloys are known to provide a unique situation, because they have large band-offset and therefore the photo-induced excitons have long life time due to carrier spatial separation. In the case of ZnSe/BeTe quantum wells carriers are accumulated enough to make excitonic complex such as trions [1]. On the other hand, in the case of CdS/ZnSe quantum wells, there exists a large lattice mismatch (~2.7 %). It is acceptable to assume that the interface roughness may cause a crucial effect on the exciton recombination processes. Ono *et al.* investigated temperature dependence of the PL in CdS/ZnSe MQWs in detail and compared with those calculated by Monte Carlo simulation modelling of an exciton hopping in the interface potentials before photo recombination [2]. These exciton trapping potentials may cause realization of a high exciton density state in strong magnetic fields [3]. This type of quantum structure can offer a possibility of studying high density exciton states such as an excitonic complex or possible occurrence of the exciton Bose-Einstein condensation (BEC). Butov *et al.* have attempted intensive studies about the effect of external magnetic field in the type-II quantum well structures [3]. They have shown that the magnetic field applied perpendicularly to the two dimensional plane can control the type-II exciton Bohr radius. They have also pointed out that the indirect exciton life time becomes longer with increasing magnetic field. These preceding researches suggested us to study the high density magneto excitons trapped in the interface roughness potentials.

2. Experiment

We investigated magneto-PL spectra of CdS/ZnSe type-II MQWs in pulsed high magnetic fields up to 53 T. The sample was fabricated on a GaAs substrate by the MBE method. Since the MQWs have a large band offset, electrons and holes are confined in the CdS and the ZnSe layer, respectively. The CdS layer with a thickness of 1.7 nm and the ZnSe layer with a thickness of 7.1 nm were stacked 20 times alternately. The CW semiconductor laser was used as the excitation light source of its wavelength 405 nm. We used the image intensified charged-coupled device array, ICCD (ANDOR TECHNOLOGY) which is attached on the polychromator of f = 250 cm. We used a long pulse magnet for measurements up to 53 T with its pulse duration about 35 ms. The PL spectra were recorded on the detector arrays in a streak mode. The measurement range of laser power was 0.02 ~ 50 W/cm^2. The details of the PL measurement used in a pulse magnetic field are presented in Ref. [4].

3. Results and Discussions

Figure 1 is typical PL spectra from CdS/ZnSe MQWs under magnetic fields. Magnetic field is applied perpendicularly to the sample plane (Faraday configuration). The PL peaks of higher and lower energy side are known as an exciton zero phonon and one phonon assisted PL line, respectively. We focused on the zero phonon line. PL intensity decreases on increasing magnetic field accompanied by slight higher energy shifts. There reported similar behavior in GaP/AlP type-II hetero structures [5], in which a model of the wave function shrinkage has been proposed. However, the reason for decrease of the intensity has not been clear yet.

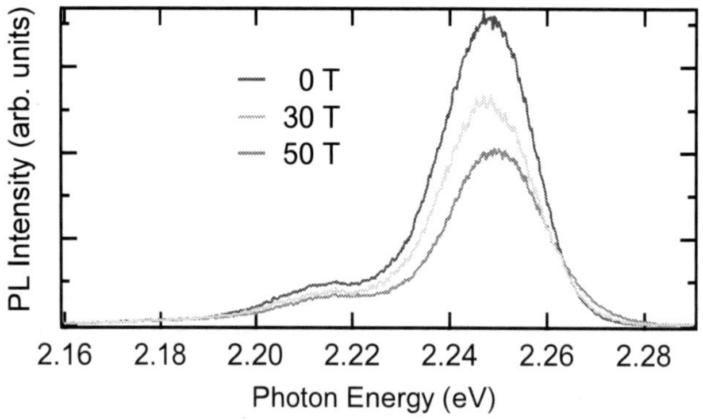

Figure 1. PL change of CdS/ZnSe MQWs in magnetic fields up to 53 T at 4.2K. Excitation density is 2.0 W/cm^2.

Figure 2 (a) shows excitation density dependence of PL peak energies at each magnetic field. In the absence of magnetic field, the shift of PL peak energy shows a logarithmic dependence on excitation density (linear line in Fig. 2 (a)). Moreover, we found that energy shifts are induced by excitation density. Figure 2 (b) is the results of the PL peak shifts at 30 and at 50 T, respectively, subtracted from those at 0 T. At each excitation density in Fig. 2 (b) the shifts in magnetic fields are considered to result from the exciton diamagnetic shift. Figure 3 shows the exciton PL peak shifts in magnetic fields at each excitation densities. We calculated a Bohr diameter of the exciton from the diamagnetic shift ΔE in Fig. 3. The exciton wave function extension is estimated from the formula

$$\Delta E = \frac{e^2}{8\mu} B^2 \langle r^2 \rangle \qquad (1),$$

where μ is the exciton reduced mass, and $\langle r^2 \rangle$ represents the spatial extent of exciton wave function, which is related to the exciton Bohr radius Δr [6]. At 0.10, 2.0, and 50 W/cm^2, Δr = 3.8 nm, 2.8 nm, 4.7 nm, respectively. This result indicates that exciton Bohr radius depends on the excitation density, and the exciton Bohr radius shows its minimum around 2.0 W/cm^2 (see Fig. 2 (b)).

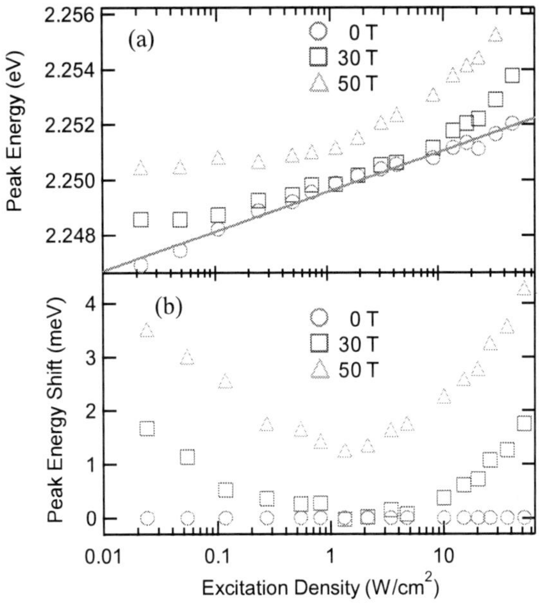

Figure 2. Excitation density dependence of (a) PL peak energies and (b) peak energy shifts from 0 T.

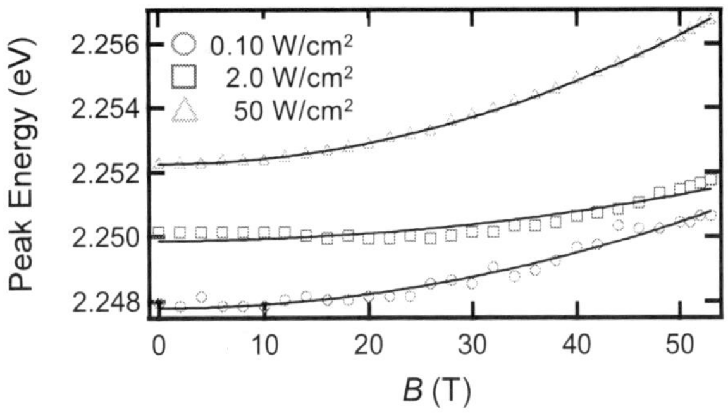

Figure 3. Magnetic field dependence of peak energy shifts at each excitation density, 0.10, 2.0, 50 W/cm^2.

The extent of exciton Bohr radius shrinkage induced by magnetic field (=diamagnetic effect) should be substantially affected by quantum confinement at the interface potentials. The diamagnetic shift of excitons at a deep potential is larger than that at a shallow potential [7]. When excitons are accumulated in the trapping potentials, the exciton self-energy increases the potentials. On increasing the exciton density, shallow potentials are easily occupied and shielded at a low density region, and then PL from the deep potentials becomes dominant, which results in the observed decrease of diamagnetic shift. Further increase of excitation density, the shielding takes place even at deeper potentials and exciton becomes less localized with widening its Bohr radius ending up with a free exciton. Our preceding study interpreted by the Monte Carlo simulation leads to the average trapping potential depth as 6 meV in our sample [2]. The self-energy is considered to exceed this potential depth above the excitation density 2.0 W/cm^2. Above this density, the exciton becomes almost free-like with a wider Bohr radius.

4. Conclusions

We performed magneto-PL measurements to study the effects of magnetic field on type-II excitons localized at the interface trapping potentials in CdS/ZnSe MQWs up to very high magnetic fields. A pulse magnet was used for measurement up to 53 T in this report. We have shown the strong excitation density dependence of the diamagnetic shifts, reflecting an extent of the exciton Bohr radius. It was shown that a large amount of excitons are accumulated up to 2.0 W/cm^2 in the deep interface potentials in this system until the exciton self-energy shields those potentials to become free-like..

References

[1] Shen R, Akimoto R and Takeyama S 2009 *Semicond. Sci. Tech.* **24** 115011

[2] Ono K, Nakajima M, Suemoto T, Makino T, Saito K, Mino H, Akimoto R and Takeyama S 2007 *Proc. of 28th Int. Conf. on the Physics of Semiconductor* (Vienna: Austria/AIP Conference Proceedings) p 467

[3] Butov L V 2004 *J. Phys. Cond. Matt.* **16** R1577

[4] Takeyama S, Watanabe K, Miura N, Komatsu T, Koike K and Kaifu Y 1990 *Phys. Rev.* B **41** 4513

[5] Uchida K, Miura N, Issiki F and Shiraki Y 2001 *Phys.* B *Cond. Matt.* **298** 310

[6] Miura N, Uchida K, Yasuhira T, Kurtz E, Klingshirn C, Nakashima H, Issiki F and Shiraki Y 2002 *Phys.* E **13** 263

[7] Grochol M, Grosse F and Zimmermann R 2005 *Phys. Rev.* B **71** 125339

Journal of Physics: Conference Series **334** (2011) 012054 doi:10.1088/1742-6596/334/1/012054

Photo induced cyclotron resonance in ZnSe/BeTe type-II quantum wells

R. Shen[1], R. Akimoto[2], and S. Takeyama[1]

[1]Institute for Solid State Physics, University of Tokyo, 277-8581, Kashiwa, Japan

[2]AIST, Ultrafast Photonics Devices Laboratory, 305-8568, Ibaraki, Japan

E-mail: shen@issp.u-tokyo.ac.jp

Abstract. We have detected a cyclotron resonance signal of the electrons induced by an external photoexcitation in an *undoped* ZnSe/BeTe type-II quantum well. The experimental spectra have also been well reproduced by the dielectric function (Drude) model. The photo-induced electron density is estimated at 1×10^{12} cm^{-2}.

1. Introduction

In ZnSe/BeTe quantum wells (QWs), due to the type-II quantum structure, a photoexcitation causes a unique situation where the electrons stay in the ZnSe layer, whereas the holes are injected into the BeTe layer. Since the energy minima for photoexcited electrons and holes lie in different layers, photoluminescence (PL) occurs as a spatially indirect (SID) optical transition. In our previous report, a positively charged exciton (composed of the spatially separated one electron and two holes) transition is a dominating mechanism for the strong SID PL in an undoped type-II QW [1]. This result also implied that the free electrons should be simultaneously created in the ZnSe layer upon photoexcitation. Furthermore, the pump-and-probe absorption spectra have evidenced the existence of free electrons in the ZnSe layer with the density higher than 10^{10} cm^{-2} [1]. In this paper, we have performed a cyclotron resonance (CR) measurement with an external photoexcitation, and investigated the photo-induced high density electron state. We have observed directly the CR signals as far-infrared absorption of the conduction electrons in the ZnSe quantum layer.

2. Experimental detail

We have investigated the *undoped* sample used in our previous report [1], which was grown on (001) GaAs substrate by molecular beam epitaxy. An asymmetric quantum structure comprises ZnSe(8 nm)/BeTe(3 nm)/ZnSe(4 nm) layers, and is sandwiched by 200-nm thick $Zn_{0.77}Mg_{0.15}Be_{0.08}Se$ barrier layers, as described in detail in Ref. 1. It was concluded there that the strong PL arises from the positively charged exciton in the ZnSe(8 nm)/BeTe(3 nm) interface, and that the free electrons are accumulated in the 8-nm ZnSe QW considering the total charge neutrality [1].

Fig. 1 shows the schematic diagram of the photo-induced CR measurement with a single-turn coil (STC) instrument. The far-infrared light sources were obtained from a CO_2 laser and an H_2O vapor laser, providing the laser wavelengths at 10.7 μm and 16.9 μm, respectively. A Ge(Cu)

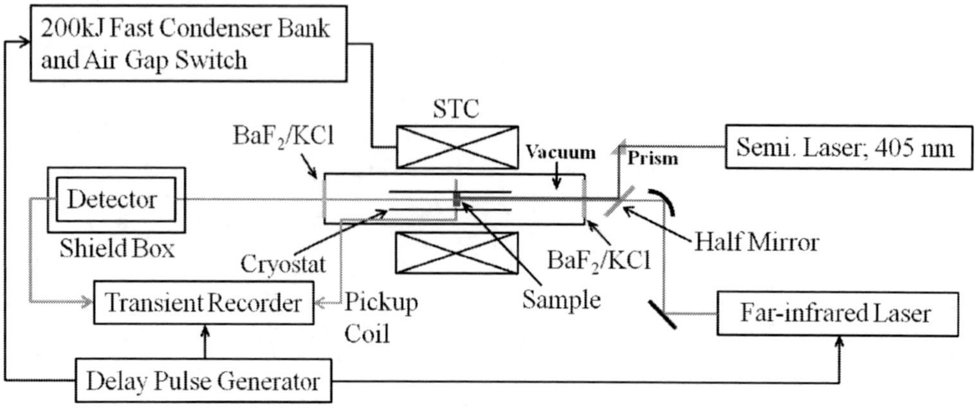

Figure 1. (Color online) Schematic diagram of the photo-induced CR measurement with the STC.

photodiode cooled down at 4.2 K and a HgCdTe photodiode cooled down at 77 K, were used as the detectors for the CO_2 laser and the H_2O vapor laser, respectively. A cw-semiconductor laser was used as an excitation light source of 405 nm wavelength. The unfocused visible light was intentionally illuminated on a bakelite mask with a hole 1.6 mm in diameter, so that the far-infrared and the visible lights should be homogeneously exposed to the same position of the sample. Furthermore, as the optical windows of the liquid He flow cryostat, Ba_2F or KCl crystals were employed, since both the far-infrared and the visible lights are transmittable. Pulse magnetic field exceeding 180 T was generated by a STC technique. The pulse duration time was about 7 μs. The magnetic field was measured by a pick-up coil wound around the sample. The signals of the optical transmission and a magnetic field were transferred to a digital recorder by the optical fibers. More details about the single-turn coil technique and the data-acquisition system are described in ref. 2.

3. Results and Discussions

Fig. 2 shows the transmission spectra of the *undoped* type-II QW at 20 K, when the probe wavelength is 10.7 μm. Every resonance peak is observed twice on rising and falling slopes of the magnetic field. The resonance peaks at around 70 T in both the spectra are identified as the transition of the residual electrons in the GaAs substrate [3]. A new broad resonance peak at the high-field side of the transmission spectra is obvious at about 160 T only in the case when applying an external photoexcitation of 405 nm wavelength (a pump light). The carrier effective mass m^* is estimated to be about 0.16 m_0 from the cyclotron frequency of the relation $\omega_c = \frac{eB}{m^*}$, where e is the elementary charge and B is the resonance magnetic field. This result is in good agreement with the value 0.15 m_0 of the free electron mass in ZnSe quantum wells [4]. Therefore, we conclude that the detected carriers are the free electrons in the 8-nm ZnSe QW, induced by the pump light.

Following the dielectric function (Drude) model, the dielectric function of the magnetic field can be described by the formula

$$\epsilon(B) = \varepsilon_\infty - \frac{\omega_p^2}{\omega(\omega - \omega_c(B) + i\frac{1}{\tau})} \tag{1}$$

where ω_p is the plasma frequency, $\omega_p^2 = \frac{n^* e^2}{m^* \epsilon_0}$, n^* is the carrier density and ϵ_0 is the vacuum

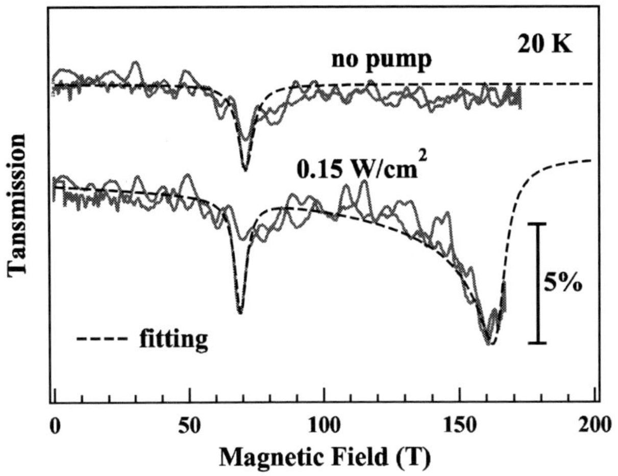

Figure 2. The pump power dependences of the CR spectra in the *undoped* ZnSe/BeTe type-II QW. The probe wavelength is 10.7 μm. The dashed lines show the calculations using eq. (2).

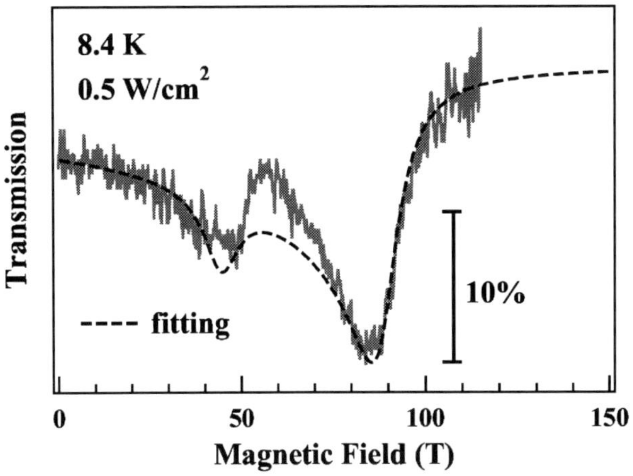

Figure 3. The photo-induced CR spectrum in the *undoped* ZnSe/BeTe type-II QW. The probe wavelength is 16.9 μm. The dashed line shows a calculated line.

permittivity, τ is the carriers scattering time. Furthermore, the cyclotron absorption is represented by

$$\sim exp\{-\frac{4\pi}{\lambda}Im(n(B)d)\} \tag{2}$$

where the refractive index $n(B) = \sqrt{\epsilon(B)}$, λ is the detected wavelength and d is the QW thickness. The experimental transmission spectra were well reproduced by eq. (2) as shown in Fig. 2. From the fitting, the density of the photo-induced free electrons is estimated at 2.7×10^{12} cm^{-2}, which is higher than that of our doped ZnSe/BeTe type-II QWs [5]. In additional, the density of the residual electrons in the GaAs substrate can also estimated to be very low amounts to 4.8×10^{12} cm^{-3}.

We have changed the wavelength of the probe light. Fig. 3 shows the transmission spectrum of the same *undoped* type-II QW at 8.4 K when the probe wavelength is 16.9 μm and the pump light power is 0.5 W/cm^2. Two resonance peaks at around 44 T and 82 T are observed, which can also be identified as the transition of the residual electrons in the GaAs substrate and the photo-induced free electrons in the 8-nm ZnSe QW, respectively. In the same way, according to eq. (2), we obtained the m^* and n^* of the photoexcited free electrons as 0.14 m_0 and 4.2×10^{12} cm^{-2}, respectively. It is obvious that the density of the photo-induced free electrons is increased with an increase in the photoexcitation power, considering in relation to the result of Fig. 2.

4. Summary

We have performed a far-infrared CR measurement with an external photoexcitation of 405 nm wavelength in an *undoped* ZnSe/BeTe type-II QW. We have detected the photo-induced CR of the free electron states at low temperatures, in two different far-infrared wavelengths. The experimental results were well reproduced by the dielectric function (Drude) model. The density of the photo-induced free electrons is estimated as higher than 10^{12} cm^{-2}. The effective mass of the high density electrons accumulated by the pumping light in the ZnSe QW was determined for the first time.

Acknowledgements

We thank A. Miyata and Professor Y. H. Matsuda for helpful comments and supports.

References

[1] R. Shen, R. Akimoto, and S. Takeyama, Semicond. Sci. Technol. **24**, 115011 (2009).
[2] S. P. Najda, S. Takeyama, N. Miura, P. Pfeffer, and W. Zawadzki, Phys. Rev. B **40**, 6189 (1989).
[3] H. Momose, H. Okai, H. Deguchi, N. Mori, and S. Takeyama, Physica E **32** 309 (2006).
[4] G. V. Astakhov, D. R. Yakovlev, V. P. Kochereshko, W. Ossau, W. Faschinger, J. Puls, F. Henneberger, S. A. Crooker, Q. McCulloch, D. Wolverson, N. A. Gippius, and A. Waag, Phys. Rev. B **65**, 165335 (2002).
[5] Ziwu Ji, H. Mino, K. Oto, R. Akimoto, K. Ono, and S. Takeyama, Semicond. Sci. Technol. **21**, 87 (2006).

Propagation length of spin waves in a conducting system

J. Gómez[1], F. Perez[1], B. Jusserand[1], G. Karczewzski[2] and T. Wojtowicz[2]

[1] Institut des Nanosciences de Paris, CNRS/Université Paris VI Campus Boucicaut, UMR 7588, 140 rue de Lourmel, 75015 Paris, France.

[2] Institute of Physics, Polish Academy of Sciences, Aleja Lotnikov 32/46, 02-668 Poland.

gomezj@cab.cnea.gov.ar

Abstract. We have studied propagating spin-flip waves excited in a two dimensional electron gas by angle resolved magneto-Raman spectroscopy. The damping rate of these excitations follows a quadratic law, as a function of q, and the group velocity is linearly dependent with q. Consequently the propagation length presents a maximum for a given q. We have estimated and analyzed the maximum propagation length showing that the spin waves are drastically attenuated in a microscopic distance, even in the intrinsic regime.

1. Introduction

Increasing interest is focused on propagating spin waves because of their potential ability to transport or modify logical information without involving charge current dissipation [1]. Obtaining the maximum coherence length of these waves is a challenge regarding this purpose.

Recent publications have highlighted the importance of the kinetic motion in the relaxation processes of the spin-flip waves (SFW) in a conducting system (see Ref. [2]). This damping mechanism is a consequence to the fact that the spin-flip single particle excitations are coupled to the SFW throughout the kinetic part of the Hamiltonian [2,3] and degrade the collective character of the spin waves.

In this work we have studied, by angle resolved electronic resonant Raman scattering (ERRS), SFW in a high mobility two-dimensional electron gas (2DEG) embedded in $Cd_{1-x}Mn_xTe/Cd_{0.8}Mg_{0.2}Te$ quantum wells. Such systems have been recently introduced as a test-bed system for spin excitations of the spin-polarized 2DEG (SP2DEG). Indeed, the *s-d* exchange coupling of conduction electron with localized electrons of the Mn atoms provide a giant Zeeman energy to the 2DEG: [4]

$$Z(B,T) = x \times N_0\alpha_e \times \langle S_z(B,T)\rangle - |g_e|\mu_B B \quad (1)$$

where N_0 is the cation sites density, α_e is the exchange coupling between the conduction electron of the well and the localized electrons on Mn impurities $N_0\alpha_e$ =−0.22 eV. $\langle S_z(B,T)\rangle$ is the thermal average spin of a single Mn atom given by the modified Brillouin function, x is the effective Mn concentration, g_e is the normal electron g factor, and μ_B is the Bohr magneton. In order to keep low alloy disorder and high electron mobility (μ~10^5 cm^2 V s), x remained below 1.2% and the electron sheet density n_{2D} ranged in between 1.5 and 4×10^{11} cm^{-2}. Depending on the Mn nominal concentration xN_0 and n_{2D}, the maximum spin-polarization degree $\zeta = (n_\uparrow - n_\downarrow)/n_{2D}$ can reach 80% for a magnetic field

B below 4 T, such that, when B is applied parallel to the quantum well plane, the Landau orbital quantization is kept negligible. Therefore, such SP2DEG is a conducting paramagnet having the spin-polarization degree of a conducting ferromagnet. Moreover, semiconductor quantum wells exhibit well-defined optical resonances which allow ERRS measurements to be performed. ERRS is powerful tool to access to the wave-vector resolution of the spin excitation spectrum [4].

Particularly, we have studied two samples, named sample A and B respectively, with similar electron density ($n_{2D} \sim 3.6\text{x}10^{11}$ cm^{-2}) and different manganese concentrations, $x = 1.13\%$ for sample A and $x = 0.22\%$ for sample B. In Fig 1a we show a typical SFW Raman spectra obtained in sample A for various Raman transferred wavevector q, the inset shows the geometrical setup with the external magnetic field applied in the plane of the quantum well. In practice, to increase the transferred wavevector we must to increase the angle, θ, of the incident laser beam with respect to the normal direction of the quantum well, as depicted in the inset.

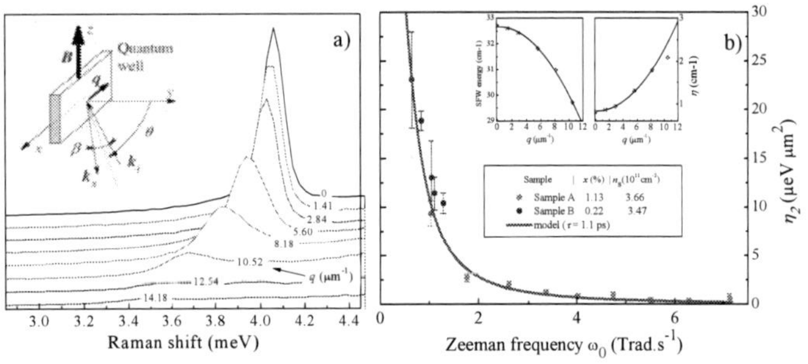

Figure 1: a) Typical SFW Raman spectra obtained for various Raman transferred wavevector q in sample A. b) η_2 parameter as a function of the Zeman frequency. Continuous line is the calculated η_2 by using the model described in Ref. [3]. The inset shows the quadratic behaviour of the energy position and line width of the Raman peak as a function of q.

The inset of Fig. 1b shows the evolution of the SFW energy ($\hbar\omega_{SFW}$) peak and the line width (η) as a function of q. Both present a quadratic behavior as explained in Ref. [3] and [5] and can be modeled by Eqs (2) and (3) respectively.

$$\hbar\omega_{\text{SFW}} = Z - \beta q^2 = Z - \frac{1}{|\zeta|}\frac{Z}{Z^* - Z}\frac{\hbar^2}{2m^*}q^2 \qquad (2)$$

$$\eta = \eta_0 + \eta_2 q^2 = \eta_0 + \frac{4E_F}{\hbar}\frac{\hbar^2}{2m^*}\frac{3\tau}{\left(\omega_0^*\tau\right)^2+1}\left[\frac{\omega_0^*}{\omega_0} - \frac{\left(\omega_0^*\tau\right)^2+\frac{1}{3}}{\left(\omega_0^*\tau\right)^2+1}\right]q^2 \qquad (3)$$

Here E_F is the Fermi energy, $Z = \hbar\omega_0$ the bare Zeeman energy, $Z^* = \hbar\omega_0^*$ the Coulomb enhanced Zeeman energy, m^* the effective electron mass, τ is the life time of the single particle excitation and η_0 is the line width at q=0. Fig. 1b shows the experimental and modeled evolution of the curvature parameter, η_2, as a function of the Zeeman frequency for the studied samples. To model η_2 we have used τ=1.1ps, as discussed in Ref. [3], this value is dominated by the disorder scattering time (τ_{dis}), with respect to the intrinsic Coulomb scattering time ($\tau_{e\text{-}e}$) contribution, $\tau^{-1} = \tau_{dis}^{-1} + \tau_{e-e}^{-1}$.

2. Propagation length of the collective SFW

To calculate the propagation length (l_{prop}) of the SFW we must use Eq. (4)

$$l_{prop} = v_g T_2 \qquad (4)$$

where v_g=dω/dq is the group velocity and T_2 is the life time of the spin wave with wavevector q. By using Eq. (2) and (3) we obtain the analytical expression for l_{prop}, see Eq. (5).

$$l_{prop} = 4\beta \frac{q}{\eta_0 + \eta_2 q^2} \qquad (5)$$

Parameters β and η_2 in Eq. (5) are modeled by the theory described in Ref. [3] and [5] and estimated by our experimental results. For fully evaluate Eq. (5) we have used the experimental value of η_0 (line width at q=0) because at the moment there is not a complete theory for describe its behavior as a function of the parameters of the sample. A representative behavior of the l_{prop} as a function of q is showed in Fig. 2a.

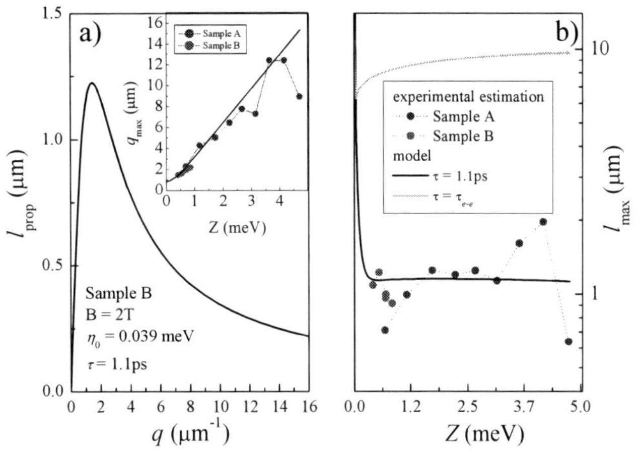

Figure 2. a) Typical propagation length as a function of q. the relaxation processes become important for high value of q limiting the propagation length of the SFW. The inset shows the evolution of the q_c as a function of the Zeeman energy. b) Maximum attainable propagation length as a function of Z. In all cases continuous lines correspond to the model.

Contrary to the intuition, the propagation length decreases for high enough value of q, this result is related to the fact that the attenuation processes become more and more important when increasing q. Such behavior, gives a particular maximum l_{max} in the curve for a given $q = q_c$, that indicates the existence of a maximum propagation length for a given spin wave with an associated wave vector q_c.

By analyzing the extremal properties of Eq. (5) we obtain the expressions for q_c and l_{max}, given by the Eqs. (6) and (7) respectively.

$$q_c = \sqrt{\frac{\eta_0}{\eta_2}} \qquad (6)$$

$$l_{max} = \frac{2\beta}{\sqrt{\eta_0 \eta_2}} \qquad (7)$$

Inset of Fig. 2a shows the evolution of q_c as a function of the bare Zeeman energy. In Fig 2b we present the maximum attainable propagation length as a function of *Z*. Both, modelled and experimental values show an almost constant behaviour (at least in the accessible experimental range) and l_{max} does not exceed 1.5 μm. Even, if the disorder scattering contribution is suppressed and we consider only the intrinsic Coulomb scattering time ($\tau_{e\text{-}e}$), the expected l_{max} is about of 15 μm, such a result evidences the effectiveness of this relaxation processes which prevents having a macroscopic propagation length.

In summary, we have estimated the propagation length of spin waves which are drastically limited by the relaxation processes originated in the delocalized character of the spin carriers. This is true even if the disorder contribution is suppressed, that means the intrinsic contribution is strong enough to damp the SFW in a short distance. We came also to the conclusion that, to make the spin wave propagate further, it needs to loose the memory of the individual electrons, say, it needs to be fully collective. Such a feature has been predicted by quantum mechanics when electrons propagate in a one dimensional channel. In such a channel, the spin is separated from the charge, it should propagate alone as a full collective mode where the individual character of electrons is lost, and therefore, no damping by the kinetic motion is expected to appear.

References

[1] V. V. Kruglyak and R. J. Hicken, J. Magn. Magn. Mater. **306**, 191 (2006).
[2] E. M. Hankiewicz, G. Vignale, and Y. Tserkovnyak, Phys. Rev. B **78**, 020404(R) (2008).
[3] J. Gómez, F. Perez, E. M. Hankiewicz, B. Jusserand, G. Karczewski, and T. Wojtowicz, Phys. Rev. B **81**, 100403(R) (2010).
[4] Cynthia Aku-Leh *et. al.* Phys. Rev. B **76**, 155416 (2007).
[5] F. Perez, Phys. Rev. B **79**, 045306 (2009).

HMF-19 IOP Publishing
Journal of Physics: Conference Series **334** (2011) 012056 doi:10.1088/1742-6596/334/1/012056

Cyclotron resonance in InMnAs and InMnSb ferromagnetic films

Y H Matsuda[1], G A Khodaparast[2], R Shen[1], S Takeyama[1], X Liu[3], J Furdyna[3] and B W Wessels[4]

[1] Institute for Solid State Physics, University of Tokyo, Kashiwa, Chiba 277-8581, Japan
[2] Department of Physics, Virginia Tech, Blacksburg, VA, 24061, USA
[3] Department of Physics, University of Notre Dame, Notre Dame, IN 46556, USA
[4] Materials Research Center, Northwestern University, Evanston, IL, 60208, USA

E-mail: ymatsuda@issp.u-tokyo.ac.jp

Abstract.
We report cyclotron resonance measurements on ferromagnetic InMnAs and InMnSb films at high magnetic fields in the megagauss range ($B > 100$ T). The cyclotron resonance transitions between the Landau levels with small indices ($N = 0, 1, 2, ..$) are clearly observed. The effective masses deduced from the resonance fields at 117 meV are $0.037m_0$ and $0.051m_0$ in InMnAs and InMnSb, respectively. They are considerably smaller than the classical band edge effective masses of heavy holes in the host semiconductors ($0.35m_0$ in InAs and $0.32m_0$ in InSb) due to the quantum effect in high-magnetic fields. The Landau levels calculated by the 8-band model, semi-quantitatively can explain the cyclotron resonance positions, indicating the itinerant holes in InMnAs and InMnSb are *p*-like holes in the valence band of the host semiconductors.

1. Introduction

Ferromagnetic semiconductors are important materials for development of spintronic devices. While effort in this area was made primarily on GaMnAs [1], other ferromagnetic III-Mn-V alloys have also been developed, including the narrow gap ferromagnetic alloys InMnAs [2] and InMnSb [3]. For example, important advances have now been made in the MOVPE growth of these latter systems [4, 5]. Investigation of the electronic structure of III-Mn-V alloys by techniques such as the cyclotron resonance (CR) can shed important light on the origin of ferromagnetism and the $p-d$ exchange interaction in III-Mn-V systems. In this work we report on CR experiments carried out on the ferromagnetic InMnAs and InMnSb films, on which clear resonance signals have been successfully observed in high magnetic fields.

2. Experiment

$In_{1-x}Mn_xAs$ films were grown by MOVPE at the Northwestern University and $In_{1-x}Mn_xSb$ films at the University of Notre Dame by low temperature MBE [3, 6, 7]. The Mn content (x), thickness (d), and hole concentration (n_p) are $x \sim 0.02$, d = 200 nm, $n_p \sim 2 \times 10^{18}$ cm^{-3} for $In_{1-x}Mn_xAs$, and $x \sim 0.02$, d=230 nm, $n_p \sim 1 \times 10^{20}$ cm^{-3} for $In_{1-x}Mn_xSb$. The Tc's are about 320 K and 10 K in $In_{1-x}Mn_xAs$ (x=0.02) and $In_{1-x}Mn_xSb$ (x=0.02) , respectively. (We use simpler descriptions, InMnAs and InMnSb, to denote the samples in the following sections). CR measurements were performed using CO_2 and H_2O lasers, providing laser radiation at 10.6,

10.7, and 16.9 μm wavelengths. Magnetic fields exceeding 100 T were generated by a single turn coil technique [8].

3. Experimental Results

Figure 1 shows the CR spectra in InMnAs and InMnSb for the excitation wavelength of $\sim$ 10 μm. The resonance fields for InMnAs are 36.0 T at room temperature (RT) and 37.5 T at 121 K. The effective masses calculated from the resonance fields are $0.036m_0$ and $0.037m_0$ at RT and 121 K, respectively. (m_0 is the mass of a free electron.) Our earlier studies on ferromagnetic

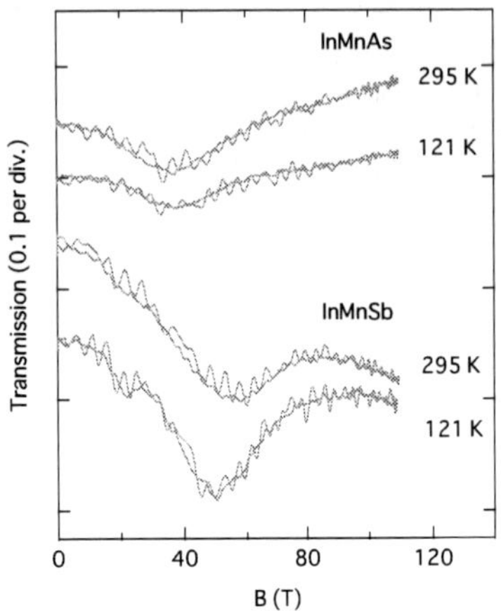

Figure 1. CR spectra for InMnAs and InMnSb films. The CR of InMnSb at 295 K, was measured at 10.6 μm; whereas, the other three resonances were measured at 10.7 μm.

InMnAs films with T_c's of $\sim$ 10 K [8], demonstrated the heavy hole (HH) effective mass ranging from $0.033m_0$ to $\sim 0.040m_0$. These significant deviations compared to the band edge HH mass of InAs ($0.35m_0$) is due to the quantum effect of the complex valence band structure [9]. The results presented here are indicating similarities in the electronic band structure of MOVPE grown InMnAs with the high T_c=320 K and the MBE grown InMnAs with the low T_c of $\sim$ 10 K.

The cyclotron mobility, μ_{CR}, was extracted from the width of the resonance peaks. We have μ_{CR}=4.8 $\times$ 10^2 cm^2/(Vs) at a room temperature and μ_{CR}=6.0 $\times$ 10^2 cm^2/(Vs) at 121 K. These values are higher than the cyclotron mobilities of the MBE grown InMnAs films reported previously [8]. Moreover, in contrast to the previous results [8], the mobility increases with decreasing temperature. Similar temperature dependence of the mobility has been reported in InAs [8]. It is likely that the impurity scattering is not a significant scattering mechanism in the MOVPE grown InMnAs, suggesting the high quality of the material.

The CR of holes in InMnSb has been observed for the first time. The measured cyclotron masses are $0.057m_0$ at RT and $0.051m_0$ at 121 K. These are much smaller than the band edge HH mass in InSb ($0.32m_0$) and similar to the observations in the InMnAs, can be explained by the quantum effects in the valence band. In addition, the observed cyclotron mass in the InMnSb is larger than that in InAs. The small Mn concentration ($x \sim 0.02$) can not modify the bandgap of the InMnSb significantly; whereas, the larger hole density in the InMnSb compared to the InMnAs, can shift the Fermi energy to a much higher level. The cyclotron mass can be enhanced when the resonance transition takes place between the Landau levels with higher indices. When we simply estimate the Fermi energy (E_f) at zero magnetic fields and 0 K using the classical band edge masses, we have E_f =16 meV for the InMnAs film and E_f =244 meV for the InMnSb film. The cyclotron resonance spectra in InMnAs and InMnSb at 16.9 μm are shown in Fig. 2. The resonance feature of InMnAs is less clear than that found at 10.6 and 10.7 μm. This fact can be due to the low mobility of the sample and the plasma absorption effect. For InMnSb, the resonance pattern is different compared to a typical single resonance peak. According to our Landau level calculations, the overlap of the light hole and heavy hole resonances can result in the unusual CR pattern.

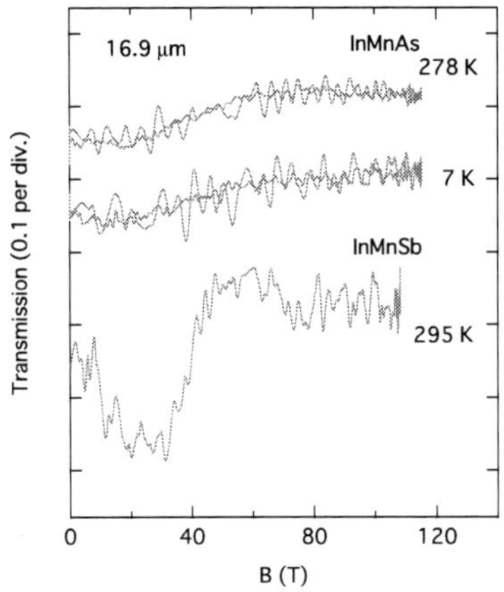

Figure 2. Cyclotron resonance spectra of InMnAs and InMnSb at 16.9 μm.

4. The Landau Level Calculations

We calculated the Landau levels for InMnAs and InMnSb and demonstrated the inter Landau level transition energies as a function of the applied magnetic field. In modeling the transitions, the modified 8-band Pidgen-Brown model [9, 10] was used, where the $p-d$ exchange interaction between the valence band holes and the Mn spins can be included in the Hamiltonian. At this stage, the number of the experimental data points is not enough to determine the exchange interaction constant. Therefore, we simply calculated the Landau levels without the $p-d$ exchange interaction; namely, calculated the Landau levels for InAs and InSb and compare the results with the experimental data.

Figure 3(a) and (b) demonstrate the calculated inter-Landau level transition energies for InAs and InSb, and the experimental data , extracted from the CR peak positions. The solid curves are the calculated transitions from N=0 to N=1 Landau levels and the dashed curves are those from N=1 to N=2 Landau levels. In Fig. 3 (a) , the experimental data for InMnAs can fit within the solid curves labeled as $b4$, corresponding to the HH (heavy hole) CR transition (N=0 to N=1). This fact is suggesting that the observed hole CR in the MOVPE grown InMnAs is due to the HH transition in the quantum limit (N=0).

In contrast to the InMnAs case, the experimental observations in the MBE grown InMnSb can not be explained by any inter-Landau level transitions from N=0 to N=1, as shown in Fig. 3 (b). The CR positions are in a better agreement with the theoretical calculations labeled as $a5'$ or $b4'$. Both curves describe the transition from N=1 to N=2 Landau levels. The $b4'$ transition mainly originates from the HH band, with small contributions from the light hole (LH), due to the Landau level mixing. The nature of the $a5'$ transition originates mainly from the LH band. Since the $b4'$ and $a5'$ levels are rather close to each other, the two resonances can overlap as a single peak. As shown in Fig. 2 for 16.9 μm excitation wavelength, the observed flat resonance in InMnSb can be due to the overlap of the two CR peaks originated from both the HH and LH bands.

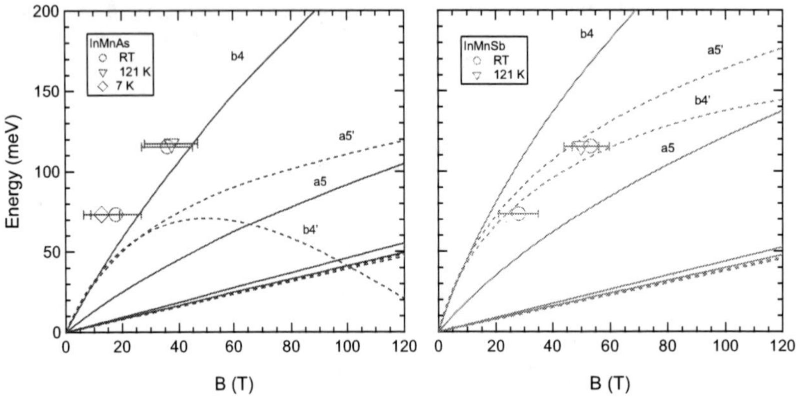

Figure 3. The calculated inter Landau level transition energies as a function of magnetic fields, for InAs and InSb, are plotted in (a) and (b), respectively. The labels such as $b4$ denote the possible Landau level transitions. The solid curves are the calculated transitions from N=0 to N=1 Landau levels and the dashed curves are those from N=1 to N=2 Landau levels.

5. Summary

The hole cyclotron resonances are clearly observed in ferromagnetic semiconductors $In_{1-x}Mn_xAs$ (x=0.02) and $In_{1-x}Mn_xSb$ (x=0.02) thin films. Both resonance are semi-quantitatively explained by the inter Landau level transitions of the host semiconductors, i.e., InAs and InSb. Therefore, the electronic structures of $In_{1-x}Mn_xAs$ (x=0.02) and $In_{1-x}Mn_xSb$ (x=0.02) are not significantly modified from that of the host materials. More detailed comparison of the cyclotron resonance of the samples with different T_c's or different Mn concentrations can provide important information on the origin of the high T_c in the MOVPE grown narrow gap III-Mn-V ferromagnetic semiconductors.

6. Acknowledgment

Supported by: NSF-DMR-0507866, AFOSR Young Investigator Program 06NE231, NSF-Career Award DMR-0846834, NSF DMR 0804479, NSF Grant DMR06-03762.

References

[1] Ohno H, Shen A, Matsukura F, Oiwa A, Endo A, Katsumoto S and Iye Y 1996 *Appl. Phys. Lett.* **69** 363
[2] Munekata H, Ohno H, von Molnar S, Segmller A, Chang L L and Esaki L 1989 *Phys. Rev. Lett.* **63** 1849
[3] Wojtowicz T, Cywinski G, Lim W L, Liu X, Dobrowolska M, Furdyna J K, Yu K M, Walukiewicz W, Kim G B, Cheon M, Chen X, Wang S M and Luo H 2003 *Appl. Phys. Lett.***82** 4310
[4] Blattner A J, Lensch J and Wessels B W 2004 it Appl. Surf. Sci. **221**155
[5] Parashar N D, Rangaraju N, Lazarov V K, Xie S and Wessels B W 2010 *Phys. Rev.* B **81** 115321
[6] Frazier M, Kini R N, Nontapot K, Khodaparast G A, Wojtowicz T, Liu X and Furdyna J K 2008 *Appl. Phys. Lett.* **92** 061911
[7] Nontapot K, Kini R N, Gifford A, Merritt T R, Khodaparast G A, Wojtowicz T, Liu X, Furdyna J K, *Appl. Phys. Lett.* 2007 **90** 143109
[8] Matsuda Y H, Khodaparast G A, Zudov M A, Kono J, Sun Y, Kyrychenko F V, Sanders G D, Stanton C J, Miura N, Ikeda S, Hashimoto Y, Katsumoto S and H. Munekata 2004 *Phys. Rev.* B **70** 195211
[9] Pidgeon C R and Brown R N 1966 *Phys. Rev.* **146** 575
[10] Kossut J 1988 *Semiconductors and Semimetals* edited by Furdyna J K and Kossut J *Academic Press, New York*, Vol.**25**, p. 183.

Cyclotron resonance in the two-dimensional metallic phase of Si quantum wells

R. Masutomi[1], K. Sasaki[1], I. Yasuda[1], A. Sekine[1], K. Sawano[2], Y. Shiraki[2] and T. Okamoto[1]

[1]Department of Physics, University of Tokyo, 7-3-1 Hongo, Bunkyo-ku, Tokyo 113-0033, Japan
[2]Research Center for Silicon Nano-Science, Advanced Research Laboratories, Tokyo City University, 8-15-1 Todoroki, Setagaya-ku, Tokyo 158-0082, Japan

E-mail: masutomi@phys.s.u-tokyo.ac.jp

Abstract. We have performed cyclotron resonance measurements on two-dimensional electrons near the metal-insulator transition. Both the cyclotron scattering time $\tau_{\rm CR}$ and the transport scattering time τ_t decrease with the electron density. The values of $\tau_t/\tau_{\rm CR}$ are close to unity, indicating that the scattering mechanism does not depend on scattering angle in the metallic phase. Short-range potential fluctuations seem to play an important role in the appearance of the metallic behavior.

1. Introduction

A metal-insulator transition (MIT) in two-dimensional (2D) systems still attracts much attention [1], since the discovery of an apparent MIT in Si metal-oxide-semiconductor field-effect transistors (Si MOSFETs) [2]. The metallic phase defies the scaling theory of localization for non-interacting 2D electron systems (2DESs) with arbitrarily weak disorder [3], which is well known as Anderson localization. In contrast, ground state of strongly correlated 2D systems is a nontrivial and fundamental issue in condensed matter physics.

Recently, theoretical calculations with the renormalization group equations suggest that the metallic phase is stabilized by electron-electron (e-e) interactions even at $T = 0$ [4]. Because the experimental data on Si MOSFETs are agreement with the theory described above, Anissimova *et al.* argue that the 2D MIT is an interaction driven and electron density N_s tuned quantum phase transition (QPT) [5]. From another point of view, the finite temperature Drude-Boltzmann theory to calculate the resistivity of 2D systems also provides good quantitative agreement with the experimental data on the metallic phase [6, 7]. The temperature dependence of the resistivity can be reproduced with the temperature dependent screening of residual impurities. Some models have been proposed to explain the 2D MIT which occurs in the region where e-e interactions are important. However, the origin of the metallic phase and its transition mechanism remain unclear and controversial.

In order to understand the 2D metallic behavior further, it is very useful to clarify dominant low-temperature scattering mechanisms. According to the Drude formula, the zero-magnetic-field dc resistivity ρ is related to the transport scattering time τ_t as $\rho^{-1} = e^2 N_s \tau_t / m^*$, where m^* is the effective mass. Since τ_t is not sensitive to small-angle scattering events, it is not identical to the single-particle scattering time τ_s [8]. Experimentally, τ_s has been deduced from

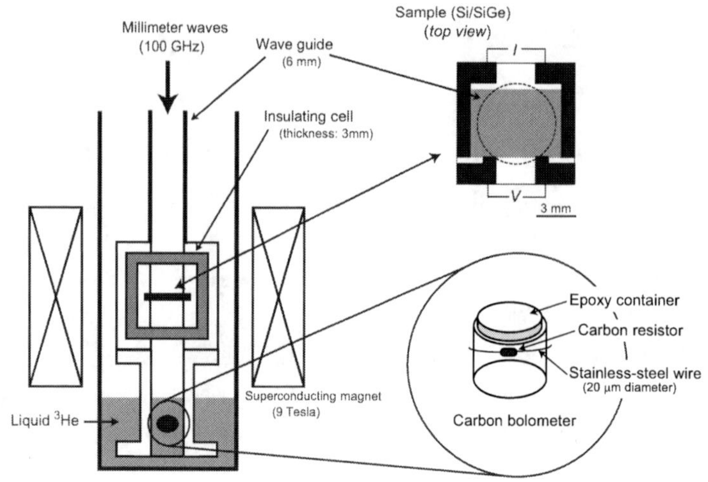

Figure 1. Schematic picture of the cyclotron resonance measurement.

the magnitude of the Shubnikov-de Haas (SdH) oscillations (τ_{SdH}) or from the linewidth of cyclotron resonance (τ_{CR}). In general, the ratio is large ($\tau_t/\tau_s \gg 1$) for long-range scattering potentials and small ($\tau_t/\tau_s \approx 1$) for short-range scattering potentials. Several experiments report large values of $\tau_t/\tau_{\mathrm{SdH}}$ in not only GaAs/AlGaAs [9, 10, 11] but also Si/SiGe [12]. However, Syed *et al.* pointed out that τ_{SdH} is strongly affected by electron density fluctuations, which cause broadening of the SdH oscillations, especially in high mobility samples [13]. In contrast, τ_{CR} is thought to be insensitive to density inhomogeneities. So far, most of the experimental studies are performed on the dc (or low-frequency limit) resistivity from which τ_t and τ_{SdH} are deduced. In this paper, we report N_s-dependence of $\tau_t/\tau_{\mathrm{CR}}$ in a high-mobility Si quantum well.

2. Experiment

Fig. 1 shows schematic drawing of the CR measurement system. This equipment consists of an insulating cell for varying the sample temperature and a carbon bolometer immersed in liquid ^{3}He. The insulating cell was made of 3-mm-thick foamed polystyrene sheets, in order to vary the sample temperature up to 10 K on the condition that the carbon bolometer was kept constant at 0.35 K. We used an Allen-Bradley carbon resistor as a sensitive detector for the CR measurements. The carbon resistor was hung with 20-μm-diameter stainless-steel wires in a sealed epoxy container (stycast 1266). Detail of the carbon bolometer is described in elsewhere [14]. We used a n-Si/SiGe heterostructure with a 20-nm-thick strained Si channel sandwiched between relaxed $\mathrm{Si_{0.8}Ge_{0.2}}$ layers [15]. The electrons are provided by a Sb-δ-doped layer 20 nm above the channel. We can control N_s by varying bias voltage V_{BG} of a p-type Si(001) substrate 2.1 μm below the channel at 20 K. The 2DES has a high mobility of 49 $\mathrm{m^2/V\ s}$ at $N_s = 2.36 \times 10^{15}\ \mathrm{m^{-2}}$ ($V_{\mathrm{BG}} = 0$ V) and $T = 0.4$ K. The use of the extremely high-mobility sample allows us to probe low N_s region, where the Coulomb interaction between electrons is sufficiently larger than the Fermi energy. The sample was mounted inside a wave-guide with a 6 mm bore inserted into a pumping ^{3}He cryostat. CR measurements have been performed in the frequency of 100 GHz using a Gunn oscillator. In order to obtain electrical transport properties of the Si/SiGe sample, Ohmic electrodes were also fabricated. The resistivity of the sample has been measured using standard low-frequency (13.7 Hz) lock-in techniques.

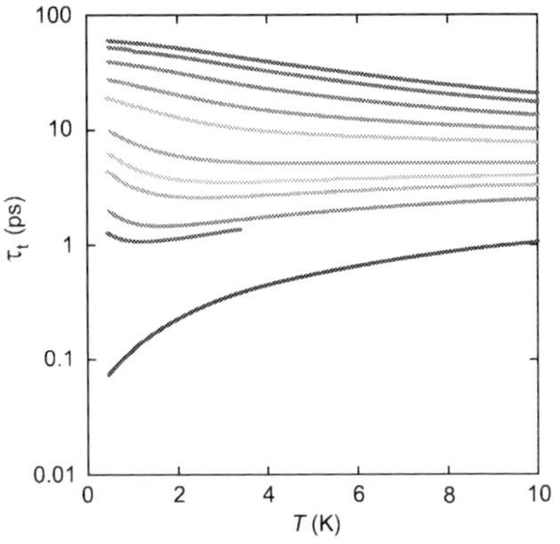

Figure 2. Temperature dependence of τ_t as a function of N_s at $B = 0$ T. The electron densities are 2.61, 2.36, 1.93, 1.51, 1.22, 0.90, 0.74, 0.66, 0.55, 0.50 and 0.44 $\times 10^{15}$ m^{-2} (from top to bottom).

3. Results and discussion

Fig. 2 shows temperature dependence of τ_t, which is given by $\tau_t = m^*/e^2 N_s \rho(T)$. Here, m^* is $0.20m_e$ [16]. While τ_t decreases with temperature for $N_s < 0.5 \times 10^{15}$ m^{-2}, τ_t increases with decreasing temperature above 0.50×10^{15} m^{-2}, i.e., $\rho(T)$ shows a metallic behavior. The critical electron density of the MIT, N_c, is almost 0.50×10^{15} m^{-2} or slightly lower. The 2D metallic phase frequently appears in large r_s regime, where r_s is the ratio of Coulomb energy to Fermi energy and is given by $\pi^{1/2}(e/h)^2(m^*/\kappa\epsilon_0)N_s^{-1/2}$. In our Si/SiGe 2DES, r_s is estimated to be 7.9 at $N_s = 0.50 \times 10^{15}$ m^{-2} ($\sim N_c$), where the relative dielectric constant κ is 11.5 for the silicon quantum well, and other parameters have their usual meanings. Interplay between *e-e* interactions and random disorder arising from charged impurities plays an important role in determining the intriguing temperature dependent τ_t.

The electron density dependence of the CR signals at 0.4 K is exhibited in Fig. 3. To avoid effect of rise in the sample temperature, the CR signals were detected with small output power of the Gunn oscillator. From the resonance magnetic field $B_c = 0.70$ T ($= m^*\omega_c/e$), m^* is estimated to be $0.20m_e$, where ω_c is the cyclotron frequency. This value is good agreement with that determined from the temperature dependence of the SdH oscillations in Ref. [16]. However, below $N_s = 0.90 \times 10^{15}$ m^{-2}, B_c slightly shifts to higher magnetic field. Although the origin is not clear at this time, it may be attributed to strong *e-e* interactions. The scattering time $\tau_{\rm CR}$ can be obtained as $\tau_{\rm CR} = B_c/(\omega \Delta B)$, where ΔB is the half width at half maximum of the CR signal. As N_s decreases, the CR signal becomes broader, that is, $\tau_{\rm CR}$ decreases.

Fig. 4 shows the electron density dependence of $\tau_t/\tau_{\rm CR}$. In the data on our Si/SiGe 2DES, the value of $\tau_t/\tau_{\rm CR}$ approaches unity as N_s decreases at 0.4 K, indicating that short-range potential fluctuations which cause angle-independent scattering play an important role in the vicinity of the MIT. Below 0.74×10^{15} m^{-2}, $\tau_{\rm CR}$ could not be obtained due to the broadening of the CR signal. Exploiting the electron spin resonance (ESR) in Si/SiGe 2DESs, Wilamowski *et al.* observed a gradual decrease in the Thomas-Fermi screening wave vector $q_{\rm TF}$ below 3×10^{15} m^{-2}, instead of the constant values of an ideal 2D system [17]. This leads to a positive feedback between loss of screening and increase in the amplitude of potential fluctuations. Thus, we consider that the MIT is induced by increase of disorder characterized by short-range (*not* long-range) potential fluctuations.

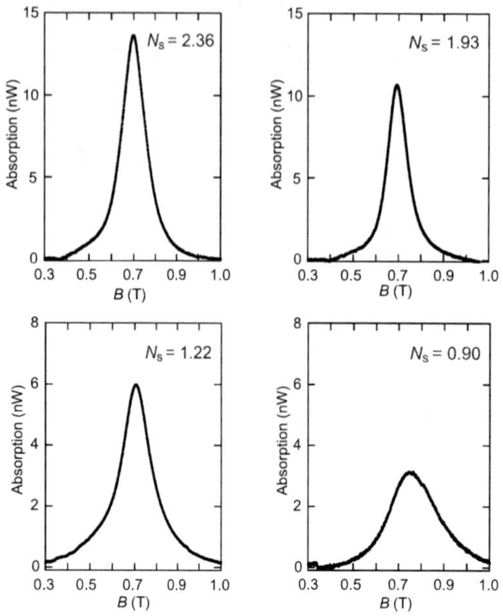

Figure 3. Cyclotron resonance signals at various N_s (T = 0.4 K). The electron density is indicated at upper side of the figures in units of 10^{15} m^{-2}.

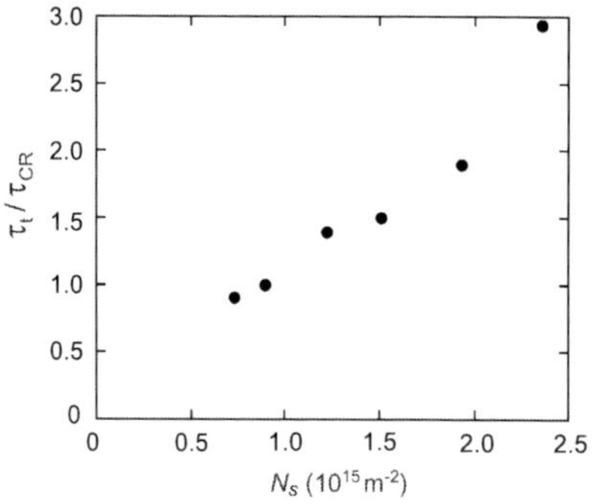

Figure 4. N_s dependence of $\tau_t/\tau_{\rm CR}$ at 0.4 K.

Acknowledgments

This work has partly supported by Grant-in-Aid for Scientific Research (B) (No. 18340080), (A) (No. 21244047), and Grant-in-Aid for Scientific Research on Priority Area "Physics of New Quantum Phases in Superclean Materials" (No. 20029005) from MEXT, Japan.

[1] For a review, see E. Abrahams, S. V. Kravchrenko, and M. P. Sarachik, Rev. Mod. Phys. **73**, 251 (2001).
[2] S. V. Kravchenko, G. V. Kravchenko, J. E. Furneaux, V. M. Pudalov, and M. D'Iorio, Phys. Rev. B **50**, 8039 (1994); S. V. Kravchenko, Whitney E. Mason, G. E. Bowker, J. E. Furneaux, V. M. Pudalov, and M. D'Iorio, Phys. Rev. B **51**, 7038 (1995).
[3] E. Abrahams, P. W. Anderson, D. C. Licciardello, and T. V. Ramakrishnan, Phys. Rev. Lett. **42**, 673 (1979).
[4] A. Punnoose and A. M. Finkel'stein, Science **310**, 289 (2005).

[5] S. Anissimova, S. V. Kravchenko, A. Punnoose, A. M. Finkel'stein, and T. M. Klapwijk, Nat. Phys. **3**, 707 (2007).
[6] S. Das Sarma and E. H. Hwang, Phys. Rev. Lett. **83**, 164 (1999); Phys. Rev. B **61**, 7838(R) (2000); Phys. Rev. B **68**, 195315 (2003); Phys. Rev. B **69**, 195305 (2004); Solid State Communications **135**, 579-590 (2005).
[7] E. H. Hwang and S. Das Sarma, Phys. Rev. B **72**, 085455 (2005).
[8] S. Das Sarma and F. Stern, Phys. Rev. B **32**, 8442 (1985).
[9] J. P. Harrang, R. J. Higgins, R. K. Goodall, P. R. Jay, M. Laviron, and P. Delescluse, Phys. Rev. B **32**, 8126 (1985).
[10] P. T. Coleridge, Phys. Rev. B **44**, 3793 (1991).
[11] B. Das, S. Subramaniam, M. R. Melloch, and D. C. Miller, Phys. Rev. B **47**, 9650 (1993).
[12] K. Ismail, M. Arafa, K. L. Saenger, J. O. Chu, and B. S. Meyerson, Appl. Phys. Lett. **66**, 1077 (1995).
[13] S. Syed, M. J. Manfra, Y. J. Wang, R. J. Molnar, and H. L. Stormer, Appl. Phys. Lett. **84**, 1507 (2004).
[14] R. Masutomi, A. Sekine, K. Sasaki, K. Sawano, Y. Shiraki, and T. Okamoto, Physica E **42**, 1184 (2010).
[15] A. Yutani and Y. Shiraki, Semicond. Sci. Technol. **11**, 1009 (1996); J. Cryst. Growth **175/176**, 504 (1997).
[16] S.-H. Song, Wei Pan, D. C. Tsui, Y. H. Xie, and Don Monroe, Appl. Phys. Lett. **70**, 3422 (1997).
[17] Z. Wilamowski, N. Sandersfeld, W. Jantsch, D. Többen, and F. Schäffler, Phys. Rev. Lett. **87**, 026401 (2001).

Journal of Physics: Conference Series **334** (2011) 012058 doi:10.1088/1742-6596/334/1/012058

High-field ESR using SQUID magnetometer

T. Sakurai[1], **R. Goto**[2], **N. Takahashi**[2], **S. Okubo**[3], **H. Ohta**[2,3]

[1]Center for Supports to Research and Education Activities, Kobe University, Rokkodai 1-1, Nada, Kobe 657-8501, Japan

[2]Graduate School of Science, Kobe University, Rokkodai 1-1, Nada, Kobe 657-8501, Japan

[3]Molecular Photoscience Research Center, Kobe University, Rokkodai 1-1, Nada, Kobe 657-8501, Japan

E-mail: tsakurai@kobe-u.ac.jp

Abstract. We have developed high-field ESR system using commercially available SQUID magnetometer equipped with the superconducting magnet up to 5 T. This is a longitudinal magnetization detection type ESR. The minimum detectable spin number is evaluated as 2×10^{13} spins/G for 105 GHz. The most advantageous points as compared with other high-field ESR are that high-field ESR can be done very easily and both microscopic and macroscopic magnetic properties can be obtained at once. Moreover, the transmission technique without using cavity enables us to make multi-frequency ESR measurement in the wide frequency region. We have succeeded in observing ESR from 70 to 315 GHz.

1. Introduction

High-field electron spin resonance (ESR) is a very powerful means to clarify local spin states of magnetic semiconductors. It has several advantages as compared with commercial X-band ($\sim$ 9.5 GHz) ESR spectrometer equipped with a water-cooled electromagnet whose magnetic field is generally limited up to about 1 T. For instance, the high-field ESR can obtain a spectrum with high resolution and observe ESR across the large energy splitting. And these advantages give us microscopic information about the magnetic center of material in detail, which can not be revealed by X-band ESR measurement. However, the high-field ESR equipments usually need large and complicated setups and its measurement can not be made so easily. Here, we demonstrate a very simple high-field ESR system using commercially available superconducting quantum interference device (SQUID) magnetometer equipped with a superconducting magnet (SQUID-ESR). This is a longitudinal magnetization detection type ESR. In this technique, the resonance is observed as the change of the longitudinal magnetization parallel to the external magnetic field. When the resonance condition is satisfied under the irradiation of the electromagnetic wave, spins are excited with the change of $|\Delta S_z| = 1$. The magnetization changes correspondingly and it is detected by the SQUID magnetometer. The advantageous points as compared with other high-field ESR system are that the high-field ESR measurement can be made much more easily and the macroscopic magnetization measurement can also be made if only the irradiation of the electromagnetic wave is stopped. The SQUID-ESR technique was first reported by Cage *et al.* [1]. In this study, we have improved the sensitivity and extended the frequency region as compared with Cage's system, as is described below in detail. We demonstrate our high-field SQUID-ESR system and several results obtained by this system.

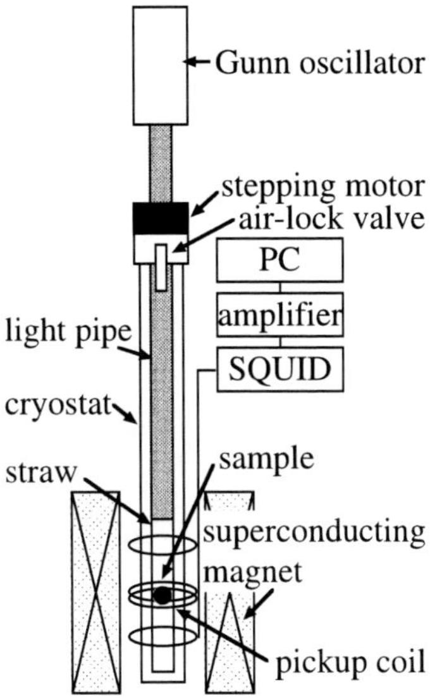

Figure 1. Schematic diagram of SQUID-ESR system.

These results will show that the SQUID-ESR is an easy but very useful tool to study magnetic properties of semiconductors.

2. Outline of the SQUID-ESR system

Our SQUID-ESR system is schematically drawn in Fig. 1. We use Magnetic Property Measurement System (MPMS) XL equipped with a superconducting magnet (Quantum Design Co. Ltd.) as the SQUID magnetometer. The magnetic field is available up to 5 T. The Gunn oscillator and its multiplier are used as the light source and the available frequency region is from 70 to 315 GHz. The sample holder rod is also used as the light pipe and the electromagnetic wave is irradiated from the top of the rod. A sample set in a straw is simply attached at the end of the rod. The other measurement procedure is the same with that of the MPMS measurement.

The ESR system using commercially available SQUID magnetometer was first reported by Cage *et al.* [1]. In Cage's system, a waveguide matched for TE_{11} mode for the frequency around 100 GHz with inner and outer diameter of 2.54 ϕ and 3.175 ϕ is used as the light pipe, while an oversized waveguide with inner and outer diameter of 5.4 ϕ and 6.0 ϕ is used in our system. In both SQUID-ESR systems, the cavity is not used and the transmission technique is adopted. Therefore, the use of a waveguide matched for the fundamental mode rather causes the loss of energy of the electromagnetic wave. As is shown later, the use of an oversized waveguide is proven to improve the sensitivity greatly. Another difference is that Gunn oscillator is used as the electromagnetic wave source in this study and it can be operated much easier than klystron which is employed in previous experiment [1]. Although its output power is generally lower than that of klystron, we have succeeded in observing ESR in the wide frequency region from 70 to 315 GHz, as is also shown later.

3. Results and discussions

We tried high-field SQUID-ESR measurement by a stainless light pipe with inner and outer diameter of 2.6 ϕ and 3.0 ϕ first for several compounds. Here we show a result of cobalt Tutton salt $(NH_4)_2Co(SO_4)_2{\cdot}6H_2O$ as an example. A single crystal with a mass of 19.2 mg is used. The magnetic ion is Co^{2+} ($3d^7$) and it is often explained as fictitious spin 1/2 with anisotropic g-value

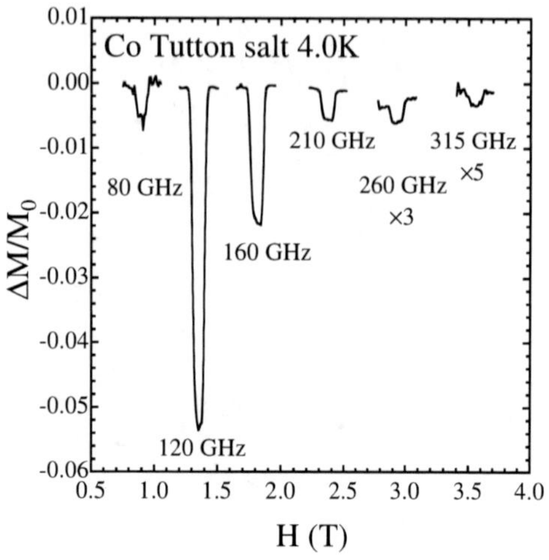

Figure 2. SQUID-ESR spectra of cobalt Tutton salt obtained at 4.0 K. M_0 is defined as $M_0 = N/2\,(g_1 + g_2)\,\mu_B S$ and ΔM is the difference of the magnetizations with or without irradiation.

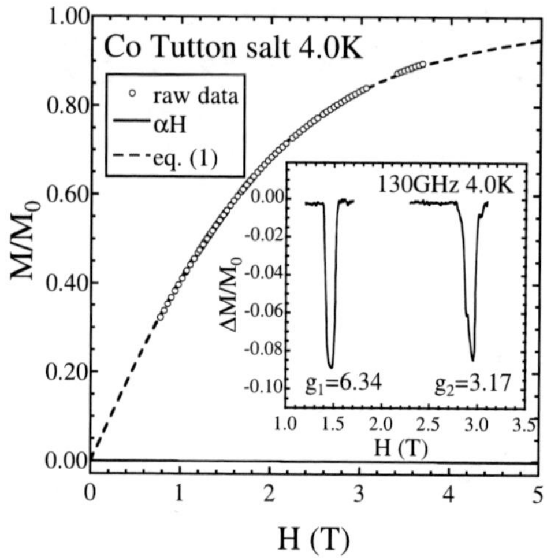

Figure 3. Magnetization curve of cobalt Tutton salt at 4.0K. The broken and solid lines are equation (1) and the second term of equation (1), respectively. The inset shows SQUID-ESR spectra obtained at 130 GHz.

when it is in nearly cubic symmetry [2]. Cobalt Tutton salt is known to have two inequivalent CoO_6 octahedra, namely two g sites [3]. The magnetic field is applied nearly parallel to the most anisotropic g principle axis for one site (site I). The measurements were done by "DC measurement" with or without irradiation of the electromagnetic wave. The magnetic field was swept by 0.3 T for one frequency with a step of 0.01 T. The magnetization was obtained by scanning 3 cm length (24 points) and by averaging two scans at each field point. Figure 2 shows the SQUID-ESR spectra of the cobalt Tutton salt at 4.0 K. The spectra are obtained by subtracting the magnetization curve obtained without irradiation of electromagnetic wave from that obtained under irradiation. We have succeeded in observing ESR in the wide frequency region form 80 to 315 GHz as shown in Fig. 2. The measurements above 140 GHz have never been done before and this is the first measurement at the higher frequency for SQUID-ESR. The signal intensity depends on the output power of the oscillator mainly. The output power tends to decrease as the frequency increases. It is 36 mW for 120 GHz and 1.5 mW for 315 GHz, for instance. A shoulder-like structure was observed for each absorption line and this is probably due to twinned crystal. From the linear fitting to the plot of frequency versus resonance field, the g-value is obtained to be $g_1 = 6.34$ for site I. And another g-value for site II is obtained to be $g_2 = 3.17$ as shown in the inset of Fig. 3. These g-values are consistent with the known values [3]. With these g-values, we can draw the magnetization curve as

$$M = \frac{N}{2}\mu_B S\,(g_1 \tanh x_1 + g_2 \tanh x_2) + \alpha H, \qquad (1)$$

where the first term is the Brillouin function with $S = 1/2$, the second term is the correction term with a fitting parameter α, $x_i = g_i \mu_B S H/(k_B T)$, N is number of magnetic ion, μ_B is Bohr magneton, k_B is Boltzmann factor, H is magnetic field and T is temperature. As shown in Fig. 3, the calculated line completely coincides with the raw data with a suitable fitting parameter α ($\alpha = -3.65 \times 10^{-8} M_0$ emu/G). Thus, it is clearly shown that the SQUID-ESR has proven to

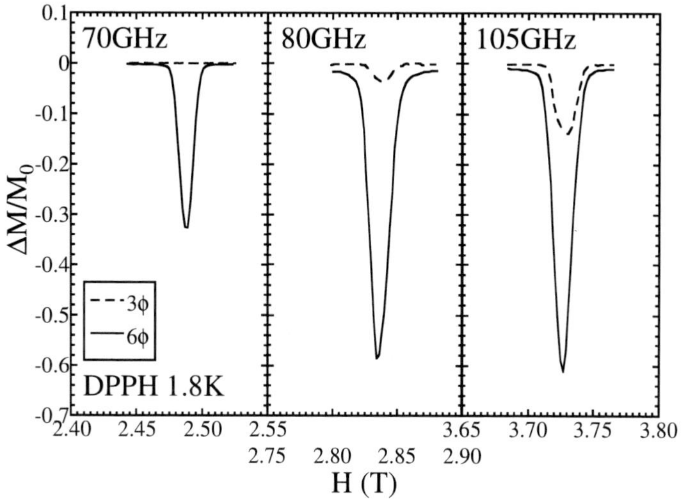

Figure 4. Comparison of SQUID-ESR spectra of DPPH. The solid and broken lines indicate those obtained by using 6 ϕ and 3 ϕ light pipe, respectively. M_0 is defined as $M_0 = Ng\mu_B S$.

be a powerful tool to obtain both magnetic macroscopic and microscopic properties.

In order to improve the sensitivity, we tried an oversized waveguide by a stainless pipe with inner and outer diameter of 5.4 ϕ and 6.0 ϕ with a handmade slide seal assembly. The results are shown in Fig. 4 with the results obtained by 3 ϕ light pipe for comparison. An organic free radical compound DPPH (1, 1-diphenyl-2-picrylhydrazyl) is used as a sample. A powder with a mass of 13 mg is used. The magnetic field was swept by 80 mT for one frequency with a step of 2 mT. As clearly seen in Fig. 4, the sensitivity improves greatly. The signal intensity increases by more than one order of magnitude for 80 GHz. Moreover, we observed ESR at 70 GHz at which ESR was hardly observed with 3 ϕ light pipe. The minimum detectable spin number is evaluated as 2×10^{13} spins/G at 105 GHz (output 52.5 mW) at 1.8 K. Thus, it is found that the oversized wave guide is very effective to improve the sensitivity for SQUID-ESR.

4. Summary

We have developed high-field ESR system using commercially available SQUID magnetometer equipped with a superconducting magnet up to 5 T. It was confirmed that the available frequency region is from 70 to 315 GHz. The minimum detectable spin number was obtained as 2×10^{13} spins/G for 105 GHz (52.5 mW) at 1.8 K by using an oversized waveguide. We found that the SQUID-ESR is a very powerful means to study magnetic properties from the macroscopic and microscopic point of view at once.

Acknowledgments

This research was partially supported by Japan Society for the Promotion of Science(JSPS), Grant-in-Aid for Creative Scientific Research (No.19GS1209), Scientific Research (C) (No. 60379477).

References

[1] Cage B and Russek S (2004) *Rev. Sci. Instrum.* **75** 4401
[2] Abragam A and Pryce M H L (1951) **A206** 173.
[3] Kuroda A, Motokawa M and Date M (1978) *J. Phys. Soc. Jpn.* **44** 1797

Intersubband population inversion and stimulated terahertz transitions between Landau levels in resonant tunneling multiple quantum well structures

Maxim P Telenkov[1], Yury A Mityagin[1], and Petr F Kartsev[2]

[1]P.N. Lebedev Physical Institute of RAS, Moscow, 119991, Russia
[2] National Research Nuclear University (Mephi), Moscow, 115409, Russia

E-mail: maxim_telenkov@mail.ru

Abstract. It is shown that a population inversion can be achieved in the system of Landau levels in cascade quantum well structures in a magnetic field under a condition of sequential resonant tunneling. This mechanism allows a wide range tuning of the emitted terahertz frequency by the magnetic field strength variation.

1. Introduction

In the present report we show the possibility to achieve a population inversion in the system of Landau levels (LL) in cascade quantum well structures in a strong magnetic field under a condition of sequential resonant tunneling, i.e. in a strong transverse electric field. The scheme of transitions is shown in Fig.1.

It was shown that in quantum wells with spacing between 1st and 2nd subbands lower than an optical phonon energy (i.e. when optical phonon scattering is suppressed) the population of 0th LL in 2nd subband exceeds that of 1st LL in the 1st subband. Therefore the stimulated emission of terahertz radiation can be achieved on the radiative transitions between these LLs.

The effect is caused by a considerable difference between intersubband and intrasubband scattering times (upper LL (2-0) is depopulated by intersubband scattering only, while the lower LL (1-1) – by intrasubband scattering to 0th level of the same subband).

In the considered system the electron-electron scattering is the most important inter-LL scattering mechanism, since other inter-LL scattering processed are strongly suppressed. So the main attention was paid to developing a model of e-e scattering in a system of LLs in quantum wells with several subbands.

The proposed mechanism allows the continuous tuning of the emitted radiation frequency by the variation of the magnetic field strength, thus giving the possibility to achieve a tunable stimulated emission.

2. Theory

Let us consider the electron kinetics in sequential resonant tunneling $GaAs\,/\,AlGaAs$ quantum well structures in quantizing magnetic field $\mathbf{B} = B\mathbf{e}_z$ (axis Z is perpendicular to the layers). In Landau gauge the single-electron spectrum is a set of confinement subbands each splitting into a series of Landau levels:

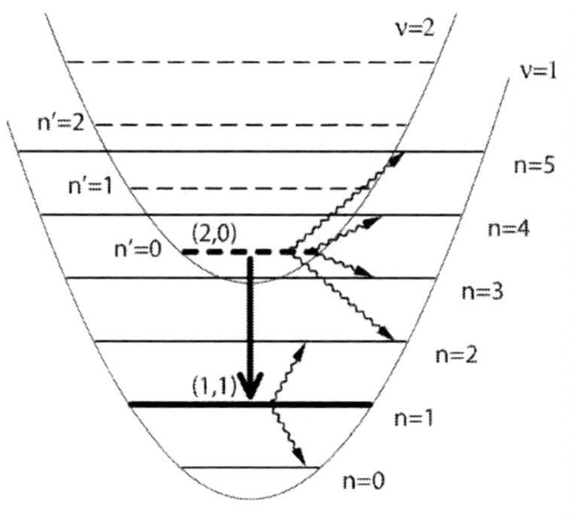

Fig.1 Proposed scheme of transitions between Landau levels in a quantum well

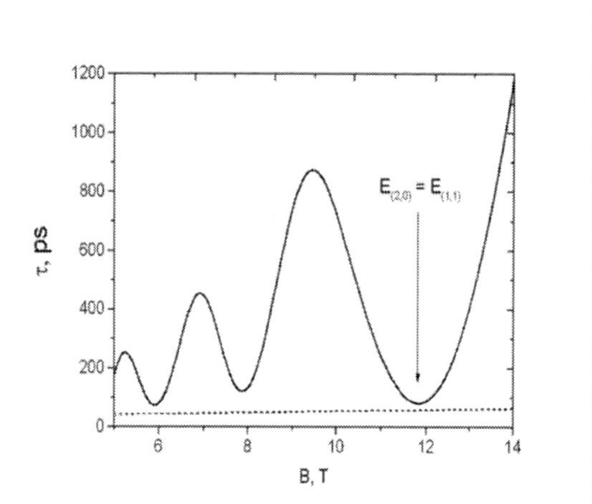

Fig.2 The calculated e-e scattering lifetimes for (2,0) (solid line) and (1,1) (dashed line) Landau levels in GaAs/AlGaAs quantum well (d_W = 25 nm)

$E=\tilde{E}_v+\hbar\omega_c(n+1/2)$ with wavefunctions $\psi(\mathbf{r})=\dfrac{\exp(ikx)}{\sqrt{L}}\cdot\Phi_n(y-k\ell^2)\varphi_v(z)$

determined by three quantum numbers – subband index v, Landau level number n and x-axis momentum component k.

Using the Fermi golden rule approximation, the electron flux to Landau state $f=(v_f,n_f)$ caused by electron-electron scattering transitions $(i,j)\to(f,g)$ is given by the expression

$$j^{e-e}_{(i,j)\to(f,g)}=\frac{N_i}{\tau_{(i,j)\to(f,g)}}, \tag{1}$$

where

$$\frac{1}{\tau_{(i,j)\to(f,g)}}=\frac{2}{L^2\alpha}\sum_{k_i}\frac{1}{\tilde{\tau}_{(i,j)\to(f,g)}(k_i)} \tag{2}$$

– total $(i,j)\to(f,g)$ transition rate,

$$\begin{aligned}\frac{1}{\tilde{\tau}_{(i,j)\to(f,g)}(k_i)}&=\frac{2\pi}{\hbar}\sum_{k_j,k_f,k_g}\left|V_{(i,j)(f,g)}(k_i,k_j,k_f,k_g)\right|^2\times\\&\times\frac{N_j}{\alpha}\cdot\left[1-\frac{N_f}{\alpha}\right]\cdot\left[1-\frac{N_g}{\alpha}\right]\cdot\delta(E_i+E_j-E_f-E_g)\end{aligned} \tag{3}$$

– scattering rate from initial state containing one electron in LL i with wave vector k_i and one electron in LL j to the final state containing electrons in LLs f and g, α - LL degeneracy, N – 2D electron concentration on the corresponding Landau state.

$$V_{(i,j)(f,g)}(k_i,k_j,k_f,k_g)=\int d\mathbf{r}_1 d\mathbf{r}_2 \psi^*_{f,k_f}(\mathbf{r}_1)\psi_{i,k_i}(\mathbf{r}_1)\frac{e^2}{\varepsilon_s|\mathbf{r}_1-\mathbf{r}_2|}\psi^*_{g,k_g}(\mathbf{r}_2)\psi_{j,k_j}(\mathbf{r}_2) \quad (4)$$

– matrix elements of electron-electron interaction.

The finite width of the LLs is taken into account by replacing the δ-function in (3) by a form-factor $F_{(i,j)(f,g)}(E_i+E_j-E_f-E_g)$, which is approximated by Lorentzian

$$F_{(i,j)(f,g)}(E)=\frac{1}{\pi}\cdot\frac{\Gamma}{E^2+\Gamma^2} \quad (5)$$

of halfwidth Γ=2 meV, typical values for quantum well structures [1,2].

3. Results

The results of the numerical calculation of the e-e scattering times are shown in Fig. 2. The calculations were carried out for GaAs\AlGaAs quantum wells of 25 nm width, with intersubband spacing ΔE_{12}=20.4 meV. The lifetime of the lower (1,1) LL was found to depend monotonically and approximately linearly upon magnetic field. On the contrary the intersubband scattering lifetime of LL (2,0) revealed an oscillating dependence on magnetic field similar to that obtained in [1,2]. The minima of lifetime correspond to the situation when LL (2,0) coincides with certain LL of lower subband or is situated exactly between them.

It is seen from Fig.2 that the intrasubband scattering time is considerably shorter than intersubband one. Within the range 5 – 10 T of magnetic field strength the lifetime of the upper (2,0) LL is always higher than that of the lower (1,1) LL, and in the intervals B=4.9÷5.6, 6.2÷7.6, 8.1÷11.0 the difference achieves the values of 3 – 10 times. It opens the possibility to achieve a population inversion for considered inter-LL transition by the resonant tunneling pumping of the upper (2,0) level.

The emission frequency $\hbar\omega = \Delta E_{21} - \hbar\omega_C$, ($\omega_C$ – cyclotron frequency) will vary within interval $\hbar\omega$ = 11.8 – 3.2 meV (ν = 95 – 26 cm-1, λ= 105 – 390 μm, f = 2.86 – 0.77 THz) for magnetic field range B = 5 – 10 T.

The important point is that the specified (2,0)→(1,1) optical transition is forbidden by selection rules when magnetic field **B** is orthogonal to structure layers. The problem can be solved by placing the structure into a tilted magnetic field where the mentioned selection rule is violated due to the mixing of in-plane and out-of-plane carrier motion similarly to that observed in resonant tunneling transitions between LLs in tilted magnetic field [3-6]. If the cyclotron energy is several times lower than intersubband spacing the following expression for the dipole matrix element can be obtained:

$$\left|\mathbf{D}_{(2,0)\to(1,1)}\right|^2=\delta_{k_1,k_2}\left|\langle\varphi_2(z)|z|\varphi_1(z)\rangle\right|^2\xi^2/2\cdot\exp\left(-\xi^2/2\right) \quad (6)$$

Here $\xi=\left[\langle z\rangle_2-\langle z\rangle_1\right]\cdot\frac{\ell_\perp}{\ell_\parallel^2}$, $\ell_\perp=\sqrt{\frac{\hbar c}{eB_\perp}}$ and $\ell_\parallel=\sqrt{\frac{\hbar c}{eB_\parallel}}$ – magnetic lengths for perpendicular ($B_\perp$) and parallel ($B_\parallel$) component of magnetic field with respect to layers, $\langle z\rangle_\nu=\langle\varphi_\nu|z|\varphi_\nu\rangle=\int dz|\varphi_\nu(z)|^2 z$ - the average z-coordinate for the states of ν-th subband. Therefore to make the dipole element nonzero there should be a difference in average coordinates of upper and lower states $\langle z\rangle_1$ и $\langle z\rangle_2$. It can be achieved by introducing asymmetric potential along the growth axis of the structure, in particular, by application of transverse electric field. In Fig. 3 the calculated according to (6) matrix element dependence on electric field for different tilt angles of magnetic field is shown.

The asymmetry can also be introduced by asymmetric well construction, say, using as an active element an asymmetric system of two or more strongly coupled quantum wells.

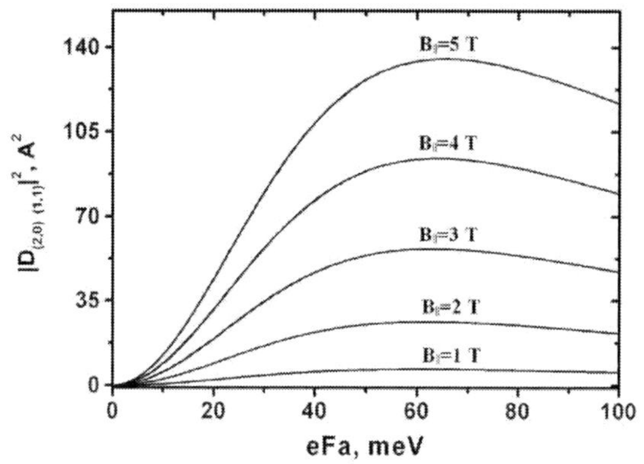

Fig. 3. The calculated dependence of dipole matrix element $\left|\mathbf{D}_{(2,0)\to(1,1)}\right|^2$ on the voltage drop per quantum well eFa in tilted magnetic field for different values of the parallel to layers component $\mathbf{B}_{\|} = 1 \div 5$ T. The calculations were made for $GaAs/Al_{0.3}Ga_{07}As$ quantum well of 25 nm width. The perpendicular to layers component of magnetic field $\mathbf{B}_{\perp} = 5$ T.

Conclusion

It was shown that due to a difference between intersubband and intrasubband e-e scattering rates a population of the Landau levels of the upper subbands can considerably exceed that of the Landau levels of the lowest one, and wide-range tunable terahertz stimulated emission on such inter-Landau-level transition may be achieved.

Acknowledgments

The work was supported by Russian Basic Research Foundation (grants No. 09-02-00671 and 08-02-92505-NCNIL) and by RF President grant for young scientists (grant No. MK-916.2009.2).

References

[1] Kempa K, Zhou Y, Engelbrecht J R, Bakshi P, Ha H I, Moser J, Naughton M J, Ulrich J, Strasser G, Gornik E and Unterrainer K 2002 *Phys. Rev. Lett.* **88** 226803-1
[2] Kempa K, Zhou Y, Engelbrecht J R and Bakshi P 2003 *Phys. Rev.* B **68** 085302
[3] Leadbeater M L, Sheard F W and Eaves L 1991 *Semicond.Sci.Technol.***6** 1021
[4] Lyo S K 1998 *Phys.Rev.* B **57** 9114
[5] Telenkov M P and Mityagin Yu A 2006 *JETP* **130** (3) 491-499
[6] Telenkov M P and Mityagin Yu A 2007 *Int.J.Mod.Phys.* B **21**(8/9) 1594

Journal of Physics: Conference Series **334** (2011) 012060 doi:10.1088/1742-6596/334/1/012060

All Optical Measurement Proposed for the Photovoltaic Hall E ect

Takashi Oka, and Hideo Aoki

Department of Physics, University of Tokyo, Hongo, Tokyo 113-0033, JAPAN

E-mail: oka@cms.phys.s.u-tokyo.ac.jp

Abstract. We propose an all optical way to measure the recently proposed " photovoltaic Hall effect ", i.e., a Hall effect induced by a circularly polarized light in the absence of static magnetic fields. This is done in a pump-probe experiment with the Faraday rotation angle being the probe. The Floquet extended Kubo formula for photo-induced optical response is formulated and the ac-Hall conductivity is calculated. We also point out the possibility of observing the effect in two layered graphene, three-dimensional graphite, and more generally in multi-band systems such as materials described by the *dp*-model.

1. Introduction

There are increasing fascinations with new types of Hall effect besides the quantum Hall effect in magnetic fields, where a notable example is the spin Hall effect in topological insulators. We have previously proposed, as another new type of Hall effect, the " photovoltaic Hall effect ", which should take place when we shine a circularly polarized light to graphene[1]. The effect is interesting since this Hall effect occurs in the absence of uniform magnetic fields. Physically, the photovoltaic Hall current originates from the Aharonov-Anandan phase (a non-adiabatic generalization of the geometric (Berry) phase) that an electron acquires during its circular motion around the Dirac points in k-space in an AC field, and bears, in this sense, a topological origin similar to the quantum Hall effect. However, unlike the quantum Hall effect, the photovoltaic Hall conductivity is in general not quantized.

We have originally proposed an experimental setup to measure the photovoltaic Hall current through the gate electrodes attached to the sample. In the present report we propose a second, all-optical setup, which employs the Faraday rotation as in the optical Hall effect discussed in refs. [2, 3], which becomes here non-linear optical measurements, since this is a "pump-probe" type experiment, where one optically probes the Hall effect that is induced in a system pumped by a circularly polarized light from a continuous laser source. This effect should grow with the intensity of the applied circularly polarized light. We also point out the possibility of observing the effect not only in graphene, but also in three-dimensional graphite, and more generally in multi-band systems such as materials described by the *dp*-lattice.

2. Kubo-formula extended for optical responses in a strong background light

So let us first propose a method to measure the photovoltaic Hall effect with non-linear optical measurements. The experimental setup is schematically displayed in Fig.1(a), where the sample is subject to a strong circularly polarized light with strength F and frequency (photon energy)

Ω. The strong external AC field changes the properties of the electronic state and the photo-induced band, or the Floquet band to be more precise, can acquire a non-trivial topological nature. This was discussed in ref. [1] as an emergence of the photo-induced Berry's curvature. Optical response can be drastically change in the presence of the photovoltaic Berry's curvature, which we show here to be captured through an optical Faraday (or Kerr) rotation measurement.

This is most easily analyzed in the Floquet formalism. In the Floquet formalism, the electrons are expressed as "dressed states", where absorption and induced-emission of photons is taken into account to all order. The strength of this method lies in the fact that one can treat these states in a similar way as in equilibrium time-independent systems and do calculations such as perturbation in a weak external probe field to obtain the linear-response formula. Our starting point to calculate the optical response is the Floquet expression for the Kubo formula extended to incorporate photovoltaic transports [1],

$$\sigma_{ab}(\boldsymbol{A}_{\rm ac}) = i\int\frac{d\boldsymbol{k}}{(2\pi)^d}\sum_{\alpha,\ \neq\alpha}\frac{[f\ (\boldsymbol{k}) - f_\alpha(\boldsymbol{k})]}{\varepsilon\ (\boldsymbol{k}) - \varepsilon_\alpha(\boldsymbol{k})}\frac{\langle\langle\Phi_\alpha(\boldsymbol{k})|J_b|\Phi\ (\boldsymbol{k})\rangle\rangle\langle\langle\Phi\ (\boldsymbol{k})|J_a|\Phi_\alpha(\boldsymbol{k})\rangle\rangle}{\varepsilon\ (\boldsymbol{k}) - \varepsilon_\alpha(\boldsymbol{k}) + i} \tag{1}$$

with $a, b = x$ or y. The DC expression can be extended to AC responses, which reads

$$\sigma_{xy}(\omega; \boldsymbol{A}_{\rm ac}) = -i\int\frac{d\boldsymbol{k}}{(2\pi)^d}\sum_{\alpha,\ \neq\alpha}\frac{f\ (\boldsymbol{k})}{\varepsilon\ (\boldsymbol{k}) - \varepsilon_\alpha(\boldsymbol{k})} \tag{2}$$

$$\times\left[\frac{\langle\langle\Phi_\alpha(\boldsymbol{k})|J_x|\Phi\ (\boldsymbol{k})\rangle\rangle\langle\langle\Phi\ (\boldsymbol{k})|J_y|\Phi_\alpha(\boldsymbol{k})\rangle\rangle}{\varepsilon\ (\boldsymbol{k}) - \varepsilon_\alpha(\boldsymbol{k}) + \omega + i} - \frac{\langle\langle\Phi_\alpha(\boldsymbol{k})|J_y|\Phi\ (\boldsymbol{k})\rangle\rangle\langle\langle\Phi\ (\boldsymbol{k})|J_x|\Phi_\alpha(\boldsymbol{k})\rangle\rangle}{\varepsilon\ (\boldsymbol{k}) - \varepsilon_\alpha(\boldsymbol{k}) - \omega + i}\right],$$

for the xy-component and

$$\mathrm{Re}\,\sigma_{xx}(\omega; \boldsymbol{A}_{\rm ac}) = \frac{\pi}{\omega}\int\frac{d\boldsymbol{k}}{(2\pi)^d}\sum_{\alpha<}[f_\alpha(\boldsymbol{k}) - f\ (\boldsymbol{k})]|\langle\langle\Phi\ (\boldsymbol{k})|J_x|\Phi_\alpha(\boldsymbol{k})\rangle\rangle|^2\delta(\varepsilon\ (\boldsymbol{k}) - \varepsilon_\alpha(\boldsymbol{k}) - \omega) \tag{3}$$

for the xx-component. Here $|\Phi_\alpha(t)\rangle$ is the α-th Floquet state, the double brakets includes a time average, $f_\alpha(\boldsymbol{k})$ the non-equilibrium distribution (occupation fraction) of the α-th Floquet state, and a positive infinitesimal. Strictly speaking, we must take into account the effect of relaxation (phonons, electrodes, etc.) to determine the non-equilibrium distribution function as a detailed balanced state with photo-absorption, photo-emission and the relaxations. Here we assume for simplicity that the non-equilibrium distribution function is given by the sudden approximation, i.e., the laser is turned on suddenly, which leads to $f_\alpha(\boldsymbol{k}) = \sum_i |\langle\langle\psi_i|\Phi_\alpha(\boldsymbol{k})\rangle\rangle|^2 f_{\rm FD}(E_i, \mu;\ _{\rm eff})$, where ψ_i and E_i are the wave function and energy eigenvalue of the system without laser and $f_{\rm FD}(E_i, \mu;\ _{\rm eff}) = 1/[\exp(\ _{\rm eff}(E_i - \mu)) + 1]$ is the equilibrium Fermi-Dirac distribution with an effective inverse temperature $_{\rm eff}$. The adiabatic approximation is an alternative candidate where the laser strength increases slowly and the distribution is given by the initial equilibrium distribution $f_\alpha(\boldsymbol{k}) = f_{\rm FD}(E_i, \mu;\ _{\rm eff})$ if i is connected to α adiabatically. A proper treatment of the relaxation can be done using the Keldysh Green's function as in ref. [1].

2.1. *Experimental feasibility*

In discussing the experimental feasibility, we can start from the relation between the induced conductivities $\sigma_{xx}(\omega), \sigma_{xy}(\omega)$ and the Faraday-rotation angle Θ_H [2], (n_0, n_s: refractive index of air and substrate)

$$\Theta_H = \frac{1}{2}\arg\left[\frac{n_0 + n_s + (\sigma_{xx} + i\sigma_{xy})/(c\varepsilon_0)}{n_0 + n_s + (\sigma_{xx} - i\sigma_{xy})/(c\varepsilon_0)}\right] \tag{4}$$

$$= \frac{1}{(n_0 + n_s)c\varepsilon_0}\sigma_{xy}(\omega) \quad (\sigma_{xy}\text{ in units of }\frac{e^2}{h}) \times 7\text{ mrad.} \tag{5}$$

This gives an estimate for the experimental precision demanded to observe the photovoltaic Hall effect through optical measurements. For example, in Fig.1 below, we estimate the photo-induced response $\sigma_{xy}(\omega; \boldsymbol{A})$ for a single-layered graphene to be around 0.2. This corresponds to a Faraday-rotation angle of Θ_H $0.2 \times 7 = 1.4$ mrad. The necessary strength of laser is around E 10^7 V/m for photon energy Ω 1 eV which is feasible with short pulse lasers.

3. Photovoltaic optical response

3.1. Single-layer graphene

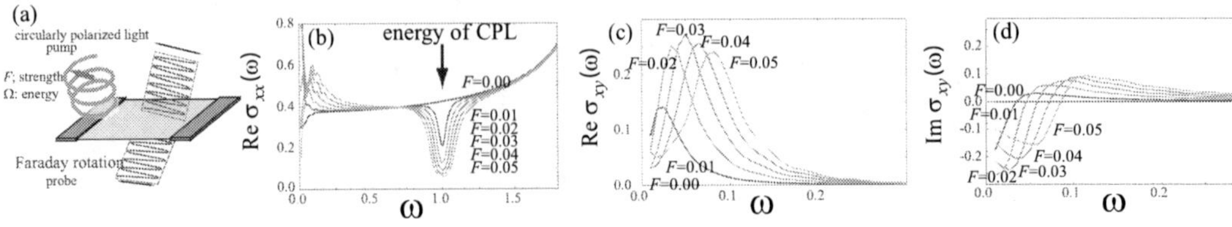

Figure 1. (a) Schematic, all-optical experimental setup to measure the photovoltaic Hall effect via non-linear optical responses. (b)(c)(d) Photovoltaic optical response in the honeycomb lattice (single-layer graphene). The optical absorption spectrum Re$\sigma_{xx}(\omega; \boldsymbol{A})$ (b) and the "photovoltaic optical Hall coe cient" $\sigma_{xy}(\omega; \boldsymbol{A})$ (c,d) are plotted for various values of strength F of the circularly polarized light. Note the different scales between (b) and (c,d). Here the energy of light is fixed to $\Omega = 1.0$ in units of the hopping integral (t 3 eV for graphene), and the effective inverse temperature $_{\rm eff} = 100$.

Using the extended Kubo formula (eqns. (2),(3)), we calculate the non-linear optical response in the presence of a strong circularly-polarized light in the honeycomb lattice. The optical absorption spectrum in Fig. 1 (b) exhibits the following: (i) At the energy of the circularly-polarized light (CPL), there is a dip in the absorption ($\sigma_{xx}(\omega)$, which is an analogue of hole burning. (ii) Near the DC limit ($\omega \to 0$), as one increases the field strength, the absorption first increases due to photo-carriers but then decreases. The decrease is due to the opening of a photo-induced gap at the Dirac point as discussed in ref. [1]. The real and imaginary parts of the photovoltaic optical *Hall* coe cient are plotted in Fig. 1 (c,d), which grow in a low-frequency region. Note that the scale of ω is different in (b) and (c,d), which is because the photo-induced ac-Hall response only takes place in the low frequency regime. The peak position of the real part shifts to higher frequency as the strength of the CPL is increased. This is important, since the experimental detection is easier for higher photon energies.

3.2. Multi-layer graphene

Photovoltaic Hall effect is not restricted to systems having a Dirac cone as in the single-layer graphene. To show this, we calculate the photovoltaic optical response in bilayer graphene (see [4] for notation and refs) described by the Hamiltonian,

$$H = \begin{pmatrix} 0 & k_x - ik_y & t_p & 0 \\ k_x + ik_y & 0 & 0 & 0 \\ t_p & 0 & 0 & k_x + ik_y \\ 0 & 0 & k_x - ik_y & 0 \end{pmatrix}. \tag{6}$$

We plot the photovoltaic optical response in Fig. 2 for this effective model. The basic features are similar to those in the monolayer graphene. The result suggests the possibility of observing photovoltaic Hall effect in multilayer graphene and graphite, since one can block-diagonalize the Hamiltonian for a multilayer system into sub-blocks of single and bi-layer graphene components as was shown in ref. [5].

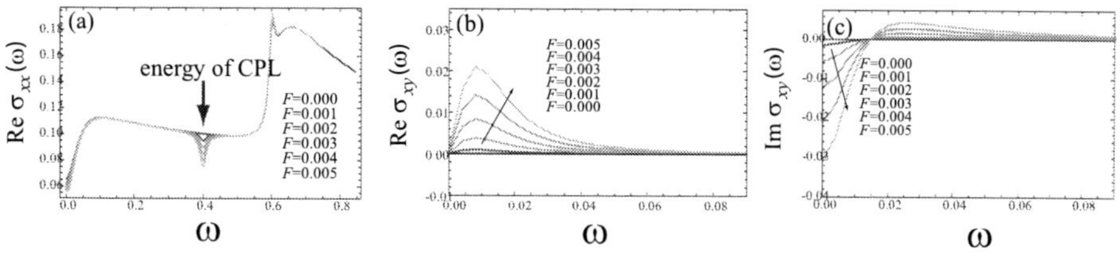

Figure 2. Photovoltaic optical response in a bilayer graphene for $\Omega = 1.0$, $t_p = 0.6$ with different scales between (a) and (b,c). Note that the input field is an order of magnitude weaker than in Fig.1.

4. Photovoltaic Berry's curvature in the *dp*-lattice

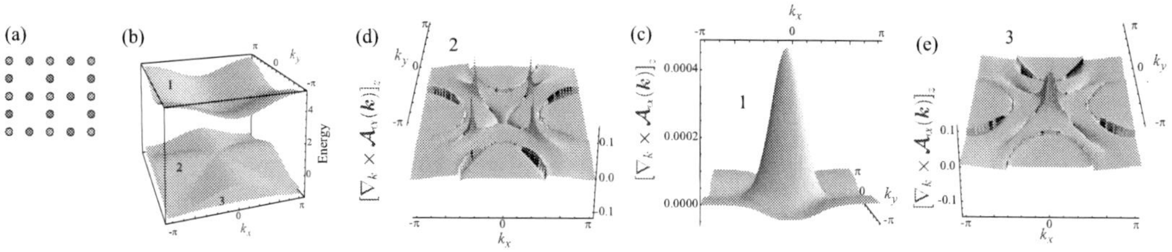

Figure 3. (a) *dp*-lattice. (b) The band dispersion in the *dp* model. (c)(d)(e) Photovoltaic Berry's curvature in the three bands (top d-band (c), and middle (d), bottom (e) p-bands) for $t_{dp} = 0.4$, $\varepsilon_d - \varepsilon_p = 4$, $F = 0.01$, $\Omega = 3.$.

We finally explore the possibility of more generally observing the photovoltaic Hall effect in materials other than graphene/graphite. For a typical multiband model, we consider here the *dp*-lattice, which is a well-studied lattice in connection with the high-Tc cuprates[7]. Here we neglect the electron-electron interaction to concentrate on the one-particle properties. The Hamiltonian is given by

$$H = \begin{pmatrix} \varepsilon_d - \varepsilon_p & -2t \sin\frac{k_x}{2} & -2t \sin\frac{k_y}{2} \\ -2t \sin\frac{k_x}{2} & 0 & 4t_{dp} \sin\frac{k_x}{2} \sin\frac{k_y}{2} \\ -2t \sin\frac{k_y}{2} & 4t_{dp} \sin\frac{k_x}{2} \sin\frac{k_y}{2} & 0 \end{pmatrix}, \tag{7}$$

where $\varepsilon_d - \varepsilon_p$ is the potential difference between the d and p bands. The photovoltaic DC Hall conductivity is expressed as Berry's curvature as [1, 6]

$$\sigma_{xy}(\boldsymbol{A}_{\rm ac}) = e^2 \int \frac{d\boldsymbol{k}}{(2\pi)^d} \sum_\alpha f_\alpha(\boldsymbol{k}) \left[\nabla_{\boldsymbol{k}} \times \boldsymbol{\mathcal{A}}_\alpha(\boldsymbol{k})\right]_z, \tag{8}$$

in terms of a gauge field $\boldsymbol{\mathcal{A}}_\alpha(\boldsymbol{k}) \equiv -i\langle\langle\Phi_\alpha(\boldsymbol{k})|\nabla_{\boldsymbol{k}}|\Phi_\alpha(\boldsymbol{k})\rangle\rangle$. Note that this expression reduces to the TKNN formula [8] in the adiabatic limit.

The band dispersion of the *dp*-lattice is shown in Fig. 3 (b) and the photovoltaic Berry's curvature for each band is plotted in (c)-(e). A peak in Berry's curvature emerges at the center of the Brillioun zone in the *d*-band, while the curvature has complex structures in the *p*-bands. This is because the two *p*-bands intersect with each other, for which the ac field induces a band mixing, while the *d*-band is separated from the *p*-bands with the separation larger than the photon energy Ω considered in the figure. The ac-response for the *dp*-lattice has not been plotted here but it also becomes finite.

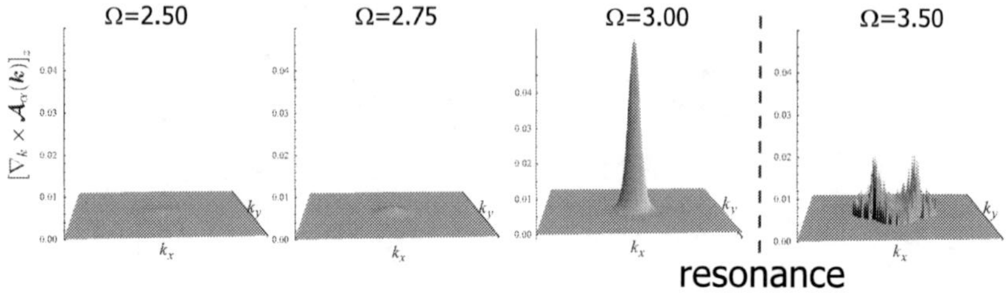

Figure 4. Photovoltaic Berry's curvature in the *d*-band of the *dp*-lattice for several values of the photon energy Ω for $t_{dp} = 0.4$, $\varepsilon_d - \varepsilon_p = 4$, $F = 0.1$.

Now, a natural question is: Is there a difference between a system with a band gap as in the *dp*-lattice when compared to a system with a massless Dirac cone as in graphene? In order to clarify this, we have calculated the photovoltaic Berry's curvature for several values of Ω in Fig. 4. When Ω is smaller than the gap, the photovoltaic Berry curvature increases as the value of Ω approaches the band gap. When Ω exceeds the gap, when a direct photo-transition becomes possible, the Berry's curvature changes into a concentric form with a negative part (not apparent in the figure). In the massless Dirac case, both the cone-like contribution and the concentric form coexist[1, 6], while in the *dp*-lattice they appear separately. Another important observation is that the magnitude of the photovoltaic Hall effect becomes large in multiband systems when Ω is slightly below the band gap, whereas in the massless Dirac case small Ω is better.

To conclude, we have studied the possibility of observing the photovoltaic Hall effect in an all-optical fashion. We have also shown that the effect is not restricted to lattices with a massless Dirac cone but is a universal phenomenon in multi-band systems. HA has been supported in part by a Grant-in-Aid for Scientific Research No.20340098 from JSPS, TO by a Grant-in-Aid for Young Scientists (B) from MEXT and by Scientific Research on Priority Area "New Frontier of Materials Science Opened by Molecular Degrees of Freedom".

References

[1] T. Oka, and H. Aoki, *Phys. Rev.* **B 79**, 081406 (R) (2009); ibid **79**, 169901(E) (2009).
[2] T. Morimoto, Y. Hatsugai, and H. Aoki, *Phys. Rev. Lett.* **103**, 116803 (2009).
[3] Y. Ikebe, T. Morimoto, R. Masutomi, T. Okamoto, H. Aoki, and R. Shimano, *Phys. Rev. Lett.* **104**, 256802 (2010).
[4] A. H. Castro Neto, F. Guinea, N. M. R. Peres, K. S. Novoselov, and A. K. Geim, *Rev. Mod. Phys.* **81**, 109 (2009).
[5] M. Koshino and T. Ando, Phys. Rev. B **76**, 085425 (2007).
[6] T. Oka and H. Aoki, *J. Phys.: Conf. Ser.* **200**, 062017 (2010).
[7] F. C. Zhang and T. M. Rice, *Phys. Rev.* **B 37**, 3759 (1988).
[8] D. J. Thouless, M. Kohmoto, M. P. Nightingale, and M. den Nijs, *Phys. Rev. Lett.* **49**, 405 (1982).

Journal of Physics: Conference Series **334** (2011) 012061 doi:10.1088/1742-6596/334/1/012061

Cyclotron resonance of two dimensional Rashba systems in InGaAs

Yasutaka Imanaka[1], **Tadashi Takamasu**[1], **Shunsaku Nitta**[2] **and Syoji Yamada**[2]

[1] National Institute for Materials Science, 3-13 Sakura, Tsukuba, Ibaraki 305-0003, Japan
[2] Japan Advanced Institute of Science and Technology, 1-1 Asahidai, Nomi, Ishikawa 923-1292, Japan

E-mail: IMANAKA.Yasutaka@nims.go.jp

Abstract. We observed an anti-crossing behavior for the magnetic eld dependence of the cyclotron resonance in a InGaAs/InAlAs two-dimensional Rashba system. Two resonances can be observed at a narrow millimeter frequency region, and this is due to the resonant coupling between di erent two spin levels because of the large zero- eld spin splitting by the Rashba e ect. The electron e ective mass is obtained as m =$0.04m_0$, which is in good agreement with the previous result at a terahertz region.

1. Introduction

Large zero- eld spin splitting has been studied intensively in various semiconductors for realizing novel spintronic applications such as the spin FET. Among narrow gap semiconductors, the InGaAs system, which has a larger spin-orbit interaction, is one of good candidates for studying the structure induced spin splitting, namely, the Rashba e ect. The zero- eld spin splitting energy for the Rashba system is estimated through the beat pattern analysis or the FFT analysis in the Shubnikov-de Haas (SdH) oscillation [1]-[4]. Cyclotron resonance (CR) is also a powerful tool for the direct observation of the spin splitting energy, recently [5]. Since the Zeeman energy becomes dominant rather than the Rashba spin splitting energy at high magnetic elds, the CR measurement at low magnetic elds is quite essential to determine the Rashba spin-orbit parameter quantitatively. In this study, hence, we measured the magnetic eld dependence of the CR in a high In-content InGaAs/InAlAs inverted heterostructure at a millimeter wave range.

2. Experiment and Sample

A $In_{0.75}Ga_{0.25}As/In_{0.75}Al_{0.25}As$ inverted heterostructure sample for the present experiment was grown on a semi-insulating GaAs substrate by the molecular beam epitaxy method. Magneto-transport experiments were performed with the van der Pauw method at high magnetic elds up to B=27T under T=1.5K. Ohmic contacts were fabricated with pure Indium soldering. The high magnetic eld is generated by the water-cooled magnet in the Tsukuba Magnet Laboratory of National Institute for Materials Science. The sweeping rate of the magnetic

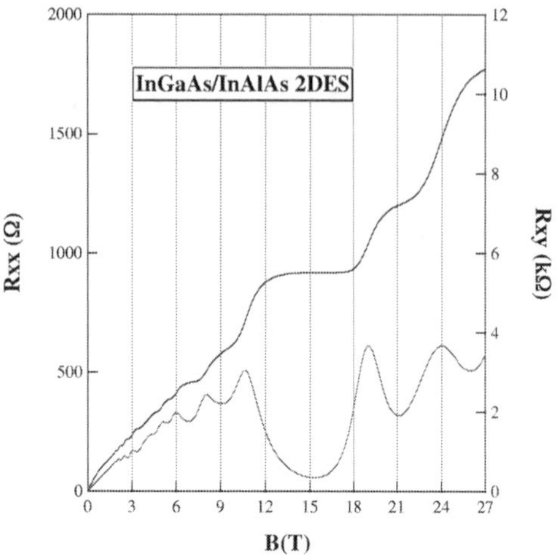

Figure 1. Magnetic eld dependence of R_{xx} and R_{xy} at high magnetic elds up to 27T. The carrier concentration of the sample is obtained as n=1.5 10^{12} cm ^{2}.

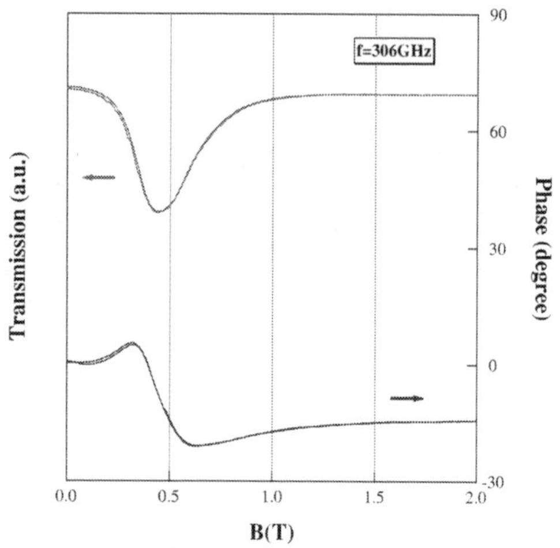

Figure 2. Typical cyclotron resonance spectra with using the MVNA : magneto-transmission (r) and phase ( ) data at f=306GHz.

eld is 1.25T/min. The SdH oscillation and the quantum Hall plateau are observed clearly at the integer lling factor in gure 1. The mobility and the carrier concentration of the sample were obtained as $\sim$20m^2/V s and n $\sim$1.5 10^{12} cm ^{2}, respectively.

CR spectra were measured with a millimeter-wave vector network analyzer (MVNA) in the range between f=150GHz and 340GHz. The magnetic eld is applied perpendicular to the two-dimensional electron layer with a superconducting magnet up to B=15T. The typical dimension of the sample is 5 5 0.5mm^3, and the backside of the sample is wedged to avoid interference e ects. More details of the MVNA is described in Ref.[6].

3. Results and Discussion

Figure 2 shows the typical CR spectra at f=306GHz. The CR measurement was performed at T=1.7K. Single Lorentzian absorption is observed around B=0.45T with the large variation of its phase. The electron e ective mass is obtained as m =0.04m_0 from the resonant magnetic eld. This value is in good agreement with the previous result of the terahertz cyclotron resonance [5]. However, in case of the frequency below f=200GHz, additional resonant peaks appear in some CR spectra. Further frequency dependence of the CR spectra below f=184GHz is shown in gure 3. Single CR peak is observed around f=172GHz, and evident additional peak appears at the higher magnetic eld side in some CR data (indicated by arrows in gure 3). The magnetic eld dependence of the additional peak is di erent from the one of the main CR peak around B=0.25T.

In order to con rm the additional resonance, a r- plot for the CR data of f=169GHz and 178GHz is shown in gure 4. The advantage for the r- plot analysis with both transmission data and phase data is known well to determine the resonant eld correctly even for weak and broad resonances in the MVNA measurement [6]. In gure 4, the dashed data at f=178GHz shows a single circle. On the other hand, the solid data at f=169GHz consists of two circles, indicating that the CR data at f=169GHz include two resonances.

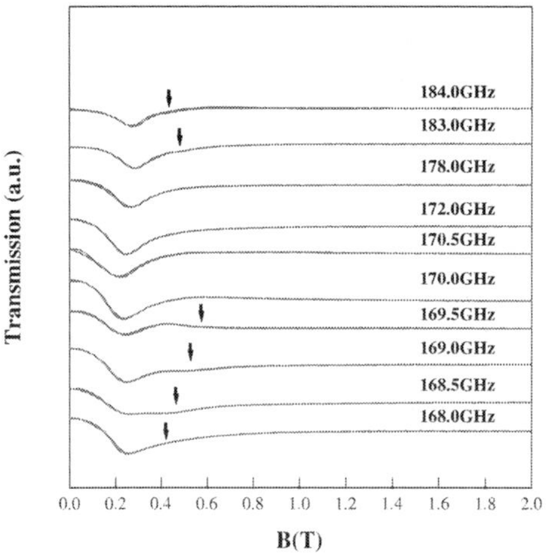

Figure 3. CR spectra between f=168GHz and 184GHz. Arrows indicate the resonant position of the additional peak.

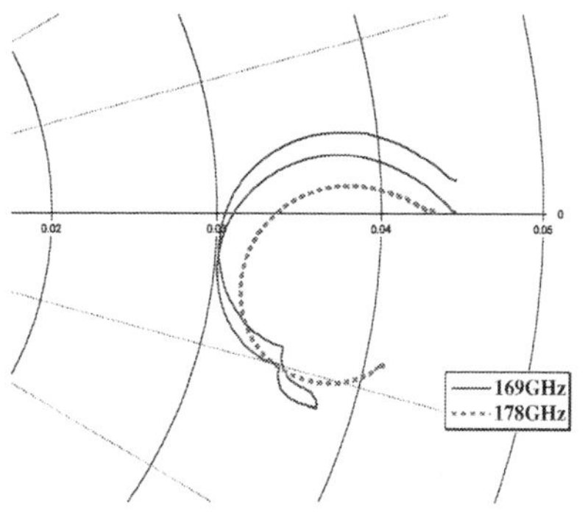

Figure 4. r- plot for the data of f=169GHz and 178GHz. The dashed data at f=178GHz shows a single circle. On the other hand, the solid data at f=169GHz consist of two circles. This indicates that the CR data at f=169GHz consist of two resonances.

Figure 5 shows the magnetic eld dependence of the CR at the frequency between f=150GHz and 340GHz. The CR energy is proportional to the applied magnetic eld except around f=170GHz. The solid line shows a calculated CR energy with m =0.04m_0, and the calculation is in good agreement with the main CR data at the wide frequency region.

The dashed curve shows the calculated result of the transition energy between di erent spin levels, N=14$^+$ and 14 from the formula:

$$E_N \;= \hbar\omega_c \left(N \quad \frac{1}{2}\sqrt{(1 \quad \frac{gm}{2m_0})^2 + N\frac{{}_R^2}{E_F\hbar\omega_c}} \right), \tag{1}$$

$$ {}_R^2 = \frac{8 \quad {}^2 m \; E_F}{\hbar^2}, \tag{2}$$

where ω_c is the cyclotron energy, which is given by $\omega_c = eB/m$, and g is the e ective g-factor. ${}_R$ and E_F are the zero- eld spin splitting energy and the Fermi energy, respectively. is the Rashba spin-orbit coupling parameter.

The dashed curve, calculated with =9 10 ^{12}eVm and $|g|$=7, explained the anomalous magnetic eld dependence of the resonance around f=170GHz qualitatively well. This Rashba spin-orbit coupling parameter, , is in fair agreement with the value estimated from the transport measurements (=12$\sim$15 10 ^{12}eVm) in the same sample. Hence, we assign that the origin of the additional peak is due to the spin- ip (SF) transition between di erent spin levels. The SF resonance was proposed theoretically by Fal'ko in the coupling of the Landau levels through the spin-orbit interaction, and an anti-crossing behavior was observed experimentally in a modulation-doped CdMnTe/CdMgTe single quantum well [8]. In the present sample, other anti-crossing points between di erent spin levels should appear at lower magnetic elds below B=1T, and two resonances were observed even at other frequency regions (not shown here). However, the carrier concentration of the present sample is too high, and the anomalous magnetic eld dependence of the additional resonance has not been understood quantitatively well yet.

Further experimental con rmation and theoretical support are needed to clarify the CR spectra at the wide range of the frequency in the InGaAs/InAlAs two-dimensional Rashba systems.

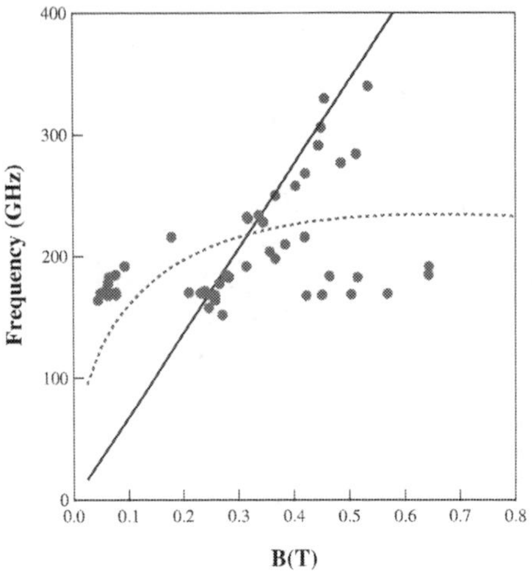

Figure 5. Magnetic eld dependence of the CR between f=150GHz and 340GHz at T=1.7K. The solid line shows the calculated result of the cyclotron energy with m =0.04m_0. The dashed curve indicates a theoretical curve of the spin- ip transition between (14, ↓) and (14, ↑) with the Rashba spin-orbit coupling constant, =9 10 ^{12}eVm and $|g|$=7.

4. Summary

We studied the CR in the InGaAs/InAlAs inverted heterostructure at the millimeter wave region. Single CR is observed at higher frequency region, and the e ective mass is obtained as m =0.04m_0 from the magnetic eld dependence of the CR. The additional resonance appears below f=184GHz at the high magnetic eld side of the CR. This resonance is probably due to the spin- ip transition between di erent spin levels. The resonant coupling of Landau levels through the large spin-orbit interaction is essential to explain the anomalous magnetic eld dependence of the CR in the two-dimensional Rashba system.

5. References

[1] J. Nitta, T. Akazaki, H. Takayanagi and T. Enoki, Phys. Rev. Lett. **78**, 1335 (1997).
[2] Th. Schäpers, G. Engels, J. Lange, Th. Klocke, M. Hollfelder and H. Lüth, J. Appl. Phys. **83**, 4324 (1998).
[3] J. Luo, H. Munekata, F. F. Fang and P. J. Stiles, Phys. Rev. B **41**, 7685 (1990).
[4] H. Choi, T. Kakegawa, M. Akabori. T. Suzuki and S. Yamada, Physica E **40**, 2823 (2008).
[5] K. Fujii, Y. Morikami, T. Ohyama, S. Gozu and S. Yamada, Physica E **12**, 432 (2002).
[6] M. Mola, S. Hill, P. Goy and M. Gross, Rev. Sci. Instrum. **71**, 186 (2000).
[7] V. I. Fal'ko, Phys. Rev. B **46**, 4320 (1992).
[8] G. Karczewski and Y. J. Wang, Physica B **298**, 402 (2001).

Acknowledgments
We wish to acknowledge the support of the sample preparation for Dr. M. Akabori. This work was supported in part by Grant-in-Aid for Scienti c Research (C) 20612011, from the Ministry of Education, Culture, Sports, Science and Technology (MEXT).

Journal of Physics: Conference Series **334** (2011) 012062 doi:10.1088/1742-6596/334/1/012062

Anisotropic spin transport affected by competition between spin orbit interaction and Zeeman effect in an InGaAs based wire

Junsaku Nitta[1], Sylvain Moulis[1], and Makoto Kohda[1,2]

[1] *Graduate School of Engineering, Tohoku University, Sendai 980-8579, Japan*
[2] *PRESTO-Japan Science and Technology, Saitama 331-0012, Japan*

E-mail address: nitta@material.tohoku.ac.jp

Abstract. Spin transport affected by competition between Zeeman effect and spin-orbit interaction (SOI) is investigated in order to check a proposed method to deduce the Rashba SOI α and Dresselhaus SOI β ratio. The experimentally obtained ratio α/β of the present sample is about 4 from angle dependence of magnetoconductance under in-plane magnetic field. The proposed method to detect the ratio by transport measurement is promising although further improvement of sample fabrication and measurement is required.

1. Introduction

The spin dynamics in solid is mainly governed by competition between the Zeeman effect and the spin-orbit interaction (SOI). In transport measurements, quantum interference such as weak anti-localization (WAL) is sensitive to the spin dynamics, and WAL analysis is often used to deduce the spin relaxation length. In III-V semiconductor heterostructures, the dominant mechanisms for SOI are the Dresselhaus SOI [1] caused by a bulk inversion asymmetry and the Rashba SOI [2] caused by a structural inversion asymmetry. When both strength of the Rashba and Dresselhaus SOIs become equal with each other, the D'yakonov-Perel' (DP) spin relaxation [3] is completely suppressed, and spin relaxation time is significantly enhanced, generating persistent spin helix state (PSH) [4], [5]. We have observed large enhancement of spin relaxation time by making the Rashba SOI parameter close to the Dresselhaus SOI parameter [6]. To realize the PSH state in semiconductor two dimensional electron gas (2DEG), precise control and electrical detection of the relative strength of the Rashba and Dresselhaus SOIs are of great importance. It is possible to determine the ratio between Rashba and Dresselhaus SOIs by using optical techniques [7-9], this is not always an option. If, e.g., the semiconductor heterostructure is covered with a top gate to tune the Rashba SOI strength, it is very difficult to carry out optical measurements. Recently, we have proposed a method to determine the relative strength of these two SOIs from transport measurements with competition between the SOIs and the Zeeman effect [10]. In this paper, we discuss the all-electrical determination of the relative strength of Rashba and Dresselhaus SOIs in an InGaAs wire structure.

2. Experiment and Discussion

Recent theories [11], [12] have predicted that a narrow wire can suppress the DP spin relaxation, and the wire can be treated as quasi-one dimensional if the wire width is shorter than the spin diffusion length. The effective magnetic field direction induced the Rashba SOI is perpendicular to the electron momentum direction, while the effective field direction due to the Dresselhaus SOI is [x, -y] direction with the electron momentum direction [x, y]. If the wire is quasi-one dimensional, the total effective field with both the Rashba and Dresselhaus SOIs is opposite for electron travelling in opposite wire directions. In this uni-axial effective field condition, the spin is conserved quantity. Therefore, no additional phase in the wave function is acquired by electrons returning to their original position and WAL is suppressed, leading to weak localization (WL). However, the uni-axial field condition is broken when a finite in-plane magnetic field is applied. If the direction of the in-plane field is in parallel to the SOIs total effective field, the uni-axial condition is preserved. Here we can expect the conductance minimum with this special in-plane field direction, which is given by the following equation [10]. The in-plane field should be the same order but sufficiently smaller than the effective field originated from the SOIs in order to obtain the accurate value of α/β ratio.

$$\theta_{\min} = \arctan\left(-\frac{\alpha\cos\phi+\beta\sin\phi}{\beta\cos\phi+\alpha\sin\phi}\right) \quad . \tag{1}$$

Here, ϕ and θ are the angles for wire and applied in-plane field directions from [100] orientation of the crystal. $\theta_{\min}$ is the special angle of in-plane field which provides the conductance minimum.

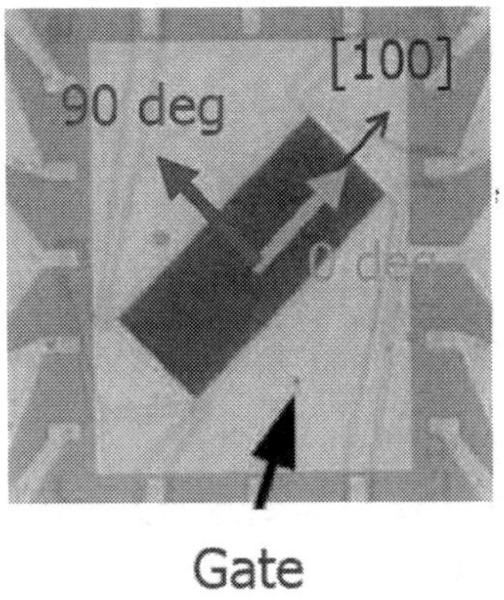

Figure 1. Photograph of fabricated sample. The 600 μm-long and 480 nm-wide 120 wires are defined along [100] direction. The wires are covered with gate electrode. In the present experiment the gate voltage was fixed at V_g = -4.5 V. Mganetoconductance was measured as a function of in-plane field angle from [100] direction.

The sample was epitaxially grown on (100) InP substrate by metal organic chemical vapor deposition. It consists of, from the top, 5 nm $In_{0.52}Al_{0.48}As$ /1.5 nm AlAs /20 nm $In_{0.52}Al_{0.48}As$ /5 nm InP for barrier /5 nm $In_{0.53}Ga_{0.47}As$ /10 nm $In_{0.7}Ga_{0.3}As$ for QW / 2.5 nm $In_{0.53}Ga_{0.47}As$ /15 nm $In_{0.52}Al_{0.48}As$ for spacer /6 nm n-$In_{0.52}Al_{0.48}As$ for carrier supply (Si doping: $N_d = 4\times10^{18}cm^{-3}$) / 200 nm $In_{0.52}Al_{0.48}As$ for buffer layer. The epitaxial wafer was processed into wires along [100] direction by electron beam lithography as shown in Fig. 1. This is because the cubic Dresselhaus SOI effect, which has four-fold symmetry, is minimized along [100] direction. The width of the wire is 480 nm and the sample consists of 120 wires to reduce the universal conductance fluctuation.

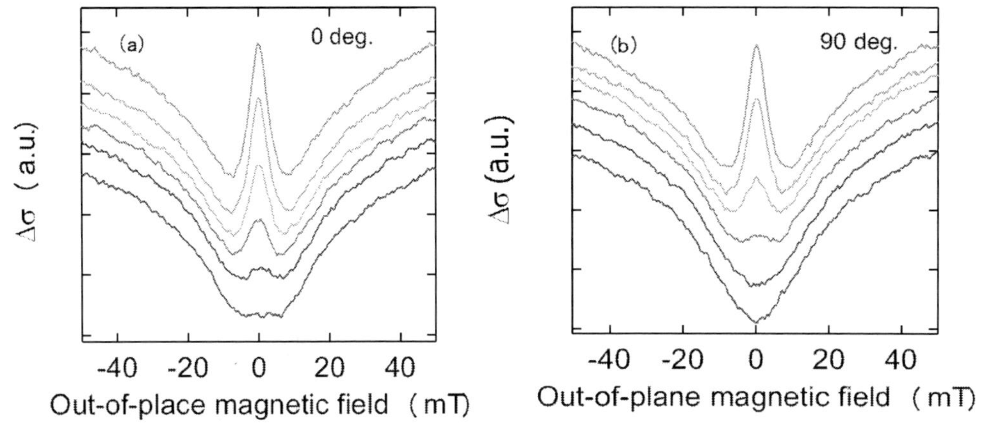

Figure 2. Magnetoconductance as a function of in-plane field strength (a) for parallel (0 deg) and (b) for perpendicular (90 deg) to the wire direction [100] at T= 1.7 K. In-plane field is increased from 0 T (top) to 1 T (bottom) with a step of 0.2 T. In-plane field showing transition from WAL to WL for parallel case is higher than that for perpendicular

We measured magnetoresitance in the fabricated wires with a fixed gate bias voltages, V_g= -4.5 V. The external magnetic field was applied perpendicular to the QW plane. The Rashba SOI parameter α was derived to be 3.9 x 10^{-12} eVm from the WAL of the magnetoresitance measurement in an ordinary Hall bar device made of the same wafer. It should be noted that the Rashba parameter is negative value because of the inverted type modulation doped heterostructure. Sheet carrier density and electron mobility are determined to be Ns= 8.7 x 10^{11} cm^{-2} and μ= 87,900 cm^2/Vs by Shubnikov de Haas oscillations and sheet resistance for the Hall bar device. All the measurements were preformed with a He^4 cryostat refrigerator at T = 1.7 K.

The magnetoconductance data as a function of in-plane field strength for parallel and perpendicular to the wires are shown in Fig. 2 (a) and (b). A transition from WAL to weak localization (WL) was observed in both cases. The transition occurred around B_{in}= 0.9 T for the parallel case while it was observed around B_{in}= 0.6 T for the perpendicular case. This anisotropy shows that the Dresselhaus SOI coexists with the Rashba SOI. The difference in the transition field indicates that the Rashba SOI field is stronger than the Dresselhaus one. This is because the Rashba SOI field direction is perpendicular to the wire while the Dresselhaus one is parallel. The applied in-plane field was decided to be 0.45 T for more detailed angle dependence measurement since the magnetoconductance data in both directions showed clear anisotropy at B_{in}= 0.45 T.

Figure 3 shows a color plot for the in-plane field angle dependence of magnetoconductance with B_{in} = 0.45 T. The angle dependence shows two-fold symmetry about 360 deg rotation. It is clear that WAL peaks are most and least suppressed at θ= 75 deg and θ= 165 deg, respectively. As expected from Fig. 3, the spin relaxation lengths obtained from WAL analysis for narrow wire [9] are maximum at θ= 75 deg and minimum at θ= 165 deg. This result suggests that the total SOI induced field is directed along θ= 75 deg. From eq. (1), α/β ratio is obtained to be about 3 – 5.4 by considering that the angle accuracy of in-plane field is ± 4 deg. The experimentally obtained Dresselhaus SOI parameter β is 0.7 x 10^{-12} – 1.3 x 10^{-12} eVm.

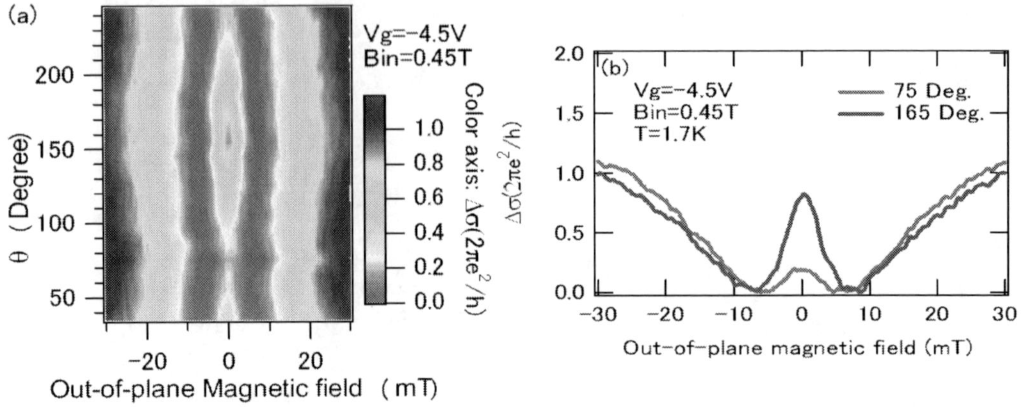

Figure 3. (a) Color plot for the in-plane field angle dependence of magnetoconductance with B_{in} = 0.45 T. (b) WAL peaks are most and least suppressed at θ= 75 deg and θ= 165 deg. This indicates that spin relaxation length at θ= 165 deg is longer than that at θ= 75 deg.

The Dresselhaus paramcter is theoretically given by $\beta = \gamma\langle k_z^2\rangle \approx \gamma(\pi / d_{QW})^2$ and is estimated to be 0.88 x 10^{-12} eVm with an assumption of cubic Dresselhaus parameter γ= 27.38 x 10^{-30} eVm^3 [13]. The experimentally obtained β is consistent with the estimated value. However, the cubic Dresselhaus parameter in GaAs has been reported in a wide range of γ= 5 x 10^{-30} – 27 x 10^{-30} eVm^3 among experiments and theories. The value of cubic Dresselhaus parameter γ should be material constant. The obtained value γ in this experiment is rather high comparing with recent data (γ= 5 x 10^{-30}eVm^3) obtained in GaAs [5]. In the present InGaAs channel is not lattice matched to the InP substrate. We have to think about the possibility of a strain induced SOI effect which is reported not to be negligible in lattice mismatch systems [14]. It should be also noted that the wire width should be 5 times smaller than the spin precession length $L_{so} = \hbar^2 / m^*\alpha$. The effective width W of the present sample is W= L_{so}/2.5. Further improvement of sample fabrication is required.

3. Summary

We have investigated spin transport affected by competition between Zeeman effect and SOIs. The angle dependence of manetoconductance under in-plane magnetic field shows anisotropy, leading to the α/β ratio. The proposed method to detect the ratio by this transport measurement is promising although further improvement of sample fabrication and measurement is required. It is possible to confirm the PSH condition by the present method. The PSH state will be realized by controlling the Rashba SOI by a gate voltage [15].

This work was partly supported by Grants-in-Aid from the Japan Society for the Promotion of Science (JSPS) and the Ministry of Education, Culture, Sports, Science and Technology (MEXT). JN acknowledges support from Strategic Japanese-German Joint Research Program.

References

[1] Dresselhaus G 1955 *Phys. Rev.* **100** 580
[2] Rashba E I 1960 *Sov. Phys. Solid State* **2** 1224
[3] D'yakonov M I and Perel' V I 1971 *Sov. Phys. Solid State* **13** 3023
[4] Bernevig B A, Orenstein J, and Zhang S-C 2006 *Phys. Rev. Lett.* **97** 236601
[5] Koralek D, Weber C P, Orenstein J, Bernevig B A, Zhang S-C, Mack S, and Awschalom D D 2009 Nature **458** 610
[6] Kunihashi Y, Kohda M, and Nitta J 2009 *Phys. Rev. Lett.* **102** 226601
[7] Giglberger *et al.* 2007 *Phys. Rev. B* **75** 035327
[8] Ganichev S D, *et al.* 2004 *Phys. Rev. Lett.* **92** 256601
[9] Meir L, et al. 2007 *Nature Phys.* **3** 650
[10] Scheid M, Kohda M, Kunihashi Y, Richter K, and Nitta J 2008 *Phys. Rev. Lett.* **101** 266401
[11] Mal'shukov A G and Chao K A 2000 *Phys. Rev. B* **61** R2413
[12] Kettemann S 2007 *Phys. Rev. Lett.* **98** 176808
[13] Winkler R 2003 *Springer Tracts in Modern Physics* **191** Berlin, Germany: Springer-Verlag *"Spin-orbit coupling effects in two-dimensional electron and hole systems"*
[14] Norman B, Trowbridge C, Stephens J, Gossard A C, Awschalom D D, and Sih V 2010 *Phys. Rev. B* **82** 081304.
[15] Nitta J, Akazaki T, Takayanagi H, and Enoki T 1997 *Phys. Rev. Lett.* **78** 1335

Magnetic field dependency of spin-splitting in $In_{0.75}Ga_{0.25}As$/ $In_{0.75}Al_{0.25}As$ two dimensional electron gas with strong Rashba spin-orbit coupling

Syoji Yamada[1†], **Shunsaku Nitta**[1], **Hiuma Iwase**[1], **Masashi Akabori**[1], **Yasutaka Imanaka**[2] **and Tadashi Takamasu**[2]

[1]*CNMT JAIST, 1-1, Asahidai, Nomi, Ishikawa 923-1292 Japan*
[2]*NIMS, 3-13, Sakura, Tsukuba, Ibaraki 305-0003 Japan*
e-mail address: shooji@jaist.ac.jp

Abstract. High-field magneto-resistance (MR) properties including Shubnikov de-Haas (SdH) oscillation and integer quantum Hall effect are observed and analyzed in high In-content InGaAs/InAlAs two-dimensional electron gas (2DEG) with strong Rashba effect. From the results, conventional longitudinal resistance oscillations with Zeeman splittng as well as quantum Hall plateaus are confirmed and a transition from zero-field spin-splitting to high-field Zeeman splitting is discussed. The two-kind spin splittings are seem to be connected via the sign changing process at finite (~3 Tesla) magnetic field, the value of which is about two times larger than that in high In-content InGaSb/InAlSb system.

1. Introduction

In 2DEG formed at narrow-gap semiconductor hetero-junction often reveals zero-field spin-splitting mainly due to Rashba-type spin-orbit interaction (SOI). This spin-splitting, Δ_R, should move to Zeeman splitting, Δ_Z, at higher fields via a particular transition process with or without of change of sign [1, 2]. Recently, we have obtained interesting results in high In-content InGaSb/InAlSb heterojunction 2DEG [3] suggesting zero spin-splitting at a finite magnetic field (~1.5 Tesla (T)) due to the transition of the sign from the zero-field to Zeeman splitting [1]. This is interpreted by the appearance of Dresselhaus-type zero-field spin splitting which is unexpectedly large in such a narrow gap material system.

However, the transition process in $In_{0.75}Ga_{0.25}As/In_{0.75}Al_{0.25}As$ system, which reveals rather stronger Rashba SOI, has not been clarified yet except the preliminary results by our group [4]. In this paper, we report longitudinal and Hall MRs in similar heterojunctions measured up to ~30 T at 1.5 K, and study the quantum Hall effect, the report of which is rare, as well as the field dependency of spin splitting in this unique material system.

2. Sample and Experiment

The measured samples are the inverted-type high electron mobility transistor (HEMT) structures. As shown in Figure 1, $d_{channel}$ = 60 nm thick InGaAs channel is located at the top most of the layered structure, which is composed of 220 nm InAlAs carrier supply layer followed by step-graded $In_xAl_{1-x}As$+AlAs+GaAs buffer layers (~1800 nm) grown on semi-insulating (001) GaAs substrate. High-field MR measurements are carried out by using a conventional superconducting magnet of 8 T at JAIST or a hybrid magnet of 27 T at NIMS at 1.5-4 K with AC or DC method. Typical sheet electron densities and mobilities of the samples at 1.5 K are n_s = 6-10$\times 10^{11}$ /cm^2 and μ_e = 1.5-2.0$\times 10^5$ cm^2/Vsec, respectively. Rashba SO coupling constant α_s determined from the beat analysis of the low field SdH oscillations are about 10-20$\times 10^{-12}$ eVm corresponding to ~8 meV spin splitting [5] at zero-field. Those μ_e and α_s values are almost comparable to the record ones in III-V compound semiconductor narrow-gap systems.

3. Results and Discussions

Figure 2 is one example of MR results: The R_{xx} (left upper panel) shows clear SdH oscillation and the R_{xy} (left lower panel) shows quantum Hall effect. However, phase change occurs at ~5 Tesla in R_{xx} oscillation and this reveals as a kink in Landau plot (right panel) for the oscillation. The two plot lines give two kind n_s values of ~7 and ~10×10^{11} /cm^2 each corresponding to the ground subband and total

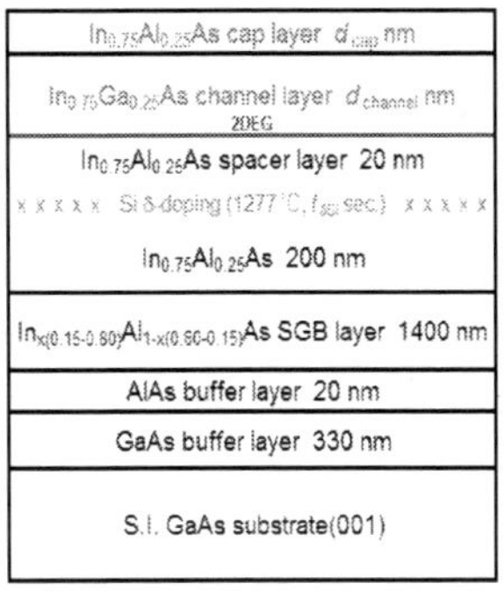

Fig. 1 Typical layered structure of $In_{0.75}Ga_{0.25}As/In_{0.75}Al_{0.25}As$ inverted HEMT structure..

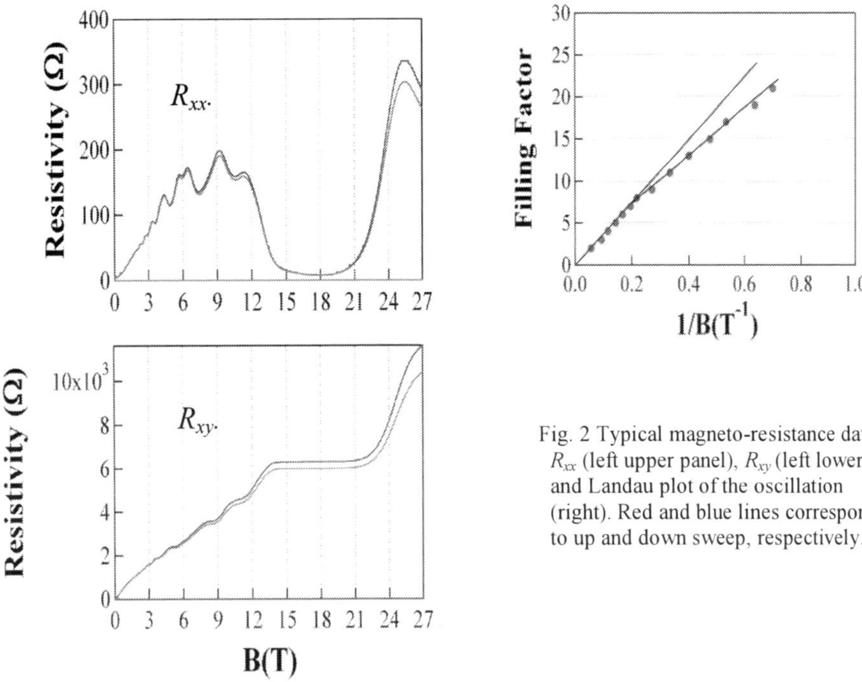

Fig. 2 Typical magneto-resistance data R_{xx} (left upper panel), R_{xy} (left lower) and Landau plot of the oscillation (right). Red and blue lines correspond to up and down sweep, respectively.

sheet electron densities, respectively. In R_{xy} trace, there observed clear and flat integer quantum Hall plateaus suggesting high quality features of our 2DEG system. However, the plateau resistance values are irregular: For example, the resistance of 18 Tesla plateau is ~6.4 kohm (~h/ie^2, $i=4$, the filling factor), while the filling factor in the Landau plot is two. This is probably due to the second subband occupation of the 2DEG in this sample, while the accurate reason for this result is not clear yet.

Figure 3 is another example of MR results: Again the R_{xx} shows clear SdH oscillation with almost no background. Since the n_s is relatively small, the oscillation is rather simple suggesting a single sub-band occupation. As seen in the figure, two types of R_{xx} were recorded depending on the different in-plane directions, [110] and [-110] of current flow. The peak amplitudes are very much different probably corresponding to the mobility anisotropy [6]. However, the amplitude ratio of 700 % is much larger than the mobility ratio of 130 %. The reason is not clear yet but this might help to investigate the origins of the in-plane mobility anisotropy itself. R_{xy} reveals distinct integer quantum Hall plateaus corresponding to the SdH oscillation, although the resistance values of which are still slightly deviated from the standard values, ~h/ie^2, i the integer.

From those MR data, we estimated magnetic field dependencies of Δ_Z and a g-factor in this 2DEG system. In this analysis, we have calculated Landau level energies as below,

$$E_{n\pm} = \hbar\omega_c[n \pm (\delta^2 + \gamma^2 n)^{1/2}] \text{ for } n = 1,2,3,,, \quad E_0 = \hbar\omega_c\delta$$

$$\delta = [1 - gm/(2m_0)]/2, \quad \gamma = (2m\alpha_s{}^2/\hbar^3\omega_c)^{1/2} = (2/\hbar^3 eB)^{1/2} m\alpha_s$$

by taking the effect of Rashba SO coupling into account and g-factor was a fitting parameter in this calculation. In other words, we determined g-factor so as to reproduce the experimental Zeeman splittings by the above calculation for spin-splitted Landau levels, $E_{n\pm}$. Here m/m_0 = 0.04, spin-orbit coupling constant α_s = 20x10^{-12} eVm are assumed and the dispersion relation of the Rashba electrons is

$$E(k) = \frac{\hbar^2 k^2}{2m} \pm \alpha_s k \quad for \quad H = \frac{\hbar^2 k^2}{2m} + \alpha_s(\vec{\sigma} \times \vec{k}) \bullet \vec{z}.$$

In Fig. 4, obtained Δ_Z and g-factor obtained for three set of Zeeman splitted peaks are plotted as functions of applied magnetic field, B. In both of the cases, non-linear dependencies on B are confirmed. Those dependencies are slightly different from that obtained formerly [4] and the magnitudes are rather smaller. Although the behaviours in very low and very high fields are still unknown yet, the data in Fig. 4 likely suggest the change of sign [1] between the two-type spin-splittings also in this 2DEG system. However, the field at which the splitting crosses zero (~3 Tesla) is found to be two times larger than that in InGaSb/InAlSb system [3].

If we discuss the origin by comparing with the InGaSb/InAlSb case, we can point out that the contribution of the Dresselhaus term is expected to be relatively small in InGaAs/InAlAs case. In fact, the contribution of the term was reported to be about one third of that of the Rashba in similar InGaAs/InAlAs system [7]. This ratio seems to give the upper limit of the Dresselhaus contribution, since we consider the rather thick quantum well in this sample, where the contribution is expected to decrease. This supports the Landau level calculation assuming only the Rashba term. In contrast, in InGaSb/ InAlSb 2DEG, the Rashba term is almost one half of the Dresselhaus one [3]. This difference could explain the difference in the zero spin splitting fields between the two material systems.

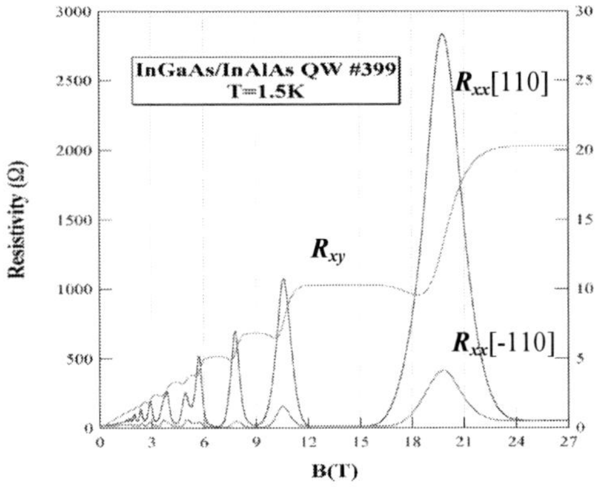

Fig. 3 Magneto-resistance data in small n_s sample: R_{xx}[110] , R_{xx}[-110] and R_{xy}. Anisotropic R_{xx}s are observed.

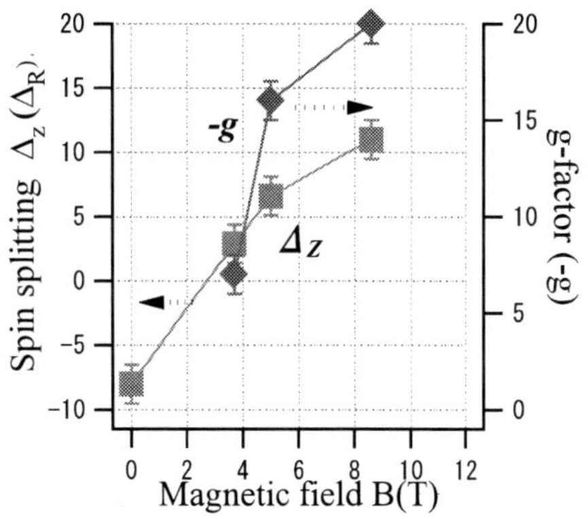

Fig. 4 Spin splittings Δ_Z and g-factor plotted as functions of applied magnetic field B. At $B = 0$, $\Delta_Z = \Delta_R$ = 8 meV.

Conclusion

We have measured high field MRs in high In-content InGaAs/InAlAs 2DEGs and investigated a variety of behaviors in longitudinal (R_{xx}) and Hall (R_{xy}) resistances with Rashba and Zeeman splittings. Clear SdH oscillations as well as corresponding (integer) quantum Hall plateaus are definitely observed and the critical sheet electron density over which the occupation of first excited subband occurs is found to be $\sim 6 \times 10^{11}/\text{cm}^2$ in this material system. The field dependencies of Zeeman splitting and g-factor are also derived by fitting calculated Landau level spectra to the spin-splitted SdH oscillation. Those results together with the zero-field splitting value seem to suggest the change of sign at finite (~3 T) field between the zero field and Zeeman spin splittings.

Reference

[1] G. Lommer et al., Phys. Rev. Lett., 60, 728 (1988).
[2] P. Pfeffer et al., Phys. Rev. B68, 035315 (2003).
[3] M. Akabori et al., Phys. Rev. B77, 205320 (2008).
[4] T. Kita et al., Physica B 298, 65 (2001).
[5] H. K. Choi et al., Physica E40, 2823 (2008).
[6] S. Nitta et al., Physica E42, 987 (2010).
[7] T.Koga, private communications.

HMF-19 IOP Publishing
Journal of Physics: Conference Series **334** (2011) 012064 doi:10.1088/1742-6596/334/1/012064

Tunable magnetic property of lateral quantum dot molecules

Fanyao Qu[1], D. R. Santos Jr.[1], R. B. Azevedo[2], V. López-Richard[3], and G. E. Marques[3]

(1) Institute of Physics, University of Brasília, Brasília- DF, 70910-900, Brazil
(2) Institute of Biology, University of Brasília, Brasília- DF, 70910-900, Brazil
(3) Departamento de Física, Universidade Federal de São Carlos, São Carlos, SP 13565-905, Brazil

E-mail: fanyao@unb.br

Abstract. We investigate theoretically magnetic properties of tunable lateral double quantum dots (DQDs) containing few electrons, subjected to spin-orbit interaction (SOI) and vertical magnetic field, using exact diagonalization (ED) method. To circumvent the cumbersome construction of matrix elements of the Hamiltonian, we adopt the modified Gaussian functions to describe atomic-like orbital states of single quantum dots. The numerical artifacts in the single-electron energy spectrum are avoided by separating the full Hilbert space into two sub-spaces. The interplay of SOI and interdot tunnel coupling leads to an oscillatory behaviour in the magnetization at low temperature. Sweeping interdot barrier voltage and adjusting interdot distance produce not only magnetic phase transition but also switches between atomic and molecular orbitals.

1. Introduction

The dependence of electronic structure of nanoscale system on their size and shape offers a great opportunity to tailor the properties of these materials [1,2]. The appearance of magnetism in otherwise nonmagnetic materials has recently been reported for semiconductor or graphene few electron quantum dots (QDs) which should not be magnetic, yet previous studies have confirmed that they may present paramagnetic response to an external magnetic field [3,4,5]. Together with the size-dependent optical and electronic properties of the QDs this magnetic behavior opens up the possibility for an extended range of technological applications and fascinating fundamental physics, which is called as "spintronics without magnetism". Spin-orbit interaction (SOI) removes the spin degeneracy of the energy levels and alters the electronic structure of QDs, acting as an "effective internal magnetic field" [6]. Theoretical studies and experimental observations have pointed out that SOI can be employed to manipulate spins as well as magnetic properties of single QDs [7].

Tunable double quantum dots (DQDs) have been fabricated in many different material systems, as for example in semiconductor heterostructures [8], semiconductor nanowires [9-11], graphenes [12] and carbon nanotubes [13]. They constitute particularly attractive blocks for practical implementation of spin-related quantum information devices due to an achievement of externally controlled interdot tunnel-coupling (TC) and electron spin-states [14-20]. Furthermore, tunneling spectroscopy measurement has demonstrated the tunability of the magnetization of DQDs [21]. Although a significant progress towards the conception of electrically addressable magnetic memories has been made experimentally and a few of theoretical studies on vertical double QDs have been performed [21], a comprehensive theory about the interplay between SOI and interdot TC controlled by in-plane gate voltages, on magnetism of lateral DQDs has been missing. This has motivated us to perform an exact diagonalization (ED) calculation in tunable lateral DQDs. We find that tuning either local interbarrier voltage or interdot distance can induce giant changes in electronic structure and spin-states of DQDs. In addition, the changes of electronic structure of DQDs are strongly mediated by SOI. Therefore, the magnetic property of DQDs can be tuned externally by in-plane local gate voltages and vertical electrical field.

2. Theoretical model

We start off by describing our model for a single particle confined in coupled InAs DQDs, subject to both Rashba SOI and the perpendicular $\vec{B}$ using the Hamiltonian $H = H_0 + H_z + H_{SO}$. In the Hamiltonian, the first term, $H_0 = \hat{\pi}^2/2m^* + V_{xy}(x,y) + V_z(z)$, is composed of kinetic energy and confinement potentials in the *x-y* plane and along the *z*-direction, where $\hat{\pi} = -i\hbar\vec{\nabla} + e\vec{A}(\vec{\rho})$ is the in-plane canonical momentum, $\vec{A}(\vec{\rho}) = \vec{B} \times \vec{\rho}/2$ is the vector potential and $\vec{\rho} = (x,y)$, e, m^*/m_0 ($\equiv m_{ef}$) and m_0 denote the in-plane coordinate, electron charge, effective and free electron mass, respectively. The second term, $H_z = g\,\mu_B\vec{\sigma}\cdot\vec{B}$, is the Zeeman coupling, where g, $\vec{\sigma}$ and $\mu_B = e\hbar/(2\,m_0)$ are the material bulk *g*-factor, Pauli spin vector and Bohr magneton, respectively. The third term, $H_{SO}^{(R)} = i\alpha(\sigma_+\pi_- - \sigma_-\pi_+)$, with $\pi_\pm = \pi_x \pm i\pi_y$ and $\sigma_\pm = \sigma'_x \pm i\sigma'_y$, describes the Rashba SOI, whose strength α can be controlled by the degree of asymmetry of the quantum well along the *z*-direction, and is measured in units of meV·Å, and $\vec{\sigma}' = \vec{\sigma}/2$. The lateral DQD's potential $V_{xy}(x,y)$, for a system defined by etched electrostatic metallic gates on the *x-y* plane, can be characterized by two localization regions and simulated by a sum of three Gaussian functions [14],

$$V_{xy}(x,y) = V_L exp\left[-\frac{(x+a)^2+y^2}{\Delta^2}\right] + V_B exp\left[-\frac{x^2}{\Delta_{b_x}^2} - \frac{y^2}{\Delta_{b_y}^2}\right] + V_R exp\left[-\frac{(x-a)^2+y^2}{\Delta^2}\right], \quad (1)$$

Where V_L (V_R) controls the depth of the left (right) side dot with line-width parameter Δ located at $x = -a$ ($x = +a$) in the $y = 0$ plane, whereas V_B tunes the height of the interdot repulsive potential barrier whose linewidth is denoted by Δ_{b_x} and Δ_{b_y}. In what follows, the physical quantities are expressed in effective atomic units, which for InAs QDs the effective Bohr radii and Rydberg energy are equal to 342.08 Å and 2.72 meV, respectively. Unless otherwise specified $V_L = V_R = V_0$, Δ= 2.5, Δ_{b_x}= 0.3, Δ_{b_y}= 2.5 and $g = -14.7$ will be adopted throughout this paper. From Eq. (1), we note that by changing the gate potentials in the DQD structure the system can be driven from a single quantum dot to a coupled double dots. We assume that only the lowest subband in the *z*-direction is populated and that the motion along in-plane and perpendicular directions are decoupled [13,14]. The eigenvalue and eigenfunctions of H_0, which is similar to an H_2^+ molecule, are obtained by ED calculation in the frame of linear combination of atomic orbital's (LCAO) in which the wavefunctions are constructed from a Gaussian type of functions (GFs) $\varphi_{n_x,n_y}^{(\pm)} = \left(\frac{x\mp a}{\lambda}\right)^{n_x}\left(\frac{y}{\lambda}\right)^{n_y} exp\left[-\frac{(x\mp a)^2+y^2}{2\lambda^2}\right] exp\left[\pm\frac{iya}{2l_B^2}\right]$ centred in each dot, where $n_x, n_y = 0,1,2,\cdots$, $l_B = \sqrt{\hbar/(m^*\omega_c)}$ and $\lambda = \sqrt{\hbar/(m^*\Omega)}$ are respectively magnetic- and effective harmonic oscillator characteristic lengths with $\omega_c = eB/m^*$, $\Omega^2 = \omega_0^2 + \omega_c^2/4$, and the phase factor in GFs is due to Gauge transformation in both left and right QDs, where $\omega_0^2 = 2\,|V_0\,|/(m^*\Delta^2)$ for $V_L = V_R = V_0$. To easily catch on the underlying physics and eventually reach a quick convergence of numerical calculation we use molecular orbital $|\Psi_k^{(\pm)}\rangle$, instead of atomic orbital GFs as basis set to construct the trial wavefunctions $|\Psi_n\rangle$, i.e. $|\Psi_n\rangle = \sum_{i=1}^{n_s} c_i^n\,|\psi_i^{(+)}\rangle + \sum_{j=ns+1}^{2\,n_s} c_j^n\,|\psi_j^{(-)}\rangle$, where $|\psi_k^{(\pm)}\rangle = \frac{|\varphi_k^{(+)}\rangle \pm |\varphi_k^{(-)}\rangle}{\sqrt{2(1\pm s_k)}}$, with the compact form for the quantum number $k = (n_x, n_y)$. Notice that s_k is the overlap integral between $\varphi_k^{(+)}$ and $\varphi_k^{(-)}$ states and n_s representing the number of GFs in each dot. Due to the non-orthogonality of the basis functions, a generalized eigenvalue problem $H^{(0)}c = E^{(0)}S\,c$ has to be handled, where $H^{(0)}$ and S, respectively, represent the matrix of the Hamiltonian H_0 and the overlap integral between basis functions, $E^{(0)}$ is the eigenvalue and c is a matrix vector whose components are the coefficients of linear combination. In a similar way, the matrix elements of total Hamiltonian can be straightforwardly derived using $|\Psi\rangle = \sum_{\sigma'_z=\uparrow,\downarrow}\sum_{j=+,-}\sum_{i=1}^{n_s} c_{i,j,\sigma'_z} c_i^n\,|\psi_i^{(j)}\rangle\,|\sigma'_z\rangle$. A convergence of the calculation is reached for $n_s = 42$, corresponding to $n_x, n_y = 0,\cdots,5$. Since $|\psi_k^{(+)}\rangle$ ($|\psi_k^{(-)}\rangle$)

possess definite parity and SOI only hybridizes the states with different parity and opposite spin, it allows us to divide the full Hilbert space into two orthogonal Hilbert subspaces. For instance, we use the hybridized spin-up (-down) symmetric and spin-down (-up) antisymmetric states to construct the Hilbert subspace I (II). In addition, it is important to remark that the Gaussian type of basis set takes great advantage over the harmonic orbitals (HOs), as usually used in the literature. For instance, when more orbitals are required to be included in basis set of HOs in order to reach higher accuracy results the brute-force numerical calculation of matrix elements becomes cumbersome. This problem is circumvented by using the set of GFs.

3. Results and discussion

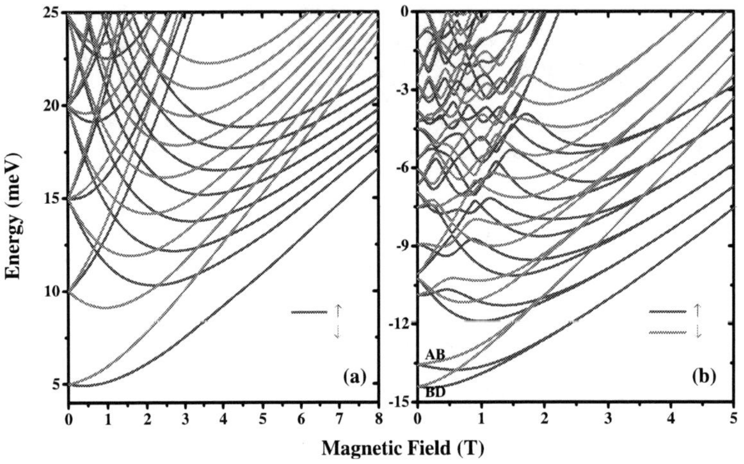

Figure 1. Energy spectrum of an electron in single parabolic quantum dot characterized by $\omega_0 = 5$ meV (a) and coupled DQD (b) defined by $V_0 = -5$, $V_B = 3.5$, $a = 1.5$, as a function of magnetic field. At zero magnetic field, they exhibit well separated electronic shells as indicated by *s*, *p*, *d*, and at high magnetic fields the Landau levels (LLs) are formed. The *BD* and *AB* represent bonding and antibonding molecular states. The blue and red lines correspond to the states with spin-up and down, respectively.

Figure 1 shows electron energy spectrum of single parabolic QD characterized by $\omega_0 = 5$ meV (a) and of a DQD defined by $V_0 = -5$, $V_B = 3.5$, $a = 1.5$ (b) as a function of magnetic field for $\alpha = 0$. It is noted that at zero magnetic field, the single QDs exhibit well separated electronic shells with the degeneracy of $2(n_+ + n_-)$, as indicated by *s*, *p*, *d*,..., where n_+ and n_- are the quantum numbers and $n_+, n_- = 0, 1, 2, \cdots$. The energy gap ΔE between consecutive shells is constant which is equal to $\omega_0 =$ 5 meV. The spin and orbital degeneracies are removed by a finite filed. Nevertheless, the accidental degeneracy presents at critical magnetic field B_c, which is determined by $B_c = m_{ef}\hbar\omega_0/\mu_B\sqrt{|g|\, m_{ef}\,(2 + |g|\, m_{ef})}$ for the first crossing. At high magnetic field, the well defined Landau levels (LLs) constructed by the harmonic orbitals with the same n_+, but different n_-, are formed. In strong contrast, at $B = 0$, the bonding and anti-bonding molecule states of DQDs, stemmed from interdot tunnelling and the hybridization of different electronic orbitals are formed, as indicated by *BD* and *AB*. For finite magnetic field, interdot coupling in a combination with Zeeman and diamagnetic shift leads to the anticrossings (crossings) due to the couplings between the states with the opposite (same) parity and spin-orientation at $B = B_c^{(ac)} (B = B_c^{(c)})$, such as, the first crossing take place at $B = 0.606$ T, where the level separation Δ_{BAB} between bonding and antibonding states of *s*- orbital is equal to the Zeeman splitting energy. With a further increasing B, the quantum confinement induced by magnetic field suppresses the splitting of *BD* and its partner *AB* states. The electronic shells constituted by closely spaced pairs of levels with opposite parity emerge. The atomic characteristic behaviour of single QDs is restored. It indicates that the competition between interdot coupling and quantum confinement induced by magnetic field makes the electronic states evolve continuously from a molecular characteristic state to an atomic-like state.

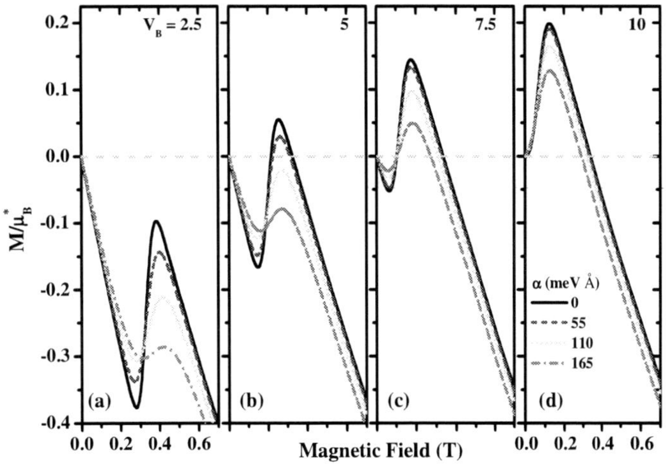

Figure 2. Impact of SOI strength α and positive interdot barrier voltage V_B on magnetization of two-electron DQD defined by $V_0 = -10$ and $a = 2$, at $T = 0.1$ K.

Figure 2 shows the dependence of magnetization on α and V_B for two-electron DQDs defined by $V_0 = -10$ and $a = 2$, at $T = 0.1$ K. Notice that the magnetization presents a "*s*"- like oscillating behavior for strong coupled DQDs ($V_B \leq 7.5$), i.e., as B increases, the magnetization decreases initially, then reaches its local minimum value, after that it abruptly increases till reaches its maximum value, eventually it decreases again. Hence, two magnetic phase transitions take place consecutively, i.e. diamagnetic to paramagnetic and then to diamagnetic. Notice also that both both α and V_B strongly affect the magnetization. For a certain V_B, such as $V_B = 5$, the SOI only very slightly impacts on the magnetization in both lower and higher magnetic fields. However, the effect is strongly enhanced in the vicinity of anti-crossing point $B = B_c^{(ac)}$. An increase in α leads to a less abrupt change of magnetization in the vicinity of $B_c^{(ac)}$ and an enlarged paramagnetic region. The underlying physics of the diamagnetic response and the "*s*"- like behavior is as follows. At $T = 0.1$ K, the ground state of the two-electron system is formed in such a way that one electron occupies the lowest spin-up BD state while the other fills up the first excited molecular state. Since the energy of the lowest single-electron state is only very slightly affected by SOI due to a large separation between BD and AB states, neither anticrossings between them nor an abrupt changes in magnetization is expected. Nevertheless, the first excited single-electron state changes its spin characteristics and slope in the vicinity of the critical field $B = B_c^{(ac)}$ due to the hybridization between the lowest spin-down BD and spin-up AB states. It induces a discontinuity in the derivative of the energy of the first excited state with respect to B, and then a "s"-like behavior of magnetization. In addition, increasing α strengths the hybridization and causes the renormalization of involved energy levels take place in an extended region. Hence, the interval of B in which system presents paramagnetic response increases. In addition to α, V_B also strongly impacts on the magnetization of two electron DQDs. For instance, V_B drives the magnetization curve move upwards, accompanied by a shift of jump position to smaller magnetic fields. In the very weakly coupled DQDs such as the DQD with $V_B = 7.5$, only one paramagnetic to diamagnetic transition remains.

ACKNOWLEDGMENTS: Authors acknowledge the Brazilian agencies: CNPq, FINATEC, CAPES, FAPES and INCT for financial support.

References

[1] S. Neeleshwar, C. L. Chen, C. B. Tsai, and Y. Y. Chen, et. al. 2005 Phys. Rev. B *71, 201307 (R)*
[2] Fanyao Qu and P. Hawrylak 2005 *Phys. Rev. Lett.* **95** 217206; Fanyao Qu and P. Hawrylak 2006 *Phys. Rev. Lett.* **96** 157201; Ramin M. Abolfath and P. Hawrylak 2006 *Phys. Rev. Lett.* **97** 186802
[3] Robert W. Meulenberg, Jonathan R. I. Lee, et. al.2009 *J. Am. Chem. Soc.* **131**, 6888
[4] G. A. Steele, G. Gotz and L. P. Kouwenhoven 2009 *Nature Nanotechnology* **4** 363; S. Moriyama, D.Tsuya, E. Watanabe, et al.2009 *Nano Lett.* **9** (8) 2891
[5] M. M. Ugeda, I. Brihuega, F. Guinea, and J. M. Gómez-Rodríguez 2010 *Phys. Rev. Lett.* **104** 096804
[6] Marian Florescu and Pawel Hawrylak 2006 *Phys. Rev. B* **73** 045304
[7] O. Voskoboynikov, O. Bauga, C. P. Lee and O. Tretyak 2003 *J. Appl. Phys.* **94** 5891
[8] W. G. Van der Wiel, S. De Franceschi., J. M. Elzerman, T. Fujisawa, S. Tarucha and L. P. Kouwenhoven 2002 *Rev. Mod. Phys.* **75** 1
[9] C. Fasth, A. Fuhrer, M. T. Bjork and L. Samuelson 2005 *Nano. Letters* **5,** 1487
[10] A. Pfund, I. Shorubalko, R. Leturcq and K. Ensslin 2006 *Appl. Phys. Lett.* **89** 252106
[11] T. Choi, I. Shorubalko, S. Gustavsson., S. Schon and K. Ensslin 2009 *New J. Phys.* **11** 013005
[12] F. Molitor, S. Droscher, J. Guttinger, A. Jacobsen, C. Stampfer, T. Ihn and K. Ensslin 2009 *Appl. Phys. Lett* **94** 222107
[13] M. J. Biercuk, S. Garaj, N. Mason, J. M. Chow and C. M. Marcus 2005 *Nano Letters* **5** 1267
[14] T. Fujisawa, D. G. Austing, Y. Tokura, Y. Hirayama and S. Tarucha 2002 *Nature* **419** 278
[15] K. Ono, D. G. Austing, Y. Tokura and S. Tarucha 2002 *Science* **297** 1313
[16] J. M. Elzerman, R. Hanson, L. H. Willems van Beveren, B. Witkamp, L. M. K. Vandersypen and L. P. Kouwenhoven 2004 *Nature* **430** 431
[17] J. R. Petta, A. .C. Johnson, J.M. Taylor, E. A. Laird, A. Yacoby, M. D. Lukin, C. M. Marcus, M. P. Hanson and A. C. Gossard 2005 *Science* **309** 2180.
[18] F. H. L. Koppens, C. Buizert, K. J. Tielrooij, I.T. Vink, K.C. Nowack, T. Meunier, L. P. Kouwenhoven and L. M. K. Vandersypen 2006 *Nature* **442** 766.
[19] D. Loss and D. P. DiVincenzo 1998 *Phys. Rev. A* **57** 120
[20] Fanyao Qu, G. I. Luiz, Victor Lopez-Richard and G. E. Marques 2009 *Appl. Phys. Lett.* **95** 083101; P. Stano and J. Fabian 2006 *Phys. Rev. Lett.* **96** 186602; L. Meza-Montes, Carlos F. Destefani and Sergio E. Ulloa 2008 *Phys. Rev. B* **78** 205307.
[21] T. H. Oosterkamp, S. F. Godijn, et. al. 1998 *Phys. Rev. Lett.* **80** 4951.

Symmetric Double Quantum Dot Energy States in a High Magnetic Field

Norman J. Morgenstern Horing[1] **and Makoto Sawamura**[2]
[1]Physics Dept., Stevens Inst. of Tech., Hoboken, N.J. 07030, USA
[2]NIMS - MANA, 1-1 Namiki, Tsukuba, 305-0044, Japan

Abstract. The dynamical Green's function and energy spectrum of a 2D symmetric quantum double-dot system on a planar host in a normal magnetic field are analyzed here, representing the two dots by Dirac delta function potentials. The proliferation of energy levels due to Landau quantization is examined in detail.

Keywords: Quantum-Dot, Double-Dot, Energy Spectrum, Magnetic Field, Landau Quantization, Green's Function

1. Introduction

Recent considerations of newly conceived quantum dot based electronic devices for the transport in nanostructured systems have often assumed that each constituent dot involved in the device would support just one single energetically accessible energy subband level in the absence of a magnetic field, and that the role of the magnetic field can be represented merely in terms of a Peierls phase factor. However, this ignores another important effect of the magnetic field in that it induces a "splintering" of the single subband energy level into a proliferation of many Landau-quantized states. The latter states may be energetically accessible making the situation much more complicated.

This work is concerned with electron states and propagation in a two-dimensional symmetric quantum double-dot system embedded in a two-dimensional host sheet subject to Landau quantization. The two dots are represented by two Dirac delta function potential terms, each of which would support just one subband state if the other were absent, if there were no magnetic field. The integral equation for the Schrödinger Green's function for this symmetric double-dot system is solved exactly in closed form in terms of the infinite sheet Green's function for two dimensional electrons subject to Landau quantization with no quantum dots. The dispersion relation for the symmetric double dot subband energies is formulated and examined by analyzing the frequency poles of the Green's function with Landau-quantization-like splintering of the levels by the magnetic field.

2. Green's Function for a Planar Symmetric Double Quantum Dot in a Normal High Magnetic Field

We consider a normal, planar double quantum dot system[1] on a two dimensional sheet in a perpendicular magnetic field, represented by a double-well potential as ($\alpha < 0$)

$$U(\vec{r}) = \alpha \sum_{\pm} \delta^{(2)}(\vec{r} - \vec{r}_{\pm}), \tag{1}$$

where $\vec{r} = x\hat{i} + y\hat{j}$ and the two potential wells defining the double dot are located at $\vec{r}_{\pm} = \pm\hat{i}d/2$ with strength α representing the product of the well depth U_0 at $\vec{r}_{\pm}$ times its area. The retarded Schrödinger Green's function for electron propagation on the sheet, $G(\vec{r}_1, \vec{r}_2; \omega)$, including the dot regions, obeys the integral equation (frequency representation)

$$G(\vec{r}_1, \vec{r}_2; \omega) = G^B_{2D}(\vec{r}_1, \vec{r}_2; \omega) + \alpha \sum_{\pm} \int d^2\vec{r}_3 G^B_{2D}(\vec{r}_1, \vec{r}_3; \omega)\delta^{(2)}(\vec{r}_3 - \vec{r}_{\pm})G(\vec{r}_3, \vec{r}_2; \omega), \tag{2}$$

where G^B_{2D} is the infinite sheet two dimensional Schrödinger Green's function for electron propagation on the sheet in a magnetic field in the absence of quantum dot potential wells. Integration with respect to $\vec{r}_3$ yields

$$G(\vec{r}_1, \vec{r}_2; \omega) = G^B_{2D}(\vec{r}_1, \vec{r}_2; \omega) + \alpha \sum_{\pm} G^B_{2D}(\vec{r}_1, \vec{r}_{\pm}; \omega)G(\vec{r}_{\pm}, \vec{r}_2; \omega). \tag{3}$$

This algebraic equation is readily solved by setting $\vec{r}_1 = \vec{r}_{\pm'}$,

$$G(\vec{r}_{\pm'}, \vec{r}_2; \omega) = G^B_{2D}(\vec{r}_{\pm'}, \vec{r}_2; \omega) + \alpha \sum_{\pm} G^B_{2D}(\vec{r}_{\pm'}, \vec{r}_{\pm}; \omega)G(\vec{r}_{\pm}, \vec{r}_2; \omega), \tag{4}$$

whence

$$\sum_{\pm} \left[\delta_{\pm',\pm} - \alpha G^B_{2D}(\vec{r}_{\pm'}, \vec{r}_{\pm}; \omega)\right] G(\vec{r}_{\pm}, \vec{r}_2; \omega) = G^B_{2D}(\vec{r}_{\pm'}, \vec{r}_2; \omega). \tag{5}$$

The inversion of this 2×2 position-space matrix equation yields

$$G(\vec{r}_{\pm}, \vec{r}_2; \omega) = \frac{D_{\mp,\mp}G^B_{2D}(\vec{r}_{\pm}, \vec{r}_2; \omega) + D_{\pm,\mp}G^B_{2D}(\vec{r}_{\mp}, \vec{r}_2; \omega)}{D_{\mp,\mp}D_{\pm,\pm} - D_{\pm,\mp}D_{\mp,\pm}}, \tag{6}$$

where we have defined

$$D_{\pm,\pm} = \delta_{\pm,\pm} - \alpha G^B_{2D}(\vec{r}_{\pm}, \vec{r}_{\pm}; \omega), \tag{7}$$

and

$$D_{\pm,\mp} = \alpha G^B_{2D}(\vec{r}_{\pm}, \vec{r}_{\mp}; \omega). \tag{8}$$

Substitution of Eqns. (6-8) into Eq. (3) yields the solution for $G(\vec{r}_1, \vec{r}_2; \omega)$, which includes the two dots located at $\vec{r}_{\pm}$ on the planar sheet. The first term on the right of Eq. (3) describes electron propagation on the sheet in the absence of the quantum dots, and the second term involving Eq. (6) describes the effect of the double quantum dot system on the electron propagation. The energy eigenstates associated with the double quantum dot arise from its frequency poles, as given by the dispersion relation

$$D_{\mp,\mp}D_{\pm,\pm} - D_{\pm,\mp}D_{\mp,\pm} = 0. \tag{9}$$

The retarded infinite sheet Green's function for a spinless 2D electron (absent dots) subject to Landau quantization is well known as[2] (magnetic field in z-direction normal to the plane of the dots; $\hbar \to 1$; c $\to 1$)

$$G^B_{2D}(\vec{r}_1, \vec{r}_2; T) = -\eta_+(T) C(\vec{r}_1, \vec{r}_2) \frac{m\omega_c}{4\pi} \frac{\exp\left[\frac{im\omega_c(X^2+Y^2)}{4\tan(\omega_c T/2)}\right]}{\sin(\omega_c T/2)}, \tag{10}$$

in position-time representation, with $T = t_1 - t_2$, $X = x_1 - x_2$, $Y = y_1 - y_2$, ω_c is the cyclotron frequency and $\eta_+(T)$ is the Heaviside unit step function. Furthermore, $\vec{A}(\vec{x})$ is the vector potential of a constant uniform magnetic field, $\phi(\vec{r})$ is an arbitrary gauge function and

$$C(\vec{r}_1, \vec{r}_2) = \exp\left[\frac{ie}{\hbar c} \int_{\vec{r}_2}^{\vec{r}_1} d\vec{x} \cdot \vec{A}(\vec{x})\right] = \exp\left[i(e/2)\vec{r}_1 \cdot \vec{H} \times \vec{r}_2 - \phi(\vec{r}_1) + \phi(\vec{r}_2)\right], \tag{11}$$

where the $\vec{x}$-integral depends on the right depends on path in general, and is taken here as a straight line. Fourier transforming from time to frequency representation, and setting $\vec{r}_1 = \vec{r}_2 = \vec{r}_\pm$ as needed in Eq. (7), ($X = 0$, $Y = 0$) with $C(\vec{r}_1, \vec{r}_2) \to C(\vec{r}_\pm, \vec{r}_\pm) = 1$, we obtain ($\omega \to \omega + i0^+$)

$$G^B_{2D}(0,0;\omega) = -\frac{m\omega_c}{4\pi} \int_0^\infty dT \frac{e^{i\omega T}}{\sin(\omega_c T/2)}. \tag{12}$$

Expanding the integrand,

$$\frac{1}{\sin(\omega_c T/2)} = 2i \sum_{n=0}^{\infty} e^{-i\omega_c(n+1/2)T}, \tag{13}$$

the T-integral may be evaluated as

$$G^B_{2D}(0,0;\omega) = \frac{m\omega_c}{2\pi} \sum_{n=0}^{\infty} \frac{1}{\omega - (n+1/2)\omega_c}. \tag{14}$$

This series diverges as $n \to \infty$, as does the T-integral of Eq. (10) for $X = Y = 0$. The divergence is an artifact of assuming a $\delta^{(2)}(\vec{r})$ potential confining the dot to a single point. It may be removed by noting that the original integral equation involves integration over a small, but finite radius "a", so we should install a "smear" radius "a" putting $X^2 + Y^2 \to a^2$ in Eq. (10), leading to

$$G^B_{2D}(0,0;\omega) \Rightarrow G^B_{2D}(a;\omega) = -\frac{m\omega_c}{4\pi} \int_0^\infty \frac{dT e^{i\omega T}}{\sin(\omega_c T/2)} \exp\left[\frac{im\omega_c a^2}{4\tan(\omega_c T/2)}\right], \tag{15}$$

or

$$G^B_{2D}(0,0;\omega) \Rightarrow G^B_{2D}(a;\omega) \Rightarrow \frac{m\omega_c}{2\pi i} \exp\left[-m\omega_c a^2/4\right] \times \sum_{n=0}^{\infty} L_n\left(\frac{m\omega_c a^2}{2}\right) \int_0^\infty dT \exp\left[i([\omega - (n+1/2)\omega_c]T)\right], \tag{16}$$

where we have expanded the last exponential on the right hand side of the T-integrand in Laguerre polynomials [3], $L_n(x)$, that embody the Landau eigenstate matrix elements on the plane in closed form. The T-integral of Eq. (16) can be carried out with the result

$$G^B_{2D}(0,0;\omega) \Rightarrow G^B_{2D}(a;\omega) = \frac{m\omega_c}{2\pi} e^{-m\omega_c a^2/4} \sum_{n=0}^{\infty} L_n\left(\frac{m\omega_c a^2}{2}\right) \frac{1}{\omega - (n+1/2)\omega_c}. \tag{17}$$

The n-series of Eq. (17) is slowly convergent[3]. As $n \to \infty$, $L_n(x) \sim n^{-1/4}$ becomes small, while the smallness of dot size "a" ensures that $L_n(m\omega_c a^2/2) \sim 1$ for lower values of n. Correspondingly, terms with index n much larger than the argument of $L_n(m\omega_c a^2/2)$,

$$n \gg m\omega_c a^2/2 \tag{18}$$

can be neglected and this yields Eq. (14) from Eq. (17), albeit the n-sum is cut off.

3. Dispersion Relation for a Symmetric Double Quantum Dot in a Normal Magnetic Field: Conclusions

The subband energy dispersion relation of Eq. (9) for a symmetric double-dot, considered with Eqns. (7), (8), and (17), takes the form (we write $\omega \leftrightarrow \omega/\omega_c$ in units of the cyclotron frequency; restore $\hbar$):

$$1 = \frac{\alpha m}{2\pi\hbar^2}\sum_{n=0}^{\infty}\frac{1}{\omega-(n+1/2)}\left[e^{-m\omega_c a^2/4\hbar}L_n\left(\frac{m\omega_c a^2}{2\hbar}\right) \pm e^{-m\omega_c d^2/4\hbar}L_n\left(\frac{m\omega_c d^2}{2\hbar}\right)\right], \tag{19}$$

from which it is clear that the energy roots occur in pairs corresponding to $\pm$. A "single pole approximation" (SPA) yields the energies, $\omega_{n\pm}$,(in units of ω_c) as

$$\omega_{n\pm} = (n+1/2) + \frac{\alpha m}{2\pi\hbar^2}\left[e^{-m\omega_c a^2/4\hbar}L_n\left(\frac{m\omega_c a^2}{2\hbar}\right) \pm e^{-m\omega_c d^2/4\hbar}L_n\left(\frac{m\omega_c d^2}{2\hbar}\right)\right]. \tag{20}$$

This SPA approximation is useful when $\alpha m/2\pi\hbar^2 < 1$, so that the denominator $[\omega/\omega_c - (n+1/2)]$ must also be small enough for the right hand side of Eq.(19) to match unity on the left. Correspondingly, the nth pair of energy levels, $\omega_{n\pm}$, is essentially determined by the single nearest frequency pole term in an approximation that neglects the other poles.

We have carried out an exact numerical analysis of the double quantum dot energies using Eq.(19) for well depths U_0=−100meV,−300mev, −600meV, with a=5nm, d=5a, ω_c = 2meV, m=0.067m_e, carrying 5,000 terms. Table 1 presents the energies, both numerically calculated and those obtained by the SPA (as well as their percent difference) for the first five energy levels with $\pm \to +$, and Table 2 does the same for $\pm \to -$. It is clear that the SPA is quite good for U_0= −100meV, less so for U_0=−300meV and worst for U_0=−600meV as one should expect from the correspondingly increasing value of the magnitude of alpha, ($|\alpha|$).

The plots of the left and right hand sides of Eq.(19) are shown in Figure 1 for U_0=−100, −300, −600meV with the Landau-quantized energy roots ($\hbar\omega_{n\pm}$) given by the frequencies at which intersections with unity occur. The right-hand-sides are plotted in blue for $\pm \to +$ and in red for $\pm \to -$.

In summary, the retarded Green's function for the Landau quantized symmetric double-dot system in a perpendicular magnetic field is given by Eqns. (4), (6), (7), (8) and ($\vec{r_\pm} = \pm\hat{i}d/2$)

$$G(\vec{r}_1,\vec{r}_2;\omega) = G_{2D}^B(\vec{r}_1,\vec{r}_2;\omega) + \alpha\sum_{\pm}G_{2D}^B(\vec{r}_1,\vec{r}_\pm;\omega)$$
$$\times\left\{\frac{(1-\alpha G_{2D}^B(a;\omega))G_{2D}^B(\vec{r}_\pm,\vec{r}_2;\omega)+\alpha G_{2D}^B(d;\omega)G_{2D}^B(\vec{r}_\mp,\vec{r}_2;\omega)}{(1-\alpha G_{2D}^B(a;\omega))^2-\alpha^2(G_{2D}^B(d;\omega))^2}\right\}, \tag{21}$$

where ($\hbar \to 1$; $c \to 1$; $X = x_1 - x_2$; $Y = y_1 - y_2$; $T = t_1 - t_2$)

$$G_{2D}^B(\vec{r}_1,\vec{r}_2;\omega) = -\frac{m\omega_c}{4\pi}C(\vec{r_1},\vec{r_2})\int_0^\infty dT e^{-i\omega T}\frac{\exp\left[\frac{im\omega_c(X^2+Y^2)}{4\tan(\omega_c T/2)}\right]}{\sin(\omega_c T/2)} =$$
$$= \frac{m\omega_c}{2\pi}C(\vec{r_1},\vec{r_2})e^{-m\omega_c(X^2+Y^2)/4}\sum_{n=0}^{\infty}L_n\left(\frac{m\omega_c(X^2+Y^2)}{2}\right)\frac{1}{\omega-(n+1/2)\,\omega_c}. \tag{22}$$

This study explicitly exhibits the splintering of the energy levels of symmetric quantum dot systems by Landau quantization in a magnetic field, in particular for quantum dots understood to have just one energetically accessible energy level in the absence of the magnetic field. Advanced concept devices must be reconsidered in the light of this proliferation of energy levels in a magnetic field as it will change the basis upon which they are intended to function. The double-dot Green's function derived here is tractable and can be used in more accurate studies of double-dot systems in a normal magnetic field.

Exact Numerical and SPA Calculated Results (with percent differences) for the First Five Symmetric Quantum Double Dot Energies $\hbar\omega_{n_+}$ for U_0=−100, −300, −600meV with $\pm \to +$ in Eqns. (19) and (20)

	U_0=−100meV			U_0=−300meV			U_0=−600meV		
$\hbar\omega_{n_+}$	Exact	Single Pole	% Error	Exact	Single Pole	% Error	Exact	Single Pole	% Error
n=0	0.47442	0.47556	0.24%	0.41573	0.42667	2.63%	0.30539	0.35334	15.70%
n=1	1.48146	1.45725	1.63%	1.44318	1.37174	4.95%	1.38436	1.24348	10.18%
n=2	2.48619	2.44346	1.72%	2.45853	2.33039	5.21%	2.41765	2.16079	10.62%
n=3	3.48945	3.43290	1.62%	3.46848	3.29871	4.90%	3.43771	3.09741	9.90%
n=4	4.49161	4.42450	1.49%	4.47491	4.27350	4.50%	4.45026	4.04701	9.06%

Exact Numerical and SPA Calculated Results (with percent differences) for the First Five Symmetric Quantum Double Dot Energies $\hbar\omega_{n_-}$ for U_0=−100, −300, −600meV with $\pm \to -$ in Eqns. (19) and (20)

	U_0=−100meV			U_0=−300meV			U_0=−600meV		
$\hbar\omega_{n_-}$	Exact	Single Pole	% Error	Exact	Single Pole	% Error	Exact	Single Pole	% Error
n=0	0.49665	0.49680	0.03%	0.48892	0.49040	0.30%	0.47382	0.48080	1.47%
n=1	1.49085	1.48807	0.19%	1.46968	1.46422	0.37%	1.42856	1.42844	0.01%
n=2	2.48680	2.47542	0.46%	2.45687	2.42625	1.25%	2.40160	2.35251	2.04%
n=3	3.48419	3.46014	0.69%	3.44930	3.38041	2.00%	3.38868	3.26082	3.77%
n=4	4.48272	4.44328	0.88%	4.44565	4.32984	2.61%	4.38479	4.15969	5.15%

Acknowledgments

This work was supported in part by DARPA grant #HR0011-09-1-0008. It is a pleasure to acknowledge computational assistance by Spencer Horton, M. Lodato and S. Donovan.

References

[1] "Quantum Wells, Wires and Dots", 2nd Ed., Paul Harrison, Wiley-Interscience (2005).
[2] N.J.M. Horing and M. Yildiz, Annals of Physics (NY) **97**, 216, Section 2; (1976).
Also, N.J.M. Horing, Annals of Physics (NY) **31**, Section 2, (1965).
[3] "Bateman Manuscript Project; Higher Transcendental Functions", Vol. 2, Ed: A. Erdelyi, et al., McGraw Hill (1953); p.189, #10.12.17 and p.199.

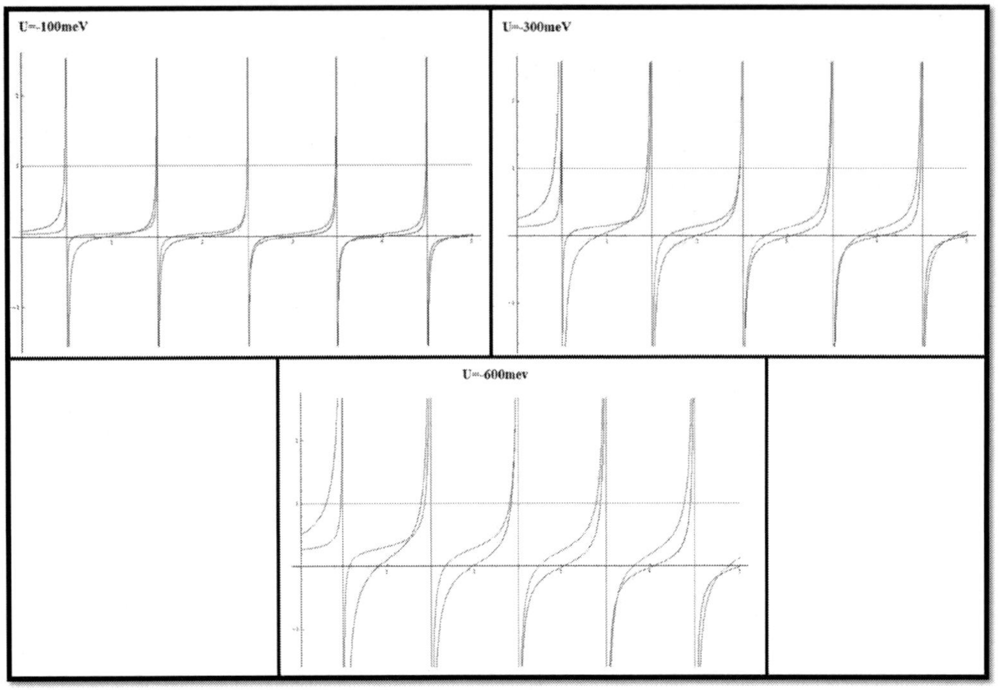

Figure 1. Plots of the left-and right-hand-sides of Eq.(19) for U_0=-100, -300, -600meV (where the function in blue is generated by taking the $\pm$ in Eqns. (19) and (20) to be positive and the red function is generated by taking it to be negative). The frequencies at which the intersections with unity occur define the Landau quantized subband energies $\hbar\omega_{n_\pm}$ of the symmetric quantum double dot system in a magnetic field. The field and material parameters are specified in the text.

Journal of Physics: Conference Series **334** (2011) 012066 doi:10.1088/1742-6596/334/1/012066

The self-consistent calculation of the edge states in bilayer quantum Hall bar

A E Kavruk, T Ozturk, A Ozturk, U Atav and H Yuksel

Selcuk University, Faculty of Science, Department of Physics, 42075 Konya, Turkey

E-mail: aekavruk@gmail.com

Abstract. In this study, we present the spatial distributions of the edge channels for each layer in bilayer quantum Hall bar geometry for a wide range of applied magnetic fields. For this purpose, we employ a self-consistent Thomas-Fermi-Poisson approach to obtain the electron density distributions and related screened potential distributions. In order to have a more realistic description of the system we solve three dimensional Poisson equation numerically in each iteration step to obtain self consistency in the Thomas-Fermi-Poisson approach instead of employing a "frozen gate" approximation.

1. Introduction

Quantum Hall effect (QHE) has continuously been a center of attraction following its discovery [1] for both theoretical and experimental physicists because of the wide range of interesting physical phenomena it involves. These phenomena are mostly explained in terms of edge channels [2-5]. Bilayer two dimensional electron systems (2DES) also present interesting physical phenomena including fractional quantum Hall effect (FQHE) [6] and hysteresis effect [7,8]. We believe that in order to understand these effects one should have a reliable knowledge of the spatial distribution of edge channels in both layers. For this purpose, like many of the theoreticians who investigate QHE [9-12], we employ self-consistent Thomas-Fermi-Poisson approach (SCTFPA) to obtain the electron density distributions and related screened potential distributions. In order to simplify the calculations, many of the researchers interested in QHE use assumptions like in-plane charges or frozen gate approximation [4,5,8-12]. We believe that solutions of the Poisson equation obtained under such assumptions do not represent the correct nature of the electrostatic structure. Thus we solve three dimensional Poisson equation (3DPE) without employing such assumptions at every step during SCTFPA.

2. Model and Calculations

2.1. The geometry of bilayer structure

The geometry of the bilayer quantum Hall bar (BQHB) considered in this study is illustrated in Fig. 1. There are two gates at the top of the structure of 150 nm width. These gates are biased with a -1.7 V potential to confine the two dimensional electron gas (2DEG) in the x direction. So that the quantum Hall bar (QHB) structure is formed by using the gates instead of etching. Under the gates there is a GaAs layer of 40 nm thickness which is called the cap. Below this layer there is a Si-doped AlGaAs layer of 10 nm thickness and followed by a 40 nm thick undoped AlGaAs layer which is called a spacer. The first 2DEG (top 2DEG) layer is located 90 nm below the surface. After that there is a GaAs layer of 30 nm thickness. At the end of this layer there is the second 2DEG (bottom 2DEG) layer at 120 nm below surface. Below this 2DEG again there is an AlGaAs spacer of 40 nm thickness followed by a Si-doped layer of 10 nm thickness, and then there is bulk GaAs. This structure actually is extended down to the substrate. However in our study the distance from the surface to the bottom is taken as 1000 nm. In the y direction the structure is taken invariant. In Fig. 1, the distance between the two gates which is denoted by w is taken 2000 nm.

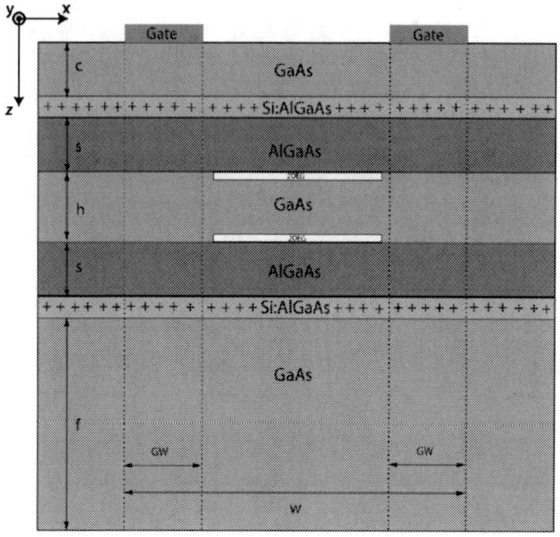

Figure 1. A schematic layout of the BQHB considered in calculations.

In this study, the average electron density and the ionized donor density which generates the 2DEG are taken equal as opposed to some theoretical studies [8-13]. The donor density of the top donor layer and the average electron density of top 2DEG are taken as $n_d^T = \bar{n}_e^T = 3.0\times10^{11}\,cm^{-2}$. In the density match situation, the density of the bottom donor layer and the average electron density of the bottom 2DEG are taken equal $n_d^B = \bar{n}_e^B = n_d^T = \bar{n}_e^T$. In the density mismatch situation, the density of the bottom donor layer and the average density of the bottom 2DEG are taken as $n_d^B = \bar{n}_e^B = 0.85\times n_d^T = 0.85\times\bar{n}_e^T$.

2.2. 3D Poisson equation & boundary conditions

We can write 3DPE inside semiconductor material as

$$-\vec{\nabla}\varepsilon\vec{\nabla}\varphi = \rho \qquad (1)$$

Here ε is the dielectric coefficient of the material and ρ is electron distribution in the 3D space. A proper solution of this equation depends on the boundary conditions which must be defined correctly. It is well known, different boundary conditions lead to different solutions of 3DPE for the same charge distribution.

The most important theoretical basis for the understanding of edge states is proposed by Chklovskii et al. [4,5]. Chklovskii et al. proposed how the incompressible strips (IS) and compressible strips (CS) are formed by the screening and edge effect. They solved the Poisson equation analytically in these studies. In order to obtain an analytical solution of PE, they make an approximation called in-plane charges assumption. In this assumption, the gates, the donors and the 2DEG are all assumed to be on the same plane in the z direction. In these studies, Chklovskii et al. did not consider electrochemical equilibrium and solved this problem using a purely electrostatic model. Lier and Gerhardts used SCTFPA to achieve electrochemical equilibrium [9]. Following this study, some further theoretical studies are made with SCTFPA [10-13], but all of them, used in-plane charges approximation. Various theoretical studies [14-18] aimed at reaching beyond in-plane approximation generally used Davies' studies [19,20] where Davies obtained analytical solutions of 3DPE to determine the confining potential on the 2DEG generated by gates on the surface of the structure under periodical boundary conditions in x and y directions. However, in these studies the confining potential due to the gates is obtained only for once and without considering the charge distributions in the 2DEG. Changes in the charge distribution of 2DEG induce a potential at the metallic gates and thus change the potential due to the gates. Ignoring this effect is called as frozen gate approximation. Considering the fact that the distance between the gates and the 2DEG is very small, we believe that changes in the charge distribution of 2DEG can effect the confining potential due to the gates. The positions and widths of compressible and incompressible strips in a bilayer structure are very sensitive to small changes in the potential, frozen gate approximation might have a substantial effect on the distribution of the edge channels.

Therefore, we numerically solve 3DPE without introducing any approximations at each step for SCTFPA [21]. The numerical solution of Poisson equation in 3D which we use in this study also has an advantage in bilayer system, because it includes all electrostatic interactions such as electron-electron interaction, interlayer charge interaction and electrostatic effects of the gates.

In this study we take y direction as invariant since we are interested only in BQHB. So we are concerned with potential distribution in just x and z directions. The boundary conditions which we use are similar to Davies' approach [19, 20]. In the z direction, at $z=0$ (on the surface) the gate potential is taken as boundary condition where there are gates, i.e. $\phi(x,y,0)=V_g$, and zero where there aren't any gates. We assume that the potential completely heals at the bottom of the sample and first derivative of the potential is assumed to vanish at the bottom (which is 1000 nm from the surface of the sample), because we maintain the overall charge neutrality in the sample, this assumption is reliable. In the x direction in the middle of the gates $\partial\phi/\partial x=0$ is taken as the boundary condition due to the symmetry of the system.

2.3. Thomas-Fermi-Poisson approximation

Thomas-Fermi-Approximation (TFA) is a widely used method to obtain density of electron distributions in 2DEG [9-15]. TFA gives density of electron distribution as

$$n_{el}(\vec{r})=\int dE\, D(E) f([E+\varphi(\vec{r})-\mu]/k_bT) \tag{2}$$

with $D(E)$ is density of state (DOS) of 2DEG, $f([E+\varphi(\vec{r})-\mu]/k_bT)=1/\left(1+e^{([E+\varphi(\vec{r})-\mu]/k_bT)}\right)$ is the Fermi function and $\varphi(\vec{r})$ is the potential function obtained from the solution of 3DPE which include potential due to the gates, donors and interaction between electrons, μ is the electrochemical potential, k_b is *the Boltzmann* constant and T is the temperature. In the absence of the magnetic field, DOS for 2DEG is $m/\hbar^2\pi$. For finite magnetic fields, DOS is Landau DOS $D(E)=\left(\frac{g_s}{2\pi\ell^2}\right)\sum_{n=0}^{\infty}\delta(E-E_n)$ with $\ell=(\hbar/m\omega_c)^{1/2}$ magnetic length, g_s is the spin degeneracy, $E_n=(n+1/2)\hbar\omega_c$ Landau energy levels and $\omega_c=eB/m$. The effective electron mass is denoted by m.

In the presence of the magnetic field we solve Poisson equation and Thomas-Fermi equation self-consistently to obtain the electron density. However the convergence is a serious problem when these equations are solved self consistently. One of the methods used to obtain convergence is Under Relaxation (UR) method. However this method is not very stable. A more stable method is Newton-Raphson (NR) method. The only disadvantage of NR method is that in order to converge this method requires an initial value which is close to the solution. So that firstly the system is solved with UR method at $T = 0$ and $B = 0$. Using this solution as the initial distribution, the electron densities of both layers are calculated self consistently with NR method for some sufficiently high temperature and zero magnetic field. Then we set the magnetic field and repeat the NR procedure for the self consistent solution. After this, the temperature is reduced slowly repeating the NR procedure at each temperature step until the system reaches to aimed temperature (T=1.4 K).

3. Result and Discussion

The ratio of the electron density and the number of Landau levels per unit area is called the *filling factor* and given by $\nu = 2\pi\ell^2 n_{el}(\vec{r})$. When this ratio is a multiple of 2 then a Landau level is completely filled and an incompressible strip is formed. In Fig. 2, the filling factors are shown as a function of the spatial coordinates for magnetic field strengths between 6.5 T and 8.5 T for a BQHB. Here average electron densities of both 2DEGs are taken as equal. In Fig. 2, the incompressible strip (IS) is indicated with yellow colour corresponds to filling factor two. In a single layer quantum hall bar structure the width of the ISs in 2DEG decrease with decreasing magnetic field [21], however in Fig.2 the width of IS in top 2DEG shows an increase with decreasing magnetic field around B~7.75 T. We believe that this anomalous behaviour is responsible for the hysteresis effect observed in BQHB. The reason of this increasing is that the screening features of the bottom 2DEG change directly because an IS appears at the centre of bottom 2DEG at this magnetic field. Although the top and the bottom 2DEGs have the same electron number, the bottom 2DEG enters to the plateau region at a lower magnetic field than the upper one. The reason of this is that the bottom 2DEG feels the confinement effect of the surface gates less than upper 2DEG. So the electrons in that layer spread a wider area and the maximum electron density in the centre remains lower than the top layer.

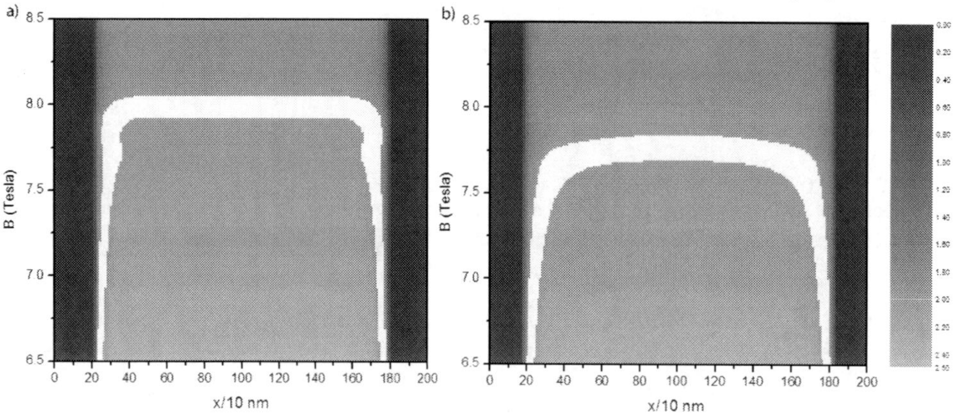

Figure 2. In match density condition, top (a) and bottom (b) 2DEGs electron distributions are shown in terms of filling factors as a function of position and magnetic field in gray scale. The color scale is given to right of the graphs. Light gray corresponds to higher filling factors. Yellow color is used to indicate ISs.

In the mismatch situation, the same investigation is done for magnetic field strengths between 6.0 T and 8.5 T. In Fig. 3, the electron distribution of the top and bottom 2DEGs are presented in terms of filling factor. Here the average electron density in the bottom layer is taken as 85% of that of the top layer. So observing of the IS in the bottom layer was rather delayed with respect to the top layer. In this situation, this layer make a screening effect for the potential which effects top 2DEG since the bottom layer consists of completely a compressible strip (CS) in a wide magnetic field interval. So around at 8.0 T, the IS which is in the middle of the top layer maintains the thickness and the location due to this flat potential.

In summary, We have calculated the spatial distributions of the edge channels for each layer in BQHB geometry using SCTFPA. We believe that we have obtained a more realistic description of the system by removing the "frozen gate" approximation.

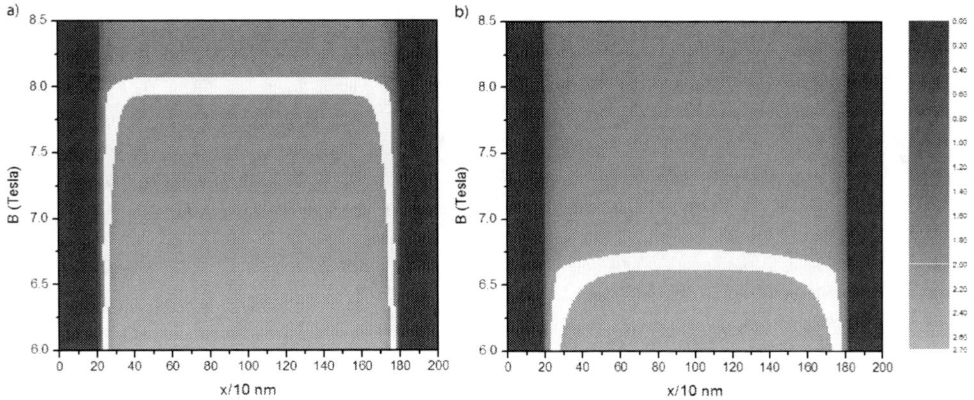

Figure 3. In mismatch density condition, top (a) and bottom (b) 2DEGs electron distributions are shown in terms of filling factors as a function of position and magnetic field in gray scale. The color scale is given to right of the graphs. Light gray corresponds to higher filling factors. Yellow color is used to indicate ISs.

Acknowledgements
This work is supported by Selcuk University BAP Grant No. 07101037. We also thank to ITAP (Institute of Theoretical and Applied Physics; Turunc/Marmaris Turkey) for allowing us to use computational facilities.

References

[1] von Klitzing K, Dorda G and Pepper M 1980 *Phys. Rev. Lett.* **45** 494
[2] MacDonald A H and Streda P 1984 *Phys. Rev.* B **29** 1616
[3] Büttiker M 1988 *Phys. Rev.* B **38** 9375
[4] Chklovskii D B, Shklovskii B I and Glazman L I 1992 *Phys. Rev.* B **46** 4026
[5] Chklovskii D B, Matveev K A and Shklovskii B I 1993 *Phys. Rev.* B **47** 12605
[6] Murphy S Q, Eisentein J P, Boebinger G S, Pfeiffer L N and West K W 1994 *Phys. Rev. Lett.* **72** 728
[7] Pan W, Reno J L and Simmons J A 2005 *Phys. Rev.* B **71** 153307
[8] Siddiki A, Kraus S and Gerhardts R R 2006 *Physica* E **34** 136
[9] Lier K and Gerhardts R R 1994 *Phys. Rev.* B **50** 7757
[10] Oh J H and Gerhardts R R 1997 *Phys. Rev.* B **56** 13519
[11] Güven K and Gerhardts R R 2003 *Phys. Rev.* B **67** 115327
[12] Siddiki A and Gerhardts R R 2003 *Phys. Rev.* B **68** 125315
[13] Siddiki A and Gerhardts R R 2004 *Phys. Rev.* B **70** 195335
[14] Siddiki A and Marquartd F 2007 *Phys. Rev.* B **75** 045325
[15] Siddiki A, Kavruk A E, Öztürk T, Atav Ü, Şahin M and Hakioğlu T 2008 *Physica* E **40** 1398
[16] Ihnatsenka S and Zozoulenko I V 2006 *Phys. Rev.* B **73** 075331
[17] Ihnatsenka S and Zozoulenko I V 2006 *Phys. Rev.* B **74** 201303(R)
[18] Ihnatsenka S and Zozoulenko I V 2007 *Phys. Rev.* B **75** 035318
[19] Davies J H 1988 *Semicond. Sci. Technol.* **3** 995
[20] Davies J H, Larkin I A and Sukhorukov E. V. 1995 *J. Appl. Phys.* **77** 4504
[21] Kavruk A E 2010 Ph.D. thesis (Turkey: Selcuk University)